HEAVY ION ACCELERATOR TECHNOLOGY

HEAVY ION ACCELERATOR TECHNOLOGY

Eighth International Conference

Argonne, Illinois   October 1998

EDITOR
Kenneth W. Shepard
Argonne National Laboratory

American Institute of Physics

AIP CONFERENCE PROCEEDINGS 473

Woodbury, New York

Editor:

Kenneth W. Shepard
Physics Division
Building 203
Argonne National Laboratory
9700 South Cass Avenue
Argonne, IL 60439
U.S.A.

E-mail: kwshepard@anl.gov

The articles on pp. 236–243, 258–266, 279–284, 384–395, 477–489, 528–535, and 566–577 were authored by U. S. Government employees and are not covered by the below mentioned copyright.

Authorization to photocopy items for internal or personal use, beyond the free copying permitted under the 1978 U.S. Copyright Law (see statement below), is granted by the American Institute of Physics for users registered with the Copyright Clearance Center (CCC) Transactional Reporting Service, provided that the base fee of $15.00 per copy is paid directly to CCC, 222 Rosewood Drive, Danvers, MA 01923. For those organizations that have been granted a photocopy license by CCC, a separate system of payment has been arranged. The fee code for users of the Transactional Reporting Service is: 1-56396-806-1/99/$15.00.

© 1999 American Institute of Physics

Individual readers of this volume and nonprofit libraries, acting for them, are permitted to make fair use of the material in it, such as copying an article for use in teaching or research. Permission is granted to quote from this volume in scientific work with the customary acknowledgment of the source. To reprint a figure, table, or other excerpt requires the consent of one of the original authors and notification to AIP. Republication or systematic or multiple reproduction of any material in this volume is permitted only under license from AIP. Address inquiries to Office of Rights and Permissions, 500 Sunnyside Boulevard, Woodbury, NY 11797-2999; phone: 516-576-2268; fax: 516-576-2499; e-mail: rights@aip.org.

L.C. Catalog Card No. 99-61607
ISBN 1-56396-806-1
ISSN 0094-243X
DOE CONF- 981087

Printed in the United States of America

CONTENTS

Preface xi
Scientific Adivsory Committee xii
Exhibitors xiii

I. RAY HERB MEMORIAL

A Retrospective of the Career of Ray Herb 3
    G. A. Norton, J. A. Ferry, R. E. Daniel, and G. M. Klody
The Canberra 14UD–a Jewel in the NEC Crown–a Living Memorial
to Ray Herb... 24
    D. C. Weisser and T. R. Ophel
The Last Large Pelletron Accelerator of the Herb Era 39
    S. Chopra, M. M. Narayanan, R. Joshi, S. Gargari, D. Kanjilal,
    S. K. Datta, and G. K. Mehta

II. ELECTROSTATIC ACCELERATORS

Voltage Limitations of Electrostatic Accelerators 47
    H. R. McK. Hyder
Gamma-Resonance Contraband Detection Using a High Current
Tandem Accelerator.. 56
    B. F. Milton, J. Beis, D. Dale, T. Debiak, E. Kamykowski, S. Melnychuk,
    J. Rathke, J. Rogers, R. Ruegg, and J. Sredniawski
Use of an ECR Ion Source in the High Voltage Terminal of the Tandem
Accelerator at JAERI 65
    M. Matsuda, C. Kobayashi, and S. Takeuchi
Fifty Years of Accelerator Based Physics at Chalk River 74
    J. W. McKay
The Ten Years' Operation of the HI-13 Tandem Accelerator at CIAE..... 80
    Y. Bingfan, Q. Jiuchang, Z. Canzhe, H. Yueming, G. Xialing, J. Yongliang,
    Z. Guilian, Y. Weimin, K. ChauXin, Y. Zhiren, S. Shengyong, L. Dezhong,
    W. Liyong, and Z. Jiazheng
Status Report of the Munich Tandem Accelerator 88
    L. Rohrer
A Foreline Oil Vapor Trap... 92
    S. M. Ferguson and E. D. Berners
A New Design of Terminal Pumping in the Lund Pelletron Tandem 94
    K. Håkansson and R. Hellborg
Replacement of a Broken Column Support in a Horizontal Pelletron 100
    K. Håkansson, R. Hellborg, and S. Uthas
Electrical Stress on Insulators of the Vivitron 104
    F. Osswald, J. Heugel, E. Jegham, N. Lahera, R. Rebmeister,
    and P. Zouloumian

III. LINACS AND BOOSTERS

Linac Boosters—An Overview 117
    D. C. Weisser and N. R. Lobanov

Superconducting Cavity Development at Legnaro 138
    A. Facco

Status of the JAERI Tandem Accelerator and Its Booster 152
    S. Takeuchi, S. Abe, S. Hanashima, K. Horie, N. Ishizaki, S. Kanda,
    M. Matsuda, I. Ohuchi, H. Tayama, Y. Tsukihashi, and T. Yoshida

Status Report in the São Paulo Pelletron-Linac Project 168
    N. Added

Construction of Superconducting RFQs at INFN-LNL 173
    G. Bisoffi, V. Andreev, E. Bissiato, M. Comunian, F. Chiurlotto,
    E. Corradin, M. Lollo, A. Lombardi, A. Pisent, A. M. Porcellato,
    T. Shirai, E. Tovo, and R. Tovo

The Non-RFQ Resonators of the PIAVE Linac 185
    A. Facco, F. Scarpa, and V. Zviagintsev

The Lead-Copper Option for Heavy Ion Accelerating Cavities:
History, Status, and Future Prospects 192
    J. W. Noé

Design of the Heavy-Ion Injector PIAVE 214
    A. Pisent

Beam Test of Niobium Sputtered QWRs and Upgrading of ALPI
Medium β Cavities 228
    A. M. Porcellato, S. Y. Stark, V. Palmieri, G. Bisoffi, A. Dainelli, M. Poggi,
    L. Bertazzo, F. Stivanello, L. Badan, A. Beltramin, L. Boscagli, D. Carlucci,
    F. Chiurlotto, S. Contran, T. Contran, M. De Lazzari, and L. Ziomi

Status of the First Batch of Niobium Resonator Production
for the New Delhi Booster Linac 236
    P. N. Potukuchi, S. Ghosh, and K. W. Shepard

Study of Acceleration Across the TTF's Zero-Crossing Velocity
in Independently Phased Linacs 244
    S. Takeuchi

A Distributed Control System Status Report of the Munich
Accelerator Control 252
    L. Rohrer and H. Schnitter

Completion of the ATLAS Control System Upgrade 258
    F. H. Munson

Status of the Linac Booster for NSC Pelletron 267
    A. Roy, P. N. Prakash, B. P. Ajithkumar, S. Ghosh, T. Changrani,
    A. Sarkar, R. Mehta, B. K. Sahu, A. Choudhury, J. Chacko, J. Anthony,
    M. V. Suresh Babu, M. Kumar, S. A. Krishnan, A. Mandal, G. O. Rodrigues,
    R. Kumar, R. K. Bhowmik, G. K. Mehta, and K. W. Shepard

Single-Gap Multi-Harmonic Buncher for NSC Pelletron 272
    A. Sarkar, S. Ghosh, P. Barua, R. Joshi, R. Ahuja, S. Rao, S. A. Krishnan,
    A. J. Malyadri, R. Kumar, S. Gargari, S. Chopra, D. Kanjilal, S. K. Datta,
    A. Roy, R. K. Bhowmik, I. R. Tilbrook, and B. E. Clifft

A Very Low Intensity Ion Beam Detector System.................................279
    G. P. Zinkann, B. E. Clifft, J. A. Nolen, R. C. Pardo, C. E. Rehm,
    and W. Q. Shen

IV. ION SOURCES

Development of ECR Plasmas for Radioactive Ion Beams 287
    R. Geller, J. L. Bouly, J. F. Bruandet, N. Chauvin, J. C. Curdy, T. Lamy,
    H. Nifenecker, P. Sole, P. Sortais, and J. L. Vieux-Rochaz

High Charge State ECR Ion Sources: Status and Developments 300
    S. Gammino and G. Ciavola

Preliminary Ionization Efficiencies of ^{11}C and ^{14}O with the LBNL
ECR Ion Sources 312
    Z. Q. Xie, J. Cerny, F. Q. Guo, R. Joosten, R. M. Larimer, C. M. Lyneis,
    P. McMahan, E. B. Norman, J. P. O'Neil, J. Powell, M. W. Rowe,
    H. F. VanBrocklin, D. Wutte, X. J. Xu, and P. Haustein

ECR Ion Source Developments at the Oak Ridge National Laboratory 321
    Y. Liu, G. D. Alton, and F. W. Meyer

A High Efficiency, Kinetic-Ejection Negative Ion Source
for RIB Generation 330
    G. D. Alton, Y. Liu, C. Williams, and S. N. Murray

A High-Intensity, RF Plasma-Sputter Negative Ion Source..................341
    G. D. Alton, R. Lohwasser, B. Cui, Y. Bao, T. Zhang, and C. A. Reed

A Multi-Sample Cs-Sputter Negative Ion Source 352
    G. D. Alton, B. Cui, Y. Bao, C. A. Reed, J. A. Ball, and C. Williams

Design and Development of the ECR Ion Source Control System
at LNL...360
    S. Canella, M. Cavenago, G. Delfitto, and G. Abrioni

The Legnaro ECR Ion Source Platform 366
    M. Cavenago, T. Kulevoy, G. Abrioni, S. Canella, and F. Cervellera

Ion Source Developments for the Relativistic Heavy Ion Collider 375
    D. B. Steski

Design Study of the Extraction System of the 3rd Generation
ECR Ion Source 384
    D. Wutte, M. A. Leitner, C. M. Lyneis, C. E. Taylor, and Z. Q. Xie

V. ACCELERATOR MASS SPECTROMETRY

Accelerator Mass Spectrometry—Big and Small..........................399
    W. Kutschera

High-Precision Measurements of ^{14}C as a Circulation Tracer
in the Pacific, Indian, and Southern Oceans with Accelerator
Mass Spectrometry (AMS) 410
    K. F. von Reden, J. C. Peden, R. J. Schneider, M. Bellino, J. Donoghue,
    K. L. Elder, A. R. Gagnon, P. Long, A. P. McNichol, T. Morin, D. Stuart,
    J. M. Hayes, and R. M. Key

Application of a Compact Microwave Ion Source to Radiocarbon Analysis 422
    R. J. Schneider, K. F. von Reden, J. M. Hayes, and J. S. C. Wills

Neutral Injection for Radioactive Ion Beams and Accelerator Mass Spectrometry 426
    A. E. Litherland, K. H. Purser, and H. E. Gove

VI. EXOTIC ION BEAMS

An Overview of ISAC 439
    P. W. Schmor

The RB Facility at KEK-Tanashi 451
    M. Tomizawa, S. Arai, Y. Arakaki, A. Imanishi, M. Okada,
    K. Niki, Y. Takeda, and E. Tojyo

Plans for Constructing a Next-Generation ISOL Facility at ORNL 466
    J. D. Garrett, G. D. Alton, R. L. Auble, C. Baktash, J. R. Beene,
    F. E. Bertrand, J. D. Fox, R. A. Gough, M. L. Halbert, J. G. Kalnins,
    Y. Liu, M. W. Ogan, F. Plasil, D. Shapira, P. T. Spampinato, J. W. Staples,
    H. Wollnik, and M. S. Zisman

An Advanced ISOL Facility Based on ATLAS 477
    J. A. Nolen, K. W. Shepard, R. C. Pardo, G. Savard, K. E. Rehm,
    J. P. Schiffer, W. F. Henning, C.-L. Jiang, I. Ahmad, B. B. Back,
    R. A. Kaye, M. Petra, M. Portillo, J. P. Greene, B. E. Clifft,
    J. R. Specht, R. V. F. Janssens, R. H. Siemssen, I. Gomes,
    C. B. Reed, and A. M. Hassanein

A Radioactive Ion Beam Facility, SIRIUS, at ISIS 490
    J. R. J. Bennett

High Power Targets for Production of Intense Radioactive Ion Beams 495
    W. L. Talbert, D. M. Drake, M. T. Wilson, J. J. Walker, and J. W. Lenz

Isobar Separators for Radioactive Ion Beam Facilities 505
    H. Wollnik and J. Garrett

The Physical Basis of the Release Curve for RIB Foil Targets 512
    J. R. J. Bennett

A Single Accelerator RIB Facility Using the Recoil Mass Spectrometer HIRA at NSC, New Delhi 523
    J. J. Das, P. Sugathan, N. Madhavan, T. Varughese, B. Kumar,
    P. V. Madhusudhana Rao, and A. K. Sinha

Beam Tests of the 12 MHz RFQ RIB Injector for ATLAS 528
    R. A. Kaye, K. W. Shepard, B. E. Clifft, and M. Kedzie

The Munich Accelerator for Fission Fragments (MAFF) 536
    O. Kester, D. Habs, M. Groß, T. Sieber, H. Bongers, A. Kolbe,
    H. J. Maier, P. Thirolf, T. von Egidy, U. Köster, E. Steichele, P. Kienle,
    H. J. Körner, A. Schempp, and U. Ratzinger

RFQ-IH Radioactive Beam Linac for ISAC 546
    R. E. Laxdal

The Radioactive Ion Beams Facility Project for the Legnaro Laboratories 559
    L. B. Tecchio

Development of a Radioactive Ion Beam Test Stand at LBNL **566**
D. Wutte, J. Burke, B. Fujikawa, P. Vetter, S. J. Freedman, R. A. Gough,
C. M. Lyneis, and Z. Q. Xie

List of Participants. **579**
Author Index. **589**

PREFACE

The Eighth International Conference on Heavy-ion Accelerator Technology was held at Argonne National Laboratory, Illinois, from October 5-9, 1998. This meeting continued the series most recently hosted at the Australian National University, Canberra in 1995 and previously held, under the name "International Conference on Electrostatic Accelerators and Associated Boosters", at Legnaro (1992), Strasbourg-Heidelberg (1989), Buenos Aires (1985), Oak Ridge (1981), Strasbourg (1977), and Daresbury (1973).

As for previous meetings, the initial topic was electrostatic accelerators. The opening session, Monday, October 5, was dedicated to the memory of the late Ray Herb of the University of Wisconsin and National Electrostatics Corporation. This was followed by a "SNEAP-like" session of informal presentations relating to electrostatic machines, which was added to this HIAT program to accommodate participants in the "Society of North-eastern Accelerator Personnel" whose 1998 meeting unfortunately could not be held as scheduled.

Continuing the trend of the last several meetings in this series, two sessions were needed to report current developments in superconducting ion accelerators. Also, many interesting recent developments were reported in a session devoted to ECR ion sources.

Several applications received attention. A session was devoted to atomic mass spectroscopy. Both because of the venue and the timeliness of the topic, two sessions and a number of poster presentations focused on the technology needed for beams of exotic, radioactive nuclei.

Regrettably, several scientists from the Nuclear Science Centre in New Delhi were unable to obtain visas from the U.S. State Department in order to present papers at this meeting. The affected papers were, to the extent possible, presented at the conference, and appear in these proceedings.

The attendees selected (with one vote per laboratory or institution represented) the Nuclear Science Centre as host for the next conference in this series, which accordingly will be held in New Delhi, India in 2001.

I would like to thank John Schiffer of Argonne and Willy Haeberli of the University of Wisconsin for their informal contributions to the Ray Herb Memorial Session. Also, I gratefully acknowledge the efforts of Margot Smith, both in organizing and conducting the conference and also in assembling these proceedings.

Kenneth W. Shepard
Argonne, Illinois

SCIENTIFIC ADVISORY COMMITTEE

J. R. Beene - ORNL, Oak Ridge, USA
N. Fazzini - CNEA, Buenos Aires, Argentina
X. L. Guan - CIAE, Beijing, China
S. S. Kapoor - BARC, Bombay, India
J. W. Noe - SUNY, Stony Brook, USA
R. Repnow - MPI, Heidelberg, Germany
P. Schmor - TRIUMF, Vancouver, Canada
S. Takeuchi - JAERI, Tokai, Japan
D. C. Weisser - ANU, Canberra, Australia

G. Ciavola - INFN-LNS, Catania, Italy
G. Fortuna - INFN-LNL, Legnaro, Italy
H. R. Hyder - Yale, New Haven, USA
G. K. Mehta - NSC, New Delhi, India
R. Rebmeister - IReS, Strasbourg, France
L. Rohrer - LMU & TU, Munich, German
D. Storm - UW, Seattle, USA
P. Thieberger - BNL, Brookhaven, USA

EXHIBITORS

AccelSoft, Inc.

Meyer Tool & Manufacturing, Inc.

Sciaky, Inc.

MegaVolt Ltd.

National Electrostatics Corp. (NEC)

Vacuum One/Leybold Vacuum Products

I
RAY HERB MEMORIAL

A Retrospective of the Career of Ray Herb

G. A. Norton, J. A. Ferry, R. E. Daniel and G. M. Klody

National Electrostatics Corp., Middleton, Wisconsin, U.S.A. 53562

Abstract. Ray Herb's career in the development of electrostatic accelerators spans 65 years. He began in 1933 by pressurizing a Van de Graaff generator, for the first time. Over the next six years, the group at the University of Wisconsin, under his direction, developed the fundamentals of equipotential rings, potential grading, corona triode control, and other basic mechanisms for the practical use of electrostatic accelerators while making fundamental contributions to experimental nuclear physics. This group held the world's record in sustaining potential difference of 4.5 MV. During World War II, he worked on radar at the Radiation Laboratory. After the war, Herb resumed his career with further fundamental contributions including metal/ceramic bonding, ultrahigh vacuum pumping, negative ion source development and metal charge carriers. The company, National Electrostatics, under his direction manufactured the accelerator which still holds the world's record for the highest sustained potential difference of 32 ±1.5 MV. Throughout his career he led teams which made the electrostatic accelerator a valuable tool for applications in a wide variety of scientific fields, well beyond nuclear physics.

INTRODUCTION

Ray Herb's active career in experimental nuclear physics, accelerator development and applications spans 65 years. His technological achievements revolutionized experimental nuclear physics. He was a rare individual who not only produced grand ideas but had the vision, understanding, capability and tenacity to pursue and develop the appropriate details to produce successful experiments and instruments. Throughout his life, Herb's personal attributes attracted good people to join his work. With a knack for judging skills and character, he molded them into highly effective teams. His contributions go beyond nuclear physics by providing the instrumentation for research, development, and production in a wide variety of fields as diverse as archeology, climate studies, biomedical research, semiconductors and many others.

Raymond George Herb was born on January 22, 1908 at the family farm just east of the small village of Navarino in central Wisconsin. He was the second youngest in a family of nine.

At a time when formal education often ended in the eighth grade, Ray Herb decided to go on to high school primarily because of football, and because at that time, boys were often expected to continue on. He had always had a fundamental curiosity about how things worked, and because of that, he decided to enroll at the University of Wisconsin taking courses in both chemistry and physics. He found the relatively clear-cut boundary

TABLE 1. Events in the Life of R. G. Herb

1908 Raymond George Herb born January 22, Navarino, Wisconsin
1926 Enrolls at the University of Wisconsin-Madison
1929 Decides to study nuclear physics
1931 Bachelor of Science, begins accelerator work
1933 First accelerator pressurized with air
1934 Ph.D. thesis work begun, ^{7}Li(p,γ) at 400 keV
1935 Ph.D. awarded, begins construction of 2.4 MV accelerator
1937 First publication describing the 2.4 MV accelerator
1939 Proton - Proton scattering paper published, 860 keV to 2.39 MeV
1940 2.4 MV accelerator upgraded to 4.5 MV
1940-1945 Radiation Laboratory (radar development)
1945 Marries Anne Williamson in Gainesville, Florida
1947 Using 4.5 MV accelerator, metal staples added to the belt
1953 First practical getter pump
1956 First practical charge exchange ion source, 20µA, of H$^-$
1959 First all metal/ceramic accelerating tube
1960 First practical metal charge carrier, with J.A. Ferry
1965 National Electrostatics Corp. is incorporated
1968 First large Pelletron order, 8 MV tandem for Brazil
1968 Receives the Tom W. Bonner Prize
1970 14 MV tandem Pelletron sold to Australian National University
1975 25 MV tandem Pelletron sold to Oak Ridge National Laboratory
1979 Voltage record of 32.0 MV $\pm$1.5 MV achieved at ORNL
1988 ORNL tandem Pelletron runs beam at 25.5 MV, 357 Ni^{+13}
1996 After overseeing the design and manufacture of over 140 accelerators, Ray Herb died, October 1

between the known and unknown in the physical sciences very satisfying. He liked the idea that matters were settled by experiment and not eloquence. In 1929, while a junior undergraduate, he heard a talk by R.H. Fowler of the Cavendish Laboratory in Cambridge. Fowler was at Wisconsin to give help on problems on temperature dependence of photo-electric emission. However, his talk discussed the experimental work of others at the Cavendish Laboratory on resonant reactions from some atomic nuclei due to alpha particle bombardment. Herb was already aware of the many important advances in atomic physics and that almost nothing was known about the nucleus. He felt, based on this talk, that a completely new field of physics was opening up. It was at that point that he decided to enter a career in experimental nuclear physics. Table 1 shows a brief outline of the major points in Ray Herb's career.

THE 1930's

In 1931, after receiving his undergraduate degree, Ray Herb and other students drove a Model T Ford from Wisconsin to California. There he met E.O. Lawrence for the first time. He saw what looked like a small tin can that Lawrence described as a particle accelerator. Herb was not impressed. On his return to the University of Wisconsin, he began working with Dr. Glenn G. Havens, a postdoctoral fellow, to build a vacuum insulated belt charged Van de Graaff type accelerator. This work was done under Professor Charles Mendenhall, who was the head of the Physics Department at Wisconsin. The vacuum container was a steel tank about 3 feet in diameter and 6 feet long. The high potential electrode was mounted on a re-entrant Textolite tube with a metal sphere open to atmosphere through which the charging belt ran. The space exterior to the sphere and its support was evacuated. The machine went to about 300 kV where it was limited by discharges from the sphere to the tank wall.

In the spring of 1933, Glen Havens left the university and, after much discussion within the Physics Department, Ray Herb decided to pressurize the vacuum container.

Herb along with D.B. Parkinson and D.W. Kerst pressurized the container with air. Measuring with a meter stick, when the flat end of the tank bowed out about 3/16 of an inch, Herb decided it was time to stop (1, 2). The potential rose to 500 kV very quickly in about 3.3 atmospheres of air. With this success, the group immediately began development of a suitable acceleration tube and the first practical pressurized accelerator.

This success led to a series of papers published from 1935 to 1940 by the group at Wisconsin headed by Ray Herb. Much of the accelerator development was done with D.B. Parkinson and D.W. Kerst with theoretical guidance for the nuclear structure program primarily by Gregory Breit. Herb always made it a point to receive guidance from theorists, whom he regarded as having an elevated position from his own experimental work. He developed a close rapport with Eugene Wigner while Wigner was working at the University of Wisconsin from 1936 to 1938. They were both politically conservative in an otherwise liberal academic environment.

FIGURE 1. Associate Professor Ray Herb in early 1940 standing outside of Sterling Hall on the campus of the University of Wisconsin in Madison. *Courtesy of Joe McKibbon.*

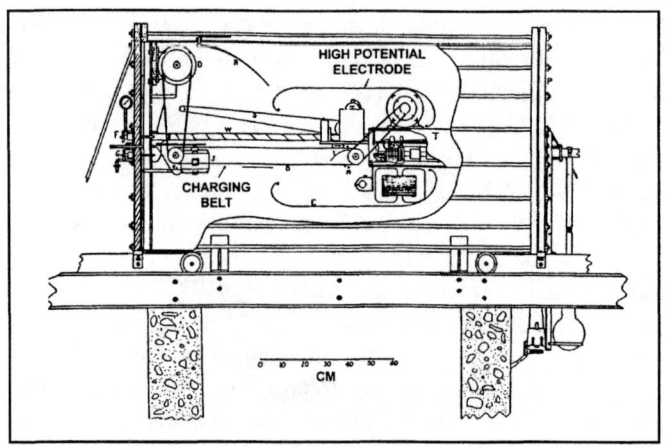

FIGURE 2. In Early 1935, Herb's first accelerator was housed in a vacuum container which was pressurized to about 3 atm. of air. This accelerator ran reliably at 400 kV and was used to produce a proton beam for Herb's Ph.D. theses, ^{7}Li(p,α). The charging belt is mounted to the left of the high voltage terminal and the accelerating tube is mounted inside a Textolite tube exiting to the right.

First Accelerator and Ph.D. Thesis

Herb's first paper was received for publication in April 1935 (3) and his second paper which contained the description of the accelerator followed immediately (4). As soon as the high pressure air experiments showed promise, they began reconfiguring the high voltage generator to accommodate an acceleration tube (Fig. 2).

The charging belt was moved out of the Textolite tube to the opposite end of the tank. This left the support Textolite tube empty to accommodate an acceleration tube. With a grant of $150.00 from the Wisconsin Alumni Research Foundation, they tried dozens of acceleration tube configurations before settling on a glass/metal assembly bonded with red sealing wax (American Express No. 2). Because of their many design attempts, the acceleration tube assembly was made separate from the Textolite support structure on its own movable platform so the acceleration tube with magnetic analyzer could be moved in and out of the high voltage generator (Fig. 3). With the tube design settled in 1934, they were able to run reliably at 400 kV. Herb then took data for his Ph.D. thesis, which was the ^{7}Li(p,α) reaction. His experiment involved both thick and thin films made by evaporating the lithium target inside the analysis endstation. His experimental techniques turned out to be superior to work done earlier by Cockroft and Walton. Already, Herb was beginning to show his talent for high precision experiments. The publication from his thesis (3) shows the many small developments that were necessary for particle counting, beam current measurement, and clean target preparation.

Other than high pressure air insulation, this accelerator was typical of the small accelerators which were just starting to be built. Both the column and tube were ungraded and little thought was given to electric field shaping. With his Ph.D. completed and his

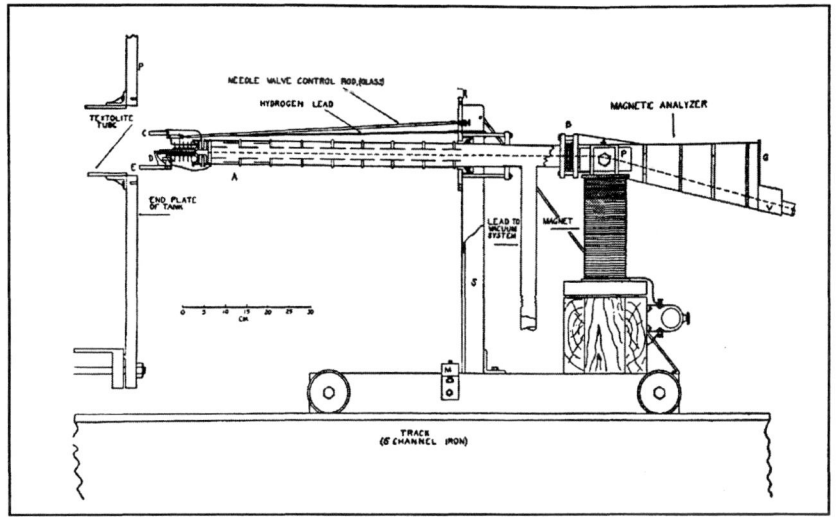

FIGURE 3. In order to facilitate the many design attempts for a successful acceleration tube, it was made self supporting independent of the pressurized vacuum chamber. The ion source, acceleration tube, and magnetic analyzer were on one rigid, movable assembly.

papers written, Herb left for the Department of Terrestrial Magnetism (DTM) of the Carnegie Institute in Washington. There he worked with Merle Tuve, Lawrence Hafstad and Odd Dahl on their large open air Van de Graaff type accelerator. This open air system operated easily at 1 MV and was used for measurements up to 1.2 MV. Herb spent the summer of 1935 building a resistance voltmeter for the accelerator.

2.5 MV Accelerator

It appears that Herb may have had more on his mind while he was at DTM than building the resistant voltmeter. He returned to the University of Wisconsin in the fall of 1935 and his third publication, "The Development and Performance of an Electrostatic Generator Operating Under High Pressure Air" was received by the Physical Review in October 1936 (5). This

TABLE 2. Fall 1935 to Fall 1936
2.5 MV Accelerator - Designed, Built and In Use

- New column design and new tank
- Potential grading on column
- Dual belt charging
- Field shaping - column rings
- Potential grading on acceleration tube
- New tube design
- Feedback voltage control
- Insulating gas mixture

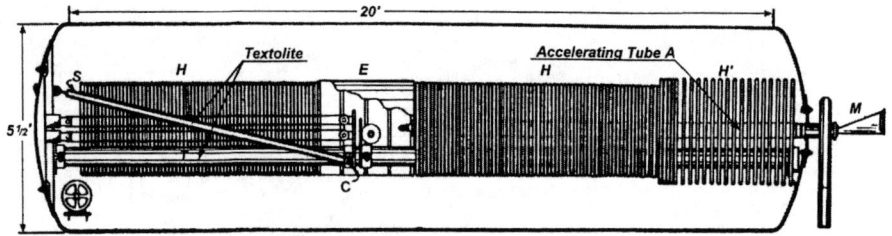

FIGURE 4. The "long tank" machine was the first electrostatic accelerator with a fully graded column and acceleration tube. The major support structure (T) is a 6" diameter Textolite tube near the base of the column, running the full length of the tank. There were two charging belts in the column to the left with the acceleration tube exiting to the right.

paper had a significant number of innovations which effectively defined the appearance and form of electrostatic accelerators to the present (Table 2).

Herb, Parkinson and Kerst obtained a new pressure vessel which was 5-1/2 feet in diameter and 20 feet long with a maximum air pressure design of 100 psig (6.8 atmospheres) (Fig. 4). They designed a column which externally looks similar to a modern day tandem accelerator, with two charging belts in one end and the acceleration tube in the other. They were very concerned about spark protection for the acceleration tube. Therefore, they decided to enclose the acceleration tube in a concentric cylinder. To best approximate a conducting cylinder, they used equipotential rings which were graded with a point to plane corona discharge system. This is the first time that field shaping was done for the entire accelerator structure.

Up until this time, all of their experience with acceleration tubes concerned straight walled glass cylinders, which were limited by flash over along the inner surface. They decided to try porcelain cylinders which were corrugated on the inside and outside as shown in Figure 5. During tests of the individual porcelain cylinders, each 2-1/2" cylinder held 100 kV without difficulty until the interior was evacuated. Therefore, they decided to add potential grading to the tube in order to prevent charge build up on the ceramic. The acceleration tube was supported by the column at 15" intervals. Herb would continue to use this design of separate potential grading for the acceleration tube and column for the rest of his career.[1]

Another innovation contained within this accelerator was feedback voltage control. They used a generating voltmeter, as they had in the small machine, which had been developed earlier (6). It was calibrated using the Li(p,γ) resonance at 440 keV. Herb had found that he could bleed charge off of the high voltage terminal by placing a single corona needle opposite the terminal at the tank wall. By amplifying the current flowing through the needle, they were able to automatically vary a charging power supply voltage to keep the electrostatic accelerator at a constant potential. This feedback system held

---

[1] In 1978, National Electrostatics developed s-series Pelletrons which have a single, tube mounted grading system. However, all Pelletrons above 5 MV maintain the dual grading system.

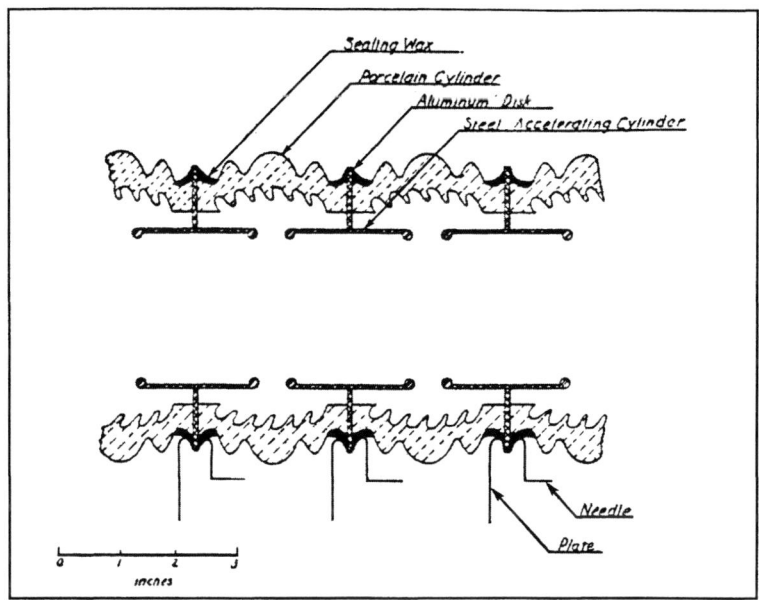

FIGURE 5. Before the long tank machine, all acceleration tubes were glass and metal. After extensive testing, a corrugated porcelain cylinder was used with a 2.5" inner diameter. The steel electrodes were bonded to the porcelain cylinders with red sealing wax and covered by an aluminum disk on the outer diameter of the porcelain cylinders. The acceleration tube was supported by the column at 15" intervals. It was immediately successful. However, at the highest voltage for this accelerator, 2.5 MV, it ran at less than half gradient of modern acceleration tubes.

the terminal potential to within 0.5% to 1%. The only periodic voltage fluctuations they saw had a frequency which was the same as that of the charging belts. They attributed this to the splice in the belt.

The new tank allowed them to reach an air pressure twice that of the small machine that had been a vacuum container. Herb had noticed while cleaning the old machine that he achieved higher voltages immediately after cleaning with CCl_4. He experimented by adding CCl_4 and other solvents. With a saturated vapor pressure of CCl_4 in air at atmospheric pressure, the breakdown potential was 1.7 times the breakdown potential of ordinary air. In the new large machine, he first used CCl_4 up to 75 lbs. and later switched to Freon.

The accelerator performed very well almost immediately. With the tank at atmospheric pressure, they were able to run at 500 kV. At 100 psig they were able to run for short periods of time at 2.5 MV and obtain reliable data at 2.16 MV. From the beginning, the acceleration tube gave no difficulties. When first mounted in the tank, it pumped down immediately to 2×10^{-6} Torr. However, the group was very conservative and conditioned up to the 2.5 MV over a period of a few months while taking data on proton elastic scattering reactions. The voltage limitation was reached when sparks occurred down the Textolite support column and belt.

This accelerator was in almost constant use supporting the experimental nuclear structure research program at the University of Wisconsin. After two years of use, the group published another paper which was received by the Physical Review in February 1938 (7). They had made no major design changes and the Textolite column supports and aluminum hoops were still in very good condition. They had not changed the potential grading corona needles throughout this time, and the needles had become very dull. They did make changes to the charging belt and to the accelerating tube which resulted in routine operation at 2.4 MV for data on proton scattering.

The original charging belts were a type of rubberized sheeting used in hospitals. They worked well, except they were frequently ripped by sparks unless thoroughly dried. The group switched to balloon fabric manufactured by the Goodyear Company.

Only minor modifications were made to the acceleration tube. These modifications involve the first five electrodes near the ion source to improve focusing. Originally, they had fixed corona gaps to provided a fixed focus. This was changed to allow a variable corona gap on the first and third sections, which allow them to vary the focus of the beam depending upon the requirements of their particular beam energy. Also, some leaks had developed in the tube as the red sealing wax became brittle. This was repaired by adding a coat of Picein wax which was a bit more plastic.

The 1938 paper also documents the fire hazard of using compressed air. Five fires over two years of operation included a wooden support fire, belt fires, power supply fires, and a small Textolite fire. The group was using a mixture of air at 100 psig with 5 kilograms of CCl_2F_2. For extinguishing fires, they permanently connected a cylinder of CO_2 to the tank, so that the tank could be flushed after the pressure had been dropped close to atmospheric pressure.

Modifications were made to their automatic voltage stabilizer by J.L. McKibben. He devised a two stage direct coupled amplifier between the corona needle opposite the terminal and the charging power supply. This allowed them to run consistently with the stability within 0.5%.

In 1936 Dr. S. Chandrasekhar suggested they look for H^-. Using the older, small machine, they reversed the polarity on the charging power supply and saw lots of x-rays and some H^- after magnetic deflection (8).

It is important to remember that all of this accelerator development was secondary to the nuclear structure program. Although the Cyclotron had surpassed the Wisconsin accelerators in energy, it could not compete with the ease and range of energy control or the low energy spread of the electrostatic accelerator. This allowed the group at Wisconsin to see much greater detail in their yield curves than was available elsewhere. Working with Gregory Breit, this group produced a series of papers from April 1935 through January 1940 investigating resonant reactions from proton scattering. They reported on the series of closely spaced resonances up to 2.6 MeV, which was the highest proton energy they could reach at the time (9, 10, 11, 12, 13, 14).

Looking at these papers, it is hard to understand when members of this group had a chance to sleep. The paper submitted in March 1937 shows yield curves for (p,γ) from below 400 keV to about 2.0 MeV for sixteen elements from carbon through lead (9). The gamma ray intensity was measured with a "quartz fiber electroscope of the Lauritsen type". The current integrator was made by Joe McKibben based on a thyratron tube. The targets were both thick and thin films, evaporated onto a target holder inside the analysis endstation. At the end of the paper, they stipulate that this was only considered to be a preliminary survey.

In May 1937 the group submitted a paper on "The Range of Protons in Aluminum and in Air" (10). This work was to be very useful for Herb's later work on the scattering of protons by protons.

Quite possibly, "The Scattering of Protons by Protons" (12) was Herb's most important paper in nuclear structure research. From January 1937 through October 1937 the group measured scattering cross sections from 15° to 45°, from 860 keV to 2.392 MeV. They constructed a hydrogen filled scattering chamber with a magnetically suppressed Faraday cup. An ionization chamber was used for scattered particle detection. Great care was taken to eliminate as many sources of error as possible. The paper describes in detail the effects of slit edge scattering and the care taken to assure the purity of all gases used. In addition, "all measurements were made several times by at least two persons." They also followed up with preliminary work on proton scattering on krypton and argon.

This work provided Gregory Breit with the necessary experimental data he needed for a thorough theoretical analysis of the short range forces between protons.

4.5 MV Accelerator

In August 1940 the Physical Review received a brief article describing the dismantling of the 2.5 MV accelerator and its replacement with a 4.5 MV accelerator in the same pressure vessel (15). The initial version of the 4.5 MV accelerator is shown in Figure 6.

The new configuration had the acceleration tube and a single belt in the same column

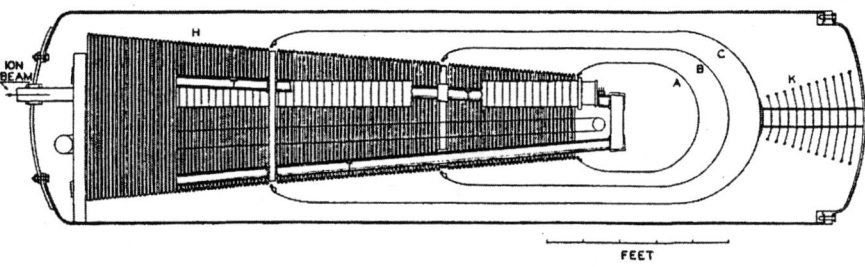

FIGURE 6. In late 1939 it was decided to redesign the 2.5 MV accelerator in the same pressure vessel. A single charging belt was used in the same column as the acceleration tube. This allowed the length of the acceleration tube to increase from 53 gaps to 62 gaps. For the first time, inner shields were used (A, B, C). For mechanical strength, insulator K supported the outer shield C with the end of the tank. However, insulator K proved troublesome, limiting voltage. For later use, at Los Alamos, inner shield C and insulator K were removed. While at the University of Wisconsin, this accelerator ran reliably at 4.3 MV. This accelerator along with a "short tank" accelerator built by Joe McKibbon were transported to Los Alamos for use in the Manhattan project.

structure. This allowed the tube to increase in length from 53, 2.5" sections to 62, 2.5" sections. To further improve the fields, two inner shields were added with a small support insulator (K). Their generating voltmeter now could only see the potential on the outer electrode (C). In order to determine the ion beam energy during their initial test, the beam was brought into air through an aluminum foil. The proton beam measured about 18 cm which, based on Herb's earlier work, indicated a proton beam energy of 3.5 MeV. Freon was added to the insulating gas and the proton beam measured 27 cm in air which indicated a terminal potential of 4.3 MV. These measurements were not done remotely. Herb merely walked up to the beam and measured it with a meter stick.

Insulator K gave significant difficulty. Meanwhile, the acceleration tube and charging belt performed very well. Therefore, the support insulator K and the inner shields were removed. This accelerator held the world's highest voltage record for about 12 years.

At a time when other institutions were attempting to build very large accelerators based on Herb's work, Herb chose to make only relatively minor changes (2). Herb felt very strongly that not enough was known about the fundamentals of electrostatics to determine if present accelerator techniques could be scaled up. It was this conservative nature which delayed his entry into the manufacture of commercial accelerators for over two decades.

THE WAR YEARS

In December 1940 E.O. Lawrence visited Ray Herb to invite him to join the MIT radiation laboratory which had just been formed in October of that year. Herb was to be the fortieth in a staff which was to grow to over 2000. He made occasional visits back to the University of Wisconsin to assist his students until they, as a group, along with the two electrostatic accelerators moved to Los Alamos to assist in the Manhattan Project.

Of this group, Joe McKibben who had just graduated in May 1940 was the most senior. McKibben's 2 MV test machine and Herb's 4.5 MV accelerator were to be in almost constant use at Los Alamos (16). The 2 MV accelerator in the short tank was converted to produce a deuteron beam for the production of radioactive iodine. The 4.5 MV machine, the long tank, was used to measure fission cross sections. During this time A. O. Hanson developed slit control for the first true energy feedback control system (17).

At the end of the war, the short tank machine was purchased by the newly formed Los Alamos Laboratory while the 4.5 MV machine was returned to the University of Wisconsin where it remained in use, with many modifications, until 1958.

At the Radiation Laboratory, Herb became a member of the laboratory's steering committee which included John Trump and Tom Bonner among its twenty-two members. Near the end of the laboratory's life, Herb served on the editorial board of the Office of Publications, which produced approximately twenty-seven technical books detailing the technical advances of the radiation laboratory.

Concerning the radar project itself, Herb was part of the ship application group which developed shipboard radar used for surface search as well as fire control and navigation aids. However, his main responsibilities were as head of the Advanced Service Base

FIGURE 7. As one of his major duties during his five years at the Rad Lab, Ray Herb was stationed in Paris heading up the Advanced Service Bases for installing radar on the continent. He reported to the head of the British Branch of the Radiation Laboratory, John Trump.

Group that installed the forward based radar installations in Europe. Herb's headquarters were in Paris which reported to the British Branch of the Radiation Laboratory headed by John Trump (Fig. 7).

Herb very rarely talked about these years. He did not consider it a significant part of his life's accomplishments. He did relate one story concerning his situation in Paris. He was part of the group that arrived in Paris very soon after the German retreat. The living conditions were rather primitive, and the food left much to be desired. The local office of IBM (truly International Business Machines) was in need of technical assistance. Herb traded expertise and material which helped IBM reestablish badly needed service and equipment. In return, Herb's group was allowed to use the commissary at IBM which had a French chef.

The Radiation Laboratory was closed on December 31, 1945. On his way back to Wisconsin, Herb married Anne Williamson. They were introduced by mutual friends in Boston while Anne was doing research in the Nutrition Laboratory at the Harvard School of Public Health. Ray Herb had taken his very first undergraduate physics course from her father, Professor Robert C. Williamson.

They traveled from the Radiation Laboratory down to Gainesville, Florida, where Professor Williamson was on the faculty at the University of Florida at the time. After their wedding, on December 26, they had a two week honeymoon in Mexico before returning to work at the University of Wisconsin.

They moved into a small home on two acres outside of Madison. Their first son, Steve, was born in 1947, followed by their first daughter, Rebecca, in 1948. Together, they raised five children including Sarah, Emily and Bill (1954 to 1962). Steve became a high energy physicist graduating from Cornell. Rebecca is a Professor of Mathematics at the University of Maryland, and Sarah is an architect in Bozeman, Montana. Emily received a fine arts degree from the University of Utah and now is active in dance and pottery. Bill received his Ph.D. in mechanical engineering from the University of Minnesota.

When not working, which was relatively rare, Herb enjoyed life in a rural setting. In the mid 1970's, the family moved approximately twenty-five miles outside of Madison

FIGURE 8. Oscar Sala and Ray Herb are shown working on the 40" radius, 90° electrostatic analyzer in 1948. This device was used in one of Herb's high precision, absolute energy experiments (20).

to a sixty acre heavily wooded lot. Recreation for Ray Herb was firewood cutting and splitting, brush clearing, camping and canoeing on the Wisconsin River. He was extremely physically active well into his mid 80's.

UNIVERSITY OF WISCONSIN

Ray Herb arrived back in Madison, Wisconsin, in mid January 1946 as a full professor. He immediately began assisting in rebuilding the nuclear physics research program at the University. He was instrumental in bringing new people to the department, including Heinz Barschall, Hugh Richards, Willy Haeberli and others, including theorists. He concentrated on increasing the precision of his nuclear scattering experiments as well as improving the equipment by pushing towards 5 MV (18).

Herb immediately began to build on the advances that occurred with the 4.5 MV accelerator while it was at Los Alamos. He built a 90°, 40" radius electrostatic analyzer (Fig. 8) with slits in order to select an ion beam with a very small energy spread. This produced an energy resolution of 1 out of 5000. His purpose was to explore very narrow resonances in nuclear reactions which were not possible with any other type of accelerator (19). His group was able to show that the width of the 985 keV ^{27}Al(p,γ) resonance was no greater than 400 eV.

This work was immediately followed by using the electrostatic deflector for absolute voltage determination (20). The potential on the 90° electrostatic deflector was produced by a stack of batteries. The battery stack produced 15 kV with 0.01% stability. This was a typical, high precision experiment designed by Ray Herb. In addition to a fully calibrated battery stack, great effort was expended to determine the precise plate separation from 0° to 90°. They investigated the Li(p,n) threshold (1.882 MeV), the strong resonance in F(p,γ) (873.5 keV) and Al(p,γ) resonance (993.3 keV). This work was done with S.C. Snowdon and Oscar Sala. Oscar Sala was to play an important role in the development of very large accelerators based on Herb's principles two decades later.

In the meantime, a small company was being formed near Boston (High Voltage Engineering Corporation) by John Trump, Denis Robinson and Robert Van de Graaff. Their first order came from the newly formed Brookhaven National Laboratory. HVEC was to provide an electrostatic accelerator to act as an injector into the first proton Synchrotron, the Cosmotron. This was to be the first accelerator to produce an ion beam in the GeV region.

Ray Herb was hired to act as consultant and to oversee the manufacture of this 4 MV electrostatic accelerator, HVEC Model D. The Cosmotron with its injector came on-line in 1952 and went off-line in January 1967, but the injector, the first commercially available electrostatic accelerator, was kept in use for radio biological research. In 1980 it was moved to Columbia University. It is still in use for radio biological purposes (21).

Herb decided not to join HVEC. As in earlier years, he was discouraged by the lack of fundamental information concerning the behavior of high fields in the vacuum of an acceleration tube. For his consulting work he was paid in HVEC stock, which came in handy many years later for the initial capitalization of National Electrostatics Corp.

Back at the University of Wisconsin, Herb began a broad based program to investigate ultrahigh vacuum techniques, high temperature metal to metal and metal to ceramic bonding, negative ion beam formation, electrostatic charging systems and electrical breakdown in vacuum (18).

In attempting to go beyond 4.5 MV, it appeared that the acceleration tube was the limitation. At that time, all acceleration tubes were sealed with rubber gaskets and organic cement and were pumped by diffusion pumps which also produce vapor in the vacuum volume. Herb felt that these were uncontrolled phenomena which would mask any fundamental limitation.

Groups under Herb's direction worked on titanium sublimation gettering combined with ionization pumping. Herb asked an undergraduate student, Dave Saxon to test various metals, when they are evaporated, for gettering affects (22). He found that titanium was far and away the best material. Two beginning graduate students, Robert Davis and Ajay Divatia, continued the work (23). They soon learned that titanium gettering only pumps chemically active gases. The background argon was not pumped. To correct this situation, they installed a center tungsten wire cathode surrounded by two concentric cylindrical wire anode grids to provide radial oscillation of electrons which would ionize inert gases and drive them into the chamber wall. The sublimated titanium would then coat the container walls, trapping the gases. This work continued and was refined to produce the pump which was known as the "Orbitron" (24). For a time, it was a commercial product of National Electrostatics Corp. (NEC), Figure 9.

FIGURE 9. "Getter Ion" Pump equipped with two pairs of ionization cells on each side of a titanium sublimator tree. A hot filament (source of electrons) was positioned at the base of each screen cylinder. A potential was maintained between the central wire and the screen cylinder.

Concerning acceleration tube development, Herb was determined to remove organic materials from the vacuum volume. He decided to investigate metals and insulators which had similar coefficients of expansion to allow high temperature bonding. The best initial results were obtained with alumina ceramic and Kovar (25). Kovar is an alloy of nickel, cobalt and iron which bonds very well to the alumina ceramic. This tube was installed in a new high gradient accelerator and was operational by 1959.

Herb continued tube development over many years, continuing well into the NEC years. For very high gradients, Herb's group developed an alumina/titanium bond which is used in all of NEC's Pelletron accelerators.

Herb's group explored the possibility of producing negative ions, which had a history dating back to before 1932 (26). The Wisconsin group noted the work by Alvarez (27) and proceeded to produce an H⁻ source (28). This source had a hydrogen gas charge exchange canal. It produced 20µAmps of H⁻. This source also produced He⁻ of up to 0.06 µAmps. Herb's group did not pursue this development much further.

As Herb continued to seek lower energy spread in his beam and better overall voltage stability, he became dissatisfied with the dielectric belt as a charge carrier. In 1947, staples were put in the standard belt. Herb reported that it gave excellent voltage stability and reasonable lifetime. However, he was dissatisfied with the manufacturing process (18). He continued developments for many years with Victor Fung, one of his students, and later with Jim Ferry, who started in this program as an undergraduate. One of the configurations was a "string of beads" which were metal charge carriers separated by an insulator. These early "string of beads" charging methods failed. Herb and Ferry began on a scheme using metal charge carriers separated by insulating spacers which were driven through a tube to the terminal. This did not seem to be working out well either.

During a visit in 1959, Joe McKibben suggested to Jim Ferry that he go back to the "string of beads" configuration. Working on his own, Jim came up with a practical solution which evolved into the present Pelletron charging chain.

During his years at the University of Wisconsin, Ray Herb officially supervised forty-four Ph.D. candidates (29). However, many students would start out with Herb learning the practical aspects of experimental nuclear physics and later switch to working with other professors in the department. Also, Dave Parkinson and Don Kerst were officially students of Gregory Breit. However, most of their work was done with Ray Herb.

In 1955, Herb was elected to the National Academy of Sciences, and in 1961 he accepted the Charles Mendenhall Professorship of Physics. He received honorary degrees from the University of Sao Paulo in 1959, the University of Basel in 1960, the University of Lund in 1993, and the University of Wisconsin (for his technological contributions outside the University) in 1988. In 1968, he was awarded the Tom W. Bonner prize in Nuclear Physics.

NATIONAL ELECTROSTATICS CORP.

By the mid 1960's, funding for equipment in nuclear physics was near an all time high while funding for studying basic electrostatics was becoming very hard to obtain. The use of electrostatic accelerators was climbing very rapidly at this time as illustrated by the number of abstracts submitted to the American Physical Society (26). It became apparent to Ray Herb that if he wished to continue to develop his ideas for electrostatic accelerators he would have to form his own organization.

Along with James A. Ferry and Theodore Pauly, Ray Herb formed the National Electrostatics Corp. on September 13, 1965 (Figure 10). In December of that year they set up shop in a warehouse just outside of Madison, Wisconsin. The stated purpose was to

FIGURE 10. Shown in the new NEC factory in 1970 are Ted Pauly, George Nelson (outside board member, banker), Ray Herb, and Jim Ferry.

manufacture and market high vacuum apparatus and high voltage accelerators. With minimal shop equipment they continued to refine the design of the titanium getter ion pumps, design and manufacture all metal valves and constructed a 2 MV electrostatic accelerator for testing.

Much of their concern shifted from the basic investigation of electrostatics into manufacturing techniques. They decided that the Pelletron accelerators would be built with a modular column which would allow relatively small design changes to go from 2 MV to 20 MV. They settled on a gradient of 1 MV per 2 feet. This remains the "official" gradient at the present time. However, the original tube design which began at NEC with thirty-three insulating gaps per MV continued to evolve.

Although there were a few small projects which came from the University of Wisconsin, there were no significant orders during the first few years. Although Herb remained a Professor at the University of Wisconsin, he had nearly exhausted all of his financial resources before the first large accelerator order was obtained.

The Department of Nuclear Physics of the Institute of Physics of the University of Sao Paulo, Brazil, had close ties with the Department of Physics at the University of Wisconsin for many years. Professor Oscar Sala was a student of Ray Herb's in the early 1950's. In Brazil, the importance of Nuclear Physics Research was well understood. A 22 MeV Betatron had already been installed. A 3.5 MeV electrostatic accelerator was built in Brazil in 1952-1954 (30). Based on the research with that first home built machine, the decision was made by the Brazilian Government to form a substantial research facility at the University of Sao Paulo.

In spite of the close ties with Herb and the University of Wisconsin, Brazil had a tough decision. The High Voltage Engineering Corporation was well established and had installed a large number of electrostatic accelerators throughout the world. By contrast, NEC had eight full time employees and two part time employees operating out of a small warehouse. NEC was proposing a 22 MeV beam energy for protons using a 4 MV electrostatic injector followed by an 8 MV vertical tandem Pelletron (Figure 11). The University of Sao Paulo decided to take the risk with this new design and this new company. The contract arrived at NEC in August 1968.

Ray Herb was now confronted with a considerable challenge. The largest accelerator he had built before ran at 4.5 MV. He recognized that this large, vertical accelerator would have to be assembled at the factory and tested before shipment. Therefore, they immediately purchased land approximately one mile from the warehouse, and construction of the building began. The original NEC building, 100 feet x 100 feet was completed in December 1968. Behind it, construction was underway for the accelerator test tower. The 126 foot high tower was completed in July 1969. The 4 MV single-ended vertical injector was mounted on a platform above the vertical 8 MV tandem Pelletron. One wall of the tower was made removable so that the 10 foot diameter, 44 foot high tank could be moved in for testing and later removed for shipment. The Pelletron Models 8UD and 4U were shipped in November 1970. The 8 MV tandem Pelletron is still in active use today.

At about the same time as the Sao Paulo order, the University of Wisconsin purchased

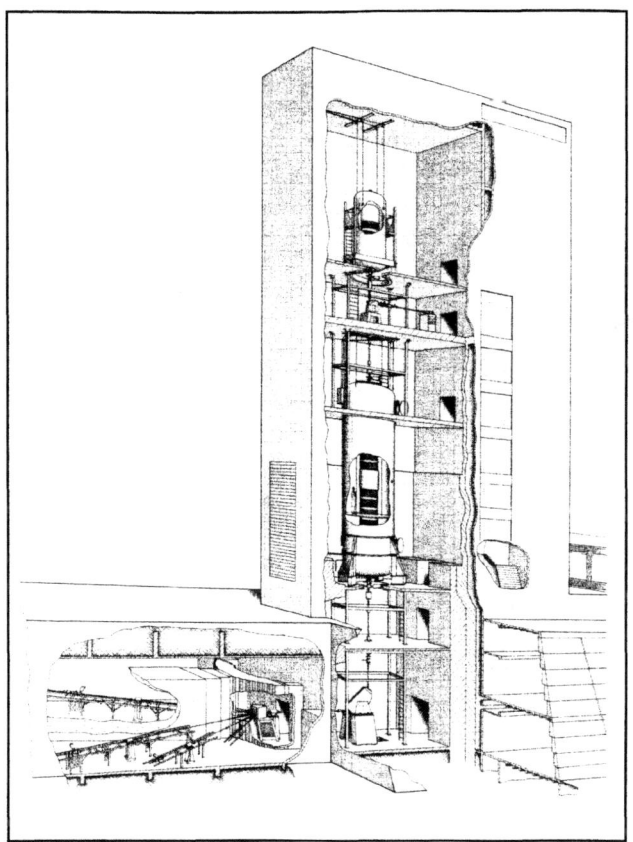

FIGURE 11. The proposed accelerator system for the University of Sao Paulo was a negative, 4 MV injector into an 8 MV tandem Pelletron. NEC expected to meet the 22 MeV beam energy by pushing the 8 MV tandem to 9 MV.

a 2 MV electron accelerator. However, the next significant challenge was to come from the Australian National University in Canberra. The Sao Paulo project had demonstrated that NEC had the ability to at least build a machine up to 8 MV terminal potential. Sir Ernest Titterton felt that NEC was a good bet to build a vertical 14 MV tandem accelerator. At the time, this was the highest guaranteed voltage from a commercial company. That order was received in December 1970. Like Sao Paulo, the ANU 14UD Pelletron continues to prosper and has run experiments, after some acceleration tube modification, as high as 15.75 MV.

The next 14UD was sold to the Weizmann Institute in Rehovot, Israel in May 1972 and was followed by a 12 MV tandem Pelletron for Tsukuba University in Japan, November 1973. Each machine was an improvement over its predecessor. The team assembled by Ray Herb continued refinements of the charging system, column structure, and acceleration tube. The next significant challenge was to come in 1975.

FIGURE 12. This photograph shows Ray Herb standing in front the high voltage terminal shell of the 25 MV tandem Pelletron manufactured for the Holifield National Laboratory in 1977. The terminal shell is 4.06 meters in diameter and 4.88 meters in height. It houses the gas/foil stripper assembly, electrostatic lenses, and the 180° magnet. *Courtesy of Oak Ridge National Laboratory.*

By 1975, funding for nuclear structure research was declining rapidly. The Oak Ridge National Laboratory had been proposing a very large accelerator system for some years. With half of their funding assured, they had to decide between two suppliers, HVEC and NEC. In their final design plan they stipulated that the accelerator must be vertical. This suited the modular construction of the NEC Pelletron. NEC submitted the lowest bid. NEC was now faced with manufacturing an accelerator with almost twice the voltage requirement of any accelerator built before. In addition, it was a folded design which had not been commercially manufactured. ((Figure 12) The folded tandem has both acceleration tubes in one column with a 180° magnet in the high voltage terminal. The system has almost 2000 beam handling parameters which required computer control.

Herb's ability to build and maintain teams was essential in building the company through the period of the manufacture of the ORNL 25URC tandem Pelletron. In March 1976 NEC also accepted an order for a 20 MV folded tandem from the Japan Atomic Energy

Research Institute. Then, in October 1977 another order was received for a vertical, straight-through 20 MV tandem from the Comision Nacional de Energia Atomica in Argentina. All three of these accelerators were required to operate at voltage levels that had never before been reliably achieved. As he had many years in the past, Ray Herb along with Jim Ferry, developed an efficient organization with competent engineers, physicists, and crafts workers to bring these projects to successful conclusions.

The first real indication that no fundamental errors had been made in the construction of the 25URC came in 1979 when the Holifield National Laboratory announced the results of the column voltage test. Without the accelerating tube installed, the 25URC ran at 32.0 MV ±1.5 MV which is the highest sustained manmade potential difference (31). Ray Herb had again reclaimed the record for the highest voltage accelerator.

From 1968 through 1980, NEC had accepted orders for twenty-two electrostatic accelerators many of which were in stages of installation or manufacture while the three very large tandem Pelletrons were undergoing design, manufacture and testing.

The original manufacturing philosophy that Herb had set up of modular construction and many common parts proved to be very valuable. Through the early 1980's, every accelerator differed from its predecessor. There were as many different models as there were customers. The flexibility of modular designs allowed NEC to survive at a time when nuclear structure research funding was rapidly decreasing and other applications were coming up. In addition, it allowed continued development and upgrades which were necessary, especially with the large Holifield accelerator.

Although the column of the 25URC went well beyond its rated voltage, the acceleration tube proved to be the limitation. NEC, along with Holifield personnel including Paul Stelson, Norval Ziegler and Charles Jones, continued acceleration tube development to produce a compressed geometry tube which was installed in the accelerator in early 1987 (32). The Holifield lab announced in 1988 that they had run an experiment with a terminal voltage of 25.5 MV (357 MeV, Ni^{+13}).

As the market place changed to lower voltage accelerators for new applications outside of nuclear physics research, this tube development proved to be very valuable. In the early 1980's NEC produced electrostatic accelerators for MeV ion implantation. They are used in a production setting. There is no time available for voltage conditioning, and the accelerator is required to run routinely at its maximum rated potential. This began the manufacture of a series of 1 MV tandem Pelletrons using the new compressed geometry tube. This trend continued into materials analysis where the accelerator would be operated by personnel with no accelerator experience. Herb oversaw the change from the manufacture of accelerators to the manufacture of complete materials analysis instruments and materials modification tools.

In the meantime, work was also progressing with electrostatic electron accelerators. Typically, the demands for electron accelerators are met by linacs where energy spread and continuous beam operation are not a necessity. However, the development of free electron lasers required a narrow energy spread, continuous electron beam in the million volt range with currents approaching 1 ampere. The group led by Dr. Luis Elias ran a series a feasibility tests at NEC. Improvements in these techniques have continued over the years for additional applications including electron cooling of antiproton beams. A DC electron beam of over 2 MeV with a continuous beam current of more than 100 milliamps has been achieved in a recirculating system (33).

Early on in this project with Dr. Elias's group, the Star Wars funding was just starting up. NEC was visited by senior officers from the three major armed forces. During a meeting, with Ray Herb present, the design for the proposed 3 MeV system was discussed. One of the officers asked if it could be scaled up to 10 MeV. The overall dimensions and difficulties were discussed in some detail. Then, another officer asked if it could be scaled up to 20 MeV and, finally up to 50 MeV. In each case, Ray Herb was ready, even with a drawing showing the overall dimensions for a 50 MeV system. His only pause came when he was asked what it would take to put it in orbit around the earth.

Ray Herb remained as Chief Executive Officer, President and Chairman of the Board of NEC until an undiagnosed illness forced him to conduct his business from home in May 1996. He made many calls every day and continued to oversee the operation of the company with Jim Ferry as Chief Operating Officer. This did not stop until just six days before his death on October 1. Ray Herb did not believe in retirement.

Throughout his life, Ray Herb was a conservative optimist. He recognized that he did not know the underlying principles for voltage conditioning or other limitations to electrostatic accelerator design. All of his advances were based on well thought out, precise experimentation. He believed very firmly in progressing step by step in order to determine what was important in the accelerator design. He always made sure that he had a firm grasp of what he did not know. One of his favorite sayings was "Life isn't dull." For him, it definitely was not.

ACKNOWLEDGMENTS

The authors would like to thank Anne Herb, Professor Hugh Richards and Professor John Cameron for reviewing and contributing to this paper. Also, the authors appreciate the information supplied by Professor Robert M. Williamson and Dr. Joe McKibben.

REFERENCES

1. Herb, R. G., "Bonner Prize Acceptance Speech," recorded by McKibben, J. L., American Physical Society Meeting, April 24, 1968.
2. Herb, R. G., "Electrostatic Accelerators in the 1930's," *Proceedings of the Third International Conference on Electrostatic Accelerator Technology*, Oak Ridge, TN, April 1981.
3. Herb, R. G., Parkinson, D. B., and Kerst, D. W., *Physical Review* **48**, 118-124 (1935).
4. Herb, R. G., Parkinson, D. B., and Kerst, D. W., *Review of Scientific Instruments* **6**, 261 (1935).
5. Herb, R. G., Parkinson, D. B., and Kerst, D. W., *Physical Review* **51**, 76-83 (1937).
6. Kirkpatrick, D., and Miyake, J., *Review of Scientific Instruments* **3**, 1 (1932).
7. Parkinson, D. B., Herb, R. G., Bernet, E. J., and McKibben, J. L., *Physical Review* **53**, 642-650 (1938).
8. Herb, R. G., private communication (1995).

9. Herb, R. G., Kerst, D. W., and McKibben, J. L., *Physical Review* **51**, 691-698 (1937).
10. Parkinson, D. B., Herb, R. G., Bellamy, J. C., and Hudson, C. M., *Physical Review* **52**, 75-79 (1937).
11. Bernet, E. J., Herb, R. G., and Parkinson, D. B., *Physical Review* **54**, 398-408 (1938).
12. Herb, R. G., Kerst, D. W., Parkinson, D. B., and Plain, G. J., *Physical Review* **55**, 998-1017 (1939).
13. Plain, G. P., Herb, R. G., Hudson, C. M., and Warren, R. E., *Physical Review* **57**, 187-193 (1940).
14. Hudson, C. M., Herb, R. G., and Plain, G. J., *Physical Review* **57**, 587-592 (1940).
15. Herb, R. G., Turner, C. M., Hudson, C. M., and Warren, R. E., *Physical Review* **58**, 579-580 (1940).
16. McKibben, J. L., private communication (1993).
17. Hanson, A. O., *The Review of Scientific Instruments* **15**, 57-63 (1944).
18. Herb, R. G., *Revista Brasileira de Fisica* **2**, 17-35 (1972).
19. Warren, R. E., Powell, J. L., and Herb, R. G., *The Review of Scientific Instruments* **18**, 559-563 (1947).
20. Herb, R. G., Snowdon, S. C. and Sala, O., *Physical Review* **75**, 246-259 (1949)
21. Marino, S. A., private communication (1998).
22. Redhead, P. A. *History of Vacuum Science and Technology*, Volume 2, New York, New York, International Union for Vacuum Science & Technology Association, 1995, Lafferty, J. M., "History of Vacuum Science: A Visual Aids Project," 97-98.
23. Herb, R. G., Davis, R. H., Divatia, A. S., and Saxon, D., *Physical Review* **89**, 897 (1953); Divatia, A. S., Davis, R. H., and Herb, R. G., *Physical Review* **93**, 926 (1954).
24. Mourad, W. G., Pauly, T., and Herb, R. G., *The Review of Scientific Instruments* **35**, 661 (1964); Douglas, R. A., Zabritski, J., and Herb, R. G., *The Review of Scientific Instruments* **36**, 1 (1965); Herb, R. G., Pauly, T., Weldon, R. D., and Fisher, K. J., *The Review of Scientific Instruments* **35**, 573 (1964); Maliakal, J. C., Limon, P. J. Arden, E. E., and Herb, R. G., *The Journal of Vacuum Science and Technology* **1**, 54 (1964).
25. Micheal, I., Berners, E. D., Eppling, F. J., Knecht, D. J., Northcliffe, L. C., and Herb, R. G., *The Review of Scientific Instruments* **30**, 855 (1959).
26. Bromley, D. A., *Nuclear Instruments and Methods* **122**, 1-34 (1974).
27. Alvarez, L. W., *The Review of Scientific Instruments* **22**, 705 (1951).
28. Weinman, J. A., and Cameron, J. R., *The Review of Scientific Instruments* **27**, 288 (1956); Windham, P. M., Joseph, P. J., and Weinman, J. A., *Physical Review* **109**, 1193 (1958).
29. Richards, H. T., private communication (1998).. Sala, O., *Revista Brasileira de Fisica* **2**, 11-12 (1972).
30. Cameron, J. R., private communication (1998).
31. Heilman, C., *The Guiness Book of World Records - 1998 Edition*, New York, Guiness Media, Inc., Bantam Books, 1998, pp. 133.
32. Raatz, J. E., Rathmell, R. D., Stelson, P. H., and Ziegler, N. F., *Nuclear Instruments and Methods in Nuclear Physics* **A244**, 104-106 (1986).
33. Larson, D. J., Anderson, D. R., Adney, J. R., Sundquist, M. L. and Mills, F. E., *Nuclear Instruments and Methods in Nuclear Physics* **A311**, 30-33 (1992).

The Canberra 14 UD - a Jewel in the NEC Crown - a Living Memorial to Ray Herb

D. C. Weisser and T. R. Ophel

Department of Nuclear Physics
Research School of Physical Sciences and Engineering
Australian National University, Canberra, ACT 0200 AUSTRALIA

I. INTRODUCTION

Ray Herb's fatherhood of Pelletron Accelerators must be judged a success. He achieved the pinnacle of fatherhood - all of his children are convinced that they are special - each of us is sure we are the jewel in the crown. It doesn't matter whether it is the biggest, the 25 URC in Oak Ridge, or the small RBS tandem in an undergraduate school or, the one surpassing its specifications - the 14 UD in Canberra. Each sparkles with its unique lustre.

The gem under the glass here is the 14 UD at the Australian National University. The many facets of Ray Herb's contributions to accelerator technology are reflected in the 14 UD's original features and indeed in some of its improvements. For like a well reared child, the 14 UD was provided with the depth of character to adapt to the challenges of growth - to survive and improve because of the pivotal developments that led to its conception and the strength of the decisions which framed its structure during gestation.

II. ELECTROSTATIC ACCELERATOR TECHNOLOGY

Many of the things that everyone takes for granted in electrostatic accelerators can be traced to a fertile period early in Herb's accelerator life that started in 1933. These and the crucial foundations of the unique features of the 14 UD and other Pelletrons were established during his "first career" at Madison. As paterfamilias, Herb nurtured an extended family of accelerator expertise at the University of Wisconsin. The NEC branch of the family, founded in 1965, fostered the commercial aspects, but never lost touch with its university roots.

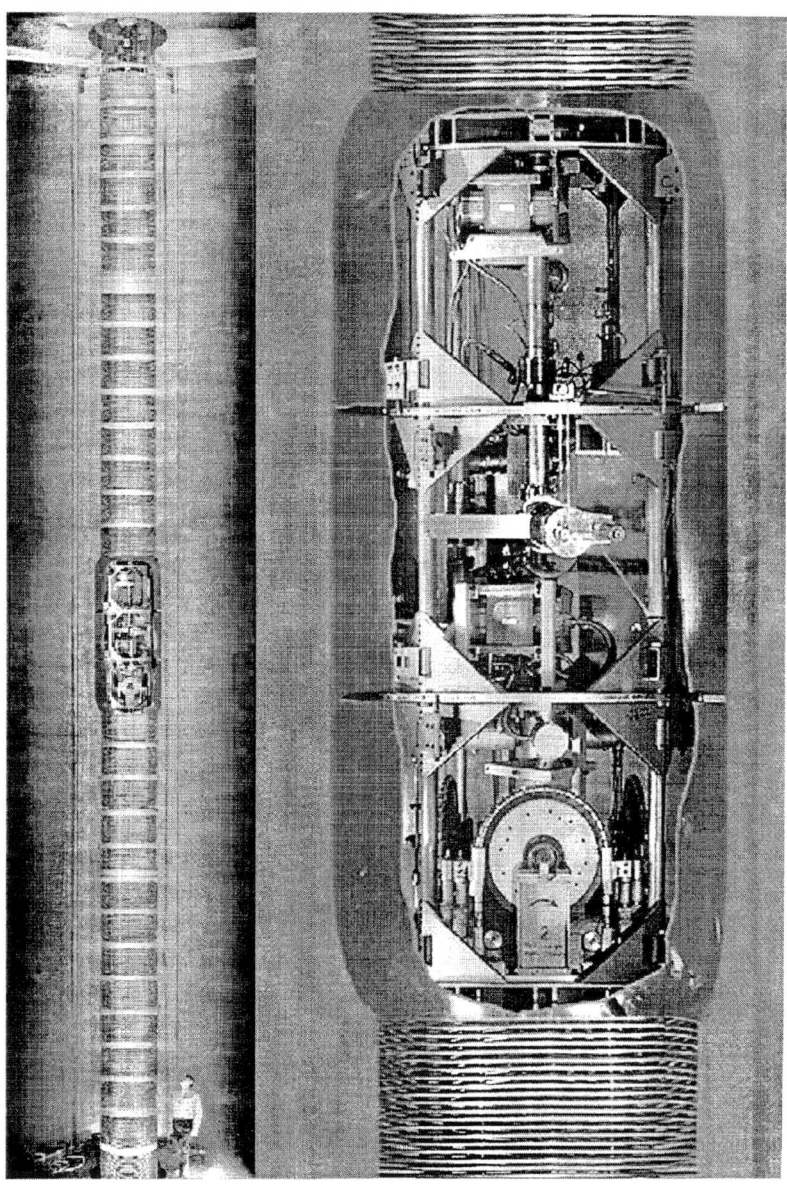

FIGURE 1. The Canberra 14 UD and a close-up view of the terminal. Both photographs are computer composites prepared by T. Thompson.

The key concepts and developments introduced over the two thirty year career spans include :-
(1)   high pressure gas insulation of electrostatic accelerators [1]
(2)   uniform voltage distribution of accelerating structures [2]
(3)   equipotential hoops [2]
(4)   corona point-to-plane voltage dividers [2]
(5)   evapor-ion pumps [3]
(6)   ceramic-titanium bonding [4]
(7)   modular tube and column structures
(8)   geometrical decoupling of tube sections
(9)   the Pelletron charging system
(10)   distributed vacuum pumping.

The latter four stemmed from work at NEC and are covered by patents. The term Pelletron is a trademark of NEC. Almost every feature of the above list is evident, either explicitly or implicitly, in Figure 1.

Interestingly, Herb's own perception of the relative importance of the developments at the University of Wisconsin and NEC differs somewhat from that given above. In 1974, his article "Pelletron accelerators for very high voltage" [5] states that the foundations were metal to ceramic bonding, the first getter-ion pump, the first ion source to provide useable negative ion beams of protons and oxygen [6] and the first negative helium beams [7]. Moreover, "the charging chain is probably not as important as the NEC accelerating tube for the practical realization of potentials above 10 MV". Modestly perhaps, those great leaps forward between 1933 and 1937 when he introduced gas insulation and uniform voltage distribution, along with equipotential hoops, were omitted. Perhaps by then he too took for granted those standard features of all electrostatic accelerators. Inclusion of the two negative ion developments is curious, though they were important to confirm the feasibility of tandem acceleration. Negative helium had in fact been reported much earlier [8]. However, a number of no less important achievements were made at Wisconsin in relation to helium beams. Intermediate energy (400 keV) stripping of He^+ to obtain high energy He^{++} beams was pioneered in the Wisconsin 4.5 MV machine [9], and in 1966, the first negative helium source using alkali metal (Cs) vapour exchange was built there [10]. Herb was involved with the former of these at least.

III. THE 14 UD -THE EARLY YEARS

The 14 UD was born out of discussions between Ray Herb and Sir Ernest Titterton (Figure 2). The ANU was granted funding late in 1969 to upgrade the Canberra facilities. NEC was considered as one possible provider from the outset. However, by Titterton's account, Herb was very reluctant to build an accelerator with a rated voltage above 10 MV. During this time, the fledgling company was completing installation of the first substantial NEC machine, an 8 UD with a 4 MV single-ended injector, for Sao Paulo. Local folk-lore in Canberra has it that Titterton finally convinced Herb to bid on a 14 MV machine during long discussions held in Madison one weekend early in 1970. During that time, the essential features of a notional accelerator were sketched out. It can be presumed with some confidence that the goal was to conceive a tandem accelerator which would surpass both the already operating HVEC MPs and the expectations of the XTU, then under test. They would have been as one from their intrinsic characters, in needing the machine to be clearly cheaper than any HVEC alternative option. The reason 14 MV was finally agreed upon is not obvious though. Later, Titterton justified it in terms of the continuity of proton energy provided by the 14 UD (0 - 28 MeV) and by a negative ion cyclotron injector installed on the EN accelerator (26 -38 MeV) at around the same time. Since the 14 UD was to be used primarily to accelerate heavy ion beams and was indeed housed with shielding consistent with only such use, the rationale in terms of proton beams is not convincing. More likely, it was simply that 14 MV was as high as Herb was prepared go after finally yielding on his earlier limit of 10 MV.

FIGURE 2. Sir Ernest Titterton (L) and Ray Herb (R) at the console of the about-to-be inaugurated (though not accepted) 14 UD accelerator (September 1973).

The ANU considered offers by three manufacturers, choosing that of NEC largely on price, but with a considerable leap of faith in the then as yet largely still to be proven NEC technology. The Sao Paulo experience had been far from unblemished, but Canberra took the gamble anyway, conscious of Herb's formidable record of achievement. In fact, it was a courageous decision by both parties. Construction of the pressure vessel and the gas handling system were to be undertaken on site by the ANU, so that NEC could only test the tube and column in air at Madison. ANU was also responsible for beam handling elements and the target area. Thus the project was the first major departure from the custom of acceptance tests prior to shipment, or at least the demonstration of a prototype of a new accelerator model, at the site of the manufacturer. Predictably, the administration at the ANU was uneasy dealing with a small, unknown company, especially when acceptance tests would not follow the usual pattern. The order was placed in December 1970, but the contract was not finally signed until May 1971.

That Herb and Titterton were right did not come at first blush. There were many such blushes along the way. As at Sao Paulo, the fundamental strengths of the NEC concepts were almost swamped by the fragility of much ancillary equipment. The acceptance tests were expected to take one NEC staff member, Robert Rathmell, three months. One physicist expanded to two when Robert Daniel joined him (Figure 3). Three months grew to twelve and both families came to Australia to console the exiles. Early progress was reasonably satisfactory. The column quickly achieved 18 MV amid dusty building site conditions. With an acceleration tube installed, the 14 UD proved capable of 14 MV and proton transmission tests were successful, though the instability of the beam on target was a matter of some contention between the locals and "the two Roberts". Generously, Herb ruled that the interpretation of "acceptable" was the province of the ANU group.

FIGURE 3. The "two Roberts" with Ray Herb (September 1973). L to R, Robert Daniel, Ray Herb and Robert Rathmell.

Concerns about the proton beam tests were soon replaced by more profound difficulties with chlorine beams. Serious loading of the charging system was evident, sometimes approaching the nominal charging limit of 100 μa for each of the three chains. The behaviour of the accelerator was erratic though. On some occasions the loading was extreme so that high voltage could only be maintained by reducing the injected current to low values; at others, the loading was much reduced and reasonable beam transmission with much higher beam intensities became possible, in fact meeting specified performance.

In recognition that the 14 UD would be used primarily with heavy ions, the terminal was large to accommodate possible focussing or charge state selection units. It is almost certain that Herb was less interested in beam optics than the attainment of high stable voltages. Injection optics seem to have been given less attention in spite of developments elsewhere to ameliorate the strong entrance focussing of an acceleration tube [11]. Somewhat serendipitously though, it became appreciated both at NEC and ANU that a shorting rod arrangement, whereby any configuration of live and shorted units could be selected, was an important beam optics tool for the accelerator. Included originally so that the corona discharge currents could be maintained at low terminal voltages without the need to reduce the SF_6 gas pressure (and perhaps also for diagnosis of troublesome units), use of the rods to maintain a constant entrance gradient meant that the injection optics would be essentially invariant with terminal voltage. Of much greater concern at NEC was the possible loading effect from the unwanted charge states in the high energy tube sections. At the outset of the 14 UD discussions, Herb and Wally Mourad proposed terminal charge state selection based on a complex three-element electrostatic deflection system. However memories of experiences with inclined field tubes provoked conservative, and undoubtedly prudent, concern at ANU about off-axis transport. Moreover, it was pointed out that terminal focussing was in many respects more important than charge state selection, and indeed essential for foil stripper operation. Over a period of time, NEC continued to put forward charge selection or enrichment schemes based on electrostatic doublet and triplet combinations, with or without steering. The ultimate scheme put forward by Mourad was a triple triplet system [12]. ANU finally opted for a single triplet. The triple triplet arrangement was still included by Herb in a notional 20 UD accelerator in 1974 [5]. By then it had been demonstrated in the Canberra 14 UD, that the only significant loading effects occurred in the low energy half of the accelerator since the extreme loading persisted even if the beam was intercepted at the terminal. Thus the loading was due to charge exchange in the low energy tube, with the consequent secondary electron production. Though pressures of 2×10^{-8} torr prevailed at the ends of the tube, it was evident that the pressure was too high within the accelerator itself. The erratic behaviour of that loading arose from another feature of the 14 UD, namely the terminal pumping that was available when gas stripping was used.

The original layout of the 14 UD included a foil stripper, similar to that of the Sao Paulo machine with a capacity of only 50 foils, located centrally within the gas stripper tube. Even before then, the EN tandem in Canberra was using foils almost exclusively [13], being one of the first accelerators to do so. NEC plainly viewed gas stripping as the preferred option. A large, multi-cartridge sublimer pump was included on each side of the gas stripper. Relatively complex rotary switching within the terminal, provided sublimer pellet selection and control of the sublimation rate. ANU pressed for provision of more foils (ultimately 280), but replacement of the foils entailed venting the entire accelerator tube. That serious deficiency was to be overcome with an ANU designed valve that sealed on each end of the stripper tube. It was vital for efficient use of the accelerator.

During the tests with the chlorine beams, the terminal pumps were generally operated before, but not during, injection of the beam. Because freshly deposited titanium continues to pump for many hours after actual sublimation ceases, the unpredictable loading was determined by the time lapse between operation of the pumps and beam acceleration. Though acceptable performance with chlorine beams could be achieved with continuous operation of the terminal pumps, NEC proposed that a pumping station be installed between units 6 and 7. This could be easily implemented because of the Herb feature of distributed power throughout the machine via rotating shafts. The formal acceptance tests were suspended for some 6 – 8 months to allow the exiles to return to Madison for R & R leave, for the ANU to improve the robustness of terminal wiring and for NEC to prepare the new pumping system. Rathmell returned at the end of 1974 to conclude acceptance tests.

IV. THE 14 UD - CONTINUOUS IMPROVEMENT

A. Vacuum Pumps

One would think that Herb would have been content with his substantial contributions to accelerator technology without taking the time to invent vacuum pumps. Growing from such innovation was a gradually more demanding, almost fierce, insistence that high voltages demanded cleaner electrodes and insulators than usually prevailed in electrostatic accelerators. Perhaps the realization that hydrocarbon-free, near ultra high vacuum was essential stemmed indirectly from the high resolution, proton capture resonance measurements he was involved with for some years, beginning in 1960/61 [14]. The aim had been to observe the so-called Lewis effect that arose at the leading edge of a resonance because energy losses of protons are discrete, rather than continuous [15]. It was found that it was extremely difficult to prepare clean enough aluminium surfaces, even with continuous evaporation.

Pump development began in 1953 with the first evapor-ion (now more commonly referred to as getter-ion) unit. In that first design, titanium wire was fed continuously onto an evaporation heater. A hot filament provided copious electrons to ionize gases not chemically reactive [3] (Figure 4). Thereafter, refinements using the orbitron principle evolved whereby electrons orbited relatively long distances around a high voltage wire [16]. In one version, the electrons were used to heat the titanium as well

[17]. By the time of the 14 UD, the pumps consisted of resistively heated titanium cartridges and companion ionizers based on the orbitron principle (Figure 5). Use of the pumps had a two-fold benefit. They provided the quality of vacuum that Herb considered essential, and at a considerable cost-saving over commercial alternatives an aspect always highly-regarded at NEC. Ionizers may have been fine in the inventor's laboratory with a skilled and patient technician nearby, but around and especially within an accelerator the life expectancy of the ionizer filaments and power supplies was short. Filament replacement was difficult with a low success rate for optimum performance.

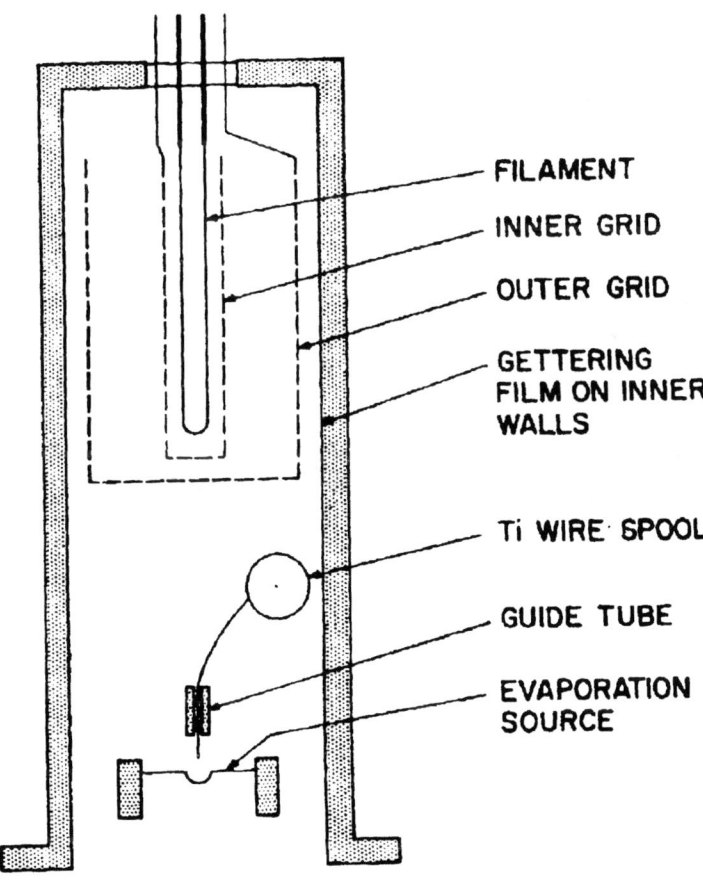

FIGURE 4. A schematic sketch of the first evapor-ion pumps.

FIGURE 5. One of the sublimer/ionizer units used initially to pump the 14 UD tube.

The reliability of the 14 UD was substantially enhanced when the ionizers were replaced with 2 and 10 litre/sec ion pumps immediately after acceptance. As users, we did not have a history invested in ionizer design. On the positive side, the sublimers proved to be robust and their use spread through the laboratory. The original sublimers remain as the main pumps at each end of the 14 UD accelerator.

B. Accelerator Tubes.

There were two major differences between NEC and HVEC accelerator tubes in the late 1960s. NEC tubes were hydrocarbon-free, exploiting diffusion bonding between high density alumina ceramic insulators and titanium electrodes. HVEC employed glass insulators glued with PVA to either aluminium, or later, stainless steel electrodes. The two designed differed as well in how they coped with the inevitable secondary electrons that were liberated when any projectile struck an electrode. HVEC relied on inclining the electric field in alternate 14 segment sections of the tube to sweep the electrons into an electrode before they gained enough energy to produce significantly more than one secondary on collision. The sweeping was augmented with transverse magnetic fields from small permanent magnets.

FIGURE 6. A column post and a module of tubes as originally installed in the 14 UD. Corona point voltage distribution units and the heated aperture assemblies are evident.

NEC in contrast inhibited the secondary emission process by inserting a titanium electrode of reduced diameter at the entrance of each tube module. Electrons produced within the tube volume were intercepted by one of these decoupling apertures, typically within two or three module lengths, so that their acceleration was geometrically limited without the need for inclined fields. Every third one of the decoupling electrodes was usually heated to about 400 degrees Centigrade during operation to maintain a low level of outgassing and thus reduce the likelihood of surface contamination of the nearby electrodes and the insulators (Figure 6).

The NEC tube suppression of secondaries proved adequate to achieve the design gradients and exceed them in the 14 UD. Indeed it soon became clear that continuous heating of any of the electrodes was unnecessary, though the ability to do so assisted initial pump-down of the tube. The space freed by removing the heated electrodes allowed the active length of the tube to be increased by 13%. Such a configuration achieved 16.7 MV after installation in the 14 UD in 1988. The resistance of the titanium to sputtering proved to be so great, that there is no apparent limit to the lifetime of these tubes. Thus Herb's original tube design provided the bedrock for further improvements (Figure 7).

C. Corona Points.

By the late 1960s, the resistor voltage grading systems had become almost universal in electrostatic accelerators, although rather expensive and unreliable. They were the cause of much lost accelerator time, "totally lacking the happy self-healing property of the corona gap" [18]. Happy perhaps with nitrogen/carbon dioxide mixtures but, as will unfold, less felicitous with SF_6, Herb returned to corona point grading for the Pelletrons, perhaps to retain their "in-house" character and reduce costs; alternatively, he had either not recognized or been convinced by advances in resistor design and application. For example, the robustness of Welwyn metal oxide resistors had been demonstrated at the critical terminal region of the EN tandem some years before in Canberra. Separate column and tube voltage grading by corona points can be seen in Figure 6.

Though SF_6 is considered self-healing, that is almost complete recombination occurs after dissociation by corona or sparking, some breakdown products are produced. Several of them are extremely corrosive in the presence of only traces of water. Somewhat ironically, it seemed that in 1982 the breakdown products in the 14 UD preferentially attacked another Herb stalwart, the Pelletron chains. During that year, some eight tank openings were brought about by chain failures. Circumstantial evidence suggested that many of the failures were due to breakdown product attack on the nylon links between the chain pellets. Certainly, deposits became manifest on the terminal spinnings and in the vicinity of the corona points (Figure 8) much more rapidly than previously, and the tank odour was more pungent, immediately after an opening.

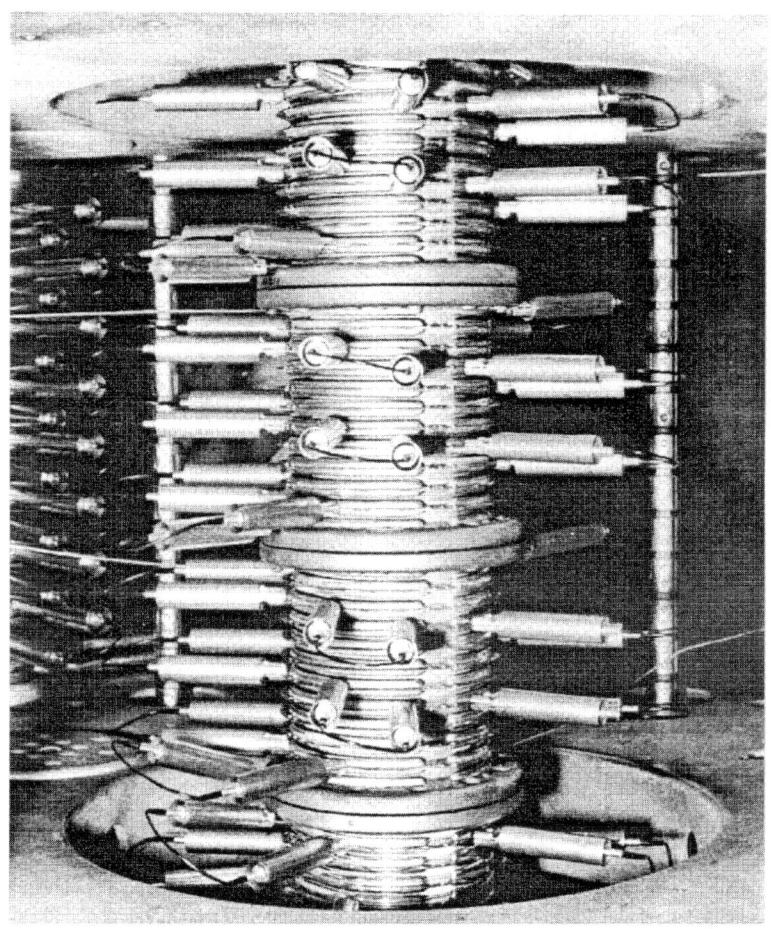

FIGURE 7. Portion of the extended tube configuration along with the resistor assemblies that replaced the corona points.

A method was devised to detect the breakdown products, allowing evaluation of the performance of the system being used to cleanse the gas [18]. Once the excessive levels of breakdown products were circumvented by substantially improving the continuous purification of the SF_6, the chains returned to their historic long lifetimes. All three chains now in the machine have run more than 40,000 hours without any problems.

By 1988, the corona points had served their purpose of bridging the period needed for the evolution of satisfactory resistor systems. Their day was done. The corona assemblies were replaced with locally-designed resistor assemblies (Figure 7) [19, 20]. Overall reliability and stability have been much improved as a result.

FIGURE 8. Typical deposits arising from breakdown products observed opposite the corona points. The colour of the deposits varied, though they were usually yellowish-white. The points shown were unusual in that those on the left unit were pistachio-green while the ones at the right were chocolate-brown.

D. Aging Column Posts

In contrast to the unlimited lifetime of the accelerator tubes, the simpler column posts, which are the strong skeleton of the machine, started to crack following an episode of reverse voltage sparking [21]. They cracked but didn't crumble and the series of failures has self terminated. Although there is no consensus on the detailed reason for the failures, there is broad agreement that it is related to excessive electric stress.

Other column post wear sites also became evident. There was spark induced erosion of the aluminium post flange where it met the last titanium spark gap electrode. It is a testament to the intrinsic robustness of the Herb concept, that these 28 year old posts can be reconditioned (Figure 9) and improved to at least double their already long service life.

V. CONCLUSION

Ray Herb fathered very special machines. He endowed them with a strength of concept and execution that has enabled them to grow and prosper. He did not get everything right every time, but even the failures were based on a drive to understand, and sometimes tripped up by a solution that was stretched just a little too far. We have all learned from these. None compromised the fundamental integrity of the NEC Pelletron. We have Ray Herb to thank for that.

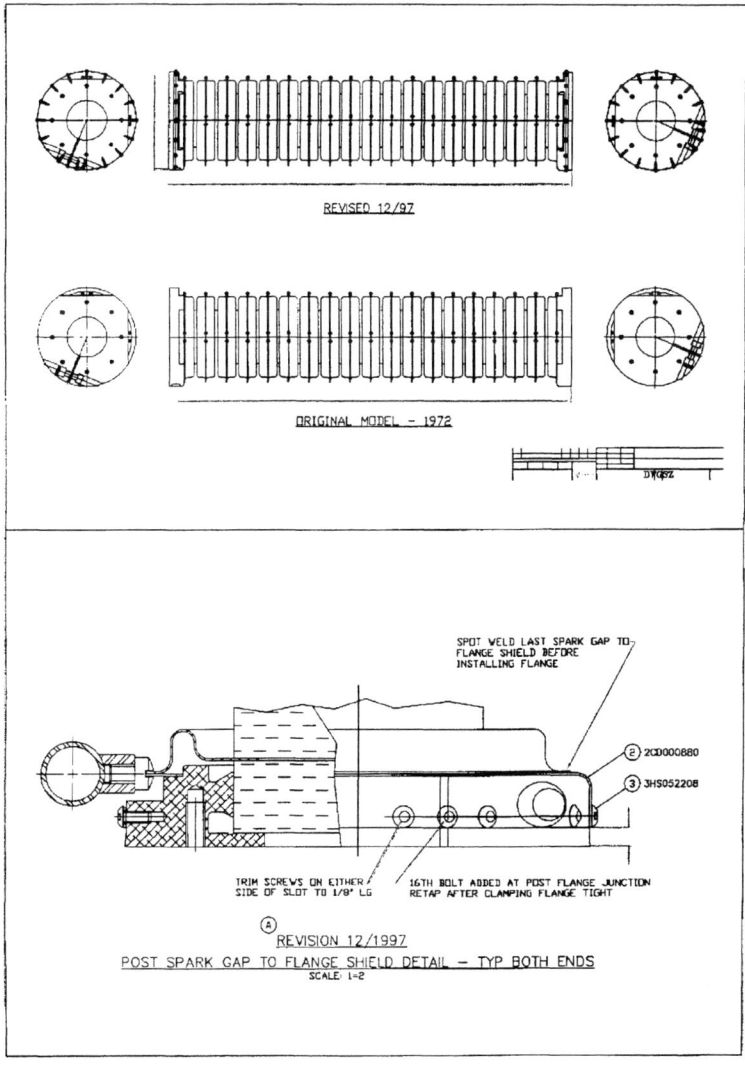

FIGURE 9. Details of a Pelletron post.

REFERENCES

1. Herb, R. G., Parkinson, D. B. and Kerst, D. W., *Rev. Sci. Instr.* **4**, 261 (1935).
2. Herb, R. G., Parkinson, D. B. and Kerst, D. W., *Phys. Rev.* **51**, 75 (1937).
3. Herb, R. G., Davis, R. H., Divatia, A. S. and Saxon, D., *Phys. Rev.* **89**, 897 (1953)
4. Michael, I., Berners, E. D., Eppling, F. J., Kneckt, D. J., Northcliffe, L. C. and Herb, R. G., *Rev. Sci. Instr.* **30**, 855 (1959).
5. Herb, R. G., *Nucl. Inst. & Meth.* **122**, 267 (1974).
6. Weinman, J. A. and Cameron, J. C., *Rev. Sci. Instr.* **27**, 288 (1956).
7. Windham, P. M., Joseph, P. J. and Weinman, J. A., *Phys. Rev.* **109**, 1193 (1958)
8. Dopel, R., *Ann. Physik* **34**, 1 (1925).
9. Bittner, J. W. *Rev. Sci. Instr.* **25**, 1058 (1954).
10. Rose, Fred A., Tolesfrud, P. B. and Richards, H. T., Post dead-line contribution, Minneapolis APS meeting, June 1966; *BAPS* **12**, 29 (1967); *IEEE Trans. NS* **14**, 78 (1967).
11. Larson, J. D., *Nuc. Inst. & Meth.* **122**, 53(1974).
12. Mourad, W. G., *Nuc. Inst. & Meth.* **121**, 333 (1974).
13. Harrison, J. S., Hill, C. J., Ophel, T. R. and Weisser, D. C., *Nuc. Inst. & Meth.* **133**, 575 (1976).
14. Walters, W. L., Costello, D. G., Skofronick, J. G., Palmer, D. W., Kane, W. E. and Herb, R. G., *Phys. Rev. Lett.* **7**, 284 (1961).
15. Lewis, H. W., *Phys. Rev.* **125**, 937(1962).
16. Douglas, R. A., Zabritski, J. and Herb, R. G., *Rev. Sci. Instr.* **36**, 1 1965). Marketed as Electro-ion pumps by Granville-Phillips Co., Boulder Colorado.
17. Orb-Ion pumps based on this method were manufactured by NRC, Newton Massachusetts
18. Bromley, D. Allen, *Nucl. Inst, & Meth.* **122**, 1 (1974). An excellent historical summary of the development of electrostatic accelerators. More specific details of the 14 UD are given by T. R Ophel, A Tower of Strength – a history of the Department of Nuclear Physics, ANU 1950-1997, (Ophel, Canberra 1998).
19. Ophel, T. R., Weisser, D.C., Cooper, A., Fifield, L. K. and Putt, G. D., *Nuc. Inst. & Meth.* **217**, 383 (1983).
20. Weisser, D. C., *US patent 418,676* (July 5 1990)
21. Weisser, D. C., Brinkley, T. A., Clarkson, G. P., Malev, M. D., Ophel, T. R. and Turkentine, R. B., *Rev. Sci. Instr.* **57**, 723 1986).

THE LAST LARGE PELLETRON ACCELERATOR OF THE HERB ERA

S.Chopra, M.M. Narayanan, R.Joshi, S.Gargari, D.Kanjilal, S.K.Datta & G.K.Mehta

Nuclear Science Centre, , P.O. Box 10502, New Delhi 110 067, India

Abstract. Prof. Ray Herb pioneered the concept and design of the tandem Pelletron accelerator in the late sixties at NEC. The 15UD Pelletron at Nuclear Science Centre (NSC), upgraded for 16MV operation using compressed geometry accelerating tubes is the last such large Pelletron. It has unique features like offset and matching quadrupoles after the stripper for charge state selection inside the high voltage terminal and consequently the option of further stripping the ion species of the selected charge states at high energy dead section, and elaborate pulsing system in the pre-acceleration region consisting of a beam chopper, a travelling wave deflector, a light ion buncher (1- 80 amu) and a heavy ion buncher (>80 amu). NSC was established as a heavy ion accelerator based inter university centre in 1985. It became operational in July 1991 to cater to the research requirements of a large user community which at present includes about fifty universities, twenty-eight colleges and a dozen other academic institutes and research laboratories. The number of users in Materials and allied sciences is about 500. Various important modifications have been made to improve the performance of the accelerator in the last seven years. These include replacement of the corona voltage grading system by a resistor based one, a pick-up loop to monitor charging system performance, conversion from basic double unit structure to singlet , installation of a spiral cavity based phase detector system with post-accelerator stripper after the analyzing magnet, and a high efficiency multi harmonic buncher. Installation of a turbo pump based stripper gas recirculation system in the terminal is also planned. A brief description of utilization of the machine will be given.

INTRODUCTION

The Tandem Electrostatic Accelerator has found widespread usage in nuclear physics research as well as in various other disciplines since the introduction of the original Van de Graaff accelerator in 1931 and the discovery of the tandem principle by Bennet in 1935 and independently by Louis Alvarez in 1951. Improved design of the charging mechanism by installing Pelletron chains and titanium ceramic diffusion bonded accelerating tubes were pioneered by the National Electrostatic Corporation (NEC), USA, founded by Prof . Ray Herb of the University of Wisconsin at Madison. This ushered in a new era in the field of big accelerators and quite a few of the presently existing large electrostatic accelerators belong to this class (e.g. 25URC machine at

Oakridge, 20 MV Tandar at Argentina, 20MV Pelletron at JAERI, Japan,and the 14UD Pelletrons at ANU, Canberra and TIFR-BARC, India). The 15UD machine at the Nuclear Science Centre (NSC), New Delhi, India can be counted among the last of such machines. We present this talk as a fitting tribute to the memory of Prof. Ray Herb, for whom this session is dedicated, and describe our experience with the machine which is the centre piece of the inter university accelerator based research facility in India.

There was a long felt need among the Indian universities to have an accelerator based facility within the university system to conduct front ranking and internationally competitive research in nuclear physics and other various disciplines. The idea found its realization when the University Grants Commission in India founded the Nuclear Science Centre as an autonomous research organization for providing such facilities with a heavy ion accelerator. The core facility was decided to be a 15UD Pelletron accelerator later upgraded to hold upto 16 MV potential by introduction of compressed geometry[1] that allowed elimination of heater units and increase of effective insulation length by about 10%. The machine was delivered by 1989 and installation was completed by June 1990. Acceptance tests were conducted subsequently and the facility was made open to users on 8th July 1991. Since then it has been running without any major breaks and have totalled upwards of 45000 chain hours. Uptime of the machine has been around 95% on the average and various modifications and upgradations have been gradually incorporated.

NOTABLE UNIQUE FEATURES

In addition to the compressed geometry tubes which was a novel featrure at the time of installation of the machine, there are a few aspects of the New Delhi Pelletron that need special mention. One of this is the charge selection capability at the terminal using an offset quadrupole[2] and a matching quadrupole situated at the terminal. There is provision for a 2nd stripper at dead section 2 (one-third way down in the high energy section) for obtaining higher energy from the accelerator. After passing through the first stripper (foil or gas) in the terminal the beam is separated into a distribution of many charge states, each of these charge states falling on the second stripper being further stripped leading to a complicated distribution at the high energy end. The off-set quadrupole triplet is shifted slightly to one side so that one of the higher charge states fall on the axis of the quadrupole. By suitable field adjustment one of the charge states is selected through the aperture and the rest of these are stopped in the body of the device. There is a certain limitation in this selection, basically due to the highest potential available for the quadrupole, expressed by the condition $E/q = 2.5$, where E is terminal potential and q the charge state. The matching quadrupole is used to optimize the acceptance of the selected charge state to the high energy side. A 40 nA ^{28}Si beam in the charge state 8^+ yielded about 80nA in the same charge state through the use of this device. This is a consequence of shifting the beam waist nearer to the high energy end resulting in improved transmission.

The pre acceleration beam pulsing system consisting of a chopper, buncher and a travelling wave deflector has been of considerable use. This allows experiments in measurement of lifetimes of nuclear states (isomers) and particle identification through time-of-flight techniques. The basic clock cycle is 4 MHz, allowing beam bursts at a repetition rate of 250ns. The beam is chopped by a pair of plates to pulses of about 20 to 40 ns width. By using double gap tube buncher, widths of the order of 1 ns for protons and about 3 ns for iodine have been obtained with intermediate values for other various beams. A set of travelling wave deflectors consisting of a series of plates operated synchronously allow the repetition rate to be cut down from 250 ns by factors of 2 all the way upto about 2 ms by throwing away intermediate packets of beam. The efficiency of this buncher is about 15%.

SIGNIFICANT IMPROVEMENTS

During the last seven years, considerable effort has been put to enhance and supplement various features of the machine. Chief among these are the following:

i) *Resistor Based Voltage Grading System*: The accelerator had corona based voltage grading system to define the potential both along the column support posts and the accelerating tube. As the Pelletron facility catered to the needs of an academic community with wide-ranging research interests, the terminal potential and ion beam species had to be varied over a large spectrum. In the corona based system, large changes of terminal potential can be accomplished only by the use of shorting rods and changes of insulating gas pressure, since the corona discharge can occur within well defined limits of gas pressure and voltage gradient.Voltage stability and corona lifetimes are also detering factors in this usage. Availability of high quality resistors gave us the opportunity to change over to a resistor based voltage grading system[3].Welwyn resistors (3 G Ω, 40 kV, $\pm$ 2%) were used. Each unit of the 15 unit machine, contained 37 tube electrode gaps and 18 column electrode gaps. A single resistance was put across each tube gap and two resistances in series across each column gap. The high value of the resistances used enables us to run the machine at high potential without overburdening the charging mechanism. A potential of 15 MV can be achieved (under no beam condition) at a typical charging current of 20 μA per chain (the limit of the chain being about 125 μA). Stability of the voltage has improved, to the tune of 2 kV in 15 MV. No shorting rod has been used for a variation of terminal potential from 2.93 MV to 15.3 MV, which are the limits of current operation. Resistance failure due to sparking has been negligible. We have lost about 15 resistances out of about 2000 in 3 years of operation.

ii) *Charging Chain Monitoring by Pick-up Rings* : Capacitive pick-up rings were installed around the Pelletron chains to observe uniformity of charge distribution on the pellets of the chains[4]. Four stainless steel electrodes of 32 mm radius were installed on each of the four column support legs at about 1.5 feet from chain inductors . Rings were mounted on nylon blocks and fixed at 10 mm away from pellets in the chain. The signal from the ring is allowed to develop across two series resistors (150 M Ω each) connected to ground. The signal derived from the midpoint of the two resistances is passed through high pass filters and amplifiers of variable gain and observed on an oscilloscope. A single pellet frequency of 380 Hz has been observed. Chain oscillations produced slower components and also the variation in the tension of chains alters the signal pattern. Monitoring of chain performance through this mechanism has been an important addition to our facility. The chains have so far run more than 45000 hours without any breakage.

iii) *Modifications to Pulsing System*: As the original pulsing system obtained from NEC showed certain shortcomings, gradually some addition and alterations were incorporated[5]. Main problem was of long term drifts in the time width of the bunched beams. The first step was to add a three phase line voltage stabilizer to give a regulation of 1%. The amplitude control circuitry for the feedback signal from the buncher pickup loop was modified to allow proper rectification and maintain better control. The travelling wave deflector circuitry was altered to reduce surge currents at the time of switching OFF and ON and also an additional interlock circuit was provided to check for proper drive at the control grid of all twelve tetrode valves. Finally to avoid phase drift due to variation in flight path of the beam through the accelerator and due to terminal voltage fluctuation a stainless steel spiral cavity was added after the analyzer magnet. The cavity resonates at a frequency of 48 MHz and has a sensitivity of 5.9 ± 0.22 μV/nA. The Stop signal for the TAC (to measure the time width) is now taken from the spiral cavity and the observed time drift has reduced from 5 ns to less than 50 ps. A typical ^{58}Ni run lasting 1 day shows a constant time width of around 2.8ns and a proton run around 700 ps.

iv) *Conversion of Double Unit Structure to Singlet* : The accelerator structure has its columns arranged in terms of double units, with each double unit consisting of six 11 electrode tubes and one shorter 8 electrode tube. In the 15 UD machine on each side there are 7 such doublets and one singlet consisting of 3 full length (11 gap) tubes and one 4 gap tube. During shorting rod operations for reduction of terminal voltage one could short only a double unit at a time. In case of problems with a unit , it could be pin pointed only upto the relevant doublet structure. Further during unit wise conditioning to bring the terminal voltage upto a high value, conditioning had to be done by 2 units at a time. We have now connected the centre point of each 8 gap tube to a corresponding casting plate through a GI wire[6]. Care has been taken to isolate vibration by a loop structure, since the connection is not from a flange but from an electrode plate. The system has functioned well. Extensive unit wise conditioning has been performed with this technique, and it has helped in unit wise trouble shooting.

FUTURE PLANS

1)*Turbo Pump in Terminal* : The vacuum in the terminal area becomes poor during gas stripper run. The titanium sublimation pumps being containment pumps take a lot of gas load. It has been planned to install 2 turbo pumps (250 litres/sec) in the terminal and eliminate the need for the titanim sublimation getter pumps.Final vacuum with this device is expected to be around 10^{-8} T even with gas stripper running at about 500 µ.

2)*LINAC*: In order to upgrade the overall beam energy and also to have beams of high atomic number accelerated above Coulomb barrier energies, a superconducting LINAC booster stage has been planned. Neccessary cryogenics, RF instrumentation etc. have already been fabricated and tested. Fabrication of superconducting resonators is underway in collaboration with Argonne National Lab in the USA. The booster will consist of 4 modules in addition to a buncher cavity.. The main modules will have eight quarter wave coaxial line superconducting niobium cavities operating at 97 MHz (with 4 watt RF power) and providing field gradients of about 3.5 MV/meter. A high efficiency (about 70%) multi-harmonic buncher (preaccelerator) to match this frequency has already been installed.[7] The superbuncher with time width of about 100 ps has already been constructed and awaits testing. The entire eight cavity LINAC module will be housed in a cryostat maintained at liquid helium temperature.. This will enable us to provide beams of upto mass number 80 at an energy of 5 MeV/nucleon. The first cryostat is already fabricated and the first LINAC module is likely to be installed by the end of 1999.

UTILIZATION

The centre, established from scratch, became operational in July 1991 with user community from many universities participating in the establishment of the facilities in the beam hall for research in nuclear structure & reaction dynamics, materials science, atomic physics, radiobiology and radiation chemistry. Experimental facilities for nuclear physics include a Gamma detector array[8] (GDA) of 12 HpGe detectors with anti Compton shields and 14 BGO multiplicity detectors, and a Recoil Mass Spectrometer called Heavy Ion Reaction Analyzer[9] (HIRA). HIRA and GDA can be coupled for tagged gamma spectroscopy measurements. A beam line has been dedicated for materials science research. It has a general purpose high vacuum experimental chamber equipped with setups for on-line or in-situ four probe measurements, luminiscence, thickness monitoring etc. This chamber is followed by two UHV chambers. In the first chamber there is provision for on-line mass analysis, and a UHV Scanning Tunneling Microscope (STM) is being installed for surface modification studies. The second UHV chamber is being commissioned for channeling studies using Goniometer. There are special facilities for ion atom collision, x-ray spectroscopy and radiobiology studies in other beam lines.

The centre has been planned as a "user facility". The user family comprises faculty and students from 50 universities, 27 colleges and 28 research institutions. The accelerator beam time is allotted by a " National Committee" which meets twice a year. - July 8 and December 18. The committee faces an increasingly difficult task in selecting proposals for beam time allotment since the demand is more than twice of what can be allotted inspite of the fact that the accelerator runs round the clock, 7 days a week, except at the time of maintenance. About 60% of the beam time is taken up by experiments in the field of basic nuclear physics. However only 25% of the users are involved in these studies. More than 60% users are involved in projects connected with materials modification in wide class of materials. Users in this area are now focussing their attention to possible dynamic studies, besides in-situ surface modification studies. Emphasis seems to be gradually shifting towards exploring possibilities of swift heavy ions provided by the Pelletron for materials engineering[10].

ACKNOWLEDGEMENTS

We are grateful to acknowledge help rendered at various stages of the development by Mr. J. Kanhiyan, Mr. K. M. Jayan, Miss Indira Iyer, Mrs. V. Jha, Mr. M. Sota and other members of NSC. We also thank NEC, USA, for help in installing the accelerator and also providing help and advice whenever required.

REFERENCES

1. D. Kanjilal, Proc. Symp. North Eastern accelerator Personnel, Kansas State Univ., World Scientific Press, 1993, p. 33.
2. D. Kanjilal, S. Chopra, M. M. Narayanan, Proc. Symp. North Eastern accelerator Personnel, AECL Research, Chalk river, World Scientific Press, 1992, p. 135.
3. M. M. Narayanan, K. M. Jayan, R. Ahuja, S. Chopra & D. Kanjilal, Ind. J. Pure & Appl. Phys. **35**, 196 (1997).
4. S. Chopra, R. Joshi, P. Barua, S. Gargari, M. M. Narayanan, D. Kanjilal, Proc. Symp. North Eastern accelerator Personnel, Rochester Univ., World Scientific Press, 1993, to be published.
5. R. Joshi, P. Barua, V. Jha, A. Sarkar, S. Ghosh, R. K. Bhowmik, Proc. Symp. North Eastern accelerator Personnel, Triangle Univ. Nucl. Lab, World Scientific Press, 1995, to be published.
6. S. Gargari, S. Chopra, R. Joshi, D. Kanjilal, S. K. Datta, Proc. Symp. North Eastern accelerator Personnel, Woods Hole Oceanographic Inst., World Scientific Press, 1996, to be published.
7. A. Sarkar et al, DAE Symp. Nucl. Phys., **41B** (1998), to be published.
8. S. C. Pancholi, R. K. Bhowmik, Ind. J. Pure & Appl. Phys. **27**, 660 (1989).
9. A. K. Sinha et al, Nucl. Instr. & Meth., **A337**, 543, (1994).
10. G. K. Mehta, Vacuum **48**, 958 (1997).

II
ELECTROSTATIC ACCELERATORS

Voltage Limitations of Electrostatic Accelerators

H.R.McK. Hyder

*A.W. Wright Nuclear Structure Laboratory, Yale University
P.O. Box 208124, New Haven, CT 06520-8124, U.S.A.*

Abstract. The history of electrostatic accelerators has been punctuated by a series of projects in which innovative designs have failed to meet the expectations of their designers. From the early, air-insulated Van de Graaffs at Round Hill to certain of the large pressurized heavy ion accelerators of the 1970s and 1980s, increases in size or changes in design and materials have not always led to the maximum voltages expected or extrapolated.

Since these failures have continued beyond childhood into a mature technology, it is reasonable to assume that the causes of voltage limitation are varied and complex. They have remained poorly understood for a number of reasons: resources for an extended program of research into breakdown and failure of electrostatic generators have always been meager, especially for large machines devoted to nuclear research; the inaccessibility of pressurized generators makes instrumentation difficult and testing slow; the calculation of transient and dynamic effects is laborious and the results difficult to verify; voltage test experiments on operating accelerators are inhibited by the significant risk of damage due to energy release on breakdown; and the total voltages (though not the local fields) achieved in many electrostatic accelerators exceed those produced in any other man-made environment.

In this review, the behavior of several generators of different designs is examined in order to assess the importance of the various design features and operating conditions that control the maximum voltage achievable in a working machine.

HISTORICAL DEVELOPMENT

The designer of a circular accelerator who specifies the radius of the ring and the strength of the magnetic field can predict particle energy with confidence. By contrast, the designer of an electrostatic machine who ventures beyond established practice enters a misty region where simple scaling is inadequate and where unexpected physical phenomena may limit the voltage, and hence the energy, which can be attained reliably.

The pioneers who built the first air-insulated accelerators were concerned with the dielectric strength of the air itself and of the insulated surfaces of the charging systems, and with vacuum breakdown in the tube through which the beam was accelerated (1). Using a pair of generators, one of each polarity, the total available voltage could in principle be doubled, but at the cost of locating a laboratory inconveniently inside a high voltage terminal. Problems with size, humidity and foreign bodies, avian and other, soon led to the use of pressurized gas and the design of electrode geometries which minimized the size and cost of the pressure vessel needed for a given voltage.

The second generation of electrostatic accelerators, beginning with Herb's first pressurized machine in 1935 (2), extended the voltage range to 12 MV, although particle acceleration was limited to 8-9 MV. In these machines, attention was paid to reliability and ruggedness because of the potential for damage due to stored energy release on sparking and because of the inconvenience of accessing the internal parts of a pressurized machine. Most were vertical, permitting a simple column structure but incurring the cost of a crane and a tall building. Several made use of multiple intershields in an effort to reduce tank size and gas inventory. High Voltage Engineering Corporation, however, used cantilevered horizontal columns without intershields for machines up to 4MV.

In 1954, Atomic Energy of Canada at Chalk River called for a 10 MeV proton accelerator. HVEC responded with a proposal which became the EN tandem, soon capable of running at 6 MV and eventually at 7MV, accelerating a wide range of light and heavy ions. Subsequent development of electrostatic accelerator technology has concentrated almost entirely on tandems, vertical and horizontal, straight and folded (3). The larger machines all use electromechanical charging systems in the form of belts, chains or ladders. Many of the smaller machines use solid state voltage multipliers, series or parallel fed, by this means eliminating the need for regular in-tank maintenance. This third generation has taken the effective voltage limit from 8-9 MV to 25 MV (4),(5), at which point it does not seem that further increases can be justified on economic grounds, even if technically feasible.

Each generation of accelerator development has presented different challenges to the goal of increased voltage. Some of the technical limitations are well understood, others result from complex phenomena difficult to analyze given the elaborate structure of the large machines. By reviewing the reported performance of machines of different types, some light can be thrown on those aspects of machine design which determine high voltage performance.

THE MACROSCOPIC ELECTRIC FIELD

The original Cockcroft-Walton set operated at modest voltage (0.8 MV) in a large room. The designers were therefore less concerned with the terminal geometry than with surface breakdown over the insulators of the rectifiers and accelerator tube. They achieved a longitudinal field of 0.44 MV/m, very close to the value of 0.46 MV/m obtained by Van de Graaff and Tuve at the Department of Terrestrial Magnetism. Van de Graaff had already achieved 1.5 MV between two 24 inch spheres, corresponding to a surface field of 2.95 MV/m, which is very close to the conventional breakdown strength of air for uniform field. Emboldened by this success, he embarked on the construction of the Round Hill generator, intended to produce 10 MV between two large (15-ft diameter) spheres (6). In fact the best voltage obtained was 5.1 MV (+2.4 MV, - 2.7 MV), corresponding to a peak field on the surface of the spheres of 1.6 MV/m. In spite of problems of high humidity – "…the hangar is about 300 yards from the sea, but during extreme high tides has been entirely surrounded by water" – the voltage limitation was probably due more to the very large surface area of the spheres, to debris on their upper surfaces and to high local fields on some of the roof girders. A famous photograph shows many discharges between the roof members and the top of the spheres and from the sphere-column junction to ground, the curved paths of these discharges showing the influence of the column potential gradient.

In 1935, Herb converted a small tank intended for vacuum insulation into a pressure vessel for air at 4 bar and obtained an increase in voltage proportional to pressure. The accelerator tube was too short to permit operation at full voltage, for which the field at the cylindrical surface of the terminal was 5.25 MV/m. The limiting factor was field enhancement due to the spherical surfaces at the ends of the terminal.

Following this success, a new accelerator was constructed in a tank designed for 8 bar with provision for a tube of increased length. Hoops were used to grade the column, which had the same diameter as the terminal, and a small quantity of freon was added to the compressed air. At 2.4 MV the field at the terminal was 7.3 MV/m. The next change was the introduction of intershields around a conical column and a further increase in the tube length. This was the 'long tank' machine of 1940 which operated at 4.5 MV, with fields on the intershields and terminal of 11-12 MV/m. A remarkable feature of this machine was the textolite tube supporting the outer intershield, provided with corona gap grading and sustaining 1.5 MV over a length of 0.86 m. The use of air at 8 atmospheres was not without hazard and four different fires are described in the paper by Herb and his collaborators (7).

This achievement prompted Van de Graaff and Trump at MIT to design a vertical 4 MV machine with two intershields, known as the "Cambridge" Van de Graaff, which was replicated in Cambridge (England), Ottawa and a number of other laboratories. The gas was either nitrogen with a small admixture of freon or, more commonly, 80% nitrogen and 20% CO_2. In spite of the increased dielectric strength resulting from the higher pressure and the addition of an electronegative component, the radial field at the design voltage did not exceed 14 MV/m. In practice, factors other than the terminal field limited the voltage achieved by these machines.

Demand for higher energies then led to the construction of the MIT-ONR and P-9 Van de Graaffs, respectively at MIT and Los Alamos (3). In both cases the design goal was set at 12 MV. At this voltage the cylindrical field on the terminal of the MIT machine was 14.7 MV/m; that on the third intershield of the Los Alamos machine was 12 MV/m. Both machines easily sustained their design voltage without tubes, but were limited to 8 MV or less with tubes.

In spite of the failure of these machines to operate at design voltage, the convenience and versatility of the electrostatic accelerator was apparent to experimenters. The demand for more and bigger accelerators then led to the commercial production of the 6 MV CN accelerator by HVEC and from that to the tandem accelerators which constitute a third generation.

The first of thirty EN tandems was supplied by HVEC to AECL in Chalk River in 1959. Originally specified as a 5MV machine, with a cylindrical field of 13.2 MV/m, similar machines later achieved voltages in excess of 7 MV when operating in SF_6. The larger FN tandem, initially guaranteed at 7.5 MV, has actually accelerated beam at

11 MV, corresponding to a cylindrical field in excess of 16 MV/m. Initial experiences with the MP tandem, which had been designed with a goal of 13 MV or higher, were unsatisfactory. Without tubes, the column would barely reach 9 MV and the limit was blamed impartially either on high fields at the ends of the bars which formed the terminal shell or on dirt. Adding radiation sources to introduce ionization and providing a screen to trap dirt at the bottom of the tank produced a modest improvement, but in these factory tests MP-0 ran at higher voltage with tubes than without (8). Changes to the terminal and the tubes were made to subsequent machines, permitting operation to at least 12 MV. Unfortunately, no controlled tests were made which would have clarified the exact importance of dirt, in the form of metallic or non-conductive particles. Whatever the cause of these early problems, MP tandems were soon limited by tube performance and further advances then followed the installation of longer tubes.

The final stage in this story is the enlargement of a horizontal MP tandem to achieve 22.5 MV, the development of very large vertical tandems for voltages of 20-30 MV and the construction of the VIVITRON, a very long horizontal tandem incorporating many radical design innovations (9).

The ESTU tandem at Yale has a central tank diameter of 25 ft, equal to that of the XTU (10). With the use of an open frame intershield, the terminal field is reduced to the low value of 12 MV/m at the rated voltage of 22.5 MV. The uniform field on the intershield bars is even less, 10.7 MV/m, but the small radius of curvature of the edges of these bars results in many regions of high local field.

The vertical NSF machine at Daresbury also had a single intershield, but the terminal and intershield fields at the maximum design voltage (30 MV) were higher, around 17 MV/m (11). However, typical operating voltages were around 20 MV, corresponding to a radial field of ~ 12 MV/m. By contrast, the vertical folded tandem (25URC) at Oak Ridge dispenses with an intershield but has a larger tank and terminal for a lower voltage rating (25MV). At this voltage, the terminal field is 13.5 MV/m. This is significantly lower than the values at which many of the smaller horizontal tandems operate.

Considering only macroscopic field strength, many tandems have operated successfully, usually in SF_6 at 6-7 bar, with radial fields at the terminal of 16-17 MV/m. Machines with intershields have achieved lower values, 10-12 MV/m being typical. However, in these machines the field at the tank wall is relatively high. In the extreme case of the VIVITRON, the use of multiple intershields produces an almost uniform radial field between column and tank. The resulting field at the terminal has the low value of 13.8 MV/m at the design voltage of 35 MV but the field at the tank wall is 5.3 MV/m when the terminal voltage is only 18 MV, i.e. about half the design rating. This is higher than the field on the wall of the Oak Ridge 25URC at the full 25 MV.

The significance of these numbers is made clear in horizontal tandems, which are very sensitive to small particles of conducting debris. Nuts, washers or even shreds of material from the charging sheaves may reduce the sparking threshold by several MV. By contrast, large pieces of insulator seem to have little or no effect. The VIVITRON recently completed several months of scheduled operation with minimal reduction in operating voltage in spite of the presence, for most of this period, of large fragments of silica-loaded epoxy resin on the floor of the tank. Some of these fragments, which came from shattered post insulators, had dimensions of 100-200 mm (12).

In horizontal tandems spark marks are concentrated towards the bottom of the terminal and adjacent column hoops, testifying to the importance of debris in initiating breakdown. Vertical tandems should be immune from radial breakdown caused by debris from the column, unless small particles are carried outwards by windage or electrostatic forces.

FINE STRUCTURE FIELDS

The field distribution around the column of a tandem departs from that between two simple concentric cylinders due to the three-dimensional curvature at the ends of the terminal shell, the geometry of the hoops surrounding the column and the gradient along the column. Near the ends of the terminal, and close to electrodes with a small radius of curvature, such as the hoops, the field is enhanced by a factor which depends on the detailed design and may be as much as 1.7 times the macroscopic surface field. On a much smaller scale, regions of even higher field may exist on the surface of sharp objects such as screw heads or hoop joints.

Two alternative approaches have been tried to optimize the electrostatic design. In the first, the terminal diameter is made larger than the column and the ends are shaped so as to minimize the field peaking while providing 'shadowing' to reduce the peak field on the nearby hoops. Examples of this approach are to be found in later versions of the MP tandem and in a single-ended Van de Graaff at Debrecen with a paraboloid terminal (13). The other approach is to make the terminal the same diameter as the column, or even smaller, and to use oval hoops near the terminal. The use of oval hoops was suggested by Boag (14), who calculated a reduction of 20-30% in radial field assuming zero axial field. However, Eastham has shown that for realistic values of axial field oval hoops offer little or no advantage (11).

The precise effect of fine structure fields on performance is not obvious. There is no doubt that the pattern of discharge marks is strongly correlated with regions of high field. Typically, the highest density of spark marks is seen on the hoops nearest the terminal. In tandems using the open "portico" type of intershield there is a high density of spark marks along the edges of the transverse or longitudinal electrodes which make up these intershields. The hoops adjacent to the end of an intershield also attract many discharges. Nevertheless a few spark marks are observed on surfaces where there is little or no field enhancement, e.g. the middle of the terminal electrode or a conventional intershield.

Breakdown may be initiated by some random event occurring on or near the tank wall in a region of relatively high field. The path of the spark may then be influenced by the existence of a region of high local field on the column or intershield and by the distance to which that high field extends into the space between the electrodes. Thus a hoop joint where the ends of the tube have edges with a radius of 0.1 mm or less may have no effect on voltage holding since the resulting field enhancement is confined to a cubic millimetre or less. The hoops themselves, having radii of 5-20 mm, perturb the field over a volume which may be a million times greater. The curved ends of a large terminal will cause a somewhat smaller field enhancement over a much larger volume still. Seen in this light, a design which reduces the peak field on the hoops at the cost of increasing the extent of a region of lower field at the ends of the terminal may not be beneficial.

The same argument might point towards the use of small diameter hoops on a close pitch. Indeed Herb suggested that the ideal method of grading a column would be to surround it with a suitable semi-conducting material, thus eliminating the hoop fine structure (15). No suitable material springs to mind. Practical limitations on the flatness of hoops and the rigidity of their fastenings place a lower limit on hoop spacing. Some difficulties have arisen in EN and FN tandems in keeping the one inch pitch, 5/8" diameter hoops uniformly spaced and able to resist unbalanced electrostatic forces arising from shorted column pitches. In MP tandems, a three-inch pitch is generally used.

Finally, it should be noted that most accelerators operate with a continuous corona discharge from the tips of the stabilizer needles. Spark damage to these points occurs occasionally if the corona mushroom is inserted too far into the tank but in normal operation the corona system is unaffected by discharges occurring elsewhere; in its immediate vicinity the density is low.

COLUMN GRADIENT AND VACUUM BREAKDOWN

Cost and convenience dictate that column length should be no greater than that needed to sustain the required terminal voltage. The actual voltage gradient may be limited by breakdown in the insulating gas, on the surface or in the volume of insulators or inside the accelerator tube.

Existing designs of Pelletrons and Laddertrons have peak fields between links which are as much as eight times the overall column gradient. Thus in order to operate with an overall gradient of 2.2 MV/m, SF_6 at a pressure of 6 bar is needed to give a 60% margin against sparking between links. Reducing this peak field would necessitate an increase in the diameter or spacing of the links. The peak fields between other electrodes, such as the hoops, are usually less and thus need not limit voltage.

Modern charging belts, when operated correctly in the appropriate field configuration, will sustain linear fields of at least 3 MV/m. The most common causes of failure are trapped moisture, changes in the composition of the material and surrounding electrodes which give rise to excessive local fields during transients.

The volume dielectric strengths of the insulators used in electrostatic accelerators are more than adequate for fields of 2-3 MV/m. Volume failure is therefore rare but may occur where flaws or voids cause local field enhancement. Alumina ceramic, which is opaque and has a high dielectric constant, sometimes suffers this type of failure since small voids may not be detectable. Surface failure may be caused by dirt or foreign bodies, over which the designer has no control, or by the absence of nearby equipotential planes, allowing surface charge distributions to build up and excessive fields to develop during transients. Conventional column structures, made up of relatively thin insulators (< 25 mm), can be effectively protected against transient stresses by multiple spark gaps. Large undivided insulators are much more vulnerable to transients, especially in large machines.

In practice, well designed column structures and charging systems, in which individual insulators sustain less than 100 kV and are surrounded by spark gaps, can be expected to operate indefinitely without failure. Processes occurring inside the accelerator tube then set the limit on column gradient.

Accelerator tubes for high gradient operation have electrodes designed to minimize the interaction between the beam and the insulators. The insulator pitch is one inch or less and the internal surface of the insulators is often convoluted. The field between adjacent dished electrodes may be as much as four times the overall field, i.e. 10-12 MV/m, but even this is well below the breakdown field for clean metals in vacuum. However, impurities and adsorbed gases give rise to small discharge currents at much lower fields and conditioning is needed to desorb these materials in a controlled way so that the resulting currents do not perturb the local gradient in such a way as to lead to instability. The energy of secondary electrons must be limited by suppression fields so that these currents do not multiply to the point where they exceed the capacity of the charging system or produce enough ionization in the gas to degrade the potential distribution. Finally, conditions on the insulator surfaces must be such that secondary emission does not result in field intensification near the cathode triple junction and cause surface breakdown. All these effects are regenerative in that a reduction in voltage across an individual pitch due to a microdischarge, to beam or electron loading, or to surface breakdown will cause a transient increase in voltage across adjacent sections. Even with the best tube designs it is difficult to control these effects at overall column gradients much over 2 MV/m. It is this that determines the length of the column.

GASEOUS INSULATION

The choice of insulating gas has important consequences beyond the primary need to allow operation at the highest voltage. The capital cost of gas, transfer plant, storage and purification is significant for all but the smallest machines. Access time for in-tank maintenance puts a premium on simple and reliable internal components. Pressure and chemical reactivity influence design and operation. Economics aside, it should always be possible to specify a gas system which will not limit the operating voltage, even though the early observation that dielectric strength is proportional to pressure is not true for most gases at pressures above a few bar.

The fortuitous observation of the dramatic effect of an electronegative gas by Herb has led to the general use of pure SF_6, a mixture of SF_6 and nitrogen, or the less expensive mixture of nitrogen and CO_2. The dielectric strength of SF_6 as a function of pressure can be expressed as $E_{max} = k.P^{0.66}$. This places a practical limit on pressure of

8-9 bar. Gas mixtures are typically used at higher pressures, 10-15 bar, and must be stored in the gas phase. Evidence as to which system is superior is lacking.

Related to the non-linear pressure dependence is the effect of electrode size. Using data due to Trump (16) and Bortnik (17), the Daresbury group calculated that the breakdown stress in SF_6 at 8 bar is only 26 MV/m for electrodes the size of the Daresbury NSF intershield, compared with 70 MV/m for small electrodes (0.1 mm^2). The cause of this effect is not fully understood, but random breakdown triggered by loose particles may be part of it.

Systems designed to operate with clean dry gas may fail to hold voltage when wet or contaminated. Although the actual dielectric strength of the gas phase may not be reduced by small amounts of water vapor, the effect of moisture on insulator surfaces and on a charging belt can degrade performance and cause damage. The presence of water vapor in SF_6 results in the formation of reactive decomposition products which are harmful to Pelletron insulators and other plastic materials. Long term good performance requires that the gas be kept clean and dry.

TRANSIENTS

Dynamic processes in electrostatic accelerators occur over a wide range of time scales, from nanoseconds for spark breakdown to tens of seconds for terminal voltage discharge. The voltage transient from a tank spark will propagate along the tank and be reflected by the unmatched impedance at the end to return inverted and almost unattenuated to damage column elements near the terminal. Attempts to incorporate matching impedances at the ends of the terminal have shown that it is possible to absorb large amounts of energy in this way, although whether this can eliminate damage to the column is not clear.

Slower transients, such as those arising during voltage ramping, can perturb the column gradient and, where intershields are installed, distort the field distribution near the intershield electrodes, causing a disproportionate increase in local fields.

Finally, an accelerator tube that is on the threshold of conditioning may become unstable and spark if beam loading or gas ionization decreases the voltage gradient over part of its length, resulting in a higher field somewhere else.

SUMMARY

1. Large machines operate at lower fields than small ones.
2. Dirt triggers breakdowns and the field on the tank wall cannot be ignored.
3. Spark patterns reveal the importance of small regions of high field.
4. Hoops and resistor frames must be rigid and planar if sparking is to be avoided.
5. Bremsstrahlung, beam loading and surface leakage all perturb the column gradient and reduce the safe voltage. Intershields are particularly affected.
6. The stored energy in large machines mandates conservative operation.
7. Components near the terminal are vulnerable to the effects of fast transients.

ACKNOWLEDGEMENTS

The author acknowledges with pleasure the benefits of valuable discussions with colleagues in many laboratories and with the staff of High Voltage-Vivirad, High Voltage Engineering Europa and National Electrostatics Corporation. Work supported in part by the United States Department of Energy under contract number DE-FG02-91ER-40609.

REFERENCES

1. Cockcroft, J.D., and Walton, E.T.S., *Proc. Roy. Soc.* **129**, 477-489 (1930).
2. Herb, R.G., Parkinson, D.B., and Kerst, D.W., *Rev. Sci. Instrum.* **6**, 261-265 (1935).
3. Bromley, D.A., *Nucl. Instr. and Meth.* **122**, 1-34 (1974).
4. Voss, R.G.P., *Revue. Phys. Appl.* **12**, 1347-1352 (1977).
5. Jones, C.M., *Revue. Phys. Appl.* **12**, 1353-1359 (1977).
6. Van Atta, L.C., Northrup, D.L., Van Atta, C.M., and Van de Graaff, R.J., *Phys. Rev.* **49**, 761-766 (1936).
7. Parkinson, D.B., Herb, R.G., Bernet, E.J., and McKibben, J.L., *Phys. Rev.* **53**, 642-650 (1938).
8. Purser, K.H., private communication, 1998.
9. Letournel, M., "New Design for an Electrostatic Accelerator", *Proc. Third Int. Conf. on Electrostatic Accelerator Technology*, Oak Ridge, 1981, pp. 247-253.
10. Hyder, H.R.McK., Baris, J., Gingell, C.E.L., McKay, J., Parker, P.D. and Bromley, D.A., *Nucl. Instr. and Meth.* **A268**, 285-294 (1988).
11. Eastham, D.A., "Calculated Field Distributions in the NSF Tandem Facility", ", *Proc. Int. Conf. on the Technology of Electrostatic Accelerators*, Daresbury, England, 1973, pp. 195-199.
12. Rebmeister, R., private communication, 1998.
13. Koltay, E., and Kiss, A., "Electrostatic Design and Acceleration Tubes in the 5 MV Van de Graaff Generator of the Atomki", *Proc. Int. Conf. on the Technology of Electrostatic Accelerators*, Daresbury, England, 1973, pp. 200-207.
14. Boag, J.W., *Proc. IEE* **100 IV**, No. 51, 63-67 (1953).
15. Herb, R.G., Parkinson, D.B., and Kerst, D.W., *Phys. Rev.* **51**, 75-83 (1937).
16. Philp, S.F., and Trump, J.G., "Compressed Gas Insulation for Electric Power Transmission", *Conference on Electrical Insulation*, N.A.S.-N.R.C., 1966.
17. Bortnik, I.M., and Vertibov, V.P., *International High Voltage Symposium*, Zurich, Switzerland, 1975, Vol. 2, p.337.

Gamma-Resonance Contraband Detection using A High Current Tandem Accelerator [1]

B.F. Milton*, J. Beis* D. Dale*, T. Debiak[†], E. Kamykowski[†] S. Melnychuk[†], J. Rathke[†], J. Rogers*, R. Ruegg*, J. Sredniawski[†]

*TRIUMF, 4004 Wesbrook Mall, Vancouver, B.C., V6T 2A3, Canada
[†]Northrop Grumman Advanced Technology and Development Center
1111 Stewart Avenue, Bethpage, NY 11714-358

Abstract. TRIUMF and Northrop Grumman have developed a new system for the detection of concealed explosives and drugs. This Contraband Detection System (CDS) is based on the resonant absorption by ^{14}N of gammas produced using $^{13}C(p,\gamma)^{14}N$. The chosen reaction uses protons at 1.75 MeV and the gammas have an energy of 9.17 MeV. By measuring both the resonant and the non-resonant absorption using detectors with good spatial resolution, and applying standard tomographic techniques, we are able to produce 3D images of both the nitrogen partial density and the total density. The images together may be utilized with considerable confidence to determine if small amounts of nitrogen based explosives, heroin or cocaine are present in the interrogated containers.

Practical Gamma Resonant Absorption (GRA) scanning requires an intense source of protons. However this proton source must also be very stable, have low energy spread, and have good spatial definition. These demands suggested a tandem as the accelerator of choice. We have therefore constructed a 2 MeV H^- tandem optimized for high current (10 mA) operation, while minimizing the overall size of the accelerator. This has required several special innovations which will be presented in the paper. We will also present initial commissioning results.

INTRODUCTION

In recent years the ease with which high explosives can be concealed and then placed on public transport such as planes has caused major concern worldwide. There is also a growing demand to reduce the transit of illegal drugs through public transport systems including the postal service. In both cases because of the required throughput and privacy requirements there is a desire for non-invasive techniques for the detection of contraband. Normal x-ray systems rely on strong

---

[1] This work was conducted under USAF contract #F0865094C0097

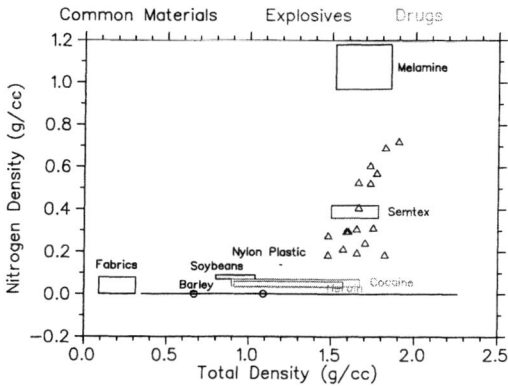

FIGURE 1. Explosives have a unique signature compared to common materials when nitrogen density is plotted against total density

density differences for identification, and have proven very successful in the detection of weapons in carry-on luggage. However while most explosives have fairly high densities, they are difficult to distinguish from other items commonly contained in luggage, particularly given the ability to shape them and/or conceal them in appropriate materials. Therefore there have been many efforts to detect contraband using methods more dependent on the exact chemical composition.

When reviewing the composition of explosives it is immediately noted that most have a large partial density of either nitrogen or chlorine when compared with most common materials. Various techniques have been developed to measure the quantities of key constituents such as carbon, oxygen and nitrogen. In fact many techniques that exploit the specific properties of these nuclei are being pursued and are reviewed elsewhere [1] [2]. Perhaps the best known of these is neutron activation.

In figure 1 the density and nitrogen partial density of a number of substances are shown. The clustering of the explosives in one part of the plot indicates how knowledge of the nitrogen partial density can lead to the identification of an explosive substance. The plot also indicates how a knowledge of both the partial nitrogen density and the total density together could be used to identify heroin or cocaine. Gamma resonant absorption (GRA) is a process that shows strong promise as a way to measure partial nitrogen (or chlorine) density.

It should be stressed that one needs to know the partial density of nitrogen in a selected volume, not the total amount of nitrogen in the overall container being inspected. For this reason it is important to have a spatially defined measurement technique, which in the case of high explosive reduces to a position resolution around 5mm. We have thus decided to combine GRA with high resolution tomography, similar to the combination of x-rays and tomography used in CAT scans. More details of the specific mission can be found in ref [3].

TABLE 1. A comparison of potential gamma resonances in carbon, oxygen, nitrogen and chlorine, using a Figure Of Merit normalized to 1000 for the $^{13}C(p,\gamma)^{14}N$ reaction.

Production Reaction	Proton Energy	Gamma Energy	FOM
$^{14}N(\gamma,p)^{13}C$	1.75	9.17	1000
$^{35}Cl(\gamma,p)^{34}S$	1.89	9.08	35
$^{35}Cl(\gamma,p)^{34}S$	2.79	8.21	76
$^{16}O(\gamma,p)^{15}N$	1.03	13.1	22
$^{12}C(\gamma,p)^{11}B$	1.39	17.23	10

GAMMA RESONANCE ABSORPTION

GRA makes use of the fact that for each element there are certain gamma energies that are strongly absorbed by the nucleus. Since the gamma energy at which this happens is specific to the target nucleus, a comparison of the absorption at the resonant energy with the absorption at an energy just off resonance will yield a measure of the sought after element. For example the reaction $^{13}C(p,\gamma)^{14}N$ has a large cross section at a gamma energy of 9.17 MeV, so by interrogating a container with gammas at this energy one could measure the nitrogen density.

From a practical point of view it is not easy to produce a near mono-energetic beam of gammas. At present the most promising approach is to use an inverse reaction to produce the gammas. For example the 9.17 MeV gammas can be produced using a beam of protons and the $^{14}N(\gamma,p)^{13}C$ reaction. Assuming favourable reaction kinematics, it is possible to find an outgoing gamma angle at which the gammas have been Doppler shifted by an amount that exactly compensates for the energy absorbed in the production reaction.

In 1994/95 the TRIUMF-Grumman team conducted an extensive study of GRA for the detection of concealed contraband. This study developed a figure of merit for evaluating various resonances. The figure of merit included factors for the strength of the resonance, the width of the resonance, gamma energy spread and other factors that affect the usefulness for imaging by tomography. The figure of merit was normalized to 1000 for the $^{13}C(p,\gamma)^{14}N$ reaction. Table 1 compares some of the considered reactions. From this the $^{13}C(p,\gamma)^{14}N$ reaction was found to be the most promising, however CDS has continued to maintain the option of using the lower energy $^{34}S(p,\gamma)^{35}Cl$ reaction, for the detection of chlorine based explosives, and the hydrochloride forms of cocaine and heroin.

One advantage of producing the required gammas using an inverse reaction is that gammas meeting the re-absorption criteria form a narrow conical fan. In the case of ^{14}N detection, the cone lies at 80.7 degrees, and is about half a degree wide. This results in a convenient inspection geometry as shown in figure 2. Gammas produced just outside this cone can be used to measure the non-resonant absorption, which can be used to normalize the resonant absorption for the calculation of nitrogen partial density, and for the calculation of total density.

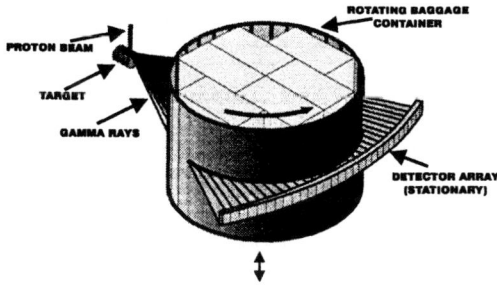

FIGURE 2. CDS detector and inspection geometry

ACCELERATOR REQUIREMENTS

For a practical inspection device there are several key elements affecting the choice of accelerator. They include image resolution, throughput, over all size and convenience of operation. The first two of these determine the number of gammas per second required and the quality of those gammas.

In order to maximize the efficiency of gamma production it is important that the energy spread of the beam be less than the target thickness. As well the lowest energy protons must have an energy at least as high as the resonance energy. Therefore the larger the proton energy spread, the higher the proton beam central energy that is required and the thicker the target that has to be used. This has a limit since we are interested in a target as thin as possible to avoid the excitation of other resonances which contribute only to the background. In addition, using too thick a target will affect the energy spread of the produced gammas, which in turn will decrease the resonance absorption signal.

The size of the beam spot on the target combines with the position resolution of the detectors to reduce the precision with which the flight path of each gamma is measured. Thus the larger the beam spot size the poorer the position resolution of the overall system. It is therefore desirable to choose a spot size that will make a negligible change in the image resolution given the resolution of the detectors.

Angular divergence of the protons at the target changes the angle of the produced gamma. Thus the proton beam's divergence creates an uncertainty as to the energy of a gamma ray even if its position at the detector is known precisely. Because of this uncertainty, some of the gamma rays that arrive at the detector on the locus of the resonant-energy conic section deviate from the resonant angle by an unknown amount up to the divergence of the beam. However these rays will not have met the resonance condition when passing through the luggage. The effect of including off-energy gammas rays in the resonant image is to dilute the resonant absorption effect which causes an increase in the resonant image noise. The larger the proton beam divergence the greater the noise added to the resonant-energy image. Simulations were used to determine the beam divergence levels that would

TABLE 2. CDS accelerator requirements for the selected production reactions

	$^{14}N(\gamma,p)^{13}C$	$^{35}Cl(\gamma,p)^{34}S$
Proton Energy (MeV)	1.75	1.89
Energy Spread (keV)	< 25	< 12
Spot Size (cm)	0.6 x 2.4	0.6 x 2.4
Divergence (mrad)	12	12
Current (mA)	10	12

cause minimal increases in the false alarm rate.

Once an image resolution, a detection level, and false alarm rate have been chosen, then throughput determines the gamma flux requirement. Calculations show that in order to achieve throughput rates in the hundreds of bags per hour for the chosen configuration, the proton beam current must be in the 10 milli-ampere range.

A summary of the beam requirements as determined in the initial design study are summarized in table 2. During the design study we performed a survey of methods for the production of protons up to 3 MeV. Accelerator options such as RFQs and cyclotrons were investigated. While RFQs looked very promising because of their compactness and high current capability they are not able to meet the other beam parameters. Electrostatic accelerators were really the only system capable of producing such high quality beams. However most electrostatic accelerators have two disadvantages for CDS, size and limited beam current. It was therefore decided to design a new tandem accelerator specifically for CDS. It should also be noted that a DC accelerator, when compared to a CW accelerator, creates a significantly lower instantaneous data rate in the detectors.

THE CDS SYSTEM

The layout of the CDS Proof-of-Principle system is shown in figure 3. This system has been designed to perform a laboratory proof of principle experiment and to allow for the testing and development of components that could be used in a commercial system.

The gamma production system begins with a high brightness H$^-$ ion source coupled by a short matching section to the tandem. The tandem is followed by a beam transport section that analyzes the beam and produces the correct beam spot parameters at the target. At present the target consists of a fixed water cooled target coated with carbon-13. However in order to use the full beam current a rotating target will be constructed. The high energy beam line has been constructed so that the resonant gamma cone lies in the horizontal plane in the region of the baggage carousel.

In order to acquire tomographic information the objects to be inspected are located on a carousel that is capable of rotating and elevating. Behind the carousel

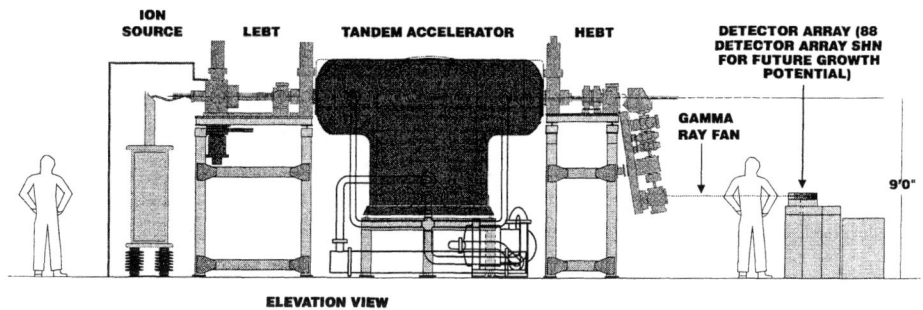

FIGURE 3. Side elevation of the CDS proof-of-principle device

is located a detection system that consists of a double deck arc of segmented BGO block detectors. The arcs subtend an angle of 53 degrees and contain 44 blocks each. Standard electronics is used to collect the phototube output which is then converted into gamma position information. Image reconstruction and interpretation is then performed in separate computers. At present we are using 7 detector blocks and offline data processing and a very simple carousel in order to assess the overall system performance. In a high throughput system the carousel would rotate and elevate simultaneously.

THE CDS TANDEM

As discussed above, the CDS requires an electrostatic accelerator capable of a very high beam current, while maintaining the smallest practical size. Using a tandem allows the terminal voltage to be halved thus reducing the demands on the voltage generator, and the amount of space required to hold voltage. The trade-off is the requirement for a stripper located in the terminal that is capable of handling the current. Since the proton beam energy at the terminal will be below 1 MeV, the energy loss in a stripping foil would be quite large. Calculations show that expected foil lifetimes would be perhaps a few minutes at the design intensities. We therefore chose to use a vapour stripper. Design of this element has concentrated on providing good differential pumping between the stripper cell, and the acceleration columns. A schematic of the system is shown in figure 4. Pressure in the acceleration columns is particularly important because interactions between the residual gas and the beam will produce free electrons that will then be accelerated and generate x-rays. Significant production of high energy x-rays could lead to voltage breakdown and constitute a radiation hazard external to the accelerator. Testing to date has shown that with the stripper operating above design pressure, the pressure in the columns is in the low 10^{-7} range.

Obviously the overall size of a tandem can be reduced by increasing the electric fields, particularly in the acceleration region. However voltage holding problems

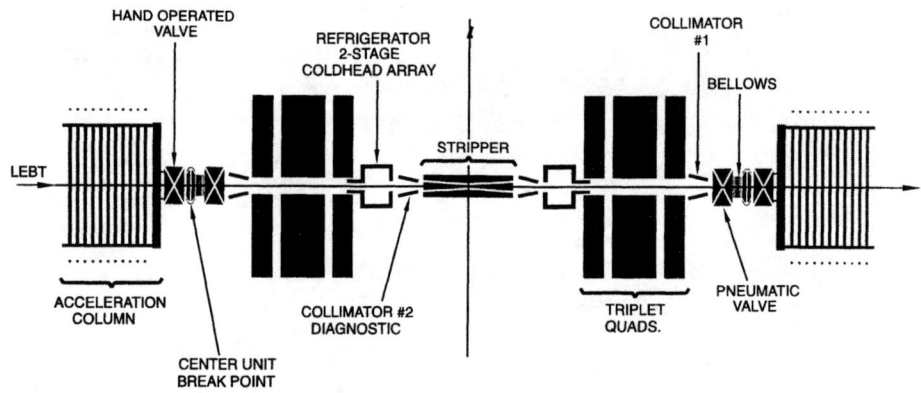

FIGURE 4. Schematic layout of the terminal vacuum system

rise steeply with electric field so we have chosen to remain at a fairly conservative 50 kV/inch in the acceleration region. By pressurizing the containment vessel to 60 PSI with SF_6 the distance between the vessel and the terminal has been reduced by factor of almost 10 over using unpressurized nitrogen (air). Keeping the terminal compact has also helped to reduce size. As well as the vacuum equipment for the stripper, the terminal contains two magnetic quadrupole triplets for matching beam between the acceleration regions and the stripper cell. Considerable effort has been devoted to making this region compact while still using commercial power supplies and controllers. Communication between ground and the terminal is achieved using fiber optics.

Of particular importance to CDS is a suitable power supply. In order to produce a 10 mA, 2 MeV beam, the power supply must be able to supply at least 20 mA at 1MV, meaning it must be capable of a power output greater than 20 kW. The preliminary CDS design study showed that most conventional power sources were incapable of providing the necessary current, and those that had produced currents in the milli-Ampere range were extremely bulky. However the design study also identified a novel power supply development at the University of Waterloo that was able to meet the CDS requirements in a compact volume. Basically this system uses a high frequency isolated core transformer to couple a large number of voltage doubler circuits that are then connected in series to develop the necessary voltage. The high frequency allows relatively small conventional components to be used, with the secondary tracks and doubler circuits mounted on normal circuit boards. This supply occupies a space less than 300 in^2 and is under 36 inches high. In the tandem the supply sits under the terminal between the support insulators and is insulated by the SF_6. In the region of the voltage generation the average field is 50 kV/in. Tests of the supply using a water load have demonstrated outputs of greater than 26 kW at 1 MV.

Fine control of the power supply voltage is achieved by varying the primary frequency, allowing rapid changes. As well since the capacitance of the doublers is small, the stored energy of the supply is small when compared to the overall tandem capacitance. Tests at 860 kV without beam have shown that even with no voltage stabilization the system shows less than a ±5 kV variation. We have implemented a system that will adjust the frequency based on the terminal voltage. This appears to take care of basic beam loading effects. However at present this system is operated at very low gain to avoid an instability in the DC power supply feeding the inverter. A rebuild of the supply should correct this and allow better stabilization of the voltage. Eventually we expect to use feedback from the high energy beamline to adjust this loop.

At present all of the system sub-components have been tested. The ion source and low energy beam transport emittance has been fully characterized at 10 mA. The tandem has undergone a series of voltage tests up to 900 kV on the terminal, including full operation of the stripper while at 865 kV. Initial tests have indicated that the beam spot at the stripper location is as specified. On the first attempt to run beam 100 μA at virtually 100% transmission was extracted. Beam current was limited to this level by target restrictions. Tests on a non-resonant target have achieved a current of 1.6 milli-amperes with no obvious limitations in the tandem systems, and nearly 100% transmission through the tandem. From observing the gamma production rate while varying the tandem voltage, the energy spread has been estimated to be σ_E=19 keV without proper voltage stabilization. Emittance measurements in the horizontal plane give an RMS emittance of 0.085 π mm-mrad (normalized). This well within that specified however limited operation time has prevented the optimization of the beam on target.

With limited beam current (due to target limitations) and almost no beam optimization a projection image of a dense object and a nitrogenous object was performed. Separation of the two objects demonstrated the basic principle of operation. A high current target has been installed, and a preliminary test indicated that it work well at beam currents up to 1 mA. A tomographic image (non-resonant) of a wedge phantom demonstrated detector resolution and the function of the image reconstruction chain. Optimization of the beam parameters at the target should increase the resonant gamma intensity and thus improve image clarity. This work is presently waiting for further funding which is expected in the coming fiscal year.

ACKNOWLEDGEMENTS

The CDS team consists of three groups and has included many people over the last 4 years. Key players at the Northrop Grumman Advanced Technology and Development Center include; J.J. Sredniawski, T. Debiak, E. Kamykowski, J. Rathke, and S. Melnychuk. At TRIUMF they have included; P. Schmor, J. Rogers, A. Altman, J. Boyd, G. Stanford, D. Dale, R. Ruegg, L. Buchman, and J. Welz. At Scientific Innovations there is J. Brondo.

REFERENCES

1. Gozani T., "Nuclear based Techniques for Cargo Inspection - A Review", Proc. of Contraband and Cargo Inspection Technology International Symposium, Washington D.C., 1992.
2. Vartsky D., Engler G., and Golberg M.B., *NIM*, **A348**, 668-691(1994).
3. J. Sredniawski, "A New Proof-of-Principle Contraband Detection System", ONDCP Conference, Nashua, N.H., October 1995.

Use of an ECR Ion Source in the High Voltage Terminal of the Tandem Accelerator at JAERI

M. Matsuda, C. Kobayashi, and S. Takeuchi

*Japan Atomic Energy Research Institute, Tokai Research Establishment
Tokai-mura, Naka-gun, Ibaraki-ken 319-1195 Japan*

Abstract. Modern electron cyclotron resonance ion source(ECRIS)s are able to produce intense beams of highly charged positive ions, of which charge states are higher than those obtained from electron stripping at the high voltage terminal of tandem accelerators. It is possible to increase beam intensity, beam energy and beam species by utilizing an ECRIS in a tandem accelerator. A small permanent magnet ECRIS has been installed in the high voltage terminal of the vertical and folded type 20UR Pelletron tandem accelerator at Japan Atomic Energy Research Institute at Tokai. Acceleration tests have been successfully carried out with beams of H^+, N^{2+}, $O^{3+,5+}$, $Ar^{6+,8+,9+}$ and $^{132}Xe^{12+,13+}$ ions.

INTRODUCTION

Electron cyclotron resonance ion sources (ECRIS) are able to produce intense beams of highly charged ions and, therefore, are used for cyclotrons and linacs as well as for experiments of atomic physics. The tandem accelerator system has been benefiting from the use of an electron stripper at the high voltage terminal. However, the most probable charge state after a foil stripper is much lower than the highest charge state of ions with an intensity of more than several $e\mu A$ from a high performance ECRIS. With respect to beam current increase, if beam current is increased the lifetime of stripper foils decreases. Especially for very heavy ions, it is impossible to obtain a stable and intense beam for a long time without foil exchange.

Use of an ECRIS is expected to open a way for a stable acceleration of high intensity beam to higher beam energy. ECRISs were, in the past, too large and too heavy to utilize in tandem accelerators. It was possible only for large vertically standing folded tandem accelerator like 25UR Pelletron tandem accelerator at Oak Ridge National Laboratory where an installation project of an ECRIS in its terminal was once considered [1]. In recent years, compact ECRISs of which plasma confinement structures are composed of permanent magnets have been developed and commercially available also. In Fig.1, charge states of ions of several eμA expected for compact permanent magnet ECRISs are compared with the most probable charge states obtained by stripping at a terminal voltage of 16MV as a function of mass number. Concerning to ions over a mass number of 100, charge states higher than 20+ can be available from an ECRIS, compared to 13+ from the stripping at a high voltage terminal of 16MV. In addition to increases of beam energy and beam intensity, use of an ECRIS clears away many problems with stripper foils, such as short lifetimes, energy straggling, emittance growth and beam intensity reduction and makes it possible to accelerate noble gas ions and alkali-metal ions. On the other hand, use of an ECRIS in a high voltage terminal has a difficult problems due to inaccessibility and operation in high pressure SF$_6$ gas and under electric surges. One needs several devices to solve these problems.

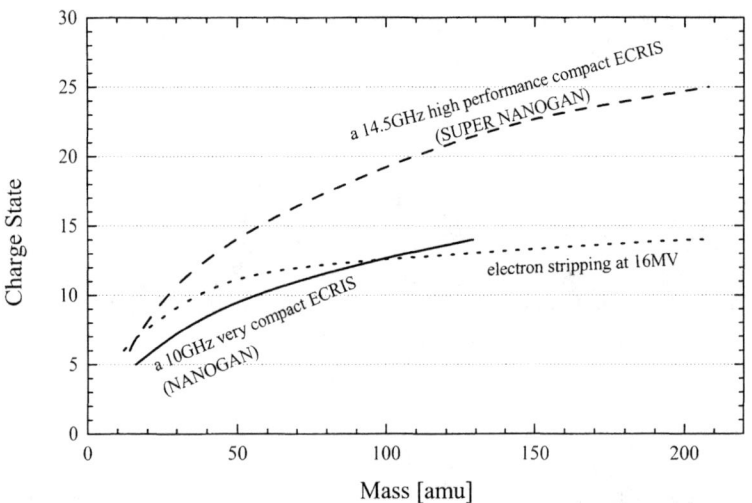

FIGURE 1. Charge state of the ions of several eμA expected for compact ECRISs are compared with the most probable charge state obtained by stripping at a terminal voltage of 16MV.

We started with a small permanent magnet ECRIS, NANOGAN [2] which works at 10GHz and with RF power of 10 to 200 W, as a preliminary step of the in-terminal ECRIS project, in order to solve many difficulties mentioned above before going to a high performance ECRIS. The JAERI tandem accelerator has been equipped with a duo-plasmatron in-terminal ion injector for pulsed proton/deuteron beams. It should be relatively easy for us to install a small ECRIS in place of the old in-terminal ion injector.

We are also aiming at increasing injection velocities of very heavy ions to the superconducting booster linac [3], of which lowest acceptance velocity is 5% of the light velocity, with this in-terminal ECRIS project. Comparison of expected energy between conventional negative ion source with foil stripping and an in-terminal ECR ion source is shown in Fig.2.

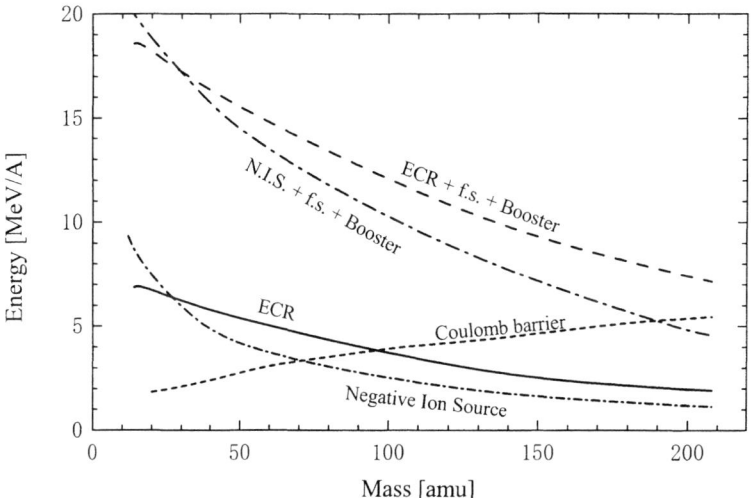

FIGURE 2. Comparison of expected energy between conventional negative ion source with foil stripping and an in-terminal 14.5GHz compact ECR ion source.

OPTIMIZATION OF OPERATING CONDITIONS

The ECRIS is placed in a severe environment; i.e. in the high pressure SF_6 gas and under an attack of electric surges from occasional high voltage sparks in the accelerator. We have considered a minimum damage to the main accelerator system in case of a trouble with the ECRIS, because the tandem accelerator is tightly scheduled to supply ion beams to its users and a fatal damage causes a accelerator vessel opening in the period of scheduled machine time. For this reason, the injection system initially needs to be simple and have minimum functions. A further improvement of performance can be made at next steps after confirming the reliability of the injection system.

Ion extraction experiments have been carried out in order to search optimum operating conditions and to minimize operating parameters. As a result of the experiments, five parameters could be reduced to three, which are the gas flow, DC bias voltage and RF power. The aim of the experiments was to obtain optimum conditions for a stable ion beam against a change in operational parameters rather than to obtain the maximum performance of the ECRIS. Extracted current of Ar ions is shown as a function of RF power in Fig.3. The beam current changes gradually, so that the beam intensity can be controlled by RF power. The RF tuner can be fixed, because the beam intensity is not sensitive to its adjustment.

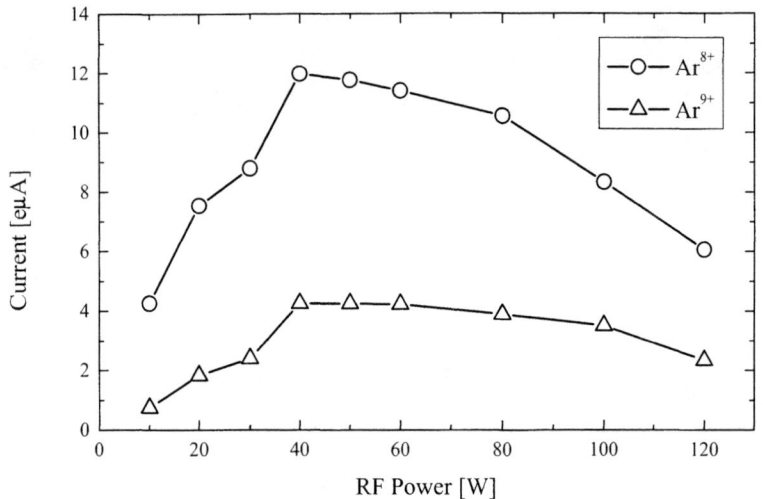

FIGURE 3. Extracted beam current of Ar ions as a function of RF power.

The support gas could be previously mixed for simplified operation. The currents of different charge states of Ar ions are seen in Fig.4 as a function of mixing ratio of oxygen gas to argon gas. The currents of highly charged ions(8+, 9+) do not depend on the mixing ratio very much, while the currents of low charge state ions decrease with increasing mixing ratio. So that, the ratio of $Ar:O_2=1:10$ was chosen, because we wanted to use an ion pump(see the next section) and to restrain the load to the pump(The total gas flow rate is fixed in the measurement for Fig.4. The flow of argon gas into the ion pump can be reduced very much.) because the pumping speed for argon gas is only about 10% compared with that for air. With an ion pump of 200l/s, the ECRIS ran stably for several months for noble gases with a flow of less than 2×10^{-5}torr·l/s. The high mixing ratio reduces unwanted low charge state Ar beams as well. The results with krypton and xenon gases were similar to argon.

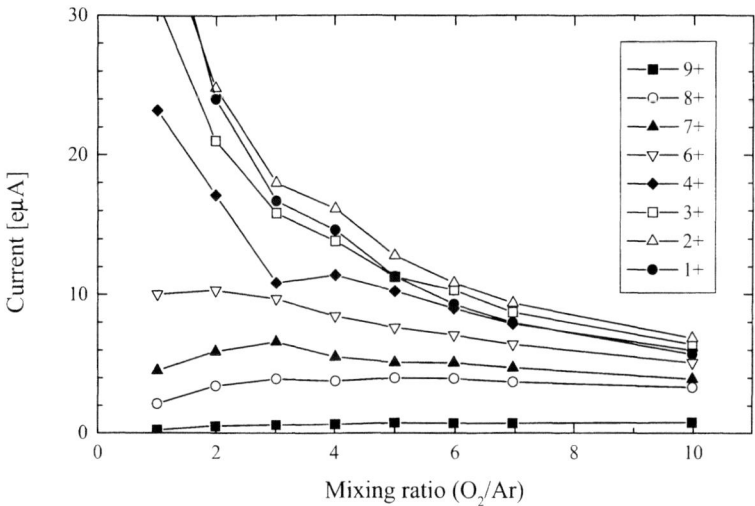

FIGURE 4. The currents of different charge states of Ar ions as a function of mixing ratio of oxygen gas to argon gas.

The gas flow, or the pressure in the plasma chamber, was a critical parameter for the condition of a stable and intense beam. It was optimum at $1.6\times10^{-4}(\pm20\%)$torr·l/s for Ar^{8+}. A calibrated leak source was employed instead of a variable gas valve, because a complicated and troublesome feedback control could be eliminated. This method was quite effective to make the ion source operation easy. A DC bias to the bias electrode was necessary to produce an intense beam. With a DC bias of -40 to -200 V, a beam was extracted with an intensity of at least ten times as intense as that without DC bias. The results of qualification experiments for three gases, Ar, Kr and Xe, mixed with N_2 or O_2 gas in above conditions, are listed in Table 1. The emittance of an Ar ion beam in the above experiments was about 10mm·mrad·MeV$^{1/2}$(80%) and the phase space profile was very complicated.

TABLE 1. The results of qualification expriments for three gases.

charge state	Extracted beam current [eµA]										
	4	5	6	7	8	9	10	11	12	13	14
Ar	12	—	11	10	7.7	1.5	—				
Kr	4.6	5.9	—	12	17	14	9.4	5.1	—	0.8	0.7
Xe		2.3	3.5	6.7	11	5.1	3.9	3.6	2.0	1.2	0.9

INSTALLATION

The JAERI tandem accelerator is a folded type machine with a 180° bending magnet in the high voltage terminal. Electric power of 10kW+15kW is available from two power generators in the terminal. A very compact permanent magnet ECRIS, NANOGAN could take the place of the old in-terminal ion injector. The layout of the in-terminal ECR ion injector is illustrated in Fig.5. An ion beam is extracted by a 30kV(at maximum) potential gap from the ECRIS, focused by an einzel lens, and then the mass and charge are roughly selected by 45° pre-analyzing magnet. The magnet is used to reduce the load to the pre-acceleration tube high voltage power supply, since the beams from the ion source amount to 2mA. An electrostatic steerer is placed just after the pre-analyzing magnet to correct the horizontal beam direction. An aperture for rough beam selection is placed just before the pre-acceleration tube. After an acceleration through the 80kV pre-acceleration tube, a desired ion beam is selected by a 45° injection magnet with a radius of curvature of about 0.3m, an electrostatic quadrupole triplet and an aperture of 4.8mm in diameter placed at about 0.4m above the accelerator tube together with a Faraday cup. Magnetic field probes are set to both bending magnets for ion beam selection.

For the RF source of 10GHz, a 200W TWT airborne amplifier with a 10GHz dielectric resonance oscillator is set in a chamber to keep it at an atmospheric pressure. The amplifier, of which power consumption is 2.2kW, is cooled by water using 180° bending magnet cooling system. The RF power is guided to the ECRIS by a wave guide including RF windows, 80kV and 30kV DC-cut elements.

The main body of the ECRIS is cooled by SF_6 gas flow which is pumped out by a diaphragm pump. The temperature of gas at the outlet was 32~35°C at RF input of 120W, which is high enough to operate the ECRIS. It can prevent a harmful temperature rise of the permanent magnets from exceeding the recommended limit of 36°C.

A 200l/s ion pump is used for the vacuum system, because it encloses a gas flow, it is fail-safe and the ECRIS works at a very small flow rate of gas supply as is described in the previous section.

There are three shield boxes of 0kV(grounded to the high voltage terminal)Deck, 80kV Deck and 110kV Deck. Source gas control circuits and a power supply for the DC bias are mounted in the 110kV Deck. Power supplies for the 30kV extraction, einzel lens, 45° pre-analyzing magnet and steerer are put in the 80kV Deck. All devices in the high voltage decks are controlled through an optical link system in communication with the 0kV Deck. The electric power is provided by means of an insulating transformer. A current source for the 45° injection magnet, cooling pump, 80kV high voltage power supply for the pre-acceleration tube and communication circuits are installed in the 0kV Deck.

The installation was carried out in April, 1998. In the first two acceleration tests, some troubles were found in power supplies, some of which were caused by high voltage sparks in the accelerator and a few of which were due to pressurized insu-

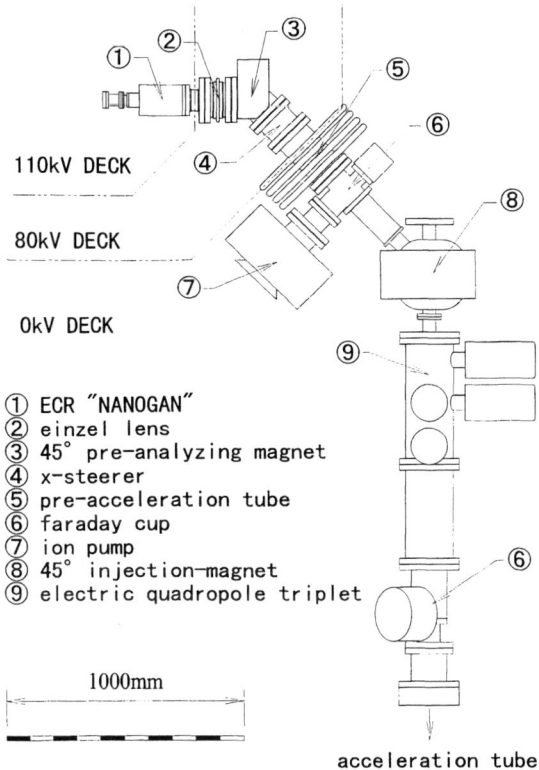

FIGURE 5. The layout of the in-terminal ECR ion injector.

lating gas. Even with these troubles, we obtained good feasibility from accelerating H^+ and Ar^{8+} ion beams. During a tank opening in August, the power supplies and their AC power lines were fixed and two gas sources were installed, which were $Ar+10\cdot O_2$ and ^{132}Xe(enriched 60%)$+10\cdot N_2$ gases. The gas flow of $Ar+10\cdot O_2$ was fixed by using a calibrated leak valve and that of $Xe+10\cdot N_2$ was controllable with a thermo-mechanical leak valve. We obtained, in the third acceleration test, a result as below.

RESULTS

The results of acceleration experiments for H^+, N^{2+}, $O^{3+,5+}$, $Ar^{6+,8+,9+}$ and $^{132}Xe^{12+,13+}$ ions are presented in Table 2. In these experiments, the extraction, pre-acceleration and terminal voltages were 15kV, 80kV and 14MV, respectively, except for Xe ions. For Xe ions, the pre-acceleration voltage was set to 50kV, because the maximum field strength of the injection magnet was not high enough to bend the ions accelerated by 80kV(The magnet was the old one which had been used for the old in-terminal proton/deuteron injector). The beam intensities for light ions H$^+$, O$^{3+,5+}$ and N^{2+} were suppressed to a large extent to their limits allowed for this facility from the point of radiation safety. The beam transmission was strongly dependent on the pre-acceleration voltage. For N^{2+} ions, the beam current after the accelerator's analyzing magnet for a 60kV pre-acceleration was over twice as much as that for a 50kV pre-acceleration. This was presumably caused by the strong gradient at the entrance of the accelerator tube. The beam profiles were very sharp at the object and image points of the analyzing magnet. The beam width of ^{132}Xe^{12+}, for example, was between 1 and 2mm(FWHM) in the bending plane. One can expect a very high quality energy beam from the ECR in-terminal injector plus electrostatic accelerator system as we know about beams from a single ended Van de Graaff.

CONCLUSIONS AND FUTURE PLANS

A compact ECRIS was installed at the high voltage terminal of the JAERI tandem accelerator. Ions of H, N, O, Ar and Xe were successfully accelerated. Now, noble gas ions are available from the accelerator, in addition to the ions normally available from the negative ion injector. We can use a high intense and high energy beam without the problems with stripper foils.

The charge states of medium mass ions, such as Xe^{12+} ions, from the very compact ECRIS are comparable to those obtained by electron stripping. A higher performance permanent magnet ECRIS is to take the place of the present ECRIS in a few years, in our in-terminal ECRIS project. Acceleration of very heavy metallic ions, such as Pb ions, are also considered in the project.

TABLE 2. The results of acceleration experiments.

	Beam Energy		Beam Current	
		After pre-acceleration	Before the acceleration tube	After the analyzing magnet
	[MeV]	[eμA]	[eμA]	[eμA]
H$^+$	14	5.5	3.1	2.3
N^{2+}	28	6.6	3.4	1.0*
O^{3+}	42	5.6	1.6	1.5*
O^{5+}	70	0.64	—	0.15
Ar^{6+}	84	10	—	2.1
Ar^{8+}	112	8.3	—	2.3
Ar^{9+}	126	0.77	0.26	0.24
^{132}Xe^{12+}	168	1.7	0.31	0.22
^{132}Xe^{13+}	182	1.3	0.26	0.17

*:limited by radiation safety

REFERENCES

1. D. K. Olsen, "Introduction to ECR Source in Electrostatic Machines" presented at the Symposium of Northeastern Accelerator Personnel, Oak Ridge, Tennessee, October, 1989.
2. P. Sortais, C. Bieth, P. Foury, N. Lecesne, R. Leroy, J. Mandin, C. Marry, J. Y. Pacquet, E. Robert and A. C. C. Villari, "Developments of compact permanent magnet ECRIS" in *Proceedings of the 12th Internatinal Workshop on ECR Ion Sources*, RIKEN, Japan, April, 1995,pp. 44-52.
3. S. Takeuchi, T. Ishii, M. Matsuda, Y. Zhang and T. Yoshida, *Nucl. Inst. and Meth.*, 1996, **A382**, pp.153-160.

Fifty Years of Accelerator Based Physics at Chalk River

John W. McKay

P.O. Box 463, Deep River, Ontario, Canada K0J 1P0

Abstract. The Chalk River Laboratories of Atomic Energy of Canada Ltd. was a major centre for Accelerator based physics for the last fifty years. As early as 1946, nuclear structure studies were started on Cockroft-Walton accelerators. A series of accelerators followed, including the world's first Tandem, and the MP Tandem, Superconducting Cyclotron (TASCC) facility that was opened in 1986. The nuclear physics program was shut down in 1996. This paper will describe some of the highlights of the accelerators and the research of the laboratory.

THE BEGINNING OF THE CHALK RIVER LABORATORIES

The origin of the Chalk River Laboratories was the Anglo-Canadian efforts in co-operation with the Manhattan Project. A laboratory was established for this purpose, at the Université de Montréal in 1942. In 1946 the effort was moved to a new site 200 kilometres north-west of Ottawa and a new establishment that later became Atomic Energy of Canada Limited (AECL), was constructed at Chalk River. The major purpose of the Laboratory as explained in he technical history of the Laboratory (1), was the exploitation of Nuclear Energy. From the beginning, basic science was a major part of the equation underpinning the development of the new reactors.

The ZEEP (Zero Energy Experimental Pile) went critical Sept 5, 1945, the first reactor outside US. It operated until 1980. ZEEP led to the NRX reactor that operated at Chalk River from 1947 to 1990 (43 years) and to the NRU reactor that began operation in 1957 and may run until 2005. Both these reactors generated the world's highest flux at the time of their inception. NRU was the first reactor to refuel under power.

In these two reactors, the Chalk River Laboratories developed the technology that went into the CANDU® reactor design. There are now 36 of these reactors now on line or under construction around the world. They are unusual devices in that they are fuelled with natural (i.e., non-enriched) uranium, use deuterium oxide moderator and refuel under power.

In addition to developing reactors, the Chalk River Laboratories were for many years, the premier Canadian facility for all branches of nuclear science. A paper published by Walter Davies in 1995 (2), summarized much of the accelerator-based study of nuclear physics at Chalk River. The nuclear physics program at the Chalk River Laboratories was shut down in 1996 and the accelerator laboratory has been closed. This paper will present some aspects of the group's work over the past fifty years.

1940's - Cockroft-Walton Accelerators - The First Observation of Direct Nucleon-nucleon Interaction

Experiments starting as early as 1946, using 150 kV and later (1949) 250 kV Cockroft-Walton accelerators, obtained as results, the unexpected and important discovery (3) that the (n,a) reaction on heavy targets was not consistent with the expected statistical behaviour, but rather with that of a direct reaction.

J.D. Cockroft was the head of the laboratory from 1944 to Sept 1946 when he returned to England to head the Harwell laboratory. He first proposed a 5 MV Van de Graaff accelerator in 1944 for the Canadian project, before the move to Chalk River. The machine was a M.I.T. design which it was agreed, was to be copied by Chalk River and Harwell. The accelerator building was finished 1948 at Chalk River but the Van de Graaff accelerator was not in reliable operation until June, 1952.

It was an impressive looking machine that towered over the researchers in contrast to the compact size of new machines such as the NEC Tandem accelerator supplied to CBAMS and other AMS laboratories.

1950's - The 3 MV Van de Graaff - The Development of Gamma-ray Correlation Studies

The Chalk River and Harwell machines ran at about 3 MV and MIT reached 3.5 to 4 MV as accelerator physicists began to explore the mysteries of high voltage. The Van de Graaff was used mostly to study the energy levels of nuclei in the 1p and 2s1d shells by a number of different reactions using proton, deuteron and helium beams. The techniques for the study of gamma-ray angular correlations were developed in this period, including the famous "Methods I and II" of Litherland and Ferguson. (4)

The control room of this era was dominated by a huge multi-channel analyser that had five channels per 19 inch rack chassis. The data recording system consisted of a pad of paper and a pencil to record the octal information from the individual indicator tubes. One day a week had to be spent checking the tubes.

This 3 MV Van de Graaff accelerator ran until the early 1960's but as researchers looked for higher energies, interest developed in the proposal for the Tandem accelerator. The world's first Tandem accelerator, EN-1, was built for AECL by HVEC and the first beam at Chalk River was seen in 1957.

1960's - The First Tandem - Nuclear Molecules, The Second Potential Well

The EN Tandem accelerator was designed to operate at 5 MV but eventually reached 7 MV. This was perhaps the first large machine to be run with pure SF6 insulation gas. Availability of beams of carbon and oxygen ions from the Tandem accelerator at energies from 6 to 35 MeV led to the discovery of quasi-molecular states in nuclear interactions.(5) Deuterons at 11.5 MeV were used in a seminal study of the ^{239}Pu(d,pf) reaction, in which clearly-defined states in the "second potential well" leading to fission isomers were identified.(6)

In the mid 1960's, the lithium-drifted germanium detector was developed at Chalk River(7) along with many innovations in electronics and data collection required to take advantage of this invention.

Success with the EN Tandem led to the construction of the 10 MV MP-3 Tandem accelerator completed in 1966. The old EN-1 was moved to l'Université de Montréal where it is still operating today. This was the machine that introduced many of us to the subtle mysteries of modern electrostatic accelerators.

1970's - The MP Tandem - Measurement of High-spin Static Quadrupole Moments

Advances in gamma-ray spectroscopy and the installation of a 10 MV Tandem accelerator culminated in a measurement of the static quadrupole moment of the high spin (J=49/2) isomer in ^{147}Gd. (8) The experimental techniques, largely pioneered at Chalk River, have been widely used throughout the world.

As the available performance of the MP Tandem gradually improved, a proposal for the next major step was made. The Superconducting Cyclotron concept originated with the Accelerator Physics Branch at Chalk River and its planning began in 1972.

One of the more intriguing subterfuges committed in this period occurred when funds to start building the machine were refused at one point. However, the accelerator physics branch did get permission to build a working model of the rf structure. The "scale" turned out to be 1:1 and I believe that "model" was still in the operating cyclotron when we finally shut down.

1980's - High-spin Gamma Spectroscopy - the 8π

Experiments on the "8π spectrometer", built as a collaborative effort between scientists from Chalk River Laboratories, McMaster University, l'Université de Montréal and the University of Toronto, began in 1985. Along with powerful analysis techniques developed at Chalk River, this instrument has been used for the last decade to make many important advances in the field of gamma-ray spectroscopy. (9)

The MP Tandem accelerator was turned around and upgraded in 1983 in preparation for the K= 520 Superconducting Cyclotron. A complete new beam-line system was designed and built to provide the spatial and time focus needed at the injection foil of the cyclotron. The first cyclotron beam, ^{127}I at 1.3 GeV, was achieved in 1985 (10).

1990's - The Superconducting Cyclotron - Hyperdeformation and Unitarity Tests in ^{10}C

This decade saw the full commissioning of the Tandem Accelerator Superconducting Cyclotron Facility (TASCC) in 1991. Many important experimental results were produced, including the first observation of hyperdeformation. (11) Measurements of branching ratios in the beta-decay of ^{10}C (12) have provided important tests of nuclear theory.

More time and effort began to be spent on applications. A major effort in accelerator mass spectrometry (AMS) of Cl-36 and I-129 built on earlier experience in the late '70's and early '80's with C-14 AMS. These two isotopes were of particular interest to AECL as products of reactor operation and to the IAEA who have developed an interest in this technique as a way to investigate what they politely refer to as "undeclared" facilities. This work is now being done at a number of AMS facilities around the world. Experiments simulating single event effects on integrated circuits developed into a major program with the discovery that very high energy particles from the cyclotron caused much more damage than lower energy particles with the same LET (linear energy transfer). The lock-up failure rate was increased by a factor of as much as 1000 in some devices. New equipment to be used in this work was under construction and we were anticipating much growth in this programme at the time we were shut down.

Studies related to reactor development were also starting to be important. These included radiolysis studies in water and heavy water and studies in simulated fuel elements.

The Conclusion of an Honourable History

The last regularly scheduled operation of the facility took place on December 19, 1996. The Canadian Government announced the termination of funding for the facility on January 16, 1997 and by end of March, all staff had been re-deployed or had left AECL.

The concept of the superconducting cyclotron was developed and patented by AECL, and after a long period of development, the machine went into regular use in 1991, operating well beyond the original design envelope. It produced 94 different beams from helium to uranium.

The MP Tandem has performed well for thirty years. Many modifications were made including installation of a chain-charging system and extended tubes. In the 1980's, the machine was rebuilt and the beam direction was reversed to facilitate injection into the cyclotron. At the end of its operation, it was capable of producing a steady beam at 16 MV and had been used in experiments at up to 15.4 MV. A 350 MeV gold beam was the highest energy used in an experiment, but we produced an identifiable beam of iodine (13+/33+) at 405 MeV.

There were many reasons for the demise of the Nuclear Physics Laboratory at Chalk River. The current Canadian government has pursued a policy of drastic programme cuts in the name of deficit reduction. They have succeeded in eliminating the deficit but the cost has been high. There have been major cuts to much basic research in Canada in Forestry, Freshwater studies in lakes, and research in all government departments. Recently, the newspapers revealed the layoff of a whole section of the Geographic Survey of Canada that had been leaders in research of the craters involved in the extinction events that killed off the dinosaurs.

AECL was directed to pursue "the CANDU business" as its prime mission. Its direct government funding has been drastically reduced and the only research to survive is that referred to as "underlying studies" related to reactor development.

The result was the end of TASCC and transfer of the Neutron and Solid State Studies branch to the National Research Council of Canada. This latter move is particularly ironic in the light of the 1994 Nobel prize awarded to Bertram Brockhouse for his work in neutron diffraction done at Chalk River. TASCC fought for other sources of support and external funding but failed to win the battle in spite of much appreciated support from all around the physics world. The neutron and solid state program will continue for a while until its neutron source, Chalk River's NRU reactor is shut down.

There are lessons that may be learned from this story. The need for political action, a skill better practised by American than Canadian scientists, is evident. It is perhaps, part of a more general teaching responsibility of scientists to the general public and hence to the politicians. Pure research motives can be mixed with the pursuit of spin-offs and applications; after all we are asking the public to spend a lot of tax dollars.

The Chalk River laboratory that became TASCC was a founding pioneer of the study of nuclear structure in the beginning and was a leader to the end. The scientists and some of the equipment have been dispersed and the Honourable History of Nuclear Physics at Chalk River has been finished. I hope that the spirit and excitement of those fifty years will be maintained by those who carry on the tradition of imagination, excitement and discovery that we were privileged to enjoy at Chalk River for half a century.

REFERENCES

(1) Atomic Energy of Canada Limited, *Canada Enters the Nuclear Age,* McGill-Queen's University Press, 1997

(2) Davies, W.G., "Nuclear Physics at Chalk River, an Historical Perspective," *Physics in Canada,* **51,** 270-272 (1995)

(3) Paul, E.B. and Clarke, R.L., *Canadian Journal of Physics,* **31,** 267 (1953).

(4) Litherland, A.E. and Ferguson, A.J., *Canadian Journal of Physics,* **39,** 788 (1961).

(5) Bromley, D.A., Kuehner, J.A., and Almqvist, E, *Physical Review,* **123,** 878 (1961).

(6) Specht, H.J., Fraser, J.S., Milton, J.C.D., and Davies, W.G., *Physics and the Chemistry of Fission,* IAEA-SM-122/128, (International Atomic Energy Agency, Vienna 1969), 363.

(7) Ewan, G.T. and Tavendale, A.J., *Canadian Journal of Physics,* **42,** 2286 (1964).

(8) Hausser, O. et. al., *Physical Review Letters,* **42,** 1451 (1979).

(9) Ward, D. et. al., *Nuclear Pysics,* **A529,** 315-362 (1991).

(10) Bigham, C.B., Davies, W.G., Heighway, E.A., Hepburn, J.D., Hoffman, C.R.J., Hulbert, J.A., Ormrod, J.H. and Schneider, H.R., "First Operation of the Chalk River Superconducting Cyclotron", *Nuclear Instruments and Methods,* **A254,** 237-51 (1987).

(11) Galindo-Uribarri, A. et. al., *Physical Review Letters,* **71,** 231 (1993).

(12) Savard, G. et. al., *Physical Review Letters,* **74,** 1521 (1979).

The Ten Years' Operation of the HI-13 Tandem Accelerator at CIAE

Yang Bingfan, Qin Jiuchang, Zhang Canzhe, Hu Yueming, Guan Xialing, Jiang Yongliang, ZhangGuilian, Yang Weimin, Kan ChauXin, Yang Zhiren, Su Shengyong, Liu Dezhong, Wang Liyong, Zhu Jiazheng

Tandem Accelerator Laboratory, Department of Nuclear Physics, CIAE, P.O.Box 275(62), BeiJing 102413, China

Abstract. A summary of the ten years' operation of the HI-13 tandem accelerator at CIAE is presented. The particular emphasis is put on the improvements on the laddertron and high voltage divider resistor system. Some statistics on the machine operation and maintenance are also given.

1. INTRODUCTION

The HI-13 tandem accelerator at CIAE has been running for more than ten years since it was put into normal operation at the beginning of 1988. The accumulated operation time and beam time is more than 40,000 h and 35,000 h respectively. The machine now is still in good technical condition with the best transmission efficiency of seventy percents and the obtainable terminal voltage of over 12 MV with beam and beam time of more than 3,500 h per year in recent years. The main improvements on the laddertron and the divider resistor system for the machine have been completed. The completely home-made ladertron has been in operation for more than 10,000 h with good performance and now it's still in use. The original "blue resistors" made by HVEC in the high voltage divider system of the HI-13 tandem accelerator has been completely replaced by the resistors with framework which are completely home-made. The performance of the framework resistor divider system is very satisfactory. To meet the requirements of the experiments, six additional beam lines were built up in the past ten years. It means that the total number of the beam lines in the laboratory now is twelve. Among them some of the new beam lines such as beam lines for atom and molecule physics studies, radioactive nuclear beam studies, material irradiation studies and AMS are playing very important roles in the laboratory. Some other improvements on the machine have also been done.

CP473, *Heavy Ion Accelerator Technology: Eighth International Conference,*
edited by Kenneth W. Shepard
© 1999 The American Institute of Physics 1-56396-806-1/99/$15.00

2. OPERATION AND MAINTENANCE

The HI-13 tandem accelerator is used in multidisciplinary studies (1). In recent years, the main scientific researches are concentrated on nuclear physics, neutron physics, radioactive beam physics, atom spectrum, nuclear data measurements and nuclear technology application studies. Up to now, more than 150 scientific projects proposed by more than 100 users from the institutes and universities all over the country have been completed.

Over the past ten years, the machine has provided more than 35,000 h beam time for the various research projects, while the operation time is about 40,000 h. Near 30 ion species such as H, D, He, Li, Be, B, C, O, F, Mg, Al, Si, P, S, Cl, Ca, Fe, Ni, Cu, Zn, Ga, Ge, Se, Br, Sr, Pd, Ag, Sn, I and Au have been accelerated to the targets with transmission efficiency of seventy percents for light ions and thirty to fifty percents for heavy ions. Fig. 1 shows the distribution of the beam time with years, and the distribution of the beam time with ion species is shown in Fig. 2. In this period, the highest terminal voltage of stable operation with beam was 12.7MV(ninety-seven percents of the rated terminal voltage of 13MV). But in recent years, the machine has to be operated at a limited terminal voltage (below 12MV) since the resistors used in the home-made resistor divider system are a mixture of two types of resistors at present which causes unbalanced high voltage distribution along the tube and column sections.

The time spent on the machine maintenance or overhaul each year was about fifty days. From year of 1988 to the end of 1997, the tank had been opened for 68 times, it means the average tank opening per year is about 6.8 times in the past ten years. Table 1 is the statistics of the tank openings. Among the tank openings, only one or two times each year was for scheduled overhaul, others were caused by unexpected troubles. They were:

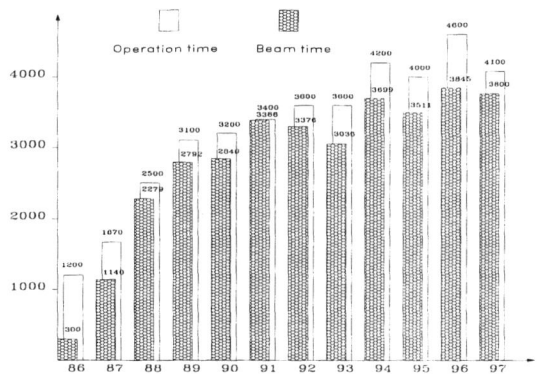

FIGURE 1. The distribution of the beam time with years

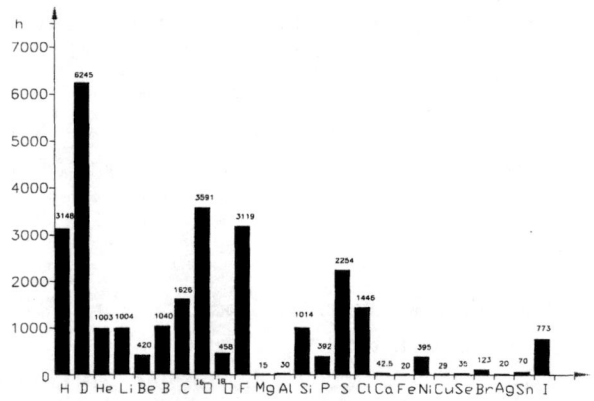

FIGURE 2. The distribution of beam time with ion species

1) Laddertron break	6 times
2) Breakdown of power supplies, cables and devices inside the tank	10 times
3) Shielding rod on the terminal and ivory bar on the equipotential rings dropped off by sparking	12 times
4) Out of control of the terminal stripper driving mechanism	5 times
5) Teeth of cog belt on the terminal generator dropped off	5 times
6) Bolts on the home-made laddertron dropped out during the test operation	5 times
7) Connection springs between resistors in the home-made resistor divider system broke during it's testing operation	6 times
8) Nuts and screws loosed or dropped out by vibration	5 times

To date, the tank was opened for more than a hundred times including the periods of machine installation, commissioning and one year testing operation. According to the calculation of SF_6 consumption, the accumulated SF6 loss in the past years was ten tons, it means each time tank opening, in our case the volume of the tank is 360 m^3, 100 Kg of SF_6 was lost.

Leak checking for the accelerator tubes during SF_6 refilling each time after tank closing is necessary since opening and closing tank probably effects the accelerator tubes mechanically even though the maintenance has not touched accelerator tube vacuum system during the tank opening. Two methods are adopted in our case: one is

Table 1. The statistics of the tank opening

Year	1988	1989	1990	1991	1992	1993	1994	1995	1996	1997
Times	8	9	4	6	8	9	7	3	8	6

leaking rate measurement at different SF$_6$ pressure in the refilling process, the other is the residual gas analysis using a quadruple mass spectrometer. Normally the results of two measurements coincide with each other. In the recent years, we find that there are always three peaks on the residual gas analysis spectrum which are at the same position as the characteristic peaks of SF$_6$ and were not observed before. To find whether it was caused by leakage or not, several tests had been carried out but no leakage was found. 1). The leaking rate measurement was quite normal. 2). Leak detecting using helium leak detector with high pressure of helium gas at all connection flanges along the tubes had shown no evidence of leakage .3). To find any possible leak anywhere along the tubes, a mixture of air and helium gas was used to pressurize the tank to the operation condition of 7.3Kg/cm^2, while two helium leak detectors were connected to the both low and high energy ends of the tank, this time no helium signal was found by detectors either. 4). Re-calibration for the residual gas analyzer with pure SF$_6$ through a leaking valve had shown that the peaks we found before were the characteristic peaks of SF$_6$ indeed. After all these experiments and careful analysis, we believe that the peaks were not caused by the leaking but from SF$_6$ which went into the tubes occasionally and were adsorbed on the wall of the tubes very firmly. During the SF$_6$ refilling of each time in recent years, the SF$_6$ characteristic peaks went up along with SF$_6$ pressure rising inside the tank for a period of time, giving a false impression of leakage, then it went down though the SF$_6$ pressure of the tank is still rising. This is because that at the first when the pressure of SF$_6$ went up, normally the temperature inside tank went up slowly too, so the SF$_6$ adsorbed on the inner wall of the tubes was released more and more, therefore the peaks went up. After some time, the temperature get into equivalence or even goes down, so the peaks get smaller and smaller.

3. THE LADDERTRON

It is well known that laddertron is very important not only to the performance of the tandem accelerator (especially to that of the terminal high voltage) but also to the machine operation efficiency. To gain good performance and long operation lifetime of the laddertron, improvements on the key-components of the laddertron such as insulating links and articulated bearings, have been carried out by several tandem accelerator laboratories in the world. In 1994, a complete home-made laddertron was put into operation at CIAE after five years' efforts (2). This new laddertron has been in operation for more than 10,000h with a highest terminal voltage of 12.7 MV and it's still in operation. It's the longest in operation lifetime among the six sets of the expendable component, such as nylon links and bushings that we have used ever since 1986 while the average lifetime of the other five sets was about 6,000 to 8,000 h. After 2,300 hours' initial operation, one nylon link on the new laddertron was broken and a electrical breakdown mark was found on another nylon link. After the replacement of these two nylon links with new one we never had the laddertron broken again. Up to now, the laddertron had been reassembled for two times to reduce the run-out of the laddertron and wipe off the dust from the links. One was done after 6,000 hours' operation, the other was made 8,200 h later. It seems to us that the nylon links could be still used for a considerable longer time if the bushings are in good condition during the operation since there is no clear evidence of wear on any side of the links.

Moisture content in the nylon material is very important to both electrical and mechanical performances of the nylon links (3). More moisture makes the links easy to be broken down, elongated and worn out. Normally, we ask factory to cast nylon material and make links in the dry season. After assembly of the laddertron we put it in a vacuum container, bake out and pump down the container at the temperature of 60 ℃ for at least two weeks to drive the moisture out from the links, then keep it in the container till we need it.

4. THE RESISTOR DIVIDER SYSTEM

The original high voltage divider resistors used on the HI-13 tandem accelerator were " blue resistors" of HVEC type, which consists of 20 small resistors of 60MΩ enclosed in the epoxy. Because of sparking, the epoxy was often broken into small pieces and dropped down to the bottom of the tank forming a new induce of sparking. On the other hand, the failure of the blue resistors had occurred more often year by year. Table 2 is the statistics of the number of the damaged blue resistors from 1984 to 1995. So we decided to develop a new resistor divider system for the HI-13 tandem accelerator. This development has been finished satisfactorily and the blue resistors have been replaced by new divider resistor section by section since the end of 1992. By the time of March 1998, the replacement was completely finished. Fig. 3 is the configuration of the new resistor dividers on the machine. Obviously, the structure of new resistor unit which consists of two 600MΩ resistors and an aluminum frame is different from the HVEC blue resistor (4). The frames form an excellent electrical shielding cage when all 72 unit of the divider resistors in a complete section of tube or column are installed protecting the divider resistors from being damaged by heavy sparking during the operation. Now all 1,152 unit of the new dividers are working well,

Table 2. Statistics of the damaged resistors

Year	1984-1989	1990	1991	1992	1993	1994	1995	Total
Damaged resistors	105	10	19	128	153	186	250	849

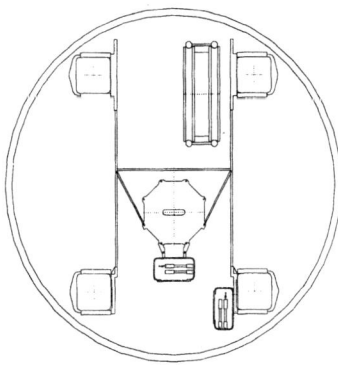

FIGURE 3. The new divider resistor on the machine

some of sections have been operating for more than five years with no one damaged after hundreds of sparking. The resistors which were made of synthetic film and installed in the earlier stage of the development are now being replaced by a kind of glass-glaze resistors since the value of the resistance of the former decreased simultaneously by a small amount after a period of operation while the resistance of the latter is very steady during the operation.

5. FACILITIES

To meet the increasing requirements of the experimental studies, a number of facilities were set up at the laboratory in the recent years.

The total number of bean lines now is twelve while in the initial stage there were only six (See Fig. 4) behind the switching magnet. The newly built Radioactive Nucleus Beam (RNB) line was put into operation at the end of 1993. Radioactive beams of ^{6}He, ^{7}Be, ^{11}C, ^{13}N, and ^{17}F have been obtained and several experiments with the beams have been done and some good results have been achieved. The purity of the secondary beams has been significantly upgraded after changing the acceptance angle of the beam line from 0° to 3°. The beam line dedicated to the atom and molecule physics studies and experimental equipment which consists of a beam-foil chamber and a 2.2 meter grazing incidence GIM-975 VUV monochromator for beam-foil spectrum study have been put into operation. Experimental studies of Highly Ionized Atom Spectroscopy with particle of S, Br, Ge, Cu, Ni, Pd, have been carried out and some phasic result have been achieved.

An automatic cycling system for isotope abundance ratio measurement has been developed for the AMS facility based on the HI-13 tandem accelerator. The main effort

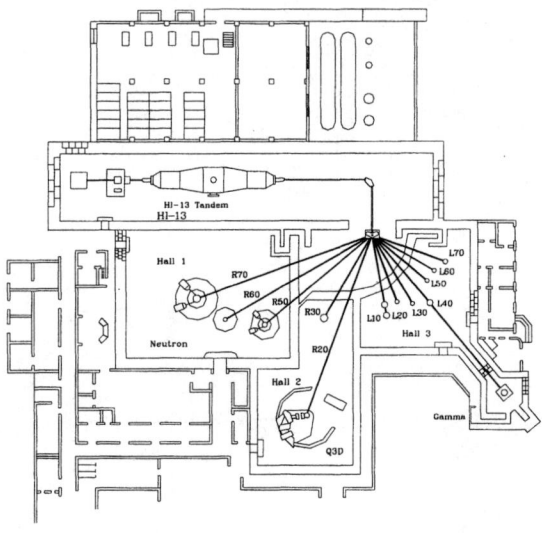

R20 Beam line for atom and molecule physics studies.   L20 Beam line for AMS.
R30 A several purpose target chamber.   L50 A secondary radioactive beam line.
L60 A heavy ion scanner for material irradiation.   L70 For material irradiation.
FIGURE 4. Twelve beam lines in the three target rooms

was focused on the cycling injection of stable isotope and radioisotope of interest into the accelerator and the measurement of stable isotopes at the outlet of the analyzing magnet. A new injector vacuum chamber with two gap lenses insulated from the deflection magnet has been constructed for beam energy modulation. A modified vacuum chamber of the analyzing magnet with two movable off-axis Farady cups near the outlet of the magnet was designed and constructed for collection of different stable isotopes such as ^{9}Be, ^{13}C, ^{27}Al, and ^{35}Cl (^{37}Cl) ions while ^{10}Be, ^{14}C, ^{26}Al, and ^{36}Cl being counted by $\triangle$E-E ionization chamber at the end of the AMS beam line.

A heavy ion scanning facility for material irradiation has been set up. It mainly consists of a magnet scanner combining with an electrical scanner and an irradiation vacuum chamber with sample transfer mechanism. Irradiation with particle of $^{32}S^{+13}$ at energy of 145MeV has been shown that the even irradiation has been obtained.

6. ACKNOWLEDGMENTS

We acknowledge the contributions of all people working in the accelerator laboratory for their every effort to the machine operation, maintenance and improvements in the past ten years.

REFERENCES

1. Guan Xialing, Qin Jiuchang et al, "The first five years' experience with the HI-13 tandem accelerator at CIAE", in *Proceedings of the sixth International conference on electrostatic Accelerators and Associated Boosters*, 1992, pp. 34-38.
2. Yang Bingfan, Zhou Junfeng, et al, "An improvement on the laddertron in Beijing HI-13", in *Proceedings of the sixth International conference on Heavy Ion Accelerator Technology*, 1995, pp. 87-88.
3. Deng yusheng, et al, "The laddertron for a 6MV tandem", in *Proceedings of the sixth International conference on Heavy ion Accelerator Technology*, 1995, pp. 101-103.
4. Zhang Guilian, Qin Jiuchang,et al, "The new voltage divider system of the HI-13 tandem accelerator", in *Atomic Energy Science and Technology*, China, 1996(3) pp. 207-213.

Status Report of the Munich Tandem Accelerator

L. Rohrer and tandem staff

*Beschleunigerlaboratorium der Universität und der Technischen Universität München,
85748 Garching, Germany*

Abstract: The Munich MP tandem accelerator is in operation since 1970. Now it is equipped with a NEC Pelletron charging system and HVEC stainless steel tubes. The negative ion injector and the high energy extension with the analyzing system were completely reconstructed. Also the instrumentation was improved continuously.
The tandem is used in an increasing degree not only for nuclear and atomic physics, but also for biological and material science work.

INTRODUCTION

The Munich MP tandem accelerator is in operation since 1970. A major conversion was made in 1975/76: a Pelletron charging system and NEC accelerator tubes were installed. The Pelletron still performs excellently. The NEC tubes were perfect with respect to beam transmission ($\geq 90\ \%$ for all ion species), but we could hardly get 12 MV terminal voltage. After the installation of extended Vivirad-High Voltage tubes in 1991 and the modification of the terminal skin in 1995 we could raise the terminal voltage to 15 MV (1).

NEGATIVE ION INJECTOR

The injector is divided into three platforms which can be independently put to high voltage: one with a polarized ion source, one with ion sources for AMS, and one with ion sources for routine operation. The polarized ion source will undergo a reconstruction at the turn of 1998/99, the hardware is already installed on a test bench (2). On the AMS platform an ultra clean source was installed. This is a sputter source with an analyzing magnet for the primary cesium beam (3). The standard injector has three ports at 45, 65, and 90 degrees of the inflection magnet. For the 90 degrees port the bending

radius of the magnet is 0.35 m and the mass energy product is 13 u·MeV. A mass resolution of 1:180 was measured. Each ion source is equipped with its own oil-free pumping system and a gate valve so that the ion sources can be changed without venting the magnet chamber. The ion species listed below are available for routine operation:
^{1}H, ^{2}H, ^{3}He, ^{4}He, ^{6}Li, ^{7}Li, ^{9}Be, ^{11}Be, ^{12}C, ^{13}C, ^{16}O, ^{17}O, ^{18}O, ^{19}F, ^{23}Na, ^{24}Mg, ^{27}Al, ^{28}Si, ^{29}Si, ^{30}Si, ^{31}P, ^{32}S, ^{33}S, ^{34}S, ^{36}S, ^{35}Cl, ^{37}Cl, ^{40}Ca, ^{48}Ti, ^{51}V, ^{52}Cr, ^{53}Cr, ^{55}Mn, ^{54}Fe, ^{56}Fe, ^{57}Fe, ^{58}Fe, ^{58}Ni, ^{60}Ni, ^{61}Ni, ^{62}Ni, ^{64}Ni, ^{59}Co, ^{63}Cu, ^{65}Cu, ^{64}Zn, ^{66}Zn, ^{69}Ga, ^{71}Ga, ^{70}Ge, ^{74}Ge, ^{78}Se, ^{79}Br, ^{81}Br, ^{92}Mo, ^{94}Mo, ^{98}Mo, ^{100}Mo, ^{107}Ag, ^{109}Ag, ^{114}Cd, ^{127}I, ^{128}Te, ^{130}Te, ^{181}Ta, ^{197}Au.

TANDEM

The charging system of our tandem is an NEC Pelletron with three chains on each side of the terminal. The Pelletron performs excellently. We had only two chain ruptures up to now, the life time being 100,000 hours of operation in the average. In recent years some of the many idler wheels were to be replaced because of bearing failures. Now we are in the process of replacing all bearings preventively: it is half finished, the other half will follow soon. The plan is to lubricate the bearings every two years, since we have good experience with this technique at the power shaft in the tank.

Also the Vivirad-High Voltage stainless steel tubes work reliably. Sometimes, especially after venting them, the tubes must be conditioned. Individual sections go up quickly in voltage to 4.3 MV - 4.5 MV. We are inclined to think that the terminal voltage is not limited by the tubes but by the tank. If the tank is clean we can go to 15 MV, but nevertheless we prefer to reduce the terminal voltage in routine operation to 13 MV - 14 MV, because we are apprehensive of damages due to occasional sparks at highest voltage.

Our terminal stripper has a gas control unit with a Pirani vacuum gauge and a piezo leak valve. The electronic part of it was replaced recently because the old one was not reliable due to insufficient voltage surge protection. Using more modern and smaller microprocessors and ADCs we gained some room for better circuits to protect the input and output lines. The new unit works without any failure.

To enable very heavy ion beams to be accelerated with the highest terminal voltage the HVEC analyzing magnet with a mass energy product of $M \cdot e/Z^2$ = 200 u·MeV was replaced in 1996 by the former Daresbury analyzing magnet with $M \cdot e/Z^2$ = 336 u·MeV (4). The different sizes of the magnets made some modifications of the analyzing system necessary. The beam lines on both sides of the magnet were completely reconstructed because the object and image focus are now on different positions, 1 m apart from the old ones. The quadrupole lens at the high energy end of the tandem was moved to the tank as near as possible. The devices formerly located between the tank and the quadrupole lens were removed. An ion pump, a faraday cup and a valve were mounted inside the tank instead. The quadrupole lens between the analyzing magnet and the switching magnet in the first target room kept its old position. To fit the beam optics to

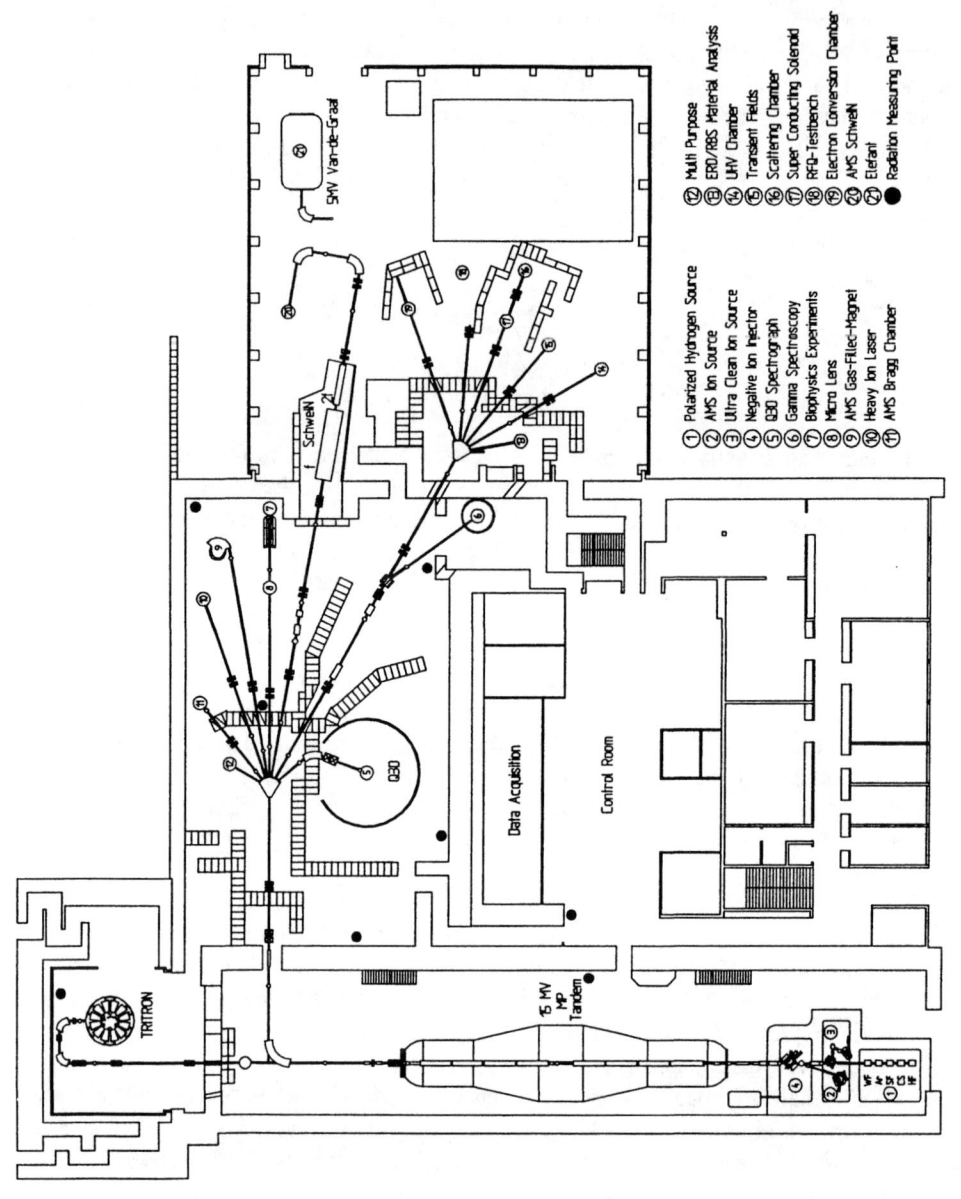

FIGURE 1. Plan of the Munich accelerator facility.

the existing beam transport system we installed an additional quadrupole doublet 3.7 m downstream from the first one.

Recently the demand on light ions with good energy stability was increasing. For light ions the energy spread in the stripper foil is quite low, so one can obtain a very stable beam energy if the terminal voltage is stable. Therefore we put the liner into operation again. This is an electrode inside the tank with capacitive coupling to the terminal. So one can apply a corrective voltage of a few kV to the terminal much faster than the corona stabilizer works, and one can increase the gain of the stabilizer. Under slit control the voltage stability is better than 1 kV_{pp} if the liner is used together with the standard corona stabilizer.

TARGET ROOMS

Some changes were made to the beam transport system and to the experimental setup. Figure 1 presents the plan of the laboratory. The AMS gas filled magnet (Pos. 9), the super conducting solenoid (Pos. 17) and the electron conversion chamber were installed. A new beam line with a vertical beam, using our former analyzing magnet, was installed for biophysics experiments (Pos. 7), and a micro lens (5) in the same beam line is under construction.

REFERENCES

1. Münzer, H. et al., Nucl. Instr. and Meth. A **382**, 78-81 (1996)
2. Hertenberger, R. et al., Rev. Sci. Instrum. 67(3), 1354 (1996)
3. Massonet, S. et al., Application of Accelerators in Research and Industry **CP392**, 795 (1997)
4. Rohrer, L., "Status report of the Munich tandem accelerator" presented at the Symposium of North Eastern Accelerator Personnel (SNEAP97), Jülich, Germany, October 13 - 15, 1997
5. Hinderer, G., G. Dollinger, G. Datzmann, and H. J. Körner., Nucl. Instr. and Meth. B **130**, 51 (1997)

A Foreline Oil Vapor Trap

S. M. Ferguson

Dept of Physics, Western Michigan University
Kalamazoo MI 49008-5151

E. D. Berners

University of Notre Dame, Physics Dept.
Notre Dame IN 46556

During my last visit to the Accelerator Lab at Notre Dame I noticed a clear plastic cylinder of alumina beads in the vacuum line between a diffusion pump and its backing pump. When I asked Ed Berners about it he pointed out that backing pumps can be an important source of back-streaming oil in a vacuum system. He said that he had checked the literature on oil-absorbing materials and found that alumina worked as well as any. He had alumina handy since it is used in the gas drying system. And alumina starts out pure white so one can easily see the oil that has been absorbed.

Alumina has two important disadvantages. First, alumina dust is very abrasive so it would cause excessive wear in a mechanical pump. One should place a filter between the alumina trap and the mechanical pump. We use the replacement cartridge from an Alcatel DFT 25 foreline dust filter; five for about $107. Second, alumina absorbs water vapor so it can take a long time to pump out the trap the first time. This can be mitigated by baking the alumina before putting it into the trap.

I changed Ed's design so that the alumina oil trap and the dust filter are contained in a single polycarbonate cylinder. The attached drawing shows our design. I don't have a more detailed drawing since this is all the detail our shop man wanted. The drawing shows the two configurations we considered for the end caps. I thought soldering brass parts together would save on labor, but our shop guy said that brass is too expensive and that it was almost as easy to put the fittings in with "O" rings.

We have been using these oil vapor traps for several months now. It is interesting to note that on some pumps the alumina is as white as the day we put it in. On other pumps we can see the oil vapor creeping up week by week and the alumina has to be replaced every few months. I don't know why pumps differ, but it is very helpful to be able to see which traps are okay and which ones need attention.

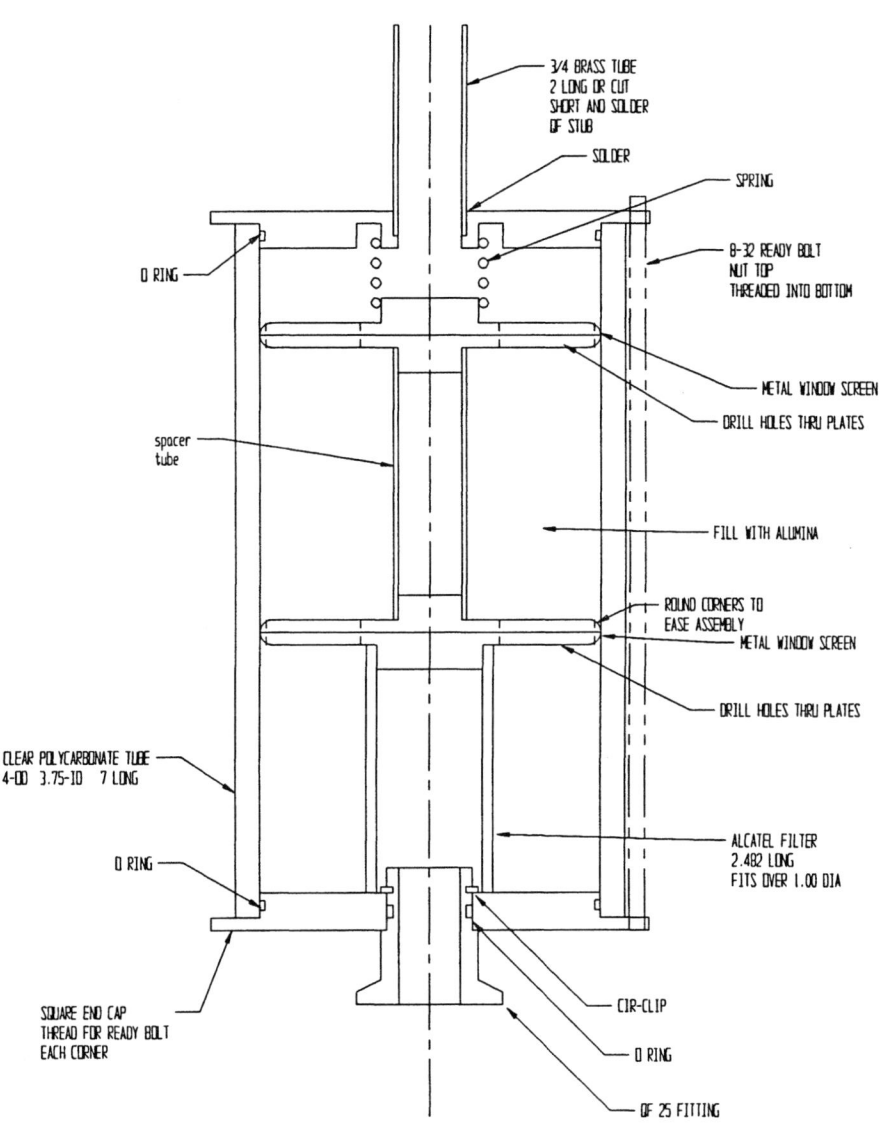

A New Design of Terminal Pumping in the Lund Pelletron Tandem

Kjell Håkansson and Ragnar Hellborg

Department of Physics, University of Lund
Sölvegatan 14, SE-223 62 LUND, Sweden

Abstract. A new design of terminal pumping has been installed in an existing 3 MV Pelletron tandem accelerator. The very limited space in the high voltage terminal (total length 600 mm and diameter 600 mm and housing equipment for two charging-chain systems and a double valve equipment, the latter to be used in connection with stripper foil replacement) and the absence of electrical power in the terminal, required an unusual design for the installation and powering of the new pumps.
Technical details of the design as well as some preliminary experience of the use of the new system, especially for accelerator mass spectrometry, will be given.

INTRODUCTION

For precise and efficient accelerator mass spectrometry (AMS) experiments the technical requirements differ to a certain extent from those for other accelerator applications. For the ion source a high and stable production efficiency of negative ions as well as a low memory effect are essential. For the electrostatic accelerator a stable terminal high voltage, a high beam transmission as well as an efficient and long-lived stripper medium are important. For the high energy analysing system a high mass- and energy-resolution is required. New dedicated AMS systems are designed on the basis of these criteria. For tandems that were originally designed for a nuclear physics programme and later on taken over for AMS, which is the case at many AMS laboratories, these requirements were not of primary concern at the initial installation. The 3 MV Lund Pelletron tandem accelerator was designed at the beginning of the 1970's. The ion beams used at that time were protons, deuterons and, to a small extend, helium ions. In connection with the introduction of the AMS programme in Lund a new ion source has been built [1], better low-energy optics have been designed [2] and several new optical units have been installed on the experimental line [3].

The requirement "an efficient and long-lived stripper medium" mentioned above can be better met with a gas stripper than with a foil stripper. During irradiation, the thickness of the foils changes in the irradiated area, and eventually the foils disintegrate. This change in thickness leads to a variation in the energy loss and in the scattering of the ions passing through the foils, and therefore to a change with time of the transmission through the accelerator system. Gas strippers have a better homogeneity compared to foils and their pressure can be kept constant with time, *i.e.* a constant stripper medium "thickness" will be obtained. In the Lund Pelletron, as well as in many other old tandems around the world, the stripper gas is pumped through the accelerating tubes using pumps outside the accelerator tank. This leads to a considerably higher pressure in the tubes. This will degrade the tube performance and lower the transmission by spreading the beam by collisions and by charge-exchange processes. This is most significant on the low-energy side since poor vacuum significantly decreases the transmission due to charge exchange from X^- to X^0 [4]. The maximum transmission through an accelerator without terminal pumping is reached at a stripper-gas pressure below that required to obtain equilibrium charge-state distribution. In order to improve the transmission using a gas stripper, the accelerator should be equipped with terminal pumping to allow the use of a higher gas density in the stripper and to minimise any effect on the vacuum in the accelerating tubes outside the gas cell.

Our work for a new stripper design including terminal pumping began as a study for transmission calculations [4], and measurements of the pressure profile in the accelerator [5] related to our AMS project. The results of these calculations and measurements clearly showed the limitation of our existing stripper design without terminal pumping. We now report on a new gas stripper, including terminal pumping, which has been designed and is now installed. The design of the new system and a test of its mechanical features, as well as a preliminary investigation of the quality of the stripper system for AMS experiments are reported.

DESIGN OF THE SYSTEM

The Stripper Housing and the Terminal Pumping

Our new installation is based on our existing stripper house. This house – the original NEC stripper house delivered with the accelerator - was slightly rebuilt some years ago to include a double valve equipment [6], to be used during foil replacement. The advantage of using the existing stripper house is obvious in the event of a failure of the terminal pumps. The accelerator can then be used for foil stripping with optical conditions very similar to those before the rebuilding. Either foil (provision for 100 foils) or gas stripping is available. The stripper canal has a total length of 400 mm and with an excess for the foil mechanism of 100 mm. The inside diameter of the canal is 8 mm. In the new design the stripper gas leaking out of the stripper canal is re-circulated by two turbo-molecular pumps (position (1) in Fig. 1) back to the stripper house. The pumps are modified Alcatel, model 5030CP. The idea of using re-circulating pumps in the terminal was reported for the first time in the beginning of the 80's [7] and has since then been used in several new installations as well as in old, rebuilt systems, *e.g.* Ref. [8]. The pumps are connected to housings (2) of high conductance on each side of the stripper house. The beam enters the first pumping housing through a limiting conductance (3) (diameter 14 mm and length 70 mm, corresponding to a conductivity of 2.9 l/s) and leaves the second pumping housing through a similar conductance (3). These short and limiting pipes (3) do not influence on the beam acceptance. However,

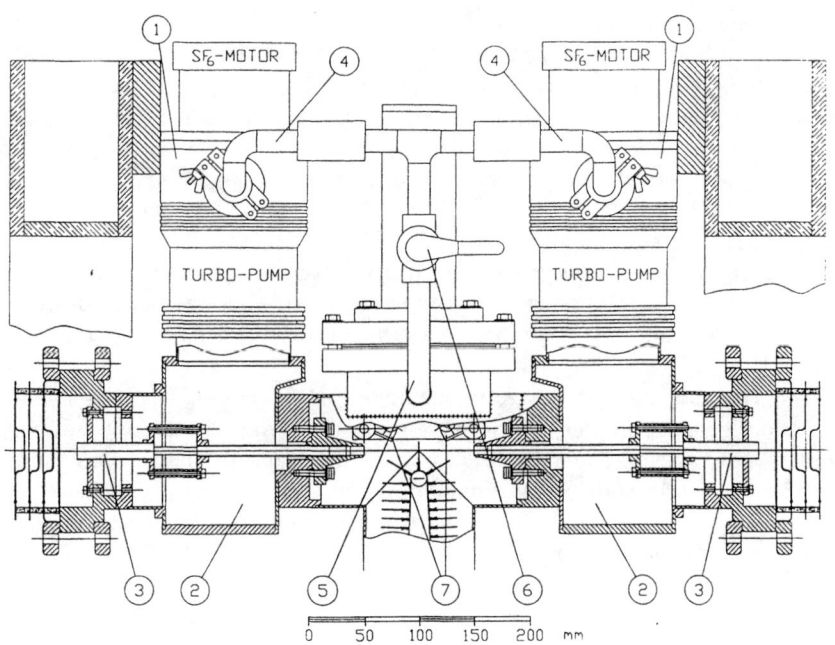

FIGURE 1. Schematic drawing of the stripper system. Legends:1-turbomolecular pumps, 2-high conductance houses, 3-limiting conductance pipes, 4- pump outlet pipes, 5-stripper house gas inlet, 6-valve, 7-double valve.

they reduce the gas flow out into the accelerating tubes considerably. The outlets of the pumps (4) are fed to the gas inlet (5) of the stripper housing. The inner diameter of these outlet lines are 16 mm. In this line a valve (6) is placed. This valve, together with the double valve equipment (7) mentioned above, gives the possibility to isolate the stripper house from the rest of the system, *e.g.* when foils are to be replaced, leak testing, etc. More than 90% of the gas is assumed to be re-circulated. New gas to replace that gas pumped away along the accelerating tubes is led in from a reservoir of about 2 litres, through a hand-operated service valve and a needle valve operated from ground via a rod. The pressure in the terminal can not be directly measured while the accelerator is in operation. However, it has been measured with the tank open during maintenance and is therefore known to be 5×10^{-3} Pa during foil stripping and in the 10^{-1} Pa and 10^0 Pa range during gas stripping without and with terminal pumping, resp.

In connection with this rebuilding of the stripper system the adjustment of the end parts of the accelerating tubes close to the terminal have been modified. All the pieces of equipment along the beam path in the terminal are machined in such a way that when bolted together they become aligned to each other. The entrance and exit of this package are aligned by our normal alignment procedure (using a telescope and apertures with 1 mm openings). As this package has well centred and steering connections for the accelerating tubes, also the tubes will be aligned. In this way a much easier way of aligning the tubes has been achieved.

FIGURE 2. A view of the high voltage terminal and part of the high energy column.

The Powering of the Terminal Pumps

The space in the terminal became very limited some years ago after a second charging chain system and the double valve equipment were installed. This together with the absence of electrical power in the terminal, required an unusual powering of the terminal pumps. The two turbo-molecular pumps have been redesigned in that the original electrical motors - mounted inside the vacuum - have been replaced by motors driven by compressed SF_6. These new motors have, of course, been placed outside the vacuum and the connection between the pumps and the motors is magnetic. In a first version ordinary, commercially available air-motors (Atlas Copco, model PIV-5020ACE50) were used. The lifetime of these motors is, however, very limited even if they are designed for continuous use. In a second version, home-made turbine motors are just now under test. In Fig. 2 a photo of the high voltage terminal after installation of the turbomolecular pumps, driven by the commercial air motors, is shown.

The compressed SF_6 for the motors is obtained from compressors – one for each pump – placed in a separate tank of about 1 m^3. In Fig. 3 a schematic drawing of the system for powering the turbo-molecular pumps is shown. Gas is transferred from the accelerator tank (containing 100% SF_6) to the compressor tank via a water cooled heat exchanger (position (B) in Fig. 3) and a pressure-reducing valve (C). The latter provides a pressure of 0.11-0.13 MPa (*i.e.* a few tenths of an atmosphere overpressure)

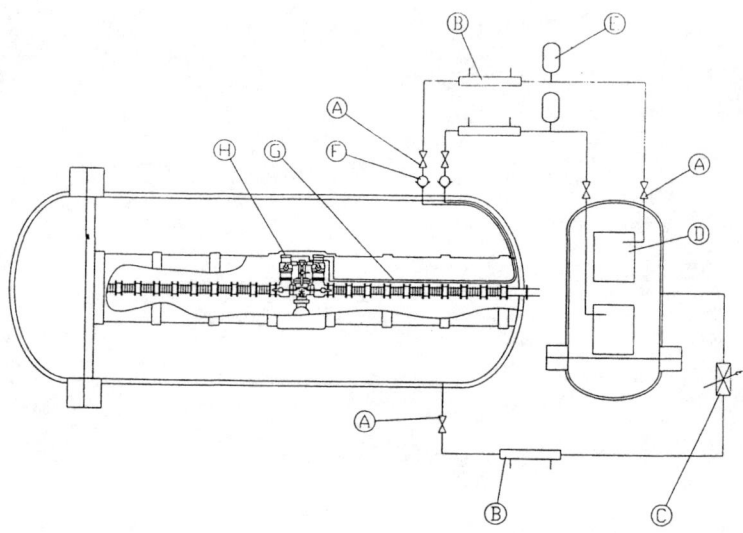

FIGURE 3. Schematic drawing of the compressor system. Legends: A-valves, B-heat exchangers, C-pressure reducing valve, D-compressors, E-buffer tanks, F-check valves, G-teflonpipes, H-motors of the turbomolecular pumps.

in the compressor tank. In this way the two compressors (D) (Atlas Copco oil-free rotating compressors, capacity 4.2 l/s at 1 MPa air, model Scroll-SF4-10) are fed with gas of a constant pressure. The outlet of the compressors (with a pressure of about 1.0 MPa) is via buffer tanks (E) with a volume of 25 litres each and water cooled heat exchangers (B) connected to a flange in the accelerator tank close to the high energy end. Between this flange and the terminal the compressed SF_6 is led through teflon hoses (G) and feeding the motors (H) of the turbo-molecular pumps. The expanding gas leaves the motors straight out into the accelerator tank. The home-made motors are built up from a one stage turbine wheel (also home-made) surrounded by replaceable nozzles. Different dimensions and numbers of the nozzles are under test, aiming at a speed of 60-75% of nominal speed (27000 r/min corresponding to 27 l/s for N_2) at a tank pressure of 0.6 MPa. A fibre optical link is installed between each pump and the outside of the tank, providing the possibility to measure the pump speed.

SUMMARY

Installation of the new gas stripper and the re-circulating pumps started eight months ago. Testing of different conditions has been done in parallel with the normal use of the accelerator. This testing will continue in order to find the optimum design of the pump motors, a correct gearing between the electrical motors and the compressors, a suitable setting of the pressure-reducing valve between the accelerator and compressor tank, as all these parameters will have an influence on the turbo-molecular pump speed. In order to increase the time between consecutive maintenance procedures, we plan to replace the ordinary steel ball bearings in the SF_6 motors with high speed bearings, i.e. similar to those used for the rotor-bearings in the turbo-molecular pumps. The turbine wheel and the magnetic coupling will be balanced – both statically and dynamically – to decrease the stress on the ball bearings. To increase the ion transmission, the N_2 stripper gas will be replaced with Ar gas [8].

The total cost of our new system, including two compressors, one compressor tank, two turbo-molecular pumps, heat exchangers, piping etc, but excluding labour costs, is relatively high, about US$ 30 000. However, the only alternative to obtain terminal pumping would be to include electrical power in the terminal. This would need an enlargement of the tank and the terminal, a much more expensive procedure.

The system has been used with a few different ion beams with excellent beam current stability and good optical quality. A 2-3 times higher ion transmission compared to the old gas stripper system and a constant stripper medium thickness compared to foil stripping is attained with the new system. The advantages for our AMS measurements using the new system will be that a higher beam current can be taken through the accelerator, a higher transmission, better quality of the beam and more stable conditions during the experiment.

Preliminary measurements of the pressure profile after the installation of the new system have been made and will be presented elsewhere as well as a report about measurement of beam transmission, mass fractionation and charge state distribution.

ACKNOWLEDGEMENT

The mechanics and electronics staff at the Pelletron accelerator T. Klemedsson, G. Larsson, S. Lundborg, G. Matthys, G. Oberlechner and S. Uthas carried out the precision work connected with the production of the new stripper system.

REFERENCES

1. Persson P., Freiman K., Hellborg R., Håkansson K., Skog G. and Stenström K., *Rev. Scientific Instr.* **69:2**, 1188 (1998).
2. Hellborg R., Håkansson K. and Skog G., *Nucl Instr Meth* **A287**, 161 (1990)
3. Hellborg R., Håkansson K. and Skog G., *The modification of an over 20-year old Pelletron to function as a high quality AMS facility,* Presented at the "31st Symp. of Northeastern Acc. Personnel" 13-15 Oct 1997, Jülich, Germany
4. Wiebert A., Erlandsson B., Hellborg R., Stenström K. and Skog G, *Nucl. Instr. Meth.* **A366**, 17 (1995).
5. Wiebert A., Erlandsson B., Hellborg R., Stenström K. and Skog G., *Nucl. Instr. Meth.*, **A364**, 201 (1995).
6. Hellborg R. and Håkansson K., *Nucl Instr Meth* **184**, 79 (1981).
7. Lee H.W., Galindo-Uribarri A., Chang K.H., Kilius L.R. and Litherland A.E., *Nucl. Instr. Meth.* **B5**, 208 (1984).
8. Bonani G., Eberhardt P., Hofmann H.J., Niklaus Th.R., Suter M., Synal H.A. and Wölfli W., *Nucl. Instr. Meth.* **B52**, 338 (1990).

Replacement of a Broken Column Support in a Horizontal Pelletron

Kjell Håkansson, Ragnar Hellborg and Sigfrid Uthas

Department of Physics, University of Lund
Sölvegatan 14, SE-223 62 LUND, Sweden

Abstract. After 23 years of use one of the column supports in the Lund 3 MV eletrostatic accelerator broke. Details of the straight-forward and quick replacement work are given.

INTRODUCTION

All standard NEC[1] Pelletron accelerators are made up of 1 MV column modules. Each module consists of four (for the biggest Pelletron accelerators more than four) insulating supports bolted between aluminium junction plates. All supports in standard Pelletrons are identical and interchangeable. These supports are made of alumina ceramic bonded to titanium metal plates utilising a metal bonding agent. They are constructed to be mechanically very strong and they are protected by spark gaps. Both ends of each support are provided with ring-shaped flanges, for mounting the support to the junction plates. These flanges are removable and fixed to the support by a screw which tightens the flange around the support.

After 23 years of use one of the twenty-four column supports in the Lund 3 MV Pelletron tandem (NEC model 3UDH) broke. The broken support was replaced with a new one in a fairly quick, straight-forward way. This replacement could be done without dismounting the main part of the column and even without breaking the tube vacuum. In this technical report, details of the replacement in a day-by-day description, and, experience of this unusual procedure for the user of an accelerator, is thus provided.

DAY BY DAY DESCRIPTION

Friday, week no. 0. When the alignment screws of the low-energy accelerating tube was loose in the junction plate no. 3 (for identification of the different units in the accelerator, see Fig. 1), a metallic noise could be heard. The reason for loosening the

---

[1] National Electrostatic Corp., Wisconsin, USA

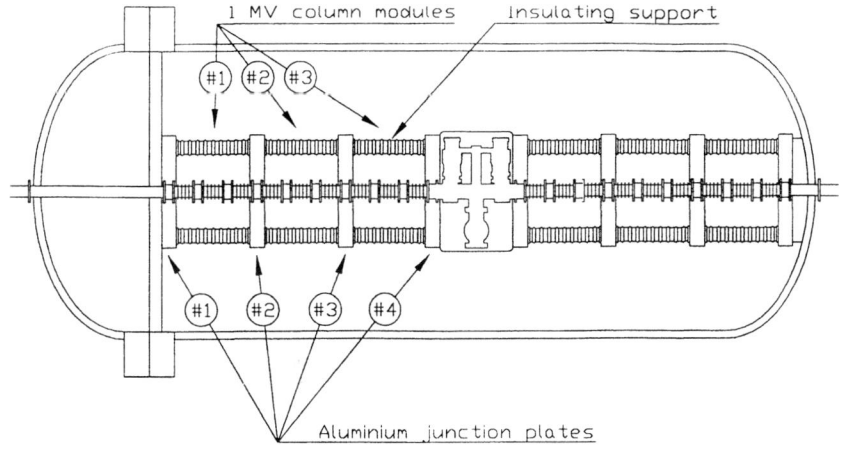

FIGURE 1. Schematic drawing of the column construction in a horizontal 3 MV Pelletron tandem.

tube alignment screws was to realign the tube in connection with maintenance work on the stripper system.

Monday, week no. 1. A piece of aluminium - 2 mm thick, 8 mm wide and in the shape of a half circle 100 mm in diameter - was found on the tank bottom. In the search for the origin of the piece of aluminium a space of 2-3 mm was noted between the flange of the lower insulating support in the column module no. 3 and junction plate no. 4. The supports are placed symmetrically in the module, and, seen along the beam direction their positions in our horizontal Pelletron are upper, lower, right hand side and left hand side, resp.

Tuesday, week no. 1. When the lower support was dismounted, the right hand side support in the same module was observed to be broken. A space of 3-4 mm could be seen in the piece of ceramic closest to the terminal. A photo of the broken support, after it was dismounted, is shown in Fig. 2. A new support (and one spare) were ordered from NEC.

Wednesday, week no. 1. The old steel beam, used 23 years ago when the column was originally positioned into the tank, was cut into a length just shorter than the length of the tank and was mounted inside the tank close to the top. Two of the ports (diameter 102 mm) on top of the horizontal tank were used to support the steel beam.

Thursday, week no. 1. Vertically threaded rods were fixed to the steel beam by plates. The lower ends of the rods were fixed to the junction plates, except to the two junction plates at ground potential. The length of the rods was adjusted to align the junction plates horizontally. After that the remaining supports in module no. 3 could be dismounted. Three of the supports in the terminal (of course not insulating and therefore of a completely different design compared to the column supports) were also dismounted, with the exception of the upper one. The tubes were pumped to check whether anything unforeseen had happened to the tubes. Good vacuum was easily obtained!

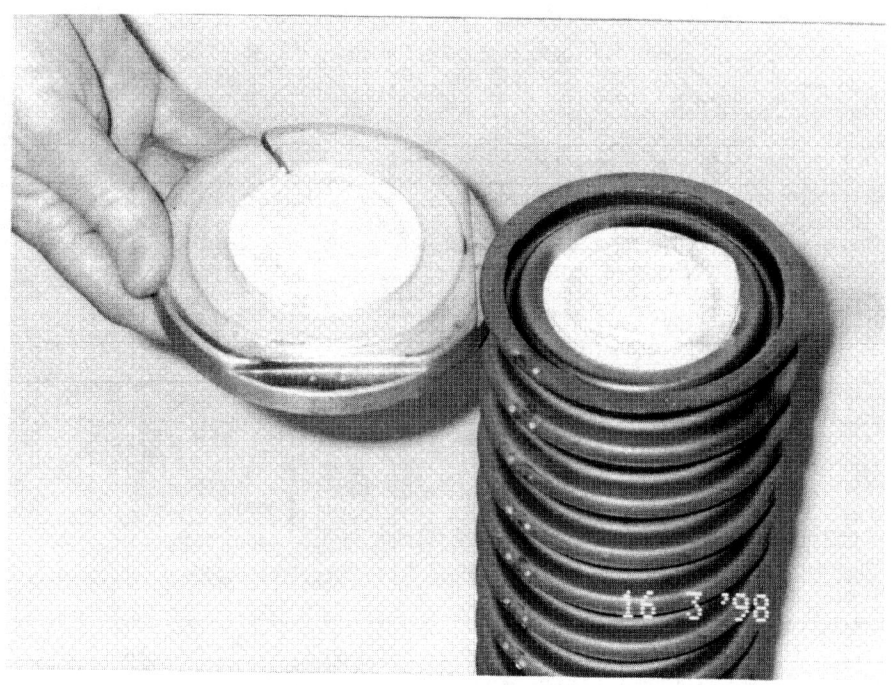

FIGURE 2. Photo of the broken insulating support.

Friday, week no. 1. Temporary supports for the terminal, 10 mm shorter than the ordinary ones, were mounted in the lower, right hand side and left hand side positions, resp. The remaining ordinary support in the terminal was dismounted.

Monday, week no. 2. One of the dismounted supports from column module no. 3 was mounted in a bench (constructed to obtain the correct distance and orientation between the flanges and also to get the flanges plane-parallel) and the pinching screws in the two flanges of the support were tightened.

Tuesday, week no. 2. The new support arrived from NEC and the tightening of the flanges of that support was checked and found to be 1.4 and 1.8 Nm. This information guided us to adjust the tightening of the support in the bench to 1.5 Nm. The other three supports were mounted in the bench and their flanges were also tightened to 1.5 Nm. The new support was installed in column module no. 3 in the lower position where the tensile stress is greatest. The rest of the supports in column module no. 3 were installed. The four supports were only lightly fixed to the junction plate no. 3 and not at all to plate no. 4. The temporary supports in the terminal were dismounted and the four newly mounted supports in the column module no. 3 were lightly tightened to junction plate no. 4.

Wednesday, week no. 2. All screws connecting the supports in column module no. 3 to the junction plates no. 3 and 4 were tightened. The right and left hand side supports in the terminal were lightly fixed to the junction plates no. 4 and 5. Before mounting the lower support in the terminal it was necessary to loosen the alignment screws of the low-energy tube (except those close to ground). The reason was that the tube forced the column sideways. The upper and lower supports in the terminal were lightly fixed to the junction plates. It was noticed that the possibility for the whole column to move axially compared to the tank was about 8 mm. To limit this distance and to press the whole column together when tightening all the flange-screws in the terminal junction plates, adjustable wedges were mounted on the high energy side between the column and the tank. After the maintenance procedures these wedges have been left in position and now limit the axial movement of the column to 1-2 mm. In this way the wedges work as a safety arrangement for the column.

Thursday, week no. 2. The horizontal alignment of the column was checked to be within ± 0.5 mm. After dismounting the vertical rods supporting the whole column the alignment was found to have increased to ±1 mm. The sideways alignment of the terminal compared to the tank was found to be ± 3mm. The vacuum in the accelerating tubes and the stripper-system was let up to atmospheric pressure, giving the possibility to align the stripper house along the optical axis. Other equipment in the terminal, such as the turbo-molecular pumps, the stripper foil mechanism etc., was mounted. During the following few days the accelerating tube was aligned along the optical axis. All corona needle plates, as well as other minor items that had to be dismounted, were remounted. The steel beam was dismounted and the ordinary maintenance work, which was interrupted when the support was found to be broken, could continue.

SUMMARY

The reason for the broken support is not known. One part of the surface of the broken piece of ceramic was white in colour and another had a yellow/brown colour. Probably the white area is fresh and is the result of this accident and the yellow/brown area is the result of previous damage and has changed colour over time. We have never noticed anything which can be connected with the initial damage.

When we discovered that one of the column supports was broken we predicted a maintenance period of several weeks. During the planning of the replacement work a relatively straight-forward way was found and this large and delicate work could be done in about a week, with an entirely satisfactory result. The description above may be of help for other accelerator users in a similar situation.

Electrical Stress on Insulators of the Vivitron

F. Osswald, J. Heugel, E. Jegham, N. Lahera, R. Rebmeister and
P. Zouloumian

Institut de Recherches Subatomiques, UMR7500 CNRS-IN2P3/ULP
Service des Accélérateurs, BP 28 F67037 Strasbourg Cedex 2, France

Abstract. The Vivitron is a large electrostatic accelerator in operation under high voltage since December 1990. Since then, the machine sustained more than 750 sparks and operated more than 17 000 hours at voltages up to 20 MV. The paper presents a review of the electric stresses applied on the insulators of that particular HV generator following the operational experience of the last years. It discusses some of the most recent results related to the electrical stress characterization i.e. calculations and data from bench, factory and field tests.

INTRODUCTION

The electrical stress characterization is an important design parameter and the key to a better understanding of the mechanisms of gas breakdown and insulator flashover or puncture occuring in the Vivitron. High Fields up to 20 MV/m applied on large areas are inducing a severe stress which have to be identified in order to improve the high voltage performance and obtain a more reliable system.

The main part of the present paper deals with the electric fields applied on the solid insulating structure and defined by calculations and measurements. Nevertheless, extrapolating field breakdown values from small scale, short time experiments and calculations to a large practical system of long expected lifetime has always been a challenge for all HV systems. Despite good results were obtained with smaller machines or low-stress designs, some of the Vivitron's limitation are related to its size and configuration. The insulating structure is composed of several dielectric materials as cast composite epoxy - fillers posts for the external part (between the porticos) and epoxy-fiberglass and Lucite boards inside the central column. The posts are composed of Biphenol-A based resin, silica and alumina fillers. The applied mechanical strength is lower than 4000 daN and the applied DC voltage is 2.3 MV under nominal conditions at 18 MV. This HV insulating technique is used with success since the seventies in compressed gas insulated cables or transmission lines but generally at a lower operational voltage. The column of the accelerator includes the charging system based on a composite nitrile rubber - polyester carcass conveyor belt. The original board and belt designs are now showing a reliability comparable to conventional systems at 18 MV. The voltage applied on the boards is 2 MV and on the belt 18 MV, the mechanical strengths are respectively 350 daN and 380 daN (tension). The central column includes the accelerating tubes made of stainless steel equipotential plates and Pyrex glass insulating gaps. The paper will not cover this aspect as the tubes suffered noticeably less damage than the other systems. The tank is filled up with sulfur hexafluoride at 0.6 to 0.7 MPa.

GLOBAL OPERATION DATA

The generator is powered since December 1990 and sustained more than 750 sparks at voltage reaching 20 MV. The spark rate is an important parameter to evaluate the stress and the induced material ageing because the conditioning is still (in some cases) a unstable process decreasing in a chronic way the voltage performance. The weak protection cannot evoid transient overvoltages which cause damages on the insulators (1). Figure 1 indicates the average number of discharge/day over the last years. This spark rate can show variations from one year to the other due to the variable status of the machine and the voltage in operation, see figures 1 and 2. We can notice that the daily spark rate decreases during the last years (1996, 1997 and 1998) and remains lower than 1/day despite the voltage in operation is increasing. This behavior can be correlated with the improvements performed during that period on the insulating structure and on the protection system against sparks (2).

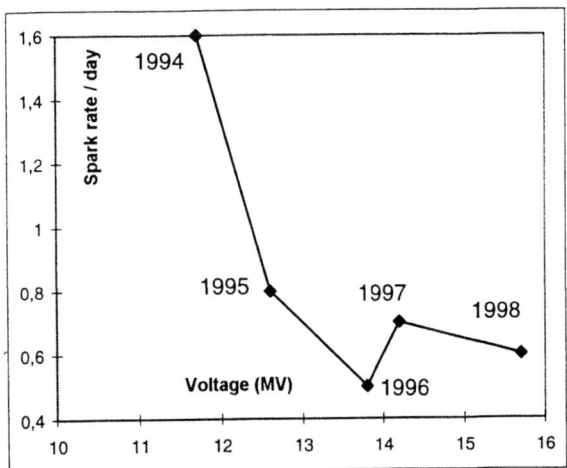

FIGURE 1. Daily spark rate averaged over a year and corresponding mean voltage.

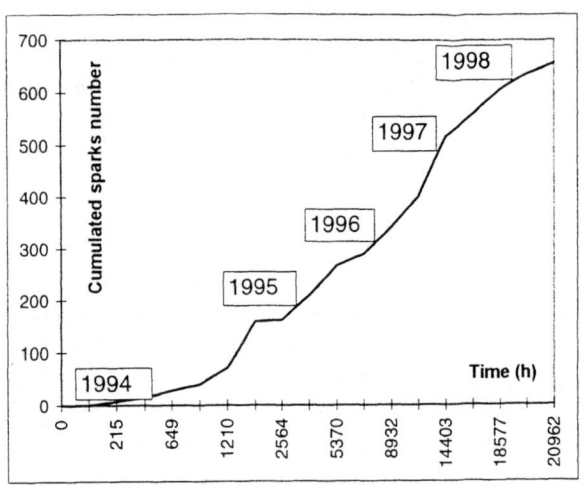

FIGURE 2. Cumulated sparks number versus cumulated time.

The damages occuring on the insulators are analyzed during the maintenance and some statistics are performed in order to complete the diagnostic i.e. to locate a weak point of the accelerator. The concentration of the damages changes from one period to the other and actually we are observing some impacts on the Lucite boards supporting the resistive potential dividers (for the column and the tubes). Other superficial tracks are sometimes noticed on the epoxy-fiberglass composite boards of the column structure, but are easily cleaned from the silicone greased surface with a tag.

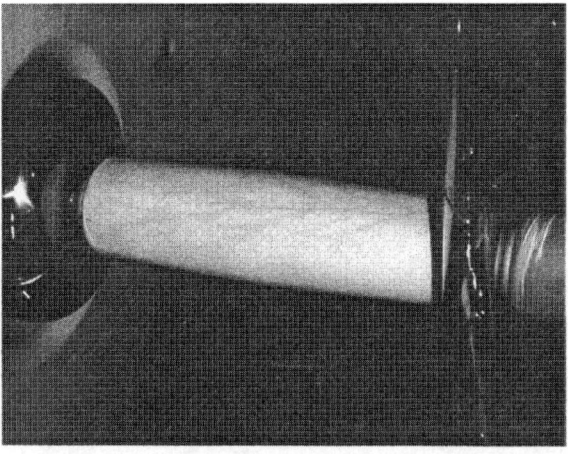

FIGURE 3. Inter-portico post insulator and superficial damage following a spark.

The insulating structure fastening the central column through the porticos is constituted by 178 long insulators (400 mm long) and 70 shorter ones (210 mm long). A few insulating posts are damaged each year. Most of them show some superficial tracks (flashover) and present several branches coming up or down, as shown in figure 3, starting sometimes from a small spot at the external interface between the dielectric material and the metallic insert (triple junction). Other rare ones (1 - 2 per year) did not sustain the electric discharge and are exploding. The multitude of very small debris (as small as dust particles) as well as larger pieces (in the order of the hand dimension) are sprayed all along the tank wall and show a typical spatial distribution, see figure 4. It is noticeable that the long posts are only showing some superficial tracks and that the short posts are merely broken when they are damaged. The HV tests performed in a CN pilot machine and our experience with the Vivitron up to 19 MV indicate that the damages on the posts are not so critical but generate more instabilities. One significant indicator of that tolerance against sparks and of the improvements is represented by the openings number/year which was 22 in 1994, 17 in 1995, 8 in 1996, 4 in 1997 and which is finally 1 up to now in 1998.

Small defects are not visible in the bulk of the material and we developed a non destructive control system composed of an ultrasonic scanner and tomographic images. We locate small flaws inside the insulators with a resolution limited to approx. 1 mm by the attenuation and the internal and interfacial reflections.

In addition of the abnormal operational conditions related to the sparks and the transient overvoltages, the insulators of the Vivitron have to sustain the standard static stress induced by the high voltage. The corresponding cumulated "dose" variation versus the time is given in figure 5 where one can see that the actual cumulated electric stress (since 1994) is in the order of 282 000 MV·h.

FIGURE 4. Broken posts sprayed on the tank wall after a spark.

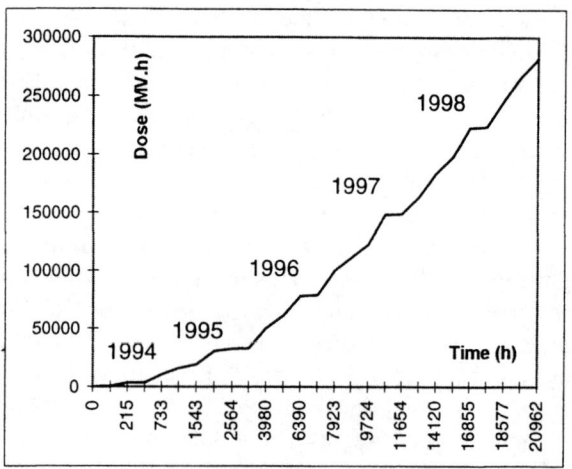

FIGURE 5. Cumulated electrostatic stress with the time.

The slope of the curve gives the average potential in operation, and starting in 1994 with a value near 12 MV, we are reaching approx. 16 MV in 1998. As mentionned, the potential corresponds to the terminal voltage which is 8 times the voltage applied on the posts (16 times on the short posts) and 9 times the voltage applied on the column boards in steady DC conditions.

LOCAL STRESS AND ANALYSES

The local electric stress as well as the dielectric properties of the materials play a key role in the global performance of a HV system like the Vivitron. With the trend to more compact, highly stressed designs and the importance of the cost and reliability, special care must be taken to perform a complete and unequivocal characterization of the electric stress whenever it is possible. Some examples of calculation, simulation and HV tests performed on insulating components of the accelerator will be presented below.

Field map inside the column

There was concern that some instabilities inside the column of the accelerator could initiate external breakdowns between the electrodes of the portico structure (3). Therefore it cannot be over-emphasized that careful observations and a good knowledge of the behavior of the column insulators in operation should be the rule to understand the actual limitation and improve the voltage performance.

It would seem that the interactions between the belt charging system (conveyor belt which follows some principles of the classic Van de Graaff accelerators), the column insulators through DC long time charging, and the beam stability (microstructure and ripple) through some coupling, are more important than initially expected. Particularly, the contact between the belt and the supporting or driving rollers (leading to some electro-aerodynamic effect on the interface in the gas) and the small but chronic corona leakage towards the surface of the insulating boards are contrary to the decoupled principle and are inducing an abnormal stress.

The electric field map inside the column represents the static and the initial conditions. It is obtained with the help of a numerical model and some calculations using the finite element method (FEM). The model represented in figure 6 is a cross section of the column and presents two planar symmetries (yoz and xoz). The boundary conditions are given by the column electrode, the accelerator tubes and their shielding bars potentials (here grounded), and the charge density on both the external sides of the belt of 10 µC/m^2 (here with an ideal infinite resistivity). At 18 MV, the average longitudinal field Ez along the column is 1 MV/m. The resultant field amplitude in figure 6 $\left(\sqrt{Ex^2 + Ey^2} \right)$ indicates some edge effect on the belt border and on the shielding bar extremity and reaches at least 1.6 MV/m (16 kV/cm). The nominal gap between the belt border and the shielding bar is 150 mm, but in running and at some locations it can reach 40 mm. The electric calculated stress applied on the lower belt surface, see figure 7, is in qualitative accordance with the preliminary measurements (4). The lateral field variation is given with the relative position from the middle (point 0) to the border of the belt (point 1, at ≅ 200 mm). Despite the charge density in this case is uniform (experimentally it is not), the field distribution is non uniform and follows the Poisson's equation and the influence of the environment.

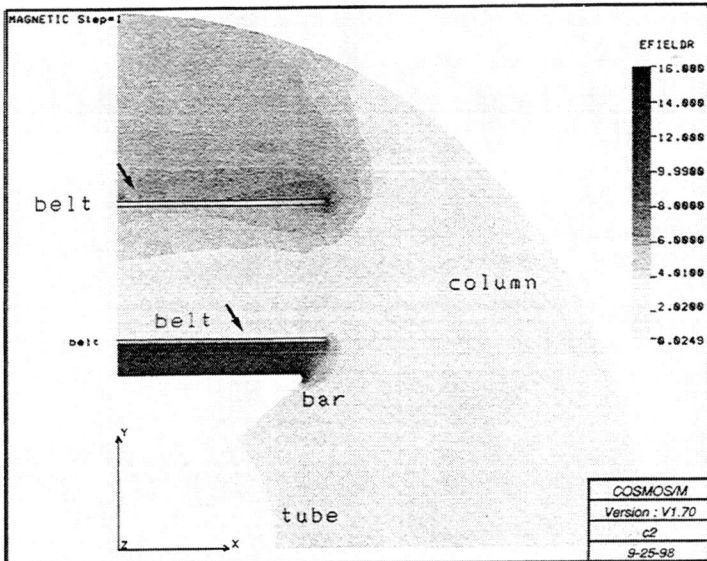

Figure 6. Electric field distribution around the belt and the tube inside the column.

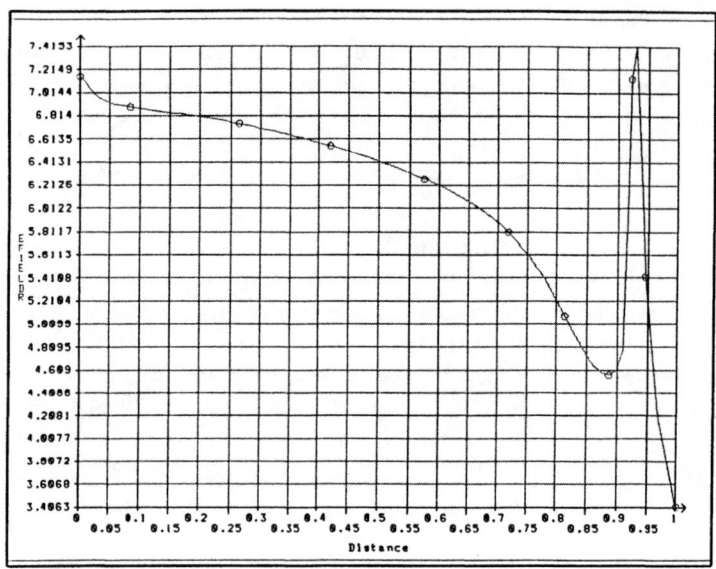

FIGURE 7. Lateral electric field variation on the belt surface (half width).

The steady state conditions used here to verify if the material and the gas can withstand the electric stress, are characterized by stable potentials and positions, and an uniform electrostatic charge density. In this case, the tangential field have to be considered if the material is contaminated, and the perpendicular component if the gas is polluted and charged with ions. The resultant DC stress have to be completed by the variation of the potentials, the migration of the charges and the movements of the belt in order to define the dynamic contribution. The capacitive coupling between the belt and the tubes have been evaluated considering the model in figure 6 i.e. the plan (belt) - rod (bar) configuration. The gap of 40 mm represents the minimum value in operation. The equivalent capacitor is in the order of 10 pF/longitudinal accelerating gap. Our actual model can integer the geometry, the potential gradient along the column and a dynamic mode. It results in a conduction current i_c given in equation (1).

A future model will have to represent the movement of the belt i.e. the displacement current given in equation (2). This dynamic component (with the time variation of the field) complets the characterization of the stress and the coupling between the belt and the environment.

$$i_c = c \frac{du}{dt} \qquad (1)$$

$$i_D = \int_s \varepsilon \frac{\partial E}{\partial t} ds = \int_s \left(\varepsilon_0 \frac{\partial E}{\partial t} + \frac{\partial P}{\partial t} \right) ds \qquad (2)$$

with P the material polarization.

Other Aspects

The insulating posts

The insulating posts are holding and supporting the central column. They are submitted to a voltage of 2.3 MV in operation at 18 MV on the terminal. The voltage is applied on the posts through the discrete electrodes of the portico structure. The ideal terminal voltage is approximated by the relation (3) in a coaxial structure with intershields :

$$V = E \sum_{1}^{8} r_{n-1} ln(r_n / r_{n-1}) \qquad (3)$$

where E is the mean outer field of the electrodes (without edge effect), r_0 the terminal radius, $r_1...r_7$ the radii of the 7 intershields and r_8 the tank radius.

The real voltage distribution along the axe of the middle of the discrete electrodes is shown in figure 8 and at 32 MV. The potential variation is indicated versus the relative position, point 0 is at the terminal, point 1 is at the tank (at 3.5 m). The discontinuities are produced by each portico from the inner to the outer face. Despite the post applied stress is not perfectly homogeneous it is in a better shape than the usual field distribution E_r given by the coaxial design without intershields :

$$V = E_r \, r ln r_8/r_0 \qquad (4)$$

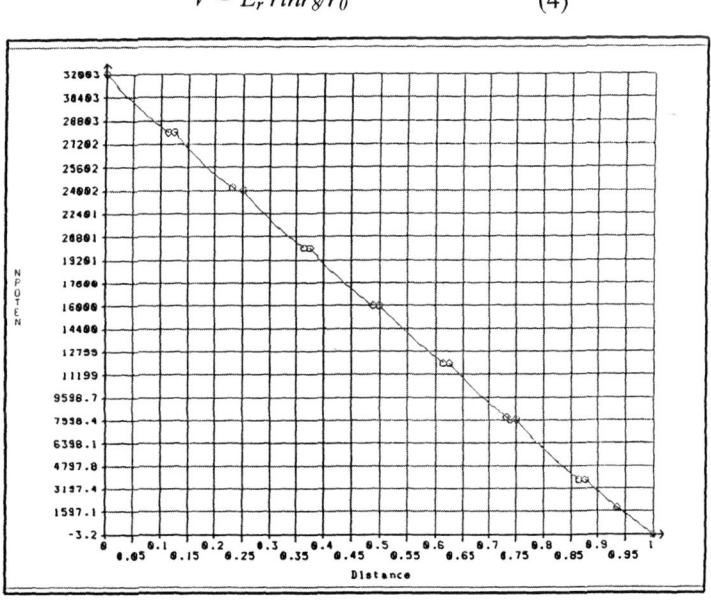

Figure 8. Voltage distribution between the terminal and the tank (relative position).

Abnormal conditions

During an abnormal condition e.g. during quasi-static perturbations and transient overvoltages the insulators can be submitted to a higher stress. The quasi-static perturbations are created by the leakage through the gas or along insulators, and the instabilities generated inside the column by the charging system and the beam. One of those perturbations were observed previously (3) and induces a stress increase of 100%. The particular conditions "overstressing" an insulator are not always leading to a spark or to a damage. In the case of a well protected and subdivided column and without post insulators, an accelerator as a MP model can support some perturbations without sparking (5).

Higher stress increases were observed with transient overvoltages occurring during a spark (6). Following our understanding, sparks are occasional events which cannot be completely avoided in electrostatic accelerators, and conditioning events occuring after a perturbation. The transients are generated in the Vivitron due to a significant traveling wave transit time from one end to the other (the length of the machine is 50 m) and create a propagation delay of the wave of approx. 150 ns (the dominant EM tramsmission mode is the TEM mode). We observe some oscillations at a frequency corresponding to the portico length. In addition, the circumferential hoop structure provides a coupling between each adjacent porticos which leads to an impedance along the portico of 15 Ω, too low to dissipate the spark energy. So, a first spark due to some failure at a particular point can lead to further sparks and overvoltages elsewhere. Despite several configurations of spark cascades are observed and recorded with cameras, the overvoltage measured along the tank and in front of the outer portico with some capacitive probes seem to be limited to the range 100-200% corresponding to a single inter-portico spark. The overvoltages create a severe stress on the insulators leading to an ageing or a failure of the structure. As they were designed by the DC level (the design of the HV insulation systems are usually determined by the transient overvoltage or lightning level occurring during the spark), an abnormal and transient stresses have been latterly integrated in the design to ensure a spark tolerant insulation, but a voltage limitation is still respected in operation to avoid failures.

The insulators ageing

Another parameter which have to be integrated into the design is the insulators ageing i.e. the long term approach. The above mentioned stresses (mechanical and electrical, in operation and during abnormal situations) have an incidence on the performance of the insulating materials and have to be evaluated with some models as the Arrhénius, the power law, the exponential or the multistresses models. Some analyses are performed with the help of the operation data and the observation of the damages during the maintenance. HV tests are actually in progress in the CN pilot machine and at voltages reaching 5 MV maintained during several weeks.

The first test program is involving :

- the ageing evaluation of the post insulators i.e. the remaining life time (after 8 years of operation) ;
- the stability under nominal conditions and during an accelerated ageing test (stress ×2) in order to estimate the influence of the post on the failure and the spark generation.

Recent results performed on a few samples indicate that for the post (long model) insulation quality is still good and that the critical level to initiate a spark between two porticos is not reached. This level is evaluated by the measurement and the monitoring of the leakage current through the post and by the resolution of the coupled problem linking the dynamic voltage distribution (and the currents) to the electric fields inside the Vivitron. It would seem that the critical level to initiate a spark is a few mC, corresponding to a few tens of µA during a few hundreds of seconds. Further tests will define the resistance of the posts against transient overvoltages.

CONCLUSION

The electric stresses applied on some insulators of an electrostatic accelerator, the Vivitron, has been reviewed and some examples are given to evaluate the influence of the electric field and the material characteristics on the global HV performance. These characteristics are given by the nominal DC values completed (or corrected) with the effects of the ageing and some quasi-static or transient overvoltages creating some severe additional stresses. Some HV tests have been performed to evaluate the physical state of the insulation and to integer the history of the materials. Some numerical simulations are indicating the behavior of the system during quasi-static and transient overvoltages i.e. abnormal conditions. More data are needed to precise these conditions e.g. during an instability inside the column, during a perturbation between porticos or on the tank wall, and so to improve the diagnostic in operation. On-line diagnostics are necessary to identify and locate temporary weak points and can help the electric stress characterization necessary to guaranty the insulation coordination (following the IEC-71 recommendations for example) and a reliable system.

ACKNOWLEDGMENTS

The authors wish to thank C.M. Cooke and S. Larigaldie for the many valuable advises and Ch. Munch, A. Tissier for their help.

REFERENCES

1. Frick, G., Osswald, F., Heusch, B., & al., «Transient voltage simulation, high voltage insulator tests and electric field calculation», 7th Int. Conf. On HIAT, Canberra, *Nucl. Instr. & Meth. in Phys. Res.* **A382**, 1996, pp. 37-50.
2. Rebmeister, R., Jegham, E., Lahera, N., & al., «Status of the Vivitron Tandem of Strasbourg», 31th SNEAP Conf., Jülich, 1997, to be published.
3. Osswald, F., Roumié, M., Frick, G., Heusch, B., *IEEE Trans. On Dielec. And Elect. Insul.* Vol.3 No 2, 1996, pp. 258-272.
4. Osswald, F. & al., «Development of the belt charging system of the Vivitron», 31th SNEAP Conf., Jülich, 1997, to be published.
5. Haas, F., Heng, G., Kempfer, C., Osswald, F., «The Strasbourg MP-10 Tandem: an accelerator used actually for nuclear physics experiments and as a test bench for the Vivitron», 5th Int. Conf. On Electrostat. Acc. And Boost, Strasbourg-Heidelberg, *Nucl. Instr. & Meth. in Phys. Res.* **A287**, 1990, pp. 18-20.
6. Cooke, C.M., Frick, G., Roumié, M., «Transient in the Vivitron», 6th Int. Conf. On Electrostat. Acc. And Boost, Padoue, 1992.

III
LINACS AND BOOSTERS

Linac Boosters- An Overview

David C.Weisser and Nikolai R.Lobanov

Nuclear Physics Department, Research School of Physical Sciences and Engineering
Australian National University, Canberra ACT 0200 AUSTRALIA

Abstract. Booster LINACS have matured into an almost standard feature of many accelerator laboratories. Some LINACS are still young and growing, some mature ones prosper through continuous improvement while others require no further development. This overview will present a profile or the current population of booster LINACS. The accomplishments of the laboratories will be surveyed since the last HIAT along with their plans for future development. Those challenges, which the entire HIAT community have in common, will be identified to attract our enthusiasm and inventiveness.

INTRODUCTION

Comprehensive reviews on Linac boosters have been made in past years. J.Delayen (1-3), K.Shepard (4) and D.Storm (5) reviewed low beta superconducting RF structures. V.Palmieri in (6) outlined some details of superconducting structures, their technological limitations and what could be required for pushing the quality of results higher. There have been eight International Workshops on RF Superconductivity. The proceedings from these workshops present detailed information on the physics, technology and applications of the field.

The general features of the linac boosters are outlined in reviews of P.Paul (7), I.Ben-Zvi (8,9), D.W.Storm (10) and H.Padamsee (11,12). The last reference also covers many of the key concepts of the RF superconductivity. Early work and historical foundations of RF superconductivity applied to heavy ion machines, are found in (4,12). G.Bisoffi in (13) reviewed the most relevant issues of the construction and conditioning techniques with particular emphasis on the experience, acquired at various laboratories, in overcoming problems such as Q-desease and frequency jitter due to both mechanical vibrations and pressure drifts in the helium cooling bath. Experience in operating the cavities before and during operation with beam is also reported.

The R&D in RF superconductivity, particularly in the field of heavy ion boosters, is adequate enough to allow the transfer of major components to industry for mass production. The essential feature required for commercial manufacturing is that there be significant reduction in manufacturing costs of all devices. G.Bisoffi in (14) will report on involvement of industry in the technology of RF superconductivity. The high fields lately achievable worldwide (particularly with bulk Nb and sputtered Nb) allow the building of much more compact machines making them accessible to large universities. The design accelerating field of 6 MV/m seems now realistic. A small engineering team need only cope with R&D while the important production could be done by industry.

New forming techniques suitable for prototype fabrication and appropriate for mass scale production are reviewed in (15). The most attractive seamless techniques are the excavation of entire cavity from bulk piece of OFHC copper (LNL/CINEL, estimated cost about U$15,000) or the still developing cold pressing technique. The last one was successfully used on half scale QWR model (15). It is not expensive (U$2,000 per cavity) and it improves the thermal conductivity of OFHC copper.

A review of RF control systems is given in (9,10,16,17). Dynamic phase control is used for small stiff resonators with small stored energy U. Resonators with high U and large dimensions employ voltage controlled reactance method (ANL, FSU (18), KSU).

This survey covers most booster linac projects which are either completed and still exist or in an advanced state of development. Taking into account that superconducting linacs are the most challenging and promising booster devices, the survey excludes circular machines and room temperature booster linac at Heidelberg, which are covered in dedicated conferences (EPAC, LINAC and Cyclotron). We do not cover topics such as cryostats, refrigeration and cryogenic engineering.

Superconducting linacs providing precision beams of heavy ions, are one of the most successful application of RF superconductivity. Many university-based laboratories have found that the capital cost of a superconducting booster is much lower than that of an equivalent voltage electrostatic tandem. Three major factors make a superconducting booster linac application economically attractive (4):

1) Low beam current requires low RF power. Low beta structures exhibit low shunt impedance. Short structures are to be used to increase velocity acceptance.
2) A booster linac is formed as an array of independently-phased modules
3) Since low frequencies are required, BCS resistance is very low. This improves thermal stability against quench and permits operation at 4.5 K and above.

Eight heavy-ion accelerator facilities are now operating, utilizing over 235 resonators. New heavy-ion accelerator facilities are being planned or under construction in Sao Paolo (Brazil), New Delhi and Bombay.

In New Delhi since HIAT95, the cryogenic facility has been commissioned which provides 600 Watts of cooling capacity @4.2K with LN_2 pre-cooling and 330 Watts without LN_2. The prototype bulk Nb QWR has been successfully tested up to 5 MV/m and achieved 4 MV/m at 7 Watts. QWRs for one module are being constructed in collaboration with ANL and cryostats have been designed. The problems to be tackled are the installation of 3 linac modules each housing 8 resonators and a superconducting solenoid. The important issue needing to be addressed is minimising the cryogenic load (19).

In Bombay installation and commissioning of Helium refrigerator (450 Watts with LN_2) and manufacturing of 12 QWRs with three cryostats is scheduled in 1999. QWRs are made of OFHC copper forgings from Hitachi Copper Japan reducing the number of E-beam welds from 4 to 2. The success rate of the fabrication process was estimated as 70%. The beam ports will be hydrogen furnace brazed. The first QWRs were lead plated using Stannostar lead acid bath (MSA chemistry) from a local supplier. Sputtered Nb is also pursued. The Nb/Cu magnetron sputtering device for QWRs is designed and tested. The goal is to commission the second part of Linac in 2000 and the first trials through the Linac, in 2002 (20).

A heavy-ion superconducting linear accelerator for radioactive beams with Nb-sputtered QWRs was proposed at the China Institute of Atomic Energy (CIAE) (21). A Nb/Cu sputtering system have been designed and manufactured in 1996.

Table 1 summarises the achieved performance of the linac boosters commissioned in various laboratories.

TABLE 1. Parameters and operating characteristics of superconducting linac boosters during the last operational year

Lab	VdG/MV	Res-Num-MV/m	Beam-MeV-pnA	R/Q	D@4.2	DB	TD
ANL	FN/9	Nb,S-48-(2.9-3.2) Nb,ID-18-(3.5-4.3)	p+ to U up to 18 MeV/u	800 225	365	254	284
SUNY	FN/8	Pb,S-26-(2.-4.0) Pb,Q-16-3.0	Zr-396-; Ge-330- Ni-423-5; Cl-303- S285-	500 250	240	83	-
UW	FN/9	Pb,Q-38-(2.6-3.0)	He-35-30; F-200-20 O/220/25; C/75/300	580 80	360	70	176
FSU	FN/9[a]	Nb,S-14-(2.0-3.0)	6Li, 7Li, Si- -0.5	150 90	300	150	300
KSU	EN/6	NbS-14-2.0-3.0	F-95- ; Si-110- C-60-	650	365	147	-
JAERI	20UR/18	Nb^{e}CuQ-44-5.0	Si-327-20; Cl-446-16 Ni-658-99; Ge-326-60 Ag-798-15; Au-912-3	400 150	-	-	-
LNL[b]	XTU/16	Pb,Q-44-2.3 Nb^{s}Cu,Q-4-4.5 Nb,Q-1-6.PIAVE Nb,Q-4-7. ALPI Nb^{s}Cu,Q-4-6.9	Zr-418; S-423; Ni-367 Ge-390 Current up to 10 pnA	10^3 200	-	30	230
ANU	14UD/16	Pb,S-10-2.0 Nb^{s}Cu,Q-2-1.5	O-160-15; Ni-320-3 F-160-7	328 60	89	28	207

Abbreviations: S-split-ring resonator; Q-quarter wave resonator; Nb^{e}Cu- Nb explosively bonded to copper; Nb^{s}Cu- Nb sputtered on copper; R/Q-He Refrigerator capacity (Watt@4.2K)/Quiescent Loss, Watt (RF off); D@4.2- days at 4.2K in 1997; DB- days with boosted beam in 1997; TD- total days with any beam in 1997; [a] upgraded with Pelletron charging system; [b] data presented for 1996 operational year.

Three laboratories reported problems running cryoplants. In the last eighteen months, 19% of the LNL scheduled beam time was lost because of faults of the screw-compressors and damage of the "cold Box" turbines caused by power failures during severe storms. The most recent major technical problem FSU has had is the failure of two of three PSI rotary screw compressors (formerly Koch) on the booster cryogenic system. SUNY had to overcome the consequences of a failure of the water "after-cooler" of the He refrigeration compressor. Water contamination of the "cold" heat exchanger stack and distribution system caused continuing problems for several months after the part was replaced.

Only UW did not report any problem with injector or booster. UW had twelve tank openings over the 1997. These were associated with tandem terminal development. In July 1997 they replaced thoriated tungsten corona points installed in March 1996. The new $13/12^{th}$ buncher worked well and solved problems with "satellite bunches".

INJECTOR

The majority of injectors are electrostatic machines made by HVEC (FN, EN, XTU). The other group uses NEC machines (14UD and 20UR). Many of these machines continue to be upgraded by the laboratories. Since the last HIAT conference, ANU and CNR New Delhi have installed gas strippers, ANU installed multi-cathode SNIC ion source.

The main developments FSU *(http://www.physics.fsu.edu)* has made in the last three years have been: installation of the Pelletron charging system to replace the belt; development of a removable terminal RF ion source, laser optically-pumped polarized alkali ion source and terminal gas stripper; installation of a 14C radioactive ion sputter source; development of a computer controlled energy scanning system for the tandem 90 degree magnet; development of an electrostatic deflector "beam switch" (22). FSU plans for a polarized Li+ source to be located in the terminal.

SUNY *(http://resonator.physics.sunysb.edu)* has completed modification of the Laddertron and expects that the choices of plastic for insulating links (polyphenylene sulphide to replace MC 901 Nylon), sleeve bearings (a solid PTFE impregnated Delrin to replace Garlock DU) and minor dimension changes have resulted in a long service life.

In 1998 the laddertron at the LNL XTU-Tandem suffered from the highest rate of failures ever observed since its first day of operation (23). The main causes of such failures were isolated and rather well understood. With the only exception of the moisture content, which has always been a concern since the first day of operation, all the other problems were directly connected with the non-standard spare laddertron links delivered by the usual supplier in 1993. Neither the nylon material, the machining tolerances, the dry bearing material, geometry nor mounting conformed to HVEC specifications. All these shortcomings add to the problem caused by the anomalous moisture contents (0.5% in weight) of the insulating links. The problems were solved using links selected from previously used laddertrons together with the correct bushings. The XTU tandem has run since February 1998 with this "emergency" solution and working at the maximum terminal voltage of 14.5 MV. Meanwhile, a completely new laddertron was assembled with new nylon links and proper bushings.

LNL in the past two years made a significant progress on PIAVE, the positive ion injector of ALPI. The new injector complex makes feasible acceleration of heavy nuclei with A>130; increases the beam intensity onto target by a factor 10; and extends the use of ALPI to rare and costly isotopes (25). The novelty of the design consists in employing two superconducting RFQs for the first acceleration stage. The rest of the acceleration is provided by eight 80 MHz bulk Nb QWRs. The beam is produced by a 14.4 GHz ECR ion source on a 350 KV platform (26). The RFQ design is based on the results of extensive MAFIA simulations combined with detailed investigation on the mechanical stability with IDEAS-S code. Currently LNL has completed development and construction of the 350 kV platform for the ECR source and associated control system; built a stainless steel prototype of the SC-RFQ1 of PIAVE (to check jigs, tuning, welding and mechanical stability); designed half scale stainless steel prototype of the SC-RFQ2; designed Nb SC-RFQ1; installed transport components of PIAVE including modified BPMs for low energy positive ion beams. In the next three years, the lab plans the commissioning of the 350 kV platform and the low energy beam line to PIAVE; commissioning PIAVE (two SC-RFQs and eight low beta QWRs) and its cryogenic system.

A major upgrade of ATLAS *(http://www.anl.gov)* has been aimed to complement the FN by an ECR source and a superconducting injector linac (Positive Ion Injector PII). This comprises a series of superconducting resonators, with beta ranging from 0.009 to 0.037. The ECR ion source is located on a 350 kV platform and can produce highly charged ions with sufficient velocity to be injected directly into the upgraded superconducting linac. Argonne has successfully produced low beta QWR Nb interdigital structures operating at 6 MV/m. In the last three years, ANL implemented a new control system, built and commissioned the new ECR2 ion source and developed a reliable method for constant-velocity-profile scaling for different beams. About 15% of all beam time now is used for radioactive beams (56Ni, 18F, 17F, 8B). In the next three years, ANL will continue development of RIB facility proposal, upgrade the ECR1 ion source to a high field design and produce light ion RIBs using both the Tandem and ECR sources.

JAERI *(http://tamalph1.tokaii.jaeri.go.jp)* installed a commercial ECR ion source in the terminal of 20UR/18 in February 1998 (24). Highly charged and intense heavy ion beams like Ar+12 can be accelerated directly from the tandem terminal and boosted by the linac.

SUPERCONDUCTING CAVITIES

The ideal superconducting cavity should satisfy the following criteria (27):
1) low RF losses in the superconducting layer independent of RF field amplitude;

2) the maximum accelerating field determined by the critical magnetic field;
3) operation at boiling point of liquid helium at atmosphere pressure;
4) the supporting structure provides mechanical stability and efficient removal of heat from the superconducting layer.

A copper substrate with deposited Pb or Nb, makes the resonator stable mechanically and reduces the risk of thermal breakdown. At ANU the temperature and power density distribution along the length of a QWR is shown on Figure 1. Power density was calculated with Superfish and temperature profile with QuickField (both Los Alamos codes). The total absorbed RF power was 5.5 Watts. The calculation was performed for typical geometry of low beta QWR: stub diameter 6 cm, 10 mm thick shorting plate and outer wall and 0.5 mm tuner plate. Figure 1 shows that the temperature rise in the tuner plate (the hottest spot of the QWR), does not exceed 0.3 K even for a resonator with a 3 mm wall. The distribution of RF power losses on the QWR surface was estimated using Superfish as following (in % to total power losses):

0.35- 0.2 mm indium gasket to tuner plate;

0.006- in indium gasket to external cut-off tubes;

1.43- in inductive coupler;

0.002- capacitive coupler;

0.004- in OFHC copper of external surface of outer wall due to RF field penetration through beam ports.

Two types of geometry are generally used in Linac boosters, (see table 1): split ring (SLR) and quarter wave resonator (QWR). Initially four development projects were completed in the sense of producing suitable accelerating structures for boosters: Argonne Nb SLR, Cal Tech Pb SLR, Karlsruhe Helix and Stanford Single-cell Nb cavities. SLRs were the only structures actually used to construct booster linacs. SLRs were built in the lead onto copper (SUNY, ANU) and solid Nb combined with explosively bonded Nb onto copper (ANL, FSU, KSU). Whereas SLRs have a longer acceleration length, QWRs have a broader transit time factor and a higher mechanical stiffness. This makes them less sensitive to microphonics and helium gas pressure variations (~1.6 Hz per 1 mBar of helium gas pressure in ANU SLRs). The complex geometrical shape of the SLR makes it suitable for electrochemical deposition of the lead layer or solid Nb combined with Nb bonded onto copper. All these factors motivated development the QWR as the superconducting structure in the recently completed booster machines (JAERI, LNL) or under construction (LNL-PIAVE booster, New Delhi, Bombay). ANU plans to upgrade its linac with 20 Nb/Cu QWR's (28). The exception is the University of Sao Paulo booster linac using ANL, Nb SLRs.

QWRs were developed both in the traditional coaxial shape with two accelerating gap (SUNY, UW, JAERI, LNL, ANU, NSF, TIFR) and in interdigital version with four gaps (ANL). There are three adopted techniques: bulk niobium (LNL), solid Nb plus composite Nb explosively bonded to copper (ANL, FSU, KSU, NSF), lead plating onto copper (SUNY, UW, LNL, ANU, TIFR) and Nb sputtering onto copper (LNL, ANU).

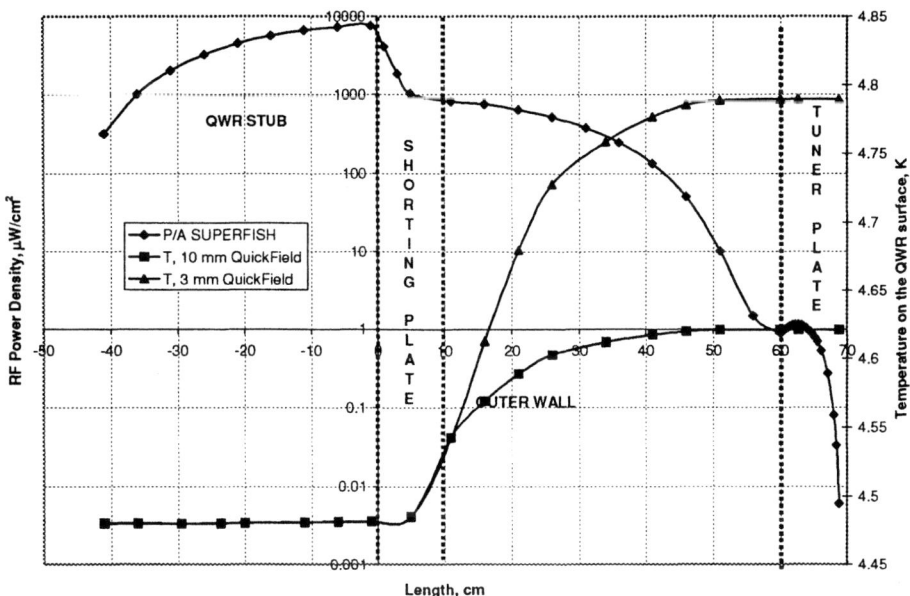

FIGURE 1. Distribution of RF power density and temperature along QWR length. Total absorbed RF power 5.5 Watts. Temperature calculated for outer wall thickness of 10 mm and 3 mm.

Surface Treatment

The careful surface treatment of the copper substrate is considered to be the most important step in any coating procedures. A review of available surface cleaning technologies and their applicability to cavity technology is given in (29). Any defect on the copper surface such as dislocations, grain boundaries or surface roughness will be reproduced by the lead or Nb films (6). A small defect on the first lead layer will act as a nucleation center for growing a microparticle with poor thermal contact with the copper substrate. The same effect may be caused by inclusion (oxides, sulfides, abrasive, grease, dust or erosion zones due to bubbles (10)) or porosity of the substrate surface (6). Delayen suggested copper plating before lead plating (1). Palmieri's group verified that copper plating of the substrate before lead plating considerably reduced the roughness of the plated lead (6). However, there is a risk of developing more surface defects due to the copper plating. A more appropriate technique to improve the quality of the substrate might be copper sputtering coating.

The influence of the oxide interface between copper and sputtered Nb was studied in (31). Removal of the oxide layer of a hydroformed cavity resulted in larger surface resistance and a larger dependence on trapped magnetic flux than in standard cavities with an oxide interface between the film and the substrate. For spun cavities the effect was minimal.

The surface roughness of the substrate has an influence on surface resistance (32). Spun cavities provided on average better results than hydroformed cavities, both in term of R_{res} and of its increase with RF field. The most apparent difference between the two kind of substrates is surface roughness. No simple mechanism explains to residual resistance in terms of roughness, although several candidates are possible viz., film crystalline structure, the effective surface area of the substrate, the effective RF path length and greater likelihood of trapping contaminants.

Lead on Copper

Lead plating was recognized as an appropriate technique providing just slightly poorer performance (BCS losses) compared to a Nb superconducting layer at low frequency. Lead plating provides fast moderate results with modest equipment and relatively low cost. In order to prevent lead oxidation and contamination during post-plating treatment, lead-tin was adopted for SUNY QWRs and the Munich cyclotron. In spite of the reported difficulties in controlling the tin concentration and lower RRR value of tin-lead compare to pure lead (50 for pure lead compare to 24 for 2% tin-lead), both laboratories demonstrated remarkable results. Today the "traditional" resonator plating chemistry is commercially no longer available, as both Shinol and fluoborates have become obsolete for economic and environmental reasons. SUNY replated six high-beta SLRs with 2 microns of Pb-Sn alloy using a modern, commercial, methane-sulfonate process (LeaRonal Solderon MHS-L) and a simple open-air procedure (33). While MHS-L is the only LeaRonal additive suitable for resonator plating, there are similar MSA products available from other companies that might work as well. The new plating chemistry and simple techniques have proven remarkably successful for three reasons: 1) consistently superior superconducting performance; 2) ease and speed of application; 3) greater safety and environmental compatibility. The improved performance makes possible routine operating fields 10-20% higher than the previous 3.0 MV/m. Two of the six SUNY resonators can be operated at accelerating gradient of 4.0 MV/m for power dissipation of 10 Watts. In addition, SUNY completed the QWR upgrade of the low beta section. Many questions about MSA-based lead-copper technology remain unanswered like: dependence of superconducting performance on tin content, film thickness, surface activation, additive concentration, metal impurities and distribution of RF power loss on the SLR surface. Nevertheless, the proven success motivated ANU to adopt MSA chemistry and to replate the first SLR in November 1998 to be followed by replating the entire ANU Linac (10 SLRs at average field of 2 MV/m, lead plated in Oxford). This might increase the booster energy gain with the same numbers of resonators by up to 100% if the best SUNY results is achieved.

LNL continues using the traditional fluoborate bath and is producing of home made Shinol. LNL experience in lead plating is supported by the high number (120) of successful plating cycles performed so far and by sixty working resonators produced for the Linac (25). Quality testing, during mass production, was applied to only 15% of total number of QWRs based on good reliability of the plating process. Recent development has addressed post-plating surface passivation (6) consisting of careful rinsing of the job with pure de-ionized water with intermediate immersion in light acid solution. The acid step dissolves oxides and modifies the lead surface. The goal is to remove all traces of fluoborate by means of subsequent dilutions.

In 1998, LNL reported that four lead plated QWRs from ALPI were taken off-line because two of their couplers failed. New cavities have a new coupler design, making them independent from the tuner mechanics. The Nb sputtering was trialed as a substitute for the electroplated lead coating in four medium beta resonators. This required minor changes in the resonator shape. Fields in excess of 4.5 MV/m at 7 Watts could be obtained off line even though the cavity geometry was not optimized for the sputtering procedure (34).

If total replacement of the installed medium beta lead plated QWRs with sputtered units is adopted, then significant improvement in ALPI performance and reduced the cryogenic power consumption are possible.

Further development Nb technology probably will make it much more attractive than lead. In spite of impressive results achieved so far with lead plating, the Nb technology offers better material with better RF performance, better stability being on exposing to air and environment friendly "dry" deposition technique. Nevertheless lead plating is still preferred for substrates of complex geometry like SLRs.

Niobium on Copper

Research programs of sputtered Nb/Cu QWRs are pursued at LNL (DC sputtering) (20) and ANU (magnetron sputtering) (28). Due to the thick copper substrate the resonator is not prone to mechanical vibrations, gas helium pressure induced frequency variations and provides efficient removal of heat from the Nb layer. The influence of coating temperature on Nb films was studied in (36). Films deposited at higher temperatures show higher RRR and lower Ar content. The results are interpreted in term of higher Nb surface mobility at higher temperature. Experimental evidence is presented in (38) that discharge gas atoms are trapped in the sputtered film. The sputtering gas concentrations are found to vary from several percent for Ne down to the ppm level for Kr and Xe. The noble gases trapped in the film affect the penetration depth, the temperature dependent losses, R_{BCS}, the losses induced by the presence of trapped fluxons, but have no significant influence on the residual resistance.

In LNL, high purity Nb films are deposited by DC diode sputtering using a negatively-biased QWR. Important results were reported in 1997 (35) for sputtering on QWRs for ALPI high beta 0.11-0.14 section. DC sputtering requires high argon pressure, up to 0.2 mBar. The bias induced ion bombardment of the growing Nb film changes in the nucleation characteristics, morphology, composition, crystalline structure and film stress. Contamination of Nb with impurities in the zones where the bias was not effective was reported. Elimination of internal sharp angles in the resonator geometry, especially in high current regions, is recommended. Higher RRR and T_c were measured when larger radius of curvature was used (39). The inner stub is straight to minimize multipactoring. The consequent higher peak magnetic field is still acceptable for Nb sputtered cavities. The end of the stub is hemispherical to minimize the surface electric field. Capacitive coupling is created by placing the exciting electrode on the stiff outer wall opposite to the end of the stub. The pick-up is located on the opposite side. This simplifies the coating procedures and removes components from the high current area. Due to relatively low deposition rate, high baking temperature in UHV is essential. A mirror electrode for the stub was introduced to eliminate film peeling at the end of stub due to electron bombardment. High argon pressure results in uniform plasma inside and outside the cylindrical cathode.

A new resonator design for high beta ALPI section, was reported. In the new QWR, all brazed joints were eliminated. This results in no trapped volumes in the brazing, improved thermal conductivity of the substrate and reduced contamination of OFHC Copper (99.99% Certified grade). The capacitive coupler was introduced to reduce the mechanical coupling to the tuner. The accelerating field in the four high beta cavities reached 6.9 MV/m at 7 Watts.

ANU Nb sputtered two QWRs of moderate Q for a superbuncher in 1992 and Time Energy Lens in 1997. The progress in R&D up to 1997 was reported in (28). The oxygen desorbed from the substrate oxide layer during sputtering was identified as the main source of Nb film contamination (40). The cathode assembly is scanned along the resonator axis providing a calculated ratio of 2.3 between maximum and minimum thickness on the resonator surface (28). Due to many other activities, there was no Nb sputtering on QWRs during the last year. Currently one QWR is being prepared (tumbling, polishing and doughnuts brazing) and the magnetron cathode is being modified for Nb sputtering in November 1998. The specific features of the modified ANU magnetron device are low discharge voltage to reduce contamination of Nb film by Ar gas and sputtering at elevated temperature to improve RRR. Modification will involve improvements of the structural stability of the cathode by introduction of spring-loaded support legs. Currently the skeleton of the device is subject to slight misalignment caused by thermal gradients.

Nb Explosively Bonded to Copper and Bulk Nb Cavities

Bulk Nb cavities are employed in ANL (SLR and Interdigital QWR), Florida State University (SLR), Kansas State University (SLR), JAERI (Nb-clad-copper and bulk Nb QWR), LNL (double wall bulk Nb QWR for the low beta sections of ALPI and PIAVE) and New Delhi (ANL developed QWR with capacitive loading and pneumatically controlled Nb bellows as a slow tuner). ANL, JAERI and New Delhi cavities have Nb-Cu resonator walls while the loop or stub and electrodes are made of bulk Nb. The resonators are similar in construction technique, material characteristics and in performance. Advantage of bulk Nb technology is a very good cavity performance. However it is quite expensive technology due to cost of E-beam welds and high purity Nb sheet (RRR up to 250). The resonator cost of LNL low beta bulk Nb QWR is about US$35,000. The E-beam welding is the main source of defects such as projections, microbubbles and craters. Welding causes compositional and microstructural heterogenities through heat affected zones. Electromagnetic hammering was claimed as technique to get a fine grain size following by annealing to re-establish the high thermal conductivity (30).

Studies have been performed on the properties of Nb thin films sputtered onto solid Nb resonators in order to determine at what treatment temperature the properties of the film merge with those of the bulk (37). Before heat treatment, the RF response was similar to that of a film coated on a copper substrate. A marked transition towards bulk-like RF behavior was observed after the 900°C treatment (sensitivity to external magnetic field). The study provides information on firing procedure using titanium as solid state getter, the production of bulk Nb cavities and their performance prior to firing and evolution of the grain size with increasing firing temperature.

Traditionally the quality control of high purity Nb concentrates on three features: purity, workability and surface quality. Another problem, which becomes more important for high gradient cavities, is inhomogeneity (cluster of foreign components, voids etc.). The eddy current method is considered as most suitable for detection of defects and inspection of welding quality (41).

At JAERI, fields of about 7 MV/m were measured off-line at 4 Watts. The Nb interdigital structures perform at the same levels. The lowest-velocity section (beta=0.009) operates at 6 MV/m. The main problem in JAERI was Q-disease (42) which was treated by increasing cooling rates from –12 K/h to –28 K/h over the hydrides precipitation temperature range of 130K to 90 K. Palmieri in (6) suggests that Q-disease may be caused by stress induced in Nb due to explosive bonding. This stress cannot be released by annealing or firing by the presence of copper and stainless steel. Tensile stress may cause the anomalous enhancement of the diffusion rate of hydrogen in Nb. Nb coated QWRs and bulk Nb double wall cavities do not suffer from Q-disease. Nb explosively bonded and bulk Nb cavities remain expensive devices. For example, the cost of New Delhi QWR was estimated as US$55,000-60,000 (43). About 25% of cost is the Nb sheet (RRR=200 for the stub and shorting plate and RRR=60-80 for outer wall). JAERI in 1997, designed low beta half-wave resonator (consisting of two quarter-wave resonant lines oscillating with opposite potentials) similar to the concept reported in (44). Due to very high cost of such Nb-clad cavities, the manufacturing was delayed (24).

Prototypes of double wall QWRs with beta range from 0.055 to 0.165 have been developed at LNL (25). The 80 MHz resonators were chosen for the low beta section of the ALPI and for high velocity section of PIAVE. The weight of such 1 m long resonators is very close to 160 MHz copper QWR allowing the use the same basic design of the cryostats as in medium and high beta sections. Tuning plates are made of lead plated OFHC copper. The average accelerating field of the first four is 7.1 MV/m at 7 Watts. The maximum peak surface electric field is about 50 MV/m and the maximum magnetic field is about 1100 G. Just slightly lower field (above 6 MV/m) was achieved in the first beta=0.047 bulk Nb QWR for PIAVE in 1997. The construction of the remaining 7 cavities will follow soon.

The highest beta = 0.28 resonator developed is the bulk Nb two gap spoke resonator operating at 11 MV/m. In low beta cavities, a record accelerating field of 18 MV/m was achieved in a 355 MHz beta = 0.12 half-wave bulk Nb resonator. The peak electrical and magnetic fields were respectively 58 MV/m and 936 G (12).

CAVITIES OPERATION

The operation of cavities will be discussed in terms of electromechanical instabilities when mechanical vibrations induce a variation in the eigenfrequency. Mechanical changes of the shape and eigenfrequency of the superconducting resonators, notably SLR and double wall bulk Nb low beta QWR, caused by pondermotive effects and/or by microphonic background (mechanical vibrations, including ground motion, vacuum pumps, refrigerator and boil off gas pressure variations) are the major contributors to amplitude and phase jitter. In all these cases, there is a source with a narrow band spectrum of noise which passes through parts of the apparatus which modify the noise signal before it interacts with the cavity. The cavity reacts with its own distinct response spectrum to the external perturbations. Problems arise when peaks in the modified noise spectrum coincide with intrinsic resonance in the cavity. The sensitivity of resonators to such perturbations defines the circulating RF power needed to hold the resonator on frequency and simultaneously affects requirements on the performance of the RF control system in terms of simplicity and price. With increasing accelerating gradients, pondermotive effects become important. In CW operation, it produces static frequency shift $\Delta f = kE_a^2$, where k is a function of the cavity shape, wall thickness and yield strength of the material. The static shift was $-0.6\ E_a^2$ Hz or 15 Hz at 5 MV/m in JAERI QWR (45).

Microphonic Perturbations

At ANU, the intrinsic resonance of the SLR were measured at room temperature with a pizoelectric sensor. Two dominant frequencies were found as previously identified by a stroboscopic technique (46). One is a torsion and bending mode of the loop, where the drift tubes move in opposite direction along the beam axis and has a frequency f1=39 Hz. The other is a bending mode of the loop where the drift tubes move in a plane perpendicular to the beam axis and has a frequency f2=45 Hz. The technique used to measure fundamental mechanical frequencies of the SLR at cryogenic temperature was described in (47). The source of microphonic vibrations was a 40 W speaker linked to an audio amplifier. The audio source for the amplifier was the spectrum analyzer providing sinusoidal or random noise signals up to 10 kHz. Sinusoidal signals produce a clear "fingerprint" in the measured frequency spectrum. Later the similar technique was used in LNL to analyze and damp mechanical modes in double wall bulk Nb QWR. Mechanical noise was excited by a powerful industrial vibrator (48).

The phase and amplitude error signals supplied by the RF controller are analyzed with a spectrum analyzer to obtain the intrinsic microphonic spectrum of the resonator. Scanning the frequency of the signal source identifies the natural mechanical resonances of the cavity. The background noise spectrum was measured with pizoelectric sensor attached to the concrete floor under 2 kg steel plate. The background noise has a spectrum consisting of a number sharp peaks (<1 Hz wide) in the 20 to 600 Hz band. The ground vibrations were re-measured with a portable vibration meter (accelerometer type). The strongest source of ground vibration was a mechanical rotary pump run at 1500 rpm (25 Hz). Accelerometer measurements on the pump show that the dominant frequency of vibration is 100 Hz. This is four times the rotation frequency of the pump which is typical for a double stage rotary vane pump. Vibrations from the rotary pump could be damped using industrial vibration dampers 6 mm thick loaded to about 3 kg/cm^2. This technique reduces the peak amplitude of vibrations by a factor of ~5 (measured 30 cm from the rotary pump toward the cryostat). Under the same arrangement, the background noise spectrum measured on the cryostat frame (figure 2). Note that the only vibration at 82 Hz is relatively strong. This noise is produced by LHe cryoplant and is transferred through LHe delivery pipes. Note that without the cryoplant running, the only 100 Hz peak is in the spectrum. The peak at 832 Hz (not shown on figure 2) is caused by 150 l/s turbo molecular pump running at 49,920 rpm. This does not coincide with a harmonics of natural resonance of SLR.

A similar problem was reported recently at KSU where the Sullair compressor was moved off the laboratory roof to a site adjacent to the laboratory in order to reduce mechanical noise. A large Kinney vacuum pump was also a source of noise. Widening the operational frequency windows on the fast frequency control modules (VCX) has greatly improved in-of-lock operation.

The spectrum in figure 3 was obtained with the phase locked resonator running at 2.2 MV/m with the amplifier delivering 70 W. The cryoplant driving the mechanical mode of 82 Hz excites the second harmonics (76 Hz and 90 Hz, mechanical quality factor of about 70) of SLR. Hopefully only tails of 82 Hz excite harmonics so it does not affect stable phase locked operation of SLR. Obviously, it is desirable that the excitation of mechanical resonance by external noise is kept to a minimum. The goal of the design of a phase stable apparatus should be that the intrinsic resonant frequencies of the device do not coincide with peaks in the noise spectrum of the environment.

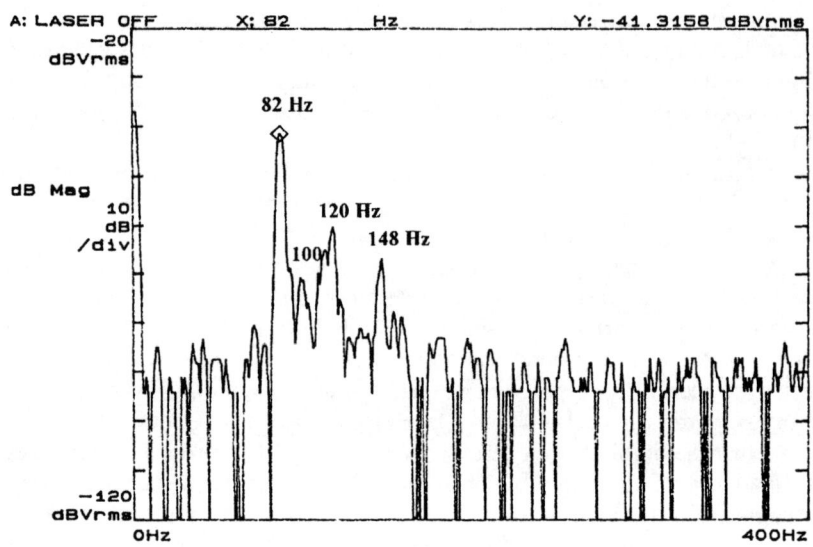

FIGURE 2. The background noise spectrum measured on the module cryostat frame.

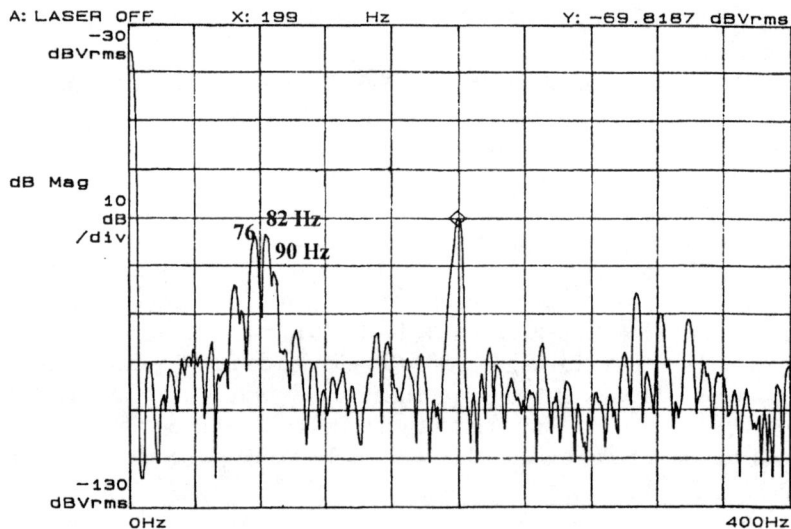

FIGURE 3. Microphonic spectrum obtained with the resonator running phase locked at 2.2 MV/m.

Slow Gas Helium Pressure Variations

The second major source of mechanical noise is slow He gas pressure variations due to cryostat refill or due to correlation between the pressure in the cryostat vessel and any pressure changes in the gas return line and in other cryostats. Sometimes fast pressure instabilities due to pressure-acoustic oscillations are observed. The effect is not important on stiff Pb/Cu QWRs (few Hz per 300 mBar pressure variations quoted at LNL, UW and ANU). The sensitivity of ANU SLR was measured to be 1.6 Hz/mbar. Full Nb cavities are slightly less sensitive to slow pressure changes. LNL reported 1 Hz/mBar for double wall bulk QWR while JAERI quoted 0.27 Hz/mbar. Both QWRs have a mechanical stiffening of the shorting plate. SUNY, ANU and JAERI facilities incorporate slow frequency tracking devices which are not in use since they introduce additional mechanical vibrations and tend to fail. The major breakthrough in ANU Linac was stabilization of GHe pressure within 25 mBar in 1998. This reduced the frequency shift to less than 40 Hz and just overcoupling was sufficient to keep the resonators locked. Figure 4 demonstrates pressure variation in the helium space in the cryostat, modulation of forward RF power delivered to the SLR and phase error signal.

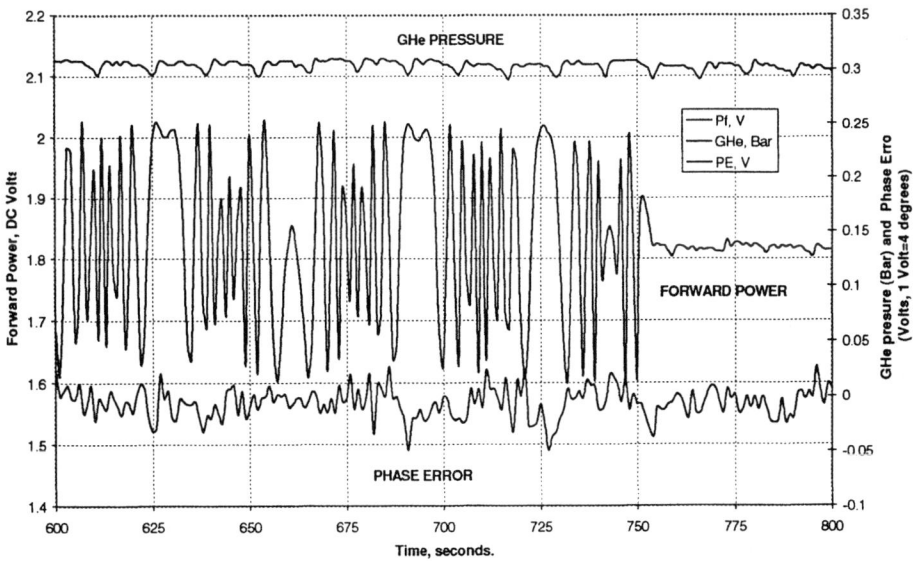

FIGURE 4. Time dependence of RF forward power (DC Volts), GHe pressure (mBar) and phase error (Volts) in two SLRs operating from 600 to 760 seconds and one only from 760 to 800 sec.

One can notice that the slow oscillation of forward power is dramatically reduced and it is correlated to gas pressure beyond 750 second of operation when the only one resonator left running. The error signal is slightly reduced as well. When more than one SLR is running, they more likely to interact thermally via common precool bar. This results in a rather complex timing pattern of forward power in figure 3 which may be described by empirical formula:

$$U = U_0 + U_m sin((4.85 + 0.97 sin(0.24t))t) , \qquad (1)$$

where U_0 is the forward power measured in DC Volts; U_m is the modulation of forward power, Volts and t is the time, seconds. Without the interaction with other resonators, the forward power is stable at $U=U_0$. Turning additional SLRs ON introduces interactions resulting in oscillation of forward power with frequency ~0.77 Hz modulated with frequency about 0.038 Hz (equation 1). The investigation of this phenomena continues in ANU.

LNL reported up to 100 mBar pressure variation in LHe reservoir corresponding to frequency drift in double wall Nb QWR up to 100 Hz. These drifts are normally in the range of 1-2 Hz per minute and may be corrected by the slow tuner. In order to lower the Q of the 42 Hz mechanical resonance, stiffening of the shorting plate similar to JAERI QWR was applied. In addition to that, a novel approach consisting of a mechanical damping of vibrations reduces their amplitude by one order of magnitude (48). These dampers are a valuable alternative to electronic fast tuners. The stability of GHe pressure in ALPI Linac is currently under investigation.

Fast GHe Pressure Variations. Thermal Acoustic Oscillations

The first evidence of the thermal acoustic oscillations in ANU Linac was observed in 1996 when the American Magnetics Helium level sensors did not provide reliable level information. The sensor relies on the heat transfer from the liquid helium to a superconducting wire, to be sufficient to keep the submerged section of the wire superconducting. The segment above the liquid, in the Helium gas space, is sufficiently warmed by an exciting current in the wire, to remain normal and so exhibit measurable resistance. The level is inferred by the resistance of the exposed portion of the wire. If there is an environmental heat load to the submerged wire which forces it to go normal, say from thermal acoustic oscillations, the device will be fooled into thinking that the liquid has disappeared. This was happening on all the Helium level detectors resulting in over-filling of the cryostats and forcing liquid Helium back into the gas return lines of the cryogen distribution system. The thermal acoustic oscillation of pressure in the He space of up to 120 mBar is shown on figure 5. This severely compromised the operation of RF control system and L'Air Liquide equipment.

The mechanism of thermal acoustic oscillations relevant to the geometry of module cryostat is described in (49). The geometry of the level sensor is a tube comprising the

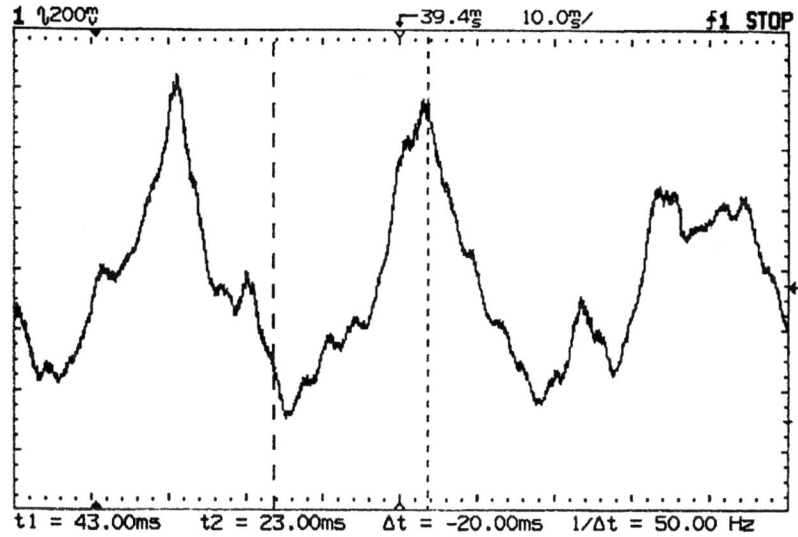

FIGURE 5. The thermal acoustic oscillations measured in GHe vessel of module cryostat. The pressure variation ranges up to 120 mBar at 110 Hz.

sensor, inside another tube, which separates the helium and the vacuum spaces. It turns out that all the heat transfer driving the oscillation takes place in a thin layer of gas on the outer tube wall. All one has to do is to intercept the acoustic wave in this thin layer as it traverses the tube. The solution was to mount an adequate baffle on the inside of the outer tube. The actual installation was further complicated by the placement of the pressure relief valve on the room temperature end of this same tube. The impedance to gas flow of the baffle compromised the safety of the system and so a piston was devises to ride on the inner tube and have sufficiently small clearance to the outer wall (0.2 mm) to be an adequate thermal acoustic baffle. If a pressure accident were to occur, the gas flow would push the piston to a larger diameter section of the assembly allowing it low impedance access to the pressure relief valve.

The second place where the thermal acoustic oscillations have been observed was return helium gas transfer tube on the L'Air valve box. Earlier the heat transfer due to oscillations produced an ice block on the room temperature fittings. The problem was solved coincidentally in 1998 after GHe pressure stabilization in ANU Linac within the range of 1.285-1.310 Bar (absolute). Due to narrower operating pressure range, the conditions in valve box no longer overlapped the oscillations instability region.

SUMMARY FEATURES AND DIRECTION THE FIELD IS GOING

Radio Frequency superconductivity has become important technology for particle accelerators. Superconducting cavities have been operating routinely for many years in a variety accelerators for low energy to medium energy for nuclear physics research. More than 235 cavities are installed and operated in state-of-art machines. There are three machines at National Accelerator Laboratories as LANL, JAERI and LNL as well as at universities such as SUNY, UW, FSU, KSU, Wuppertal and ANU.

The university-based superconducting heavy ion facilities have been short of funds for many years. KSU reported that running superconducting booster under limited budget was only possible due to the exceptional quality of the technical support staff and their dedication to their tasks. FSU is looking for wider applications of boosters and associated technology. SUNY is proposed to promote the small university lab where good science can be done which does not fit large US national labs.

Many laboratories promote stronger commitment to support SNEAP as a viable resource for communication of problems and solutions (KSU). The cooperation between booster community regarding technical exchange has been tremendous and a great boon (SUNY, LNL, ANU).

Several installations (LNL, ANL, JAERI) require further development and study of high charge state ECR ion sources. A few ion sources have been installed in the terminals of tandems. For example an ECR source in the terminal at JAERI, and FSU has a terminal ion source of polarized Li+. BARC is transforming their machine into folded Van de Graaff. UW built a terminal ion source of intense beams of low energy 3He and is being modified to make all Helium and Hydrogen isotopes.

Many new applications are coming which require development of low intensity ion beam diagnostics. LNL is preparing a system for the measurement of transverse emittance and a RF system for longitudinal beam emittance. KSU is upgrading beam diagnostics for cold gas jet targets which require a minimum of two beam scanners per beam line. Development there is undergoing of electronics to "lock" the beam onto the target based on beam position information provided by scanners.

Intense studies improved the understanding of sputtered superconducting Nb films resulting in the outstanding performance of LNL Nb/Cu cavities showing that this technology can lead to successful series fabrication. New surface analysis techniques will improve understanding and lead to reduce contamination effects, field emitters and defects. Development of very low beta structures will require study of mechanical stability both QWR and RFQ. This will demand novel E-beam welding techniques. The prospects of other thin film superconductors, such as NbTiN or Nb_3Sn, are considered. RF application of high temperature superconductors seems to be going well for communication applications. However, high T_c cavities still seem to be out of reach.

More installations have already required industrial scale production methods, pushing down cost and at the same time assuring quality control. This becomes crucial for university-based laboratories. A great step toward cheaper cavity production was made at LNL where entire QWR can be excavated or cold pressed without brazed or welded joints.

Implementation of booster automatic set up is on agenda. LNL reported developing a spread sheet for the on-line study of the longitudinal and transverse dynamics of Linac booster.

Booster RF electronics is in continuous progress. KSU will upgrade VCXs with the latest design from ATLAS. LNL collaborate with ANU on testing of the prototype RF control system for ANU Linac upgrade.

RIB production becomes a hot topic. A radioactive beam proposal is in preparation at ANL ATLAS. This field requires development of very low beta accelerating structures for RIB pre-acceleration. Radioactive safety issues related to RIB production need careful study.

Accelerating structures for high intensity proton beams (RFQ, drift tube linacs and high beta structures) for various neutron spallation sources are example of advanced applications of RF superconductivity in particle accelerators. This presents new technical challenges for mechanical stability and thin film production of Nb/Cu cavities.

ACKNOWLEDGEMENTS

We are grateful to a large number of colleagues for corresponding and providing data. The authors acknowledge the essential help of A.Danielli, T.Gray, E. Myers, D.Storm, R. Lefferts and G. Zinkann.

REFERENCES

1. Delayen J., in *Proceedings of the 2nd Workshop on RF Superconductivity*, (H.Lengeler ed., CERN Geneva, 1984), 1984, p.195.
2. Delayen J., in *Proceedings of the 3rd Workshop on RF Superconductivity*, (K.W.Shepard ed., ANL Argonne, 1987), ANL-PHY-88-1, 1987, p.469.
3. Delayen J., in *Proceedings of the 4th Workshop on RF Superconductivity,* (Y.Kojima ed., KEK Tsukuba, 1989) KEK89-21,1989, p.249.
4. Shepard K., *Nuclear Instruments and Methods* A **382**, 125-131 (1996).
5. Storm D.W., in *Proceedings of the 6th Workshop on RF Superconductivity,* (R. Sindelin ed., CEBAF Newport News, 1993) CEBAF-93, 1993, pp.216-235.
6. Palmieri V., in *Proceedings of the 7th Workshop on RF Superconductivity,* (Gif-sur-Yvette), 1996, pp.237-258.
7. Paul P. and Sprouse G., *Comments Nucl.Part.Phys.* **11-5** (1983), pp.217-229.
8. Ben-Zvi I., *Particle Accelerators* **9** (1979), pp.169-186.
9. Ben-Zvi I., *Nuclear Instruments and Methods* A **287**, 216-223 (1990).
10. Storm D.W., *Nuclear Instruments and Methods* A **328**, 213-220 (1993).
11. Padamsee H., *IEEE Transaction on Applied Superconductivity,* **5**(1995) pp.828-833.
12. Padamsee H., Knobloch J., Hays T., *"RF Superconductivity for Accelerators"*, New York, 1998

13. Bisoffi G., in *Proceedings of the 8th Workshop on RF Superconductivity*, (V.Palmieri ed., Abano Terme 1997) to be published
14. Bisoffi G., in *Proceedings of the Applied Superconductivity Conference (ASC98)*, Palm Desert (CA), to be published
15. Palmieri V., "Seamless Cavities", in *Proceedings of the 8th Workshop on RF Superconductivity*, (V.Palmieri ed., Abano Terme)1997, to be published
16. Graf H.D., in *Proceeding of the 5th Workshop on RF Superconductivity*, (DESY M-92-01, Hamburg, Germany), 1992, p.317.
17. Simrock S.N., in *Proceedings of the 6th Workshop on RF Superconductivity*, (R. Sindelin ed., CEBAF Newport News, 1993), 1993, pp.294-306.
18. Myers M. et al, *Nuclear Instruments and Methods in Physics Research* **B40/41**, pp.904-907(1989).
19. Roy A., private communication
20. "Superconducting LINAC Booster", TIFR-BARC Joint Project, Technical Report, 1997
21. Chen Jiaer, Zhao Kui, "R&D on RF Superconductivity in China", *in Proceedings of the 8th Workshop on RF Superconductivity*, (V.Palmieri ed., Abano Terme), 1997, to be published
22. Myers M., private communication.
23. "The Laddertron Charge System of the LNL XTU-TANDEM. Problems, Remedies, Perspectives", LNL internal report, 1998 (unpublished).
24. Takeuchi S., private communication
25. Fortuna G. et al, in *Proceedings of the 18th International Linear Accelerators Conference*, vol2., CERN, Geneva, 1996, pp.905-909.
26. Cavenago M. et al, *Review of Scientific Instruments*, **69-2**, pp.659-661(1998).
27. Weingarten W., in *Proceedings of the 7th Workshop on RF Superconductivity*, (ed., Gif-sur-Yvette, October), 1996, pp.129-142.
28. Lobanov N. and Weisser D., in *Proceedings of the 8th Workshop on RF Superconductivity*, (V.Palmieri ed., Abano Terme 1997) to be published
29. Kneisel P. and Lewis B., *in Proceedings of the 7th Workshop on RF Superconductivity*, (ed., Gif-sur-Yvette), 1996, pp.311-327; Schirm K. et al, in *Proceedings of the 7th Workshop on RF Superconductivity*, (ed., Gif-sur-Yvette), 1996, pp.461-465.
30. Singers W., Tesla Workshop on Seamless Cavities, DESY, Hamburg, 1995
31. Benevitti C.et al, "On the Role of the Interface Between Niobium Films and Copper RF Resonators", in *Proceedings of the 8th Workshop on RF Superconductivity*, (V.Palmieri ed., Abano Terme),199, to be published
32. Benvenuti C. et al, "Niobium Sputter-Coated Copper Resonators", *in Proceedings of the 8th Workshop on RF Superconductivity*, (V.Palmieri ed., Abano Terme), 1997, to be published
33. Lipski A., Noe J. and Realmuto C., *"An Environment-Friendly Method for Resonator Surface Fabrication"*, presented at the SNEAP-30, 1996, to be published
34. Porcellato A.M. et al, "Upgrading of the ALPI cavities and beam tests", in *Proceedings of HIAT98 Conference*, to be published.
35. Stark S. et al, "Niobium Sputter-Coated QWRs", in *Proceedings of the 8th Workshop on RF Superconductivity*, (V.Palmieri ed., Abano Terme), 1997, to be published
36. R.Russo and S.Sgobba "Influence of Coating Temperature on Nb Films", Proc.8th Workshop on RF Superconductivity, (V.Palmieri ed., Abano Terme 1997) to be published
37. Benvenuti C. et al, "Structural and RF Properties of Nb Films Deposited Onto Annealed Nb Resonators", in *Proceedings of the 8th Workshop on RF Superconductivity*, (V.Palmieri ed., Abano Terme), 1997, to be published
38. Benvenuti C. et al, "Properties of Copper Cavities Coated with Nb Using Different Discharge Gases", in *Proceedings of the 8th Workshop on RF Superconductivity*, (V.Palmieri ed., Abano Terme), 1997, to be published
39. Palmieri V. et al, *Nuclear Instruments and Methods in Physics Research* **A382**, pp.112-117(1996).
40. Malev M., *Nuclear instruments and Methods in Physics Research*, **A382**, pp.161-166(1996)
41. Singer W. et al, "Diagnostic of Defects in High Purity Niobium", *in Proceedings of the 8th Workshop on RF Superconductivity*, (V.Palmieri ed., Abano Terme), 1997, to be published

42. Takeuchi S. and Matsude M., "First Three Years Operational Experience With the JAERI Tandem-Booster", in *Proceedings of the 8th Workshop on RF Superconductivity*, (V.Palmieri ed., Abano Terme), 1997, to be published
43. Roy A., private communication.
44. Delayen J. and Mercereau J., *Nuclear Instruments and Methods in Physics Research* **A257**, pp.71-76(1987).
45. Takeuchi S. et al, *Nuclear Instruments and Methods in Physics Research* **A382**, pp.153-160(1996).
46. Delayen J. et al, *IEEE Trans.Nucl.Sci* **NS-24-3**, p.1759(1970).
47. Lobanov N. et al, "Vibration Studies of the SLR", presented at the HIAT95, Canberra, 1995, (unpublished).
48. A.Facco, "Mechanical Mode Damping in Superconducting Low Beta Resonators", in *Proceeedings of the 8th Workshop on RF Superconductivity*, (V.Palmieri ed., Abano Terme), 1997, to be published
49. Weisser, D., "Superconducting Linac Booster at the ANU Commencing Physics Operations", in *Proceedings of the SNEAP XXX*, (Karl F. von Reden and Robert J. Schneider ed.) World Scientific, pp. 61-75 (1996).

Superconducting Cavity Development at Legnaro

A. Facco

INFN-Laboratori Nazionali di Legnaro
Via Romea, 4, I-35020 Legnaro (Padova) Italy

Abstract. 10 years after the construction at LNL of its first superconducting resonator, the laboratory has become one of the leading centers in the development of low beta superconducting cavities, including superconducting rfqs. The technologies of bulk niobium, niobium sputtering and lead plating have been used and some of the innovations introduced in the field during the last decade have been developed at LNL. The present technology allows producing cavities that can reach more than 50 MV/m peak surface electric field, both in test cryostats and after installation in the linac. New techniques have been developed to minimize the resonators construction cost, like spinning of multicell cavities, modular design of quarter wave resonators and niobium sputtering on copper: these technologies have reached a high level of performance and reliability. A new technique of mechanical damping was developed for stabilization of large size quarter wave resonators. The lead plating technique was enriched by the stabilization of the lead surface against oxidation. The construction of the first superconducting rfq to be used in an accelerator facility, which required original solutions to the new kind of problems emerging from this project, is at an advanced stage. Recent results obtained at LNL show that the goal of operating superconducting low beta resonators at 8 MV/m in cw mode can be presently reached; future low beta linacs could be designed with an accelerating field which is about two times higher than the one required in the past, with a significant cost reduction that could make these machines affordable also for small laboratories.

INTRODUCTION

The development of superconducting resonators at LNL started in 1987, in the framework of the ALPI project; the aim was the construction of a superconducting linac booster for heavy ions (1). Taking advantage from the considerable experience already accumulated by other laboratories (2,3,4,5), we chose the lead plated quarter wave resonators (QWR's) technology for the first phase of the construction; for its low cost and its well known techniques, it appeared to be the most convenient way to enter the field of superconducting cavities in the view of a future upgrading. At the same time a laboratory for research and development in niobium sputtering was created, to acquire the capability to replate copper resonators by means of this new and promising technique; moreover, a program on bulk niobium quarter wave resonators started, with the aim of developing, at moderate cost for industrial production, high performance cavities for heavy ion linacs (6). In this frame the Legnaro laboratories had the possibility to gain experience in the all three technologies, with the perspective of applying them in different stages of the linac construction: copper-lead cavities for the medium-β, 160 MHz resonators; copper-niobium for the 160 MHz, high-β cavities where the geometry was favorable for sputtering; bulk niobium for the large size, 80 MHz low-β resonators.

During 1996 the PIAVE project (7) required a new step forward for the construction of a positive ion injector for ALPI, and the construction of 2 superconducting RFQ's, together with a new type of low-β quarter wave cavities, has started.

LEAD PLATED RESONATORS

The activity on copper-lead plated resonators started with tapered QWR's (8). The first problem to be faced in the initial activity was the difficulty of overcoming multipacting; the switch to cylindrical cavities allowed to contain the problem within reasonable limits and gave the possibility to test the resonators. Since then, the technique was developed and about 120 plating cycles have been done; various plating procedures were tested before fixing the parameters, as widely described elsewhere (9,10).

The resonators are made of OFHC copper; no electron beam welding (EBW) is used, the beam ports and the stainless steel flange of the liquid helium inlet are connected to the cavity by brazing in vacuum. The resonators body can be obtained, at about the same cost, either by machining a full copper cylinder or by brazing together two parts, one of which being the outer conductor cylinder; in the first case a higher thermal stability can be obtained, but material defects that cause the rejection of the substrate can be discovered only after the completion of the mechanical construction.

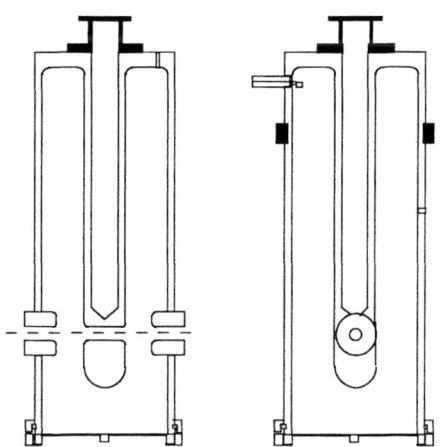

FIGURE 1. Sketch of the CU-Pb medium-β resonators

The main innovation introduced at LNL is the passivation of the Pb surface (11). The procedure, consisting of a light acid solution applied for 8 seconds to the fresh plated surface and followed by a very careful water rinsing, protects the Pb surface from oxidation for weeks in normal atmosphere. Since it is very difficult to avoid exposing the lead plated surface to oxygen during the installation of resonators in a cryostat, the Pb passivation allows to reduce the risk of Q degradation; it is a simple and attractive alternative to lead-tin plating, successfully used in other laboratories to prevent oxidation (12). Lead-tin, even if characterized by a slightly higher critical temperature in comparison with lead, does not give real advantages in terms of rf performance and its plating is not easily reproducible (13); after a few experiments at LNL it was abandoned for Pb passivation.

An unusual application of lead plating and passivation is found in the bulk niobium resonators tuning plates, made of OFHC copper for better cooling and exposed to low rf current. The coupling of lead and niobium, in this case, gave good results: the resonators could reach very high field and no sign of performance limitations appeared that could be attributed to the Pb surface. The treatment have been applied successfully to all tuning plates of the bulk niobium cavities and to the last two lead plated cavities installed in the linac in 1995.

At present, 48 Cu-Pb resonators are installed in Alpi. The average accelerating field is 2.65 MV/m at 7 W and the best resonators can work at 3.1 MV/m. They undergo periodical helium conditioning at 100 W; this procedure was implemented in our linac since the beginning and it can be executed automatically.

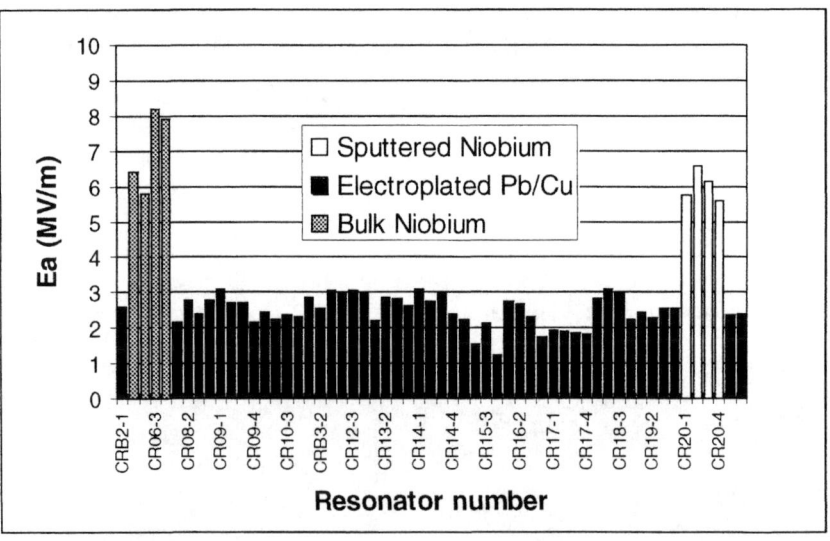

FIGURE 2. The accelerating field at 7 W of the low-, medium- and high β resonators installed in ALPI and calibreted by means of ion beam acceleration.

Through many years of operation, these resonators have maintained their performance and they have always worked reliably. Since 1995 their installation has ceased.

The main advantages of lead plated cavities are the well-proven reliability, the low cost of the substrate and the relatively cheap plating technology.

The main disadvantage is the low performance, which requires the installation of many resonators and related equipment.

At present, the gradual replacement of all the Pb resonators with Nb-sputtered ones has started; however, since this operation will require years, the expertise in Pb plating is being maintained alive at LNL.

NIOBIUM SPUTTERED RESONATORS

The first laboratory dedicated to sputtering technologies was created at LNL in 1988, with the aim of further developing and applying to quarter wave resonators the techniques successfully used at CERN for the LEP multicell cavities (14). The first QWR, after a first period of study on samples, was sputtered in 1991, and in 1993 the newly designed high-β prototypes (round corners geometry and external beam ports) could reach 6 MV/m at 7 W. The first cryostat (CR20 with 4 high-β resonators) was installed in ALPI in 1995 and in 1988 the first sputtered medium-β cavities, originally plated with lead, were successfully tested and installed in the linac (15).

The quarter wave resonators for sputtering are made of copper OHFC 99.99% certified grade. Even if brazed substrates could be sputtered successfully, for best performance the cavity must be machined from a full piece. This improves the thermal conductivity of the substrate and prevents copper contamination during brazing; it is necessary to start from Cu cylinders free from defects or inclusions that otherwise would be detected only at the end of the machining work. The shape of the high-β resonators, free of sharp angles, was studied together with the cathode shape in order to optimize the sputtering process; the beam ports are external and connected to the resonator by means of screws. Differently from the other LNL cavities, the coupler of the high-β ones is capacitive, located in a lateral position not far from the tuning plate; thus, the substrate geometry is free from holes which could cause bad niobium deposition in the high rf current region.

The substrate preparation after machining is a very important step and includes tumbling, high pressure water rinsing (HPR), electropolishing, chemical polishing, baking in high vacuum. The sputtering technique is the DC biased diode, widely described elsewhere (16); HPR, followed by ethanol rinsing and drying with nitrogen gas, are being done before installing the resonators in their cryostats.

All the production sequence can be considered well established and the final result reliable; the best resonator reached 8 MV/m at 7 W and had maximum peak fields of 53 MV/m and 1150 G. The on-line average field at 7 W of the resonators recently installed in the CR20 cryostat is 6 MV/m. These results demonstrate that the great possibilities of this technology have been extended to QWR's and that nowadays a high performance heavy ion linac can be done with sputtered cavities.

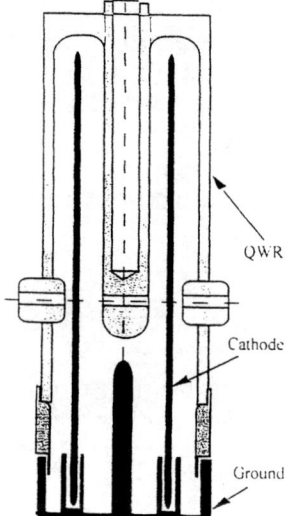

FIGURE 3. the dc-biased sputtering configuration.

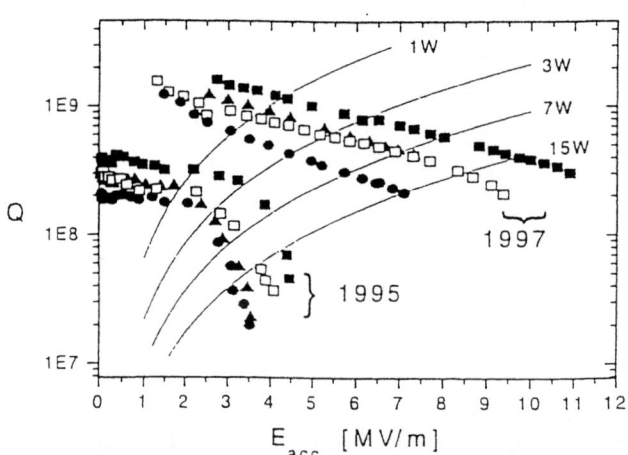

FIGURE 4. Q vs. E_a curves of the niobium sputtered high-β resonators recently installed in the ALPI cryostat n.20 (laboratory test), in comparison with the curves (on-line test) of the first sputtered cavities installed in 1995.

The old medium-β cavities geometry, because of its sharp corners and the presence of beam ports, is less favorable to sputtering; nevertheless, a reliable procedure was set up to recycle the existing lead plated cavities with minimal mechanical changes to the old substrate. The aim is bringing them to 4-5 MV/m at 7 W or reduce to 3 W the rf losses at 3 MV/m in order to lower the cryogenic load. The laboratory test of the first cryostat has shown that both goals have been achieved. A detailed report on this work can be found in these proceedings (17).

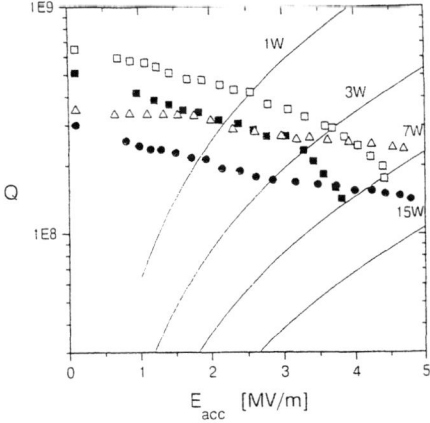

FIGURE 5. Q vs. Ea curves of the sputtered medium-β cavities.

The major difficulties of the sputtering technology in QWRs are related to the long time needed to develop a cathode suitable for the required resonator geometry; moreover, not all geometries seem to be suitable for sputtering. This technology, at present, still needs to be managed in a laboratory environment.

The main advantages are the high rf performance, the low sensitivity of sputtered resonators to earth magnetic field and helium pressure fluctuations, the absence of oxidation problems.

The cost of the machined qwr copper substrate is much lower than the cost of a finished bulk niobium resonator of the same shape. The cost of the substrate preparation and sputtering, still managed in a laboratory environment, is difficult to evaluate; in the view of a large production, however, the initial investment for setting up a sputtering laboratory and for developing a cathode could be widely compensated by a lower cost of a resonator.

In the future plans of LNL there is the sputtering of all the 45 lead plated cavities of ALPI in order to bring the linac performance to the design levels with less resonators than originally planned.

BULK NIOBIUM RESONATORS

The first LNL niobium cavity was developed in 1987 in collaboration with the Weizmann Institute and CERN. Most of the characteristics of the following Legnaro quarter wave resonators were already present: all niobium structure, cylindrical shape with no tapering, absence of "donut-shaped" drift tube. The simple and symmetric design was studied to reduce the construction cost, to avoid multipacting and to allow for high temperature vacuum treatment. In 1990 (before the HPR age) an improved 160 MHz, β_0=0.11 cavity could give 4.7 MV/m at 7 W (18); it was the first resonator installed in ALPI, lately removed due to a cryogenic accident.

In 1992 a 80 MHz, β_0=0.055 prototype, and a 240 MHz β_0=0.17 one, constructed using the material recycled from an old resonator, were built and tested with good results (19); since then we have tested different methods of resonator treatments and started applying HPR.

In 1994 the first lot of 6 low-β niobium resonators was constructed and the first one, that reached 6.3 MV/m at 7 W, was installed in ALPI (20). Four of the remaining cavities could be properly installed only this year. In 1996 a new 80 MHz prototype, with the drift tube geometry modified to reach the optimum velocity β_0=0.047 for the PIAVE project, was designed, produced and successfully tested (21).

All these resonators, having a similar design but a different length in order to work at different frequency and match different beam velocity, could be built with similar components and construction procedure; in principle, all frequencies between 80 and 240 MHz could be covered, and all v/c between 0.03 and 0.3. This modular design is especially suitable for building a heavy ion linac with low, medium and high beta sections. The needs of the ALPI project, however, prompted us to concentrate our efforts on the large size, 80 MHz resonators; the results of this work, however, could be extended to all resonators of the same type. We have produced until now 14 low-β cavities for ALPI and one for PIAVE; five of them are presently in the linac and 7 more are under construction.

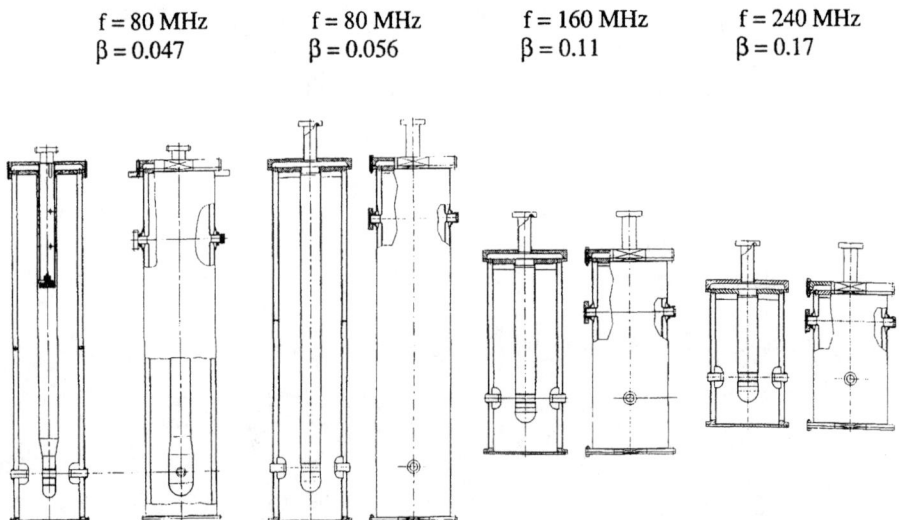

FIGURE 6. The bulk niobium quarter wave resonators developed at LNL.

The resonators are made of RRR≥150 niobium, except for the parts not exposed to rf , in normal grade material; the inner and outer conductor cylinders are made of 2 mm thick niobium sheet and the removable tuning plate is made of lead-plated copper. All parts are characterized by a simple geometrical design so that they can be easily machined and the construction cost can be reduced; moreover, the construction procedure includes the EBW of many niobium components in parallel, to limit the number of openings of the welding machine vacuum tank. The required maximum surface roughness after machining is 10 μm. The surface treatment, performed at CERN and consisting of the standard 2:1:1 chemical polishing for niobium, is usually able to guarantee high Q_0; the only treatment that is usually done at LNL is a 100 bar water rinse, followed by a pure ethanol rinse, to remove residual particles from the high field regions. The passivated Pb-Cu tuning plate, insensitive to oxygen but strongly deteriorated by any contact with ethanol, is being mounted with some delay, when the resonator is dry. In the absence of a clean area, particular care is devoted to prevent dust contamination; a protection is mounted on the beam ports and removed just before closing the cryostat. After a cycle of experiments we have realized that a high temperature thermal treatment is not necessary for these resonators. We have never observed, until now, any "Q-disease"(22) related to the resonator cooling speed.

In addition to the first prototype, we have tested successfully 7 resonators of the production lots; one further resonator, rejected because it was affected by a welding defect, was repaired by the company. The off- and on-line measurements show that all the cavities tested until now reach between 6 - 8 MV/m at 7 W. Some of the resonators

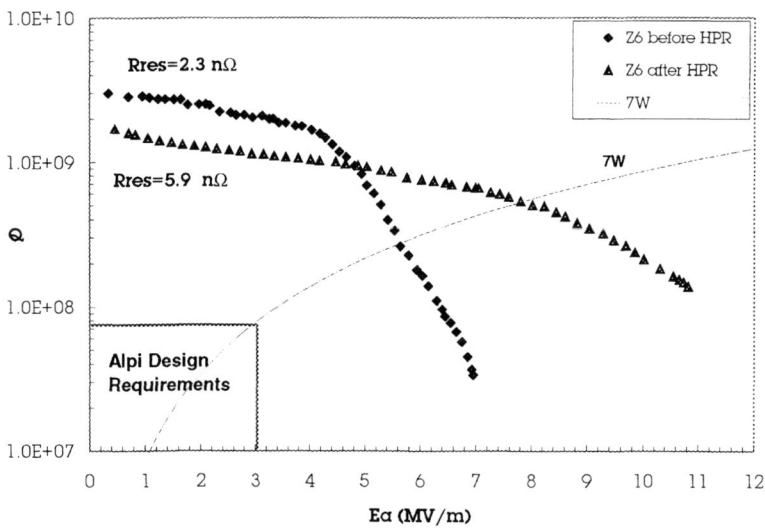

FIGURE 7. Q vs Ea of the bulk niobium cavity n. 6, before and after high pressure rinsing.

have been tested many times, sometimes together with fast tuners prototypes; their performance, after storage in air and dust contamination, could always be recovered by means of HPR.

The lowering of the low field quality factor, while strongly reducing field emission at high field, is a typical effect of HPR in our cavities. The highest Q of 3×10^9, corresponding to a residual surface resistivity of about 2 nΩ, was obtained in resonator n.6 before the HPR; field emission, in this test, started at 4 MV/m. The same resonator, after HPR and helium conditioning, had a Q_0 below 1.8×10^9 but no strong field emission appeared up to 11 MV/m (Fig.7). The most remarkable result was obtained on-line in the cryostat n. 6, recently mounted in the linac, housing 4 resonators produced in 1994 and previously tested many times in the laboratory: even if two resonators had Q_0 below their typical value of 1.5×10^9, field emission practically disappeared in all cavities after 1h of helium conditioning at 150W; the accelerating field, averaged in all of them, was 7.1 MV/m at 7 W, and 8 MV/m at 11 W. In all resonators the maximum field was about 11 MV/m, limited by quenching originated from a hot spot: this corresponds to more than 50 MV/m and 1100 G peak fields (23). To run the four resonators at 7W would produce more than 5 MV acceleration in one cryostat; to run them at a much higher field, however, would be possible at higher rf power (Fig 8).

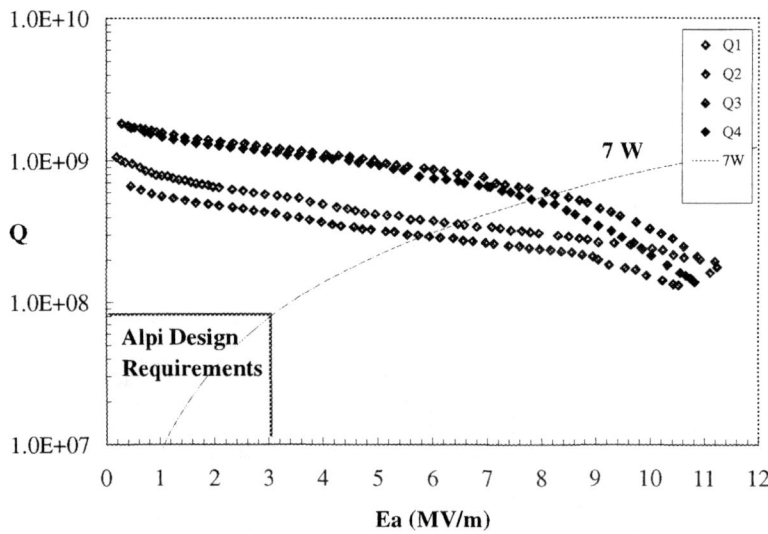

FIGURE 8. On-line performance of the bulk niobium resonators installed in cryostat n. 6.

Our rf system, however, was originally planned for operating the resonator at 3 MV/m and cannot guarantee a sufficiently wide bandwidth at more than 5 MV/m in low-β cavities; to work safely at 8 MV/m we should change our 80 MHz amplifiers. In our particular case, the main limitation to the field level in operation is given by anomalous helium pressure changes in our cryogenic system. While the first resonator prototype did not show to be sensitive to pressure fluctuations (20), the following ones of the production lot have shown a variable response, up to 1 Hz/mbar. When the cryogenic system is in stable operating conditions, the helium pressure is slowly changing (say, below 10 mbar/minute), and the mechanical tuner driven in a feedback loop allows for stable operation; in particular conditions, however, some sudden changes of pressure may appear (sometimes up to 200 mbar in two minutes), causing phase unlocking. In the preliminary testing we have been able to do until now, the lock at 4 MV/m could be maintained for days; in stable pressure conditions, we expect the cavities to operate at 5 MV/m. We are presently working for smoothing the pressure jumps in our cryogenic system and for increasing the speed of the mechanical tuner; the JAERI solution of reinforcing the top plate (24) is also being considered. In the low-β section of our linac, anyhow, a gradient higher than 5 MV/m could be hardly used without beam losses, since the beam line was designed for 3 MV/m and, at that time, the high field values presently obtained were not expected for low-β cavities.

Mechanical mode damping of large-size resonators

A general problem in large size superconducting resonators, which does not appear in the short, medium-β quarter wave ones, is the presence of low frequency (say, below 150 Hz) mechanical modes. These modes, easily excited by environmental vibrations, produce oscillations of the resonator frequency and make the phase lock to an external reference difficult. The stiffening of the resonators by increasing their wall thickness usually decreases the mechanical mode eigenfrequencies without solving the problem. Niobium and copper are equivalent from this point of view, being the values of the Young modulus approximately the same for both materials.

Whenever possible, resonators should be designed so to have only mechanical modes above 150 Hz, as it was aimed at with the four-rod 80 MHz superconducting rfq under construction at LNL.

Since in λ/4 cavities the lowest mechanical mode frequency is inversely proportional to the square of the inner conductor length, a 80 MHz resonator has modes with frequency four times lower than in the corresponding 160 MHz one. Electronic fast tuners (25) are able to correct for frequency changes of the order of hundreds of Hz, at the price of introducing some limitation to the resonator performance and some complication in the control system. To avoid these drawbacks, we have developed a mechanical damper to be installed inside the resonator (26); this device dissipates the energy of the most dangerous mechanical modes and keeps the resonators vibration amplitude to tolerable levels in normal conditions. The damper does not influence the resonator performance and can be a valuable alternative to fast tuners. The resonators installed in our linac and equipped with these dissipators have obtained a clear benefit from them: while in normal noise conditions the background frequency noise seemed to be slightly increased, when applying artificially low frequency vibrations to the cryostat the damped resonators have shown a much higher stability (23).

We have also developed a low power fast tuner (27); in this device a variable capacitance, regulated by 12 high voltage pin diode switches, is mounted outside the cryostat and connected to a coupler by a rf cable. The tuning range is ±12 Hz, enough for our resonators, but it was never used after the introduction of the dissipators.

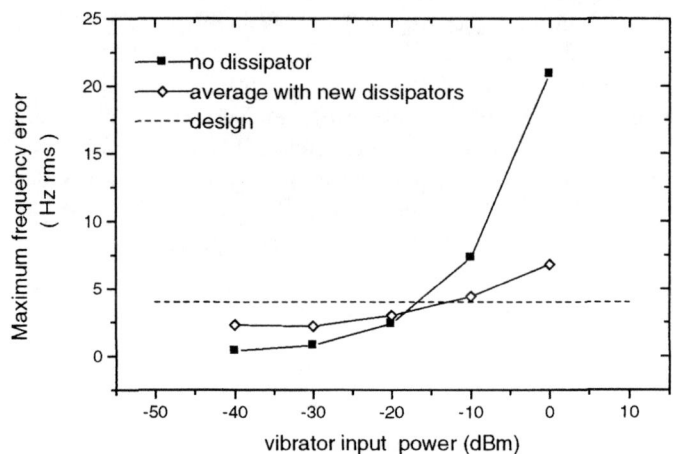

FIGURE 9. Maximum frequency error in cavities with and without mechanical damper in linac cryostats; the vibration was induced artificially in the frequency range 5-150 Hz, at different vibrator power levels.

The main drawbacks of bulk niobium cavities are their cost, lower than in the past but still higher than in the copper ones, and their sensitivity to helium pressure fluctuations which require either a stable helium pressure or an adequate mechanical tuning system.

The main advantages are represented by their very high performance, their light structure which allows for large size resonators, the possibility of building resonators of any shape, the absence of oxidation problems and in the possibility of recovering high performance by means of high pressure water rinsing; and, last but not least, the possibility of purchasing this technology from industry.

OTHER DEVELOPMENTS

In addition to quarter wave resonators, other technologies are being developed at LNL. The first superconducting RFQ for heavy ions for the PIAVE linac (28) is under construction and it will be probably the first one of this type in operation. The completely new characteristics of the resonator required original solutions in design and construction technique; a report on this work can be found in the present proceedings (29).

Another important development is the construction of the first seamless multicell cavity (30). A nine-cell, 1.5 GHz cavity for electrons has been produced by spinning a circular disk of niobium by means of a dismountable mandrel. The machining required only one day, with no need of welding and intermediate annealing. In comparison with the standard construction technique, which includes 19 electron beam welds in critical regions and many intermediate frequency adjustments, this is a technological breakthrough that could open new perspectives in the construction of electron linacs. Another application, of interest to heavy ion resonators technology, is the cold forming of seamless copper tubes and quarter wave resonators.

POSSIBLE APPLICATIONS IN FUTURE LINACS

To evaluate how the use of high field cavities could influence the design of heavy ion accelerators, we have done an approximate calculation of the total cost (except for taxes, physicists and personnel needed to assemble the linac building blocks) of a typical 40 MV booster built according to the new technological possibilities and using only average resonators parameters measured on line (8 MV/m at 11 W). In our hypothesis the linac is made of "PIAVE-like" modules, containing 8 QWRs in two cryostats, two magnetic doublets and one diagnostics box per module; vacuum, cryogenic system, electronics, rf, control systems, buildings and plants.

We found that, by using the present technology of niobium, either bulk or sputtered, the total cost could be reduced to about 50% of that of a similar lead-copper

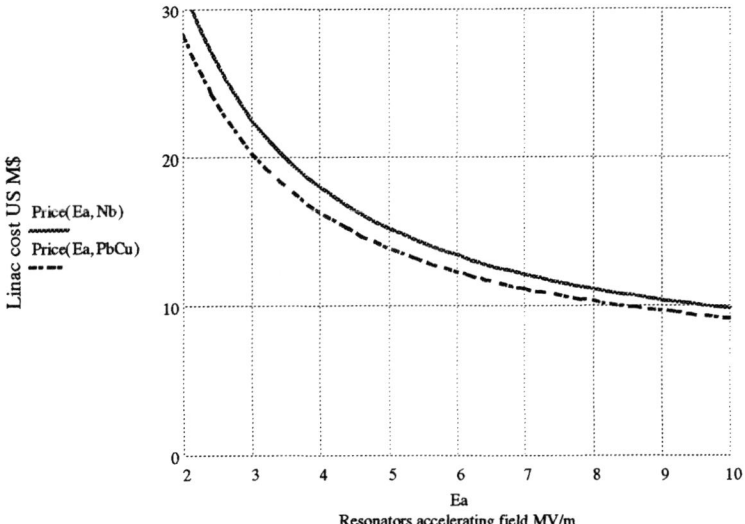

FIGURE 10. Cost estimate vs. operation accelerating field in a 40 MV superconducting booster.

linac working at 2.65 MV/m (fig.10). The resonators performance is a much more important parameter in comparison with the resonators price: to reach the same linac cost with lead plated resonators, they should be able to work at about 7 MV/m.

By purchasing most of the linac building blocks from industry, this project could be managed by a relatively small number of persons.

CONCLUSIONS

The LNL have developed, during the last 10 years, the technologies of lead plating, niobium sputtering and bulk niobium for superconducting low-β cavities. The present achievements allow the production of quarter wave resonators with remarkable on-line performance, i.e. about 50 MV/m peak electric field and 1100 G peak magnetic field. The LNL superconductive niobium rfq design requirement of 25 MV/m peak field, which could appear to be somehow ambitious a few years ago, can now be considered a realistic goal. More than 5 MV energy gain per unit charge at 7 W/cavity (or 6.2 MV at 14 W/cavity) in one, 1m long cryostat with 4 quarter wave, β_0=0.055 bulk niobium resonators have been demonstrated on-line; similar results have been obtained with β_0=0.14 niobium sputtered cavities. Further improvements in resonators design and surface treatment seem to make the goal of working at 10 MV/m feasible. Moreover, a careful mechanical design and the new technique of vibration damping in resonators extend the possibility of working at very high field with low frequency, large size cavities. The design of future linacs could take advantage from these new possibilities; high gradient resonators allow to reduce the linac size and construction cost drastically. Both technologies of bulk and sputtered niobium are able to provide high performance resonators at a lower cost than in the past; the bulk technology can already be purchased from industry.

This strong decrease of their cost could make high gradient superconducting linacs the ideal postaccelerators for radioactive beam facilities.

ACKNOWLEDGMENTS

I am grateful to G. Bisoffi, S. Canella, D. Carlucci, A. Lombardi, V. Palmieri, A. Pisent, A. Porcellato, J.S. Sokolowski, S. Stark, V. Andreev, V. Zviagintsev and all the LNL Accelerator Division for the clarifying discussions and the documentation kindly provided.

REFERENCES

1. G. Fortuna et al., *Nucl. Instr. and Meth.* **A287** (1990) 253.
2. L.M. Bollinger, *Annu. Rev. Nucl. Part. Sci.* **36** (1986) 475.
3. J.N. Noe, *Rev. Sci. Instr.* **57** (1986) 757.
4. I. Ben-Zvi, B.V. Elkonin, J.S. Sokolowski and I. Tserruya, *Nucl. Instr. and Meth.* **A254**, 306.

5. D.W. Storm et al., *Nucl. Instr. and Meth.* **A287** (1990) 247.
6. A.Facco and J.S. Sokolowski, *Nucl. Instr. and Meth.* **A328** (1993) 275-278.
7. A. Lombardi et al., "The new superconducting Positive Ion Injector for the Legnaro ALPI booster", Proc. of the 18th Int. Linear Accel. Conf., Geneva, Switzerland, 1996, p.125-128.
8. I. Ben-Zvi and J.M. Brennan, *Nucl. Instr. and Meth.* **A212** (1983) 73.
9. G. Fortuna et al., *Nucl. Instr. and Meth.* **A238** (1993) 236.
10. A.M. Porcellato et al., Proc. Of the Symposium of North Eastern Accelerator Personnel, World Scientific 1990, p.259-275.
11. S. Gustafsson et al., Proc. of the 7th Workshop on Rf Superconductivity, Gif-sur-Yvette, France, 1995, ed. B. Bonin, p.589.
12. U. Trinks, P.Schutz, Proc. of the 6th Workshop on Rf Superconductivity, CEBAF Newport News, USA, 1993, ed. R.M. Sundelin, p. 167.
13. V. Palmieri, Proc. of the 7th Workshop on Rf Superconductivity, Gif-sur-Yvette, France, 1995, ed. B. Bonin, p. 237-258.
14. C. Benvenuti, "Superconducting coatings for accelerating cavities: past, present and future", Proc. of the 5th Workshop on Rf Superconductivity, Hamburg 1991.
15. S. Stark et al., "Niobium Sputter-coated QWRs", Proc. of the 8th Workshop on Rf Superconductivity, Abano, Italy, 1997, ed. V. Palmieri (to be published).
16. V. Palmieri et al., *Nucl. Instr. and Meth.* **A382** (1996) 112-117.
17. A. Porcellato et al. ,"Upgrading of the ALPI cavities and beam test", These proceedings.
18. I. Ben-Zvi, B. Elkonin, A. Facco and J.S. Sokolowski, Proc. of the 2nd European Particle Accel. Conf., Nice, France, 1990, p. 1103.
19. A. Facco, J.S. Sokolowski, , I. Ben-Zvi, E. Chiaveri and B.V. Elkonin, Proc. of the 1993 Particle Accel. Conf., Washington, USA, 1993, p. 849-851.
20. A.Facco and J.S. Sokolowski, *Nucl. Instr. and Meth* **A382** (1996) 107-111.
21. A. Facco, F. Scarpa and V. Zviagintsev, "The non-rfq resonators of the PIAVE linac", These proceedings.
22. M. Shibata, T. Ishii, S. Kanazawa, S. Takeuchi and T. Yoshida, Proc. of the 6th Workshop on Rf Superconductivity, CEBAF Newport News, USA, 1993, ed. R.M. Sundelin, p. 895.
23. A. Facco, V. Zviagintsev, S. Canella, A. Porcellato and F. Scarpa, "On-line performance of the LNL mechanically damped superconducting low beta resonators", Proc. of the 1998 European Particle Accelerator Conf., Stockholm, Sweden, 1998 (to be published).
24. S. Takeuchi and M. Matsuda, "First three year operational experience with the JAERI tandem-booster", Proc. of the 8th Workshop on Rf Superconductivity, Abano, Italy, 1997, ed. V. Palmieri (to be published).
25. N. Added, B.E. Clift, K.W. Shepard,"Upgraded phase control system for superconducting low velocity accelerating structures", Proc. of the 16th Int. Linear Accel. Conf., Ottawa, Ontario, 1992.
26. A. Facco, "Mechanical mode damping in superconducting low-β resonators", Proc. of the 8th Workshop on Rf Superconductivity, Abano, Italy, 1997, ed. V. Palmieri (to be published).
27. V. Zviagintsev, V. Andreev and A. Facco ,"A room temperature, low power fast tuner for superconducting resonators", Proc. of the 8th Workshop on Rf Superconductivity, Abano, Italy, 1997 (to be published).
28. A. Pisent, "Design of the Heavy Ion Injector PIAVE", These proceedings.
29. G. Bisoffi, "Construction of Superconducting RFQ's at INFN-LNL", These proceedings.
30. V. Palmieri, "Seamless cavities: the most creative topic in RF superconductivity", Proc. of the 8th Workshop on Rf Superconductivity, Abano, Italy, 1997, ed. V. Palmieri (to be published).

Status of the JAERI Tandem Accelerator and Its Booster

Suehiro Takeuchi, Shinichi Abe, Susumu Hanashima, Katsuzo Horie,
Nobuhiro Ishizaki, Susumu Kanda, Makoto Matsuda, Isao Ohuchi,
Hidekazu Tayama, Yoshihiro Tsukihashi and Tadashi Yoshida

Japan Atomic Energy Research Institute, Tokai Research Establishment
Tokai, Naka, Ibaraki 319-1195 Japan

Abstract The JAERI tandem accelerator has been operating effectively since 1982 and heavy ions from the tandem has been boosted in energy by the superconducting independently phased booster linac for the experiments at higher bombarding energies since 1994. Present operating status of the tandem accelerator and its booster, status of their use for experiments and an on-going improvement project with an ECR ion source are described.

INTRODUCTION

The tandem accelerator at JAERI Tokai has been delivering various heavy ions to its twelve target beam lines for experimental researches on nuclear physics, atomic and solid-state physics for 16 years. The tandem accelerator is a model of 20UR Pelletron made by National Electrostatic Corp.(NEC) in U.S.A., which was commissioned at a terminal voltage of 18.5 MV with beams in 1982. In early time operations for nuclear physics experiments, the accelerator was operated at a voltage as high as 18 MV, having a risk of high voltage sparks which would give a damage to electronic devices in the high voltage terminal and two shorted sections in the column. In the recent years, it has been operated under 17 MV to have much fewer chance of their damages and accelerator tank opening. There has not been significantly long down times due to serious troubles.

The layout of the accelerator facility is shown in Fig. 1. A superconducting booster linac is placed on the line extended straight from the analyzing magnet. The booster construction was planned as soon as the tandem accelerator started beam delivering service, in order to give enough bombarding energy to medium and very heavy ions to overcome Coulomb barrier energy against a similarly heavy target nucleus. The superconducting(s.c.) technology which was being successfully developed at Argonne National Laboratory was introduced to the development of new superconducting quarter wave resonators matched to the booster of JAERI tandem accelerator[1,2]. After a successful development of s.c. resonators and prototype acceleration units, the construction of the booster started in 1988[3-5]. The booster was commissioned in 1994 and has been contributing to higher bombarding energy experiments[6].

The recent status of the use of the beams from the tandem accelerator and its booster is described after the description of the present status of the accelerators.

The lifetime problem of foil strippers against heavy ions, which was once solved by finding a method of long lived carbon stripper foils[7-10], is growing as very heavy ions are now often accelerated and injected into the booster. We are trying to increase beam intensity and energy using an ECR ion source at the high voltage terminal. This on-going improvement project is presented in the last section.

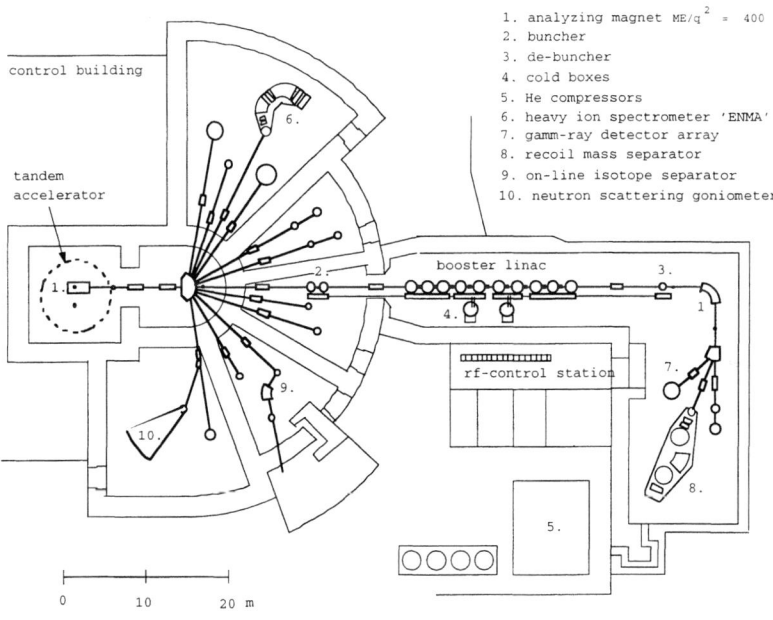

FIGURE 1. Layout of the JAERI tandem accelerator and its booster

TANDEM ACCELERATOR

Negative Ion Injectors

The tandem accelerator is illustrated in Fig. 2. It is equipped with two negative injector decks which are usually operated at 200 kV. One is the originally built deck which has four injection lines with four different ion sources. The other one was built later in order to alternatively prepare the ions for the next day during the operation. In these years, we have been using two cesium sputter negative ion sources(NEC-SNICS-2) on the old deck and another SNICS-2 on the new deck. Many ion species are available from only the single kind of sources(Table 1).

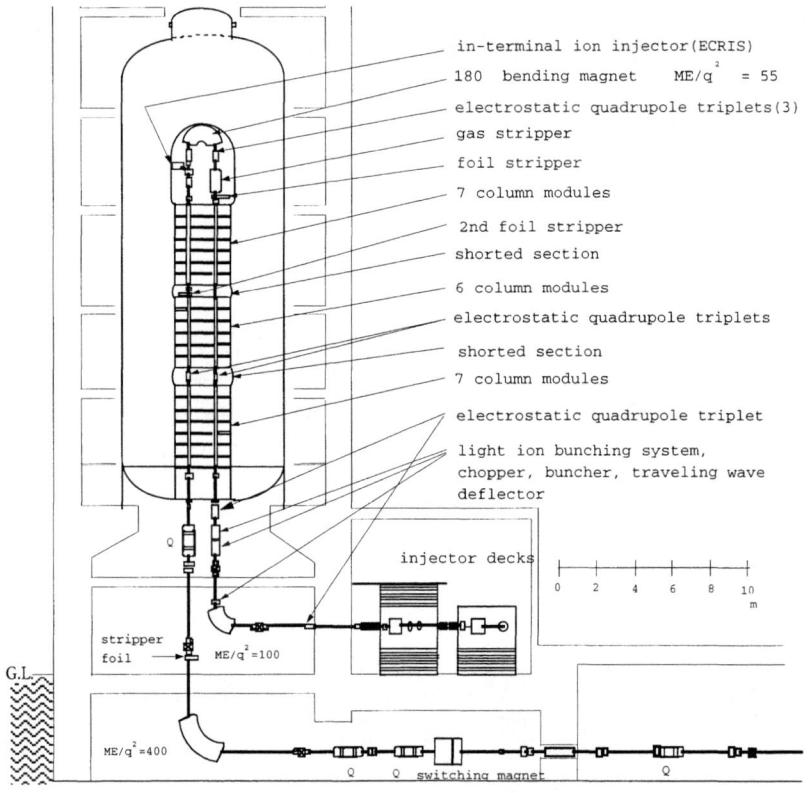

FIGURE 2. Side view of the JAERI tandem accelerator.
The low energy and high energy accelerator tube lines are situated in the plane perpendicular to the horizontal beam lines in the real facility.

TABLE 1. Ion Species available from the Cesium Sputter Negative Ion Sources

Ion	Material	Form	Current(mA)	Ion	Material	Form	Current(mA)
Li	Li_2O	Li	5.7	Fe	Fe	Fe	3.0
B	$B + O_2$	BO	21	Co	Co	Co	5.0
C	graphite	C	150	Ni	Ni	Ni	22
O	oxide	O	10-50	Cu	Cu-rod	Cu	15
F	fluoride	F	10-50	Ge	Ge	Ge	8.0
Na	Na_2O	Na	2.2	Se	CdSe	Se	2.6
Al	Al-rod	Al	10	Br	RbBr + Al	Br	25
Si	Si	Si	90	Zr	ZrO_2	ZrO	5
P	P	P	21	Ag	Ag	Ag	20
S	NiS, Ce_2S_3	S	23, 7	I	KI + Al	I	37
Cl	$NiCl_2$	Cl	55	Pt	Pt	Pt	20
Sc	Sc	Sc	5.4	Au	Au	Au	70

Accelerating Structure

The tandem accelerator is a vertically standing folded type Pelletron accelerator[11]. The accelerator column including the high voltage terminal is 16 m high and is placed in a high pressure vessel of 8.2 m in diameter and 26 m in height. A hundred percent SF_6 gas of about 5.5 bar is used as the insulating gas. The insulating column consists of twenty 1MV-column modules of 2.7 m in diameter and 0.6 m in height. They are divided into three sections by placing two shorted sections. The low energy and high energy accelerator tubes are laid in parallel at a distance of 1.42 m. The electron stripped positive ions are bent back by a 180 degree bending magnet. In the lower shorted section, electrostatic quadrupole triplets are placed in the both acceleration lines. In the upper shorted section, second stripper foils are mounted in an foil exchanger box on the high energy acceleration line.

Accelerator Tubes, Pellet chains, and Corona Needle Potential Dividers

NEC-standard 17.3 cm long accelerator tubes($2 \times 3 \times 20$ tubes) with heater plates have ever been used since the construction of the accelerator, except a few replacements. The voltage rates per module are presently in the vicinity of 0.90 MV for the cases that a group of 7(6) column modules are under high voltage tension. The full column voltage reaches to a voltage between 16.6 and 17.3 MV during the conditioning usually done.

Three sets of pellet chains(two chains for each set) were installed[11], but now only one set is in use. Most of the chains used lived longer than the recommended replacement time of 30,000 h. There was only once a broken chain during operation, which was due to a stop of oiling to the chain.

Open type corona needles have been used for dividing the potentials along accelerator tubes and insulating column. Enclosed type corona needle potential dividers were installed first but exchanged soon to the open type ones because of heavy damage of needles by high voltage sparks and decomposed corrosive gas of SF_6. Shorting rods are used in case of a very low terminal voltage.

Rotating Shaft Power Delivering System

Electric power is delivered by two rotating shafts of insulating rods which are driven by synchronous motors placed under the column bottom and drive power generators of 10 kW and 15 kW at the high voltage terminal and 3 kW and 5 kW in the upper and lower shorted sections, respectively. A 500 W generator for heater plates in the accelerator tubes is embedded in every cast plate between column modules, but now not connected to the rotating shaft because of disuse of the heater plates. The driving mechanism has been changed from the beginning one. The shafts are now directly coupled to the motors and the generators in the high voltage terminal are rotated through gear-up transmissions. The generators in the shorted sections are still belt driven. In the AC power lines to the motors, programmable frequency inverters are inserted to make a soft start.

The shafts are composed of two column modules long rod insulators of 14 cm in diameter and flexible couplings with ball bearings which are mounted on the cast plates. In recent years, bearings needed to be frequently replaced. In every maintenance period which comes in every four months, check and replacement work is carried out inevitably as well as for regular maintenance items of pellets of charging chains, corona needles and so on.

In the High Voltage Terminal

The high voltage terminal is equipped with a stripper foil exchanger which can mount 240 foils, a gas stripper with 0.9 m long canal of 9.5 mm in diameter which is retractable from the beam axis and placed in a large getter pump chamber, an achromatic 180 degree bending magnet of a radius of curvature of 0.71 m and of mass energy product of 55 amu-MeV, two electrostatic quadrupole triplets placed symmetrically before and after the bending magnet, an in-terminal positive ion injector with its injection magnet and an electrostatic quadrupole triplet in front of the high energy accelerator tube. A duoplasmatron ion source has been ever situated in the injector till 1997. It was recently replaced by a compact ECR ion source(see the section of "ON-GOING IN-TERMINAL ECR ION INJECTOR PROJECT"). Four sets of a Faraday cup and a variable aperture are placed at the beam waist points. Four 100 l/s ion pumps are used for the beam line. They are still pumping well without any change of elements, but their power supplies were changed to newly designed ones. A 200 l/s ion pump is put to the positive ion injector in place of an old getter pump.

Shorted Sections

In the lower one of two shorted sections, electrostatic quadrupole triplets are put on the both accelerator tube lines. In the upper one, there are a foil exchanger for the second stripper and a Faraday cup in the high energy tube line. Four 100 l/s ion pump are used for these tube lines in the shorted sections.

Maintenance and Repair Work

There has not been significantly long down times due to serious troubles. The accelerator has been regularly opened and maintained for 1 - 1.5 months in every four or five months. Beam times have been continuing on a 24 hours a day and 7 days a week basis in a three to four months long operation period. Frequencies of tank opening including unexpected opening in the past 15 year are shown in Fig. 3. Frequencies of what sort of devices were repaired in the past 13 years are shown in Fig. 4. The repair frequency was very low in the 1990s, because of fewer operations above 17 MV.

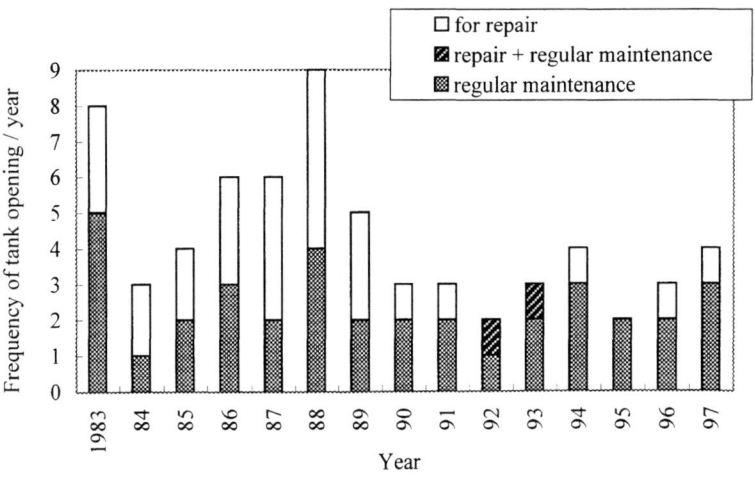

FIGURE 3. Frequency of tank opening for repair and maintenance.

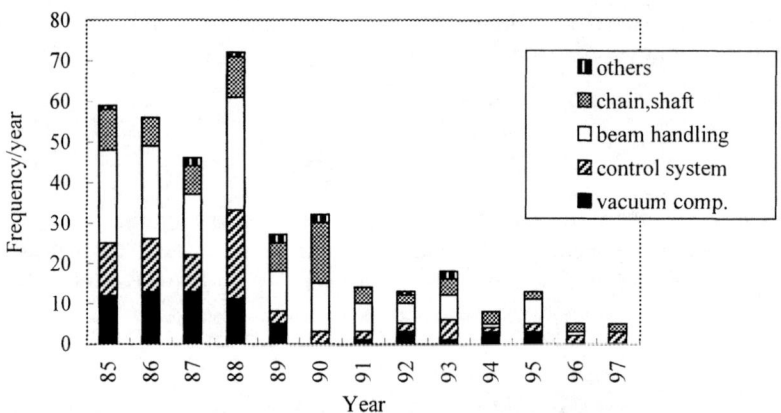

FIGURE 4. Frequency of repair in different kinds of accelerator components.

Stripper Foils

About 210 carbon stripper foils of about 5 μg/cm² are mounted on the foil exchanger in the terminal. The foils are prepared by an arc-discharge method specially developed for long-lives against heavy ion bombardment, supported by collodion film and slackened by shrinking the frames. The arc-discharge deposition method of long-lived stripper foils was developed in late 1970s by S. Takeuchi et al[7-10]. Their lives for heavy ions up to medium mass are long enough to operate the tandem accelerator stably at a moderate injection beam intensity. Foils were loaded into the exchanger once a year. The frequency was, however, recently doubled due to an sharp increase of beam times with very heavy ions. The state of a foil in use is monitored on screen by a TV camera. Foils of 10 μg/cm² are used at the second and post tandem foil exchangers.

Communication System

All the beam line devices are controlled by a digital communication system through a CAMAC system. The central computer was replaced by a parallel processing computer developed at JAERI(see section of "CONTROL SYSTEM"). The control /monitoring data to/from the devices in the accelerator column are transmitted on infra-red light. CAMAC crates and modules work in the terminal and shorted sections to send signals to many devices. The signals between the CAMAC modules and devices in the high voltage decks for the in-terminal ion injector are transmitted through optical fibers. Malfunctions of CAMAC components were greatly reduced after 1990 together with other components(Fig. 4). It can be understood that they were mainly due to high voltage sparks, because of operations under 17 MV in 1990s.

Injection and High Energy Beam Transport Lines

The injection and analyzing magnets are double focusing 90 degree magnets with radius of curvature of 1.0 m and 1.8 m and mass energy product of 85 and 400 amu.MeV, respectively. A pulsing system for light ions and heavy ions were installed in the injection beam line. The fundamental bunch frequency is 4 MHz. The frequency can be reduced at a multiple of 1/2 by a traveling wave deflector. The light ion buncher has been used for protons. The heavy ion buncher has not been used for experiments and is removed now. The system is not used for the booster operation, either. A switching magnet is placed to deliver a beam to one of 12 extended beam lines(Fig. 1). On the high energy beam line and all the extended beam lines, magnetic quarupole triplets, Faraday cups, beam profile monitors, double slits, valves and ion pumps are put at their proper places.

SF_6 Gas Handling System

Pure SF_6 gas is used in the accelerator vessel. In case of opening the vessel, SF_6 gas is transferred from the accelerator vessel of 1200 m^3 to three storage tanks of 30m^3 by using two compressors of 150 HP and two 600 m^3/h rotary pumps. The flow rate at the beginning is 1200 m^3/h and the gas of 0.55 MPa is evacuated from the accelerator vessel into vacuum of less than 1 kPa in 10 hours. In case of gas filling, liquid SF_6 from the storage vessels is vaporized and transferred into the accelerator vessel in 5 hours. During accelerator operation, SF_6 gas in the accelerator vessel is circulated through a cooler. The system is equipped with a gas purifier, but it has not been used because the gas is purified in case of liquefaction. The gas handling high pressure system is inspected every year to ensure its safety.

SUPERCONDUCTING BOOSTER

The booster is an independently phased linac composed of 46 superconducting quarter wave resonators of an optimum velocity of 0.1c, which can accelerate any ions faster than a velocity of 0.05c(c = light velocity in vacuum)[3,4]. A continuous ion beam from the tandem accelerator is bunched into a 130 MHz pulsed beam by a set of double drift harmonic buncher which consists of two 130 MHz QWRs and two 260 MHz QWRs. One for each frequency is enough for operation. The capture efficiency is 60 %. Forty 130 MHz QWRs are used for acceleration, which are housed in 10 linac units(cryo-modules). A CW beam is obtained after the beam is de-bunched and compressed in energy spread by one of the two 130 MHz QWRs in the de-bunching unit and analyzed by a double focusing 90 degree magnet with a mass energy product of 400 amu-MeV.

Superconducting Quarter Wave Resonators

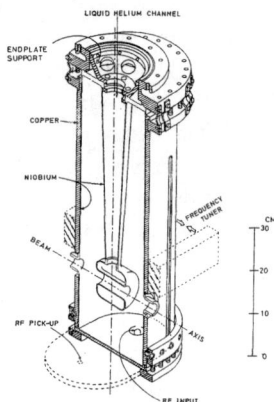

FIGURE 5. Cut-away view of a superconducting quarter wave resonator

The QWRs are made of solid niobium and niobium-clad copper sheets(Fig. 5). An average accelerating field gradient of 6.5 MV/m was obtained at rf input of 4 W from their off-line performance testing using a test cryostat[5]. The corresponding acceleration energy gain is 1 MeV/charge for the optimum injection velocity of 0.1c, because the reference acceleration length is 0.15 m. The average field gradient was lowered to 4.6 MV/m when the on-line resonators are cooled down by refrigerators because of their slow cooling speed of about -10 K/h(Fig. 6 upper case). It was due to hydrogen pollution during the electro-chemical surface treatments and precipitation of niobium-hydrides on the niobium surfaces at a temperature of 130 - 90 K during the cool-down[5,12]. A severe degradation happened to approximately 10 resonators. The on-line performance could be improved by cooling the resonators as fast as -20 to -40 K/h by means of sequential switching of entire cold helium flow to one of a few groups of cryomodules over the range of 130K and 90K[13]. An average of 5.4 MV/m was obtained by the sequential cooling(Fig. 6 lower case), which satisfied our designe value of 5 MV/m. The sequential cooling is applied to a case that operations at very high energy is scheduled in a period that the resonators are kept cold.

Electron field emission is still not so strong, but sometimes happened to some resonators and dropped their field gradients to about 3.5 MV/m level. It can be easily relieved by high power rf pulse processing.

Recently, a big air leak happened due to a sharp and strong ion beam hit on beam line bellows while the resonators were cold. As a result of warming up the resonators to the room temperature, it was found that there were no appreciable degradation in their performances, fortunately.

There have been no troubles with the resonator control components such as rf input couplers and frequency tuners, and any cryomodule has never been opened since 1994.

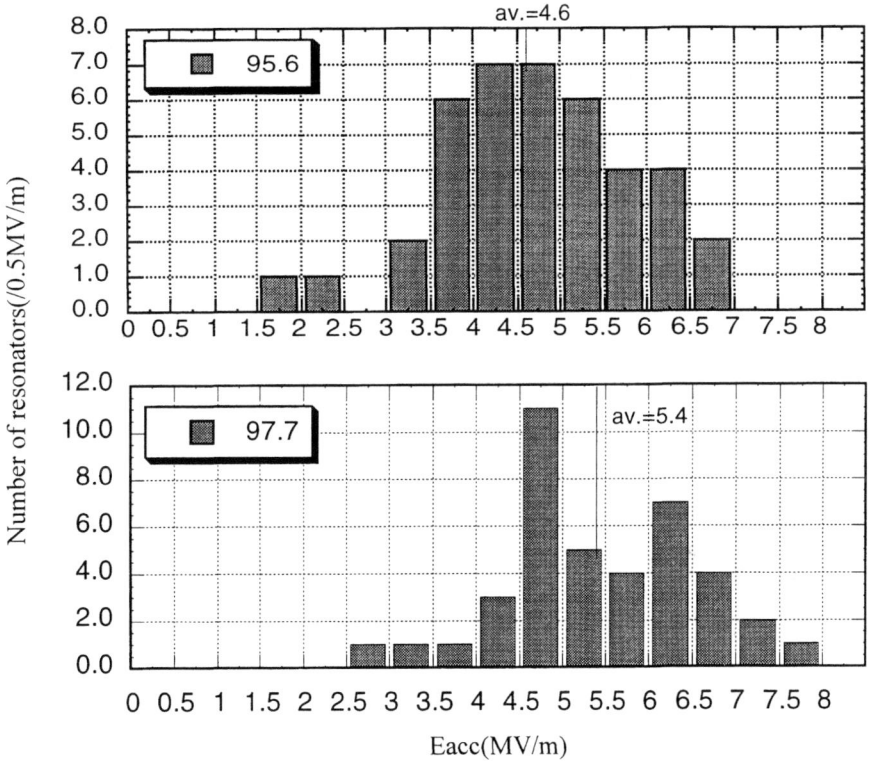

FIGURE 6. Histograms of accelerating field gradients. Upper case is obtained after normal precooling without sequential fast precooling and lower case with sequential fast precooling.

Beam Acceleration

Every resonator is controlled and can be phase-locked in a strongly coupled self-exited resonance loop which includes phase and amplitude feed back control. Every resonator phase is set up to a synchronous phase after knowing the beam bunch phase at the resonator, in an independently phase linac. In this booster, three normal conducting QWR are placed after the third, sixth and tenth cryomodules to detect the beam bunch phases. The time of arrival of the bunched beam from the resonator under phase setting can be measured as a function of resonator phase by sweeping the resonator phase and measuring the rf phase of the rf signal from a detector at the same time[6]. From the measured curve, one can find the resonator phase which corresponds to a desired synchronous phase to the beam. The data taking is automatically done with a small measurement and control station. This method is quickly done and sensitive to the beam current or usable even below 1 nA. One can set all the resonators and quadrupole doublet lenses in 1.5 hours, although the resonator phases are presently manually set at the resonator control stations.

The beam transmission was 40 -100 % at the commissioning test[6] and improved to some extent by a re-alignment of the linac units and quadrupole doublet lenses. Heavy ion beams were frequently transported through the booster without acceleration for some experiments in the booster target room. In that case, 100 % of the beams passed through the booster. The booster is equipped with baffle apertures of 16 mm in diameter at the beam entrance ports of all the linac units in order to prevent spread beams from hitting resonators with a bore diameter of 25 mm. Beam handling onto the axis is crucial in the present situation. Further improvement is wanted, including investigations of longitudinal beam spread.

The heavy ions accelerated for experiments in 1997 and 1998 are listed in Table 2.

TABLE 2. Ion Species Accelerated by the Booster in 1997 and 1998

Ions	Beam energy(MeV) in	out	Average acc. field gradient(MV/m)	No. of resonators
$^{58}Ni^{11+}$	180	635	3.7	25
$^{58}Ni^{13+}$	215	270	2.9	11
$^{60}Ni^{11+}$	180	283-294	3.6-3.6	21-23
$^{60}Ni^{12+}$	195	312	2.5	28
$^{68}Zn^{22+}$	185	573	3.2	40
$^{74}Ge^{22+}$	205	625	3.7	37
$^{76}Ge^{22+}$	195	365-648	3.7-3.8	37-39
$^{82}Se^{24+}$	190	743	4.3	40
$^{90}Zr^{12+}$	205	340-470	3.5-3.9	27-40
$^{127}I^{28+}$	205	580-686	2.9-3.3	38-40
$^{197}Au^{25+}$	340	823	4.3	38

Cryogenic System

The booster is equipped with two identical refrigeration systems(Sulzer's Model TCF-50 with modifications) with actual refrigeration powers of 275 W for a liquid helium 4.5 K loop and 1.57 kW for a helium gas 80K loop. The 80K loops are used for shielding of the transfer lines of 4.5K loops and the resonators. The acceptable thermal load from resonators is 24-28 W/cryomodule or 120-140W/refrigerator on the stable operation. The quiescent loss is about 75W. The total load to the system is stabilized as to balance with the refrigeration power against a fluctuation of rf load by a 400W heater situated at the sump in the cold box[6]. The heater consumes as much as 60 - 80 W at the maximum external load as long as a stable condition is held, under the present situation. With respect to the pressure stability, usual fluctuation is a few or several mbar and a pressure change at a moderately unstable condition is within 30 mbar. It is satisfactory to the resonator operation, because the pressure sensitivity of the s.c. QWRs is 0.27 Hz/mbar.

CONTROL SYSTEM

The control computer installed together with the tandem accelerator has been replaced by a multiprocessor-based parallel processing computer developed at the JAERI tandem accelerator[14]. The conceptual block diagram of the control system is illustrated in Fig. 7. The system is divided into main- and sub- host systems using work stations, a central system, serial highway drivers, a CAMAC serial highway

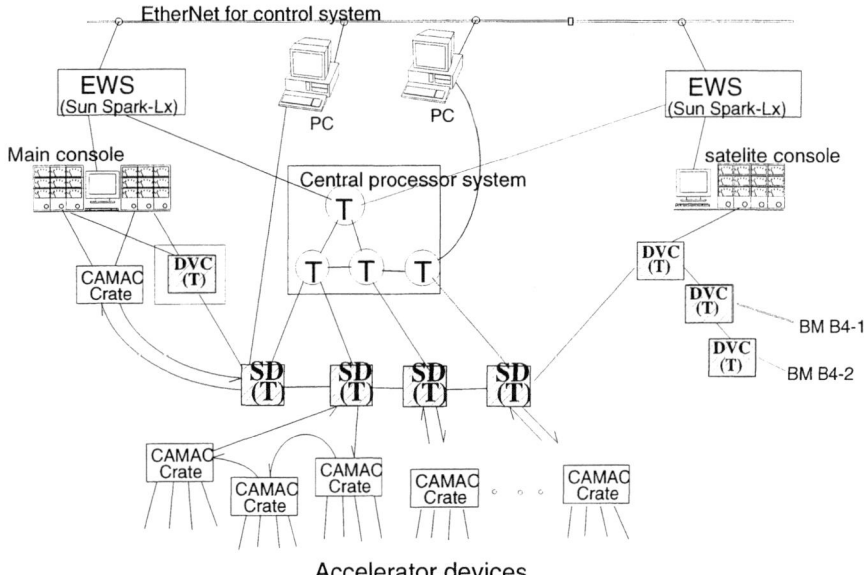

FIGURE 7. Conceptual block diagram of the control system for the JAERI tandem accelerator and its booster. T: Transputer, SD: serial highway driver, DVC: Device controller.

system and two personal computers. Multi-processors of four Transputers of Model T805(INMOS Limited) with local memories of 2 Mbytes are used as the multiprocessors in the central system. The transputer is a 32-bit micro-processor which supports parallelism at the hardware level and has communication links to other transputers. These features enable us to construct a multiprocessor system and to get large computing power easily at small costs. The serial highway drivers are the front end processors of the central system to the CAMAC system. Transputers are also used in the driver. The CAMAC serial highway system is almost the same as the old one. The control system was developed to include the booster's beam line device controls. Two bending magnets for the booster work under transputer driven device controllers. They have been introduced to establish a field-sensing-feedback field control with NMR field meters.

STATUS OF THE UTILIZATION

Experimental apparatuses at the ends of 12 extended beam lines(Fig. 1) are a gamma-ray goniometer for nuclear structure studies, irradiation chambers for studies on irradiation effects in materials, a heavy ion spectrometer of $QM_1D_1M_2D_2M_3$ type for nuclear reaction studies, a multi-purpose scattering chamber, irradiation chambers for irradiation effects in materials at low and high temperatures, an on-line isotope separator for nuclear chemistry, a neutron scattering goniometer and so on. In the booster target room, there are a recoil mass separator, a gamma-ray detector array and an irradiation chamber.

The beam time for experiments has been constantly about 200 days a year (Table 3). Statistics of the beam times in different fields in 1997 are shown in Table 4. Statistics of accelerated ions in the two terms of 1985 to 1987 and 1995 to 1997 are shown in Fig. 8 as a function of atomic number of the ions. The trend has shifted to heavier ions since the booster started operation. The use of the booster was limited at about 20 - 25 % among the whole beam time in the present user's situation. Accelerated ions are already shown in Table 2. The fraction is expected to increase because other 25 % of the whole beam time was used without energy boost for the experiments in the booster target room in these years.

TABLE 3. Operational Status in 1997

State	Total time(day)	Percentage(%)
Beam time for experiments	234	64
Regular maintenance	74	20
Days off	47	13
Down time for repair	9	2.5
Others	1	0.3

TABLE 4. Beam Time in Different Research Fields in 1997

Research field	Beam time(days)	Percentage(%)
Heavy ion nuclear physics	127	54
Solid state physics		
atomic and molecular physics	44	19
Nuclear chemistry	28	12
Neutron physics	14	6
Irradiation effect in materials	13	6
Detector development	4	2
Radiation chemistry	2	1
Accelerator development	2	1

(a)
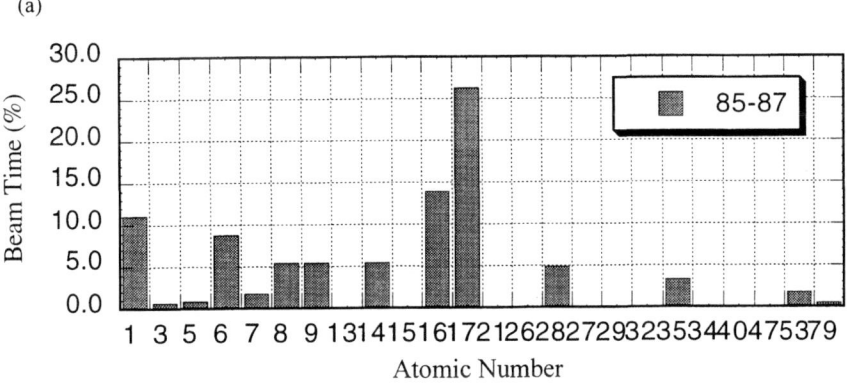

(b)
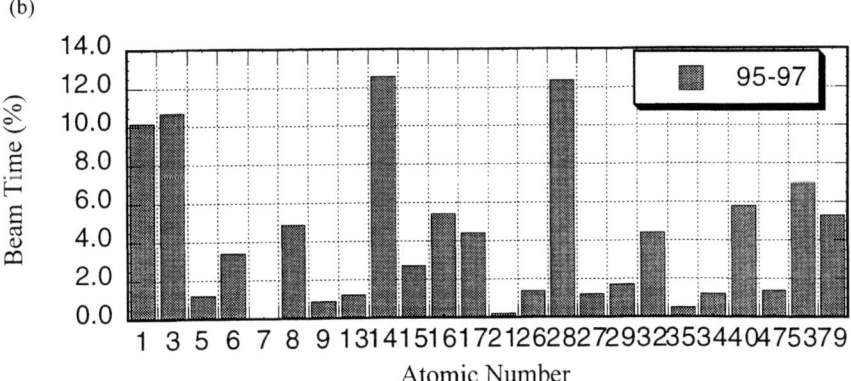

FIGURE 8. Statistics of accelerated ions in the two terms of (a) from 1985 to 1987 and (b) from 1995 to 1997.

ON-GOING IN-TERMINAL ECR ION INJECTOR PROJECT

Trend of the demands from users is moving toward heavier ions as is shown in Fig. 8. The beam intensity is limited by use of stripper foils. Foil's life is very limited for very heavy ions. The booster's resonators do not accept low velocity ions slower than 0.05c. For that case, the second stripper situated in the tandem accelerator column is used and naturally decreases the beam intensity to one-third or less. Therefore, we need a system which do not use stripper foils unless very strong long-lived thin stripper foils are developed. Use of an ECR(electron cyclotron resonance) ion source in the high voltage terminal is a promising option, nowadays.

A decade ago, use of an ECR ion source was proposed for the HHIRF tandem accelerator at ORNL[15,16]. At that time, a compact ECR ion sources with 100% permanent magnet plasma confinement system were not available except a large one such as NEOMAFIOS developed at Grenoble. Today, a very compact ion source, NANOGAN(10GHz) which was developed at GANIL is commercially available, which can be installed in our tandem(Fig. 2)[17]. From an ECR ion source, one can accelerate noble gas ions which are impossible to accelerate in tandem accelerators. We, then, started a project of beam intensity increase and energy increase utilizing ECR ion sources.

We tried a NANOGAN first[18]. The installation work has been carried out. We recently succeeded in a test of acceleration of ions of H^+, O^{3+}, N^{2+}, $Ar^{6+,8+,9+}$ and $^{132}Xe^{12+,13+}$. The beam quality seemed very high, because there was no energy straggling in the stripper foil. The details of this work are reported elsewhere in the proceedings. We are planing to accelerate metallic ions such as Pb^{25+} and to use a more powerful 14.5 GHz compact ECR ion source, in the future.

CONCLUSION

The JAERI tandem accelerator has been used stably for experiments without big troubles for 16 years. Major improvements were the addition of a superconducting booster, the addition of a second injector, the renewal of the control system with a multiprocessor-based parallel processing system using transputers and the replacement of the in-terminal ion source from a duoplasmatron to a compact ECR ion source. The acceleration energies were increased two to four fold by the booster. Noble gas ions are now available from the ECR ion source. Although a use of an ECR ion source in a tandem accelerator means a way back to a single ended electrostatic accelerator, we are free from the problems with stripper foils and it is a promising way to higher beam intensity and higher beam energy.

REFERENCES

1. Shepard. K. W., Takeuchi, S. and Zinkann, G. P.,"DEVELOPMENT OF SUPERCONDUCTING NIOBIUM ACCELERATING STRUCTURES FOR HEAVY IONS", *IEEE Trans. Magn.* MAG-21, 1985, pp. 146-148
2. Takeuchi, S., "Status of Work on Superconducting Quarter Wave Resonators at JAERI", *Proceeding of the third Workshop on RF Superconductivity*,Argonne, Illinois, 1987, pp. 429-434.
3. Takeuchi, S., Ishii, T. and Ikezoe, H., "Niobium Superconducting Quarter-wave Resonators as a Heavy Ion Accelerating Structure", *Nucl. Instr. and Methods* A281, 1989, pp.426-432.
4. Takeuchi, S., Ishii, T., Ikezoe, H. and Tomita, Y., "DEVELOPMENT OF THE JAERI TANDEM SUPERCONDUCTING BOOSTER", Nucl. Instr. and Methods A285, 1990,pp.257-262.
5. Ishii, T., Shibata, M., Takeuchi, S., "Construction of the JAERI tandem booster", *Nucl. Instr. and Methods* A328,1993,pp231-235.
6. Takeuchi, S. Ishii, T., Matsuda, M., Zhang, Y. and Yoshida, T., "Acceleration of heavy ions by the JAERI tandem superconducting booster", *Nucl. Instr. and Methods* A382, 1996, pp. 153-160.
7. Takeuchi, S. Kobayashi, C., Satoh, Y., Yoshida, T., Takekoshi, E. and Maruyama, M., "A PREPARATION METHOD OF LONG-LIVED CARBON STRIPPER FOILS", *Nuclear Instr. and Methods* 158, 1979, pp.333-338.
8. Takeuchi, S. and Kanazawa, S., "CARBON DEPOSITION APPARATUS TO PRODUCE LONG-LIVED STRIPPER FOILS FOR THE JAERI TANDEM ACCELERATOR", *Nucl. Instr. and Methods* 197, 1982, pp.267-272.
9. Takeuchi, S. and Kanazawa, S., "CARBON STRIPPER FOILS LESS SHRINKABLE AGAINST HEAVY ION BOMBARDMENT", in *Proceedings of the 11th World Conference of the Insternational Nuclear Target Development Society,* Seattle, Washington, 1982, pp.1-17.
10. Takeuchi, S. and Kanazawa, S., "SERVICE LIVES OF LONG-LIFE CARBON STRIPPPER FOILS IN THE FIRST THREE YEARS OPERATION OF THE JAERI TANDEM ACCELERATOR", *Nucl. Instr. and Methods* A249, 1986, pp. 133-136.
11. Maruyama, M.,"The JAERI Tandem Accelerator Facility", in *Proceedings of the Third International Conference on Electrostatic Accelerator Technology*, Bejin, 1981, pp.17-22.
12. Aune, B., Bonin, B., Cavedon, J. J., Jillard, M., Godin, A., Henrio, C., Leconte, Ph., Safa, H., Veyssiere, A., and Zjlberajch, C., "DEGRADATION OF NIOBIUM SUPERCONDUCTING RF CAVITIES DURING COOLING TIME", in *Proceedings of the 1990 Linear Accelerator Conference*, Albuquerque, New Mexico, 1990, pp. 253-255.
13. Takeuchi, S. and Matsuda, M., "First three year operational experience with the JAERI tandembooster", to be published in *Proceedings of The Eighth Workshop on RF Superconductivity*, Padova, Italy, 1997.
14. Hanashima, S., "Status of Control System for the JAERI Tandem Accelerator", in *Proceeding of the 11th Symposium on Accelerator Science and Technology*, Harima Science City Garden, Hyogo, Japan, 1997,pp.519-521.
15. Olsen, D. K., "Proposal for an ECR source in the HHIRF Tandem Terminal", *private communication*, 1990.
16. Olsen, D. K., "Introduction to ECR Sources in Electrostatic Machines", in *Proceedings of Symposium of Northeastern Accelerator Personnel*, Oak Ridge, Tennessee, 1989.
17. Sortais, P. Bieth, C., Foury, P., Lecesne, N., Leroy, R., Mandin, J., Marry, C., Pacquet, J. Y., Robert, E., Villari, A.C.C., "Developments of compact permanent magnet ECRIS", in *Proceeding of the 12th INTERNATIONAL WORKSHOP ON ECR ION SOURCES*, RIKEN, Japan, 1995, pp.44-52.
18. Matsuda, M., Takeuchi, S. and Kobayashi, C., "Application of an ECR ion source to the JAERI tandem accelerator", in *Proceedings of the 11th Symposium on Accelerator Science and Technology*, Harima Science City Garden, Hyogo, Japan, 1997, pp.165-167.

Status Report in the São Paulo Pelletron-Linac Project

N. Added

Laboratório Pelletron, Instituto de Fisica,
Universidade de São Paulo, SP, Brazil

Abstract. This work describes the present situation of Pelletron-Linac Project and the steps we have accomplished in the last two years.

INTRODUCTION

The Pelletron-Linac Project was created in order to improve the maximum projectile energy for the experiments performed in our laboratory. We are installing a new superconducting accelerator as a post-accelerator after our present tandem. In this work I will describe briefly the present situation of each item of this project.

BUILDING

The building has been the major problem in this project. Funds for construction should have come from the State Government, but negotiations were never finalized. So we were obliged to look for funds in other agencies (federal and state). Because of the value (US$ 3 M) it was impossible to get money from just one agency, so we decided to divide the project into several steps and to obtain funds for each one independently. This strategy is working well and we were able to start construction in 1996.

Basically the construction was divided into four steps: groundleveling, foundations and concrete walls, electric and hydraulic installations, and architecture. The first two steps are complete as can be seen in Fig 1.

For the third step, the most expensive, we have just received US$ 1 M from FAPESP. The liberation of these funds will allow us to re-initiate the construction in the second semester of 1998.

The last step, which we call architecture, will be used to close the building. In other words, painting, floors, roof sealing, windows, everything to finish it. Funds for that will come from University budget and from a Pronex project. We hope to complete these two last steps of construction by the end of the second semester of 1999.

CP473, *Heavy Ion Accelerator Technology: Eighth International Conference,*
edited by Kenneth W. Shepard
© 1999 The American Institute of Physics 1-56396-806-1/99/$15.00

FIGURE 1. Present view of the new accelerator building

BUNCHING SYSTEM

Our bunching system can be divided into three parts: pre-buncher, chopper and phase detector. The mechanical pieces were installed in the beam line during 1996 and 1997. During the conditioning process of the pre-buncher grids we found a problem with the connections to the copper cones. To solve this problem we had to modify the connection between the molybdenum grids and copper cones. It is important to note that there is no easy way to bond molybdenum with other materials and that we have done several tests to find the best solution. This solution was to braze the grids to a thin copper ring with a special silver material in an oven where we could control the internal temperature. Finally, these rings were welded to the cones using regular solder.

In a first phase, we have tested the components of the RF bunching system independently in an offline experimental setup. Basically, these tests have consisted in evaluating the Q factor for each circuit after adjusting the drive and pick-up probes of each component in order to get the correspondent resonant frequency. Typically, unloaded Q values were about 600 for the pre-buncher (for the first three harmonics) and chopper and about 1500 for the phase detector. These values were reproduced after the installation in the accelerator beam line. Recently we have started to use beam time to check and calibrate the bunching system control eletronics. Using ^{16}O and ^{35}Cl beams we have generated time spectra between the bunched beam and a surface barrier silicon detector mounted in the 30B scattering chamber. In these preliminary tests we have used only the pre-buncher.

The results of these tests show good agreement with our predicted values. We have performed several tests in order to find the best conditions for the parameters of the system. The best value we have obtained for the time resolution was about 1.5 ns for ^{16}O (40 MeV) e ^{35}Cl (67,5 MeV) beams. We hope to improve this value after installing the phase shifter controller, the last piece of eletronics needed to complete the eletronic circuit loop for the bunching system.

We have also performed tests to verify the chopper efficiency for cleaning the events from non-bunched beam particles. Results for this test were excellent.

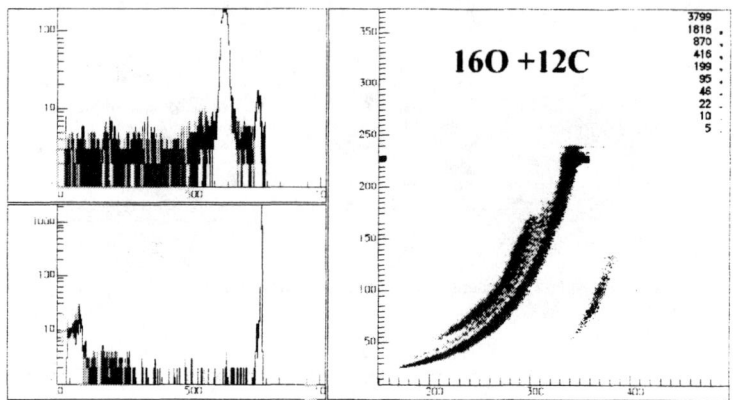

FIGURE 2. a) Time of flight and energy spectrum for ^{35}Cl + ^{197}Au at the energy of 67.5 MeV. Tac spectrum shows the bunching efficiency. b) Two dimentional spectrum energy versus time of flight showing the mass separation for a distance of 35 cm between target and silicon detector

RESONATORS

All the split-ring resonators are completely assembled and each one has been tested at liquid helium temperature at least once. As we can see in table 1, most of them achieve maximum electric field over 3 MV/m. Exceptions are the two high beta resonators H-59 and H-62, both with problems at 1 MV/m. We hope to identify the problems with them in a future cold test, where we will use special thermometry in order to look for defects in the welding joints of the arms which could be generating the heat responsible for the electric field limitation.

TABLE 1. Maximum electrical field for Brazilian resonators after the first cold test

Low β	Electrical Field (MV/m)	High β	Electrical Field (MV/m)
L - 24	5.0	H - 56	3.0
L - 25	5.0	H - 57	3.5
L - 26	4.5	H - 58	4.3
L - 27	3.2	H - 59	1.0
		H - 60	4.5
		H - 61	5.6
		H - 62	1.0
		H - 63	3.5
		H - 64	3.5

If we were forced to start the operation of the superconducting accelerator, we could use one of these resonators as a rebuncher, because in this position the electric field is always under 1MV/m. In other words, there would be only one resonator to limit the operation. Ten of the fourteen resonators are already in Brazil, waiting for final installation on the beam line.

CRYOGENIC SYSTEM AND CRYOSTATS

Our cryogenic system was planned to work with two plants: a liquid helium plant and a liquid nitrogen plant. The liquid helium plant from Cryogenic Consultant Inc. is rated at 600W/150l/h. Although it was purchased some time ago, it has not yet been installed because of the delay in the building construction. Our liquid nitrogen plant consists of two 50 l/h Linit units from Linde Cryogenics. They arrived in Brasil in December 1997 and were commissioned in June 1998. Both nitrogen plants are running perfectly, producing liquid above the nominal values. In order to install these plants we have worked to close an area in the new building close to the compressor room. All items of the helium plant were also installed in the same area and we hope to start this equipment during 1999.

Finally, to complete this description we should describe the cryostats situation. The resonators will be mounted in four cryostats. Two of them will have just one resonator mounted (superbuncher and rebuncher) and the other will accommodate 6 and 7 resonators. Funds for the larger cryostats were obtained from Finep in 1997. After consulting some companies we have decided to build the cryostats with an Italian company (CINEL), that has worked on the Legnaro accelerator. This company is supposed to start manufacturing our cryostats before the end of year. The situation for the smaller cryostats is undefined because we are requesting funds for them.

FIGURE 3. LINIT50 Liquid Nitrogen Plant

FINAL COMMENTS

It is still necessary to obtain some funds in order to complete this project. Besides the small cryostats, we will have to purchase vacuum pumps, beam diagnostic stations, superconducting solenoids and control eletronics for the resonators. We have estimated that we will spend about US$ 600K more to finalize the project. However, we are confident that we will complete these projects before the end of 1999. If we have no more problems with construction we hope to have beam out of the new accelerator by the second semester of 2000.

Meanwhile, some improvementes were implemented in our target area. Two projects were installed: a 1 m diameter scattering chamber (with a 14 triple telescope hodoscope) under supervision of prof. Alejandro Szanto Toledo's group and a mini-ball for particle (11 plastic scintillators) and gama detection (6 GeHP, 4 Anti-Compton and 8 NaI) under supervision of prof. Roberto Ribas' group. About US$ 1.5 M were used for project and installation of these projects. Finally, profa. Alinka Lépine's group has a US$ 400 K project to design a twin solenoid system for our accelerators based in the design of Notredame Laboratory.

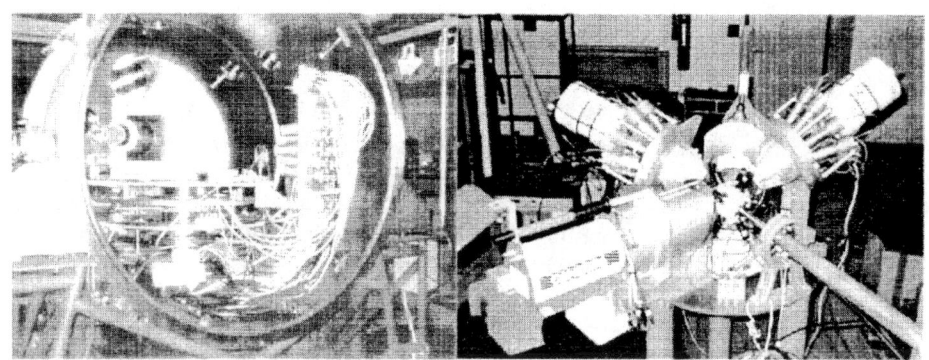

Figure 4. *Left:* View of 1 m scattering chamber and hodoscope. *Right:* Mini-ball for particle and gama detection.

Construction of Superconducting RFQs at INFN-LNL

G. Bisoffi°, V. Andreev°, E. Bissiato°, M. Comunian°,
F. Chiurlotto°, E. Corradin*, M. Lollo°, A. Lombardi°, A. Pisent°,
A.M. Porcellato°, T. Shirai[†], E. Tovo*, R. Tovo˙

°INFN – Laboratori Nazionali di Legnaro, Legnaro, Italy
*Dipartimento di Ingegneria Meccanica, Università di Padova, Padova, Italy
[†]NSRF-ICR, Kyoto University, Kyoto, Japan
˙Dipartimento di Ingegneria Meccanica, Università di Ferrara, Ferrara, Italy

Abstract. The tests on the stainless steel prototype for one of the two superconducting RFQs (SRFQs) for PIAVE (SRFQ2), the being built heavy ion injector for the Legnaro booster, were completed in summer 1998, while the construction of the first niobium resonator started in February 1998 and is expected to be completed by April 1999. The structure, resonating at 80 MHz, is 0.8 m long and 0.76 m in diameter. All technological aspects connected with the construction of the SRFQs, and the corresponding tests on the stainless steel model, are reviewed: development of the parts, assembly sequence, electron beam welding (EBW) steps, rough and fine adjustment of the resonant frequency, bead-pull measurements, characterization of the mechanical vibration modes, frequency change due to cooling down and chemical etching tests. The updated development of SRFQ1 is briefly reviewed.

INTRODUCTION

The realization of the first full niobium superconducting (s.c.) RFQ (SRFQ2) is close to completion. It is the second RFQ on the beam line of PIAVE (1), the new being built injector of the s.c. booster ALPI at INFN-LNL. The whole layout comprises two SRFQ's in a single cryostat, followed by eight Quarter Wave Resonators (QWRs, β_{opt}=0.047). The 80 MHz resonant frequency, plus the typical issues related to the construction of a superconducting resonator (2) imposed the design of a four-rod structure, with 90°-apart stems (3). These, together with the stiffening structure of the outer tank, were carefully optimised in shape so that the large size resonator (~0.8 m in length and in diameter) was made rigid towards possible mechanical vibrations. The novelty of the structure, with respect in particular to the complexity of the mechanical construction and the electron-beam welding (EBW) joints, suggested to devote a substantial effort in building a full scale stainless steel model first. This allowed to face a number of technological issues before proceeding with the considerable investment on the full Nb version.

Fig. 1 shows a picture of the completed full scale stainless steel model of SRFQ2, during RF tests. In the following chapter the major construction issues of the mechanical construction are reviewed.

FIGURE 1. A photo of the stainless steel model of SRFQ2 is shown. In its s.c. version, the resonator itself is made out of high purity Nb (RRR>250), whereas the external stiffening cage plus the end-flanges are made out of Ti

All experimental tests performed on the stainless steel model are then reported: bead-pull measurements around the beam axis are compared to beam dynamics requirements and the fine tuning range is compared to MAFIA (4) simulations. Electromechanical stability is discussed: the frequency response spectrum of the cavity to mechanical excitation is compared with FEM simulations; the characteristics of a slow frequency tracking system (5) and the possible use of a fast tuner (6) are analysed. The resonant frequency increase, due to the cooling down of the resonator to liquid nitrogen temperature, is reported and a procedure of efficient chemical etching of the Nb resonator internal surface is proposed.

The final chapter reviews the tests performed on the Al model of SRFQ1, the first resonator on the beam line but the second in the construction sequence. Table 1. shows the main parameters of the two SRFQs.

TABLE 1. Main characteristics of SRFQ1 and SRFQ2 (design parameters are shown in italics, while the others are obtained with a model developped with the code MAFIA and are shown for a Cu room temperature structure of the same geometry).

	SRFQ1	SRFQ2
RF Frequency [MHz]	*80*	*80*
Vane voltage [kV]	*148*	*280*
Vane length [mm]	*1378*	*746.1*
Cell number	*42.6*	*12.4*
Ave. aperture [mm]	*8*	*15.3*
Modulation	*1.2 - 3*	*3*
Stored energy [J]	2.1	3.6
Capacitance [pF/m]	142	135
Max. surface field [MV/m]	*25.5*	*25.5*
Max. magnetic field [Gauss]	249	241
Shunt impedance[MW] (Cu)	0.28	0.32
Power Loss [kW] (Cu)	79.6	133
Q value (Cu)	13200	13400
Field bump [%]	0.3	0.5

MECHANICAL CONSTRUCTION

Machining of parts

Fig.2 represents a quarter of SRFQ2 and it can help in the explanation of the resonator construction sequence.

All parts are machined independently and then joined by EBW. The prototype is made out of AISI-304 stainless steel, while the s.c. resonator is made out of Nb (RRR>250) and pure Ti for the stiffening cage.

The four parts of each stem (detail A in fig.1), the cylinders on the rear of electrodes (detail B in fig.1) and tank quarters (detail C in fig.1) are obtained by "drop hammer forming". The modulated vanes (detail D in fig.1) are milled with a specially machined tool (stainless steel: HSS+Co5%) the curvature of which corresponds to the transverse radius of curvature of the vane and the radial rake angle is ~ 30° for efficient removal of the shaving, while the optimum milling machine velocity was found to be 30 m/min.

The Ti external stiffening cage (detail E in fig.1) features a 6 mm thick reactor grade niobium (RRR > 30) insertion, electron-beam-welded on the tank side. The Nb

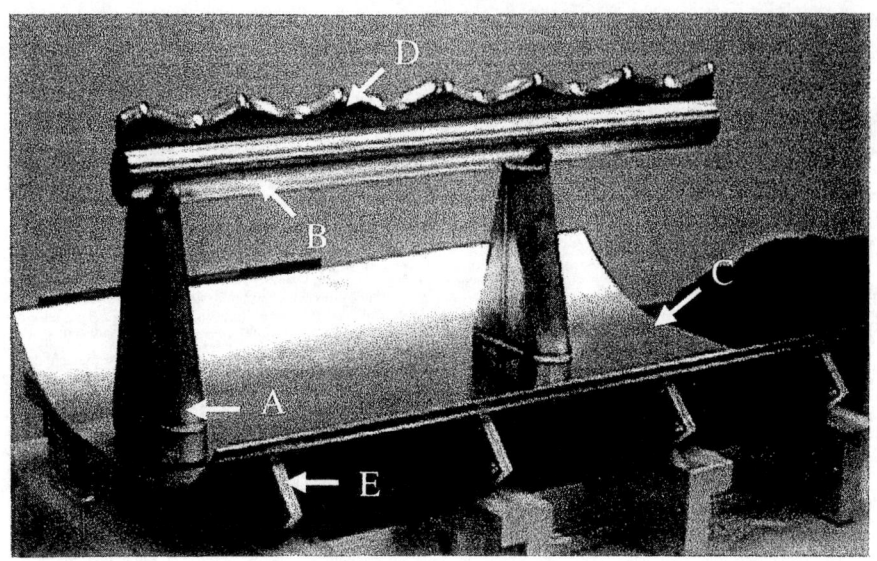

FIGURE 2. The picture of a quarter of SRFQ2, taken during assembly of the resonator, helps in following the description of the construction of the various parts (A. stem, B. cylinder on the rear of electrodes, C. quarter of tank, D. modulated vane, E. Ti stiffening cage) to which the text refers.

insertion is in turn tag-welded to the tank, thus avoiding Ti diffusion into the high RRR Nb of the tank itself ($T_{fusion,Ti}$ = 1672 °C vs $T_{fusion,Nb}$ =2468 °C). Ti was theoretically estimated to diffuse by only 0.2÷0.4 mm beyond the common EBW molten zone (~1 mm deep) (7), but the timescale of the project did not allow to verify this estimation experimentally and the reactor grade Nb insertion was thought to be a safe practical solution.

Titanium is the ideal material for the stiffening cage. Beside being very similar to Nb from the point of view of the overall thermal contraction between 300 and 4 K (thus minimizing mechanical deformations during cooldown), it features about the same Young modulus of Nb (E_{Ti} = 105 GPa vs E_{Nb} = 100÷110 GPa) but half its density (ρ_{Ti} = 4.5 g/cm^3 and ρ_{Nb} = 8.66 g/cm^3): this is particularly important since, for equivalent geometries, mechanical vibration eigenfrequencies scale with $(E/\rho)^{1/2}$.

Electron Beam Welding

The technology of EBW is the most appropriate and best investigated joining method of high RRR Nb parts, since the seam quality can be made excellent. Moreover the purity of the metal from metallic inclusions or carbon, oxygen, hydrogen and nitrogen contaminations is very well preserved.

The electron beam welding company (Zanon spa, Schio, Italy) uses a 150 kV electron gun with a maximum power of 30 kW. The gun voltage is kept constant and the beam current and the weld velocity are adjusted, according to the desired weld penetration depth. In our case, where all but one of the welds are performed from the inside of the RF surface, beam current and weld velocity are adjusted so that a penetration of 2.5-2.8 mm is obtained on the 3 mm thick Nb pieces: this gives good weld uniformity and a stiff joint; moreover we thus minimize the risk of voids or grooves along the seam, which would be present if the penetration were full. Depending on the particular geometry of the joint, the electron beam current ranges between 28.5 and 34 mA and the weld velocity is 12 mm/s.

Smoothness and uniformity of the joint largely depend on either the electron beam focusing parameters or the pattern and size of beam oscillation along the welding path. In our case an elliptical oscillation of the electron beam (a = 2.6 mm and b = 2.0 mm being the ellipse radii) at a frequency of ~9 kHz was found to give a very homogeneous weld while minimizing the natural fluctuations of the molten region, thus reducing both the voids along the seam and the number of sudden sprays of niobium, blown out of the weld. This method, mastered by Zanon SpA under the supervision of J. Brawley from T. Jefferson Laboratory and in collaboration with INFN-LNL, was found to be superior to simple beam defocusing. Greater care has only to be taken in switching on/off of the beam at the two extremes of the welding path: at the seam ends it is necessary to add Nb dummy pieces, where the beam current is increased/decreased in 10-20 mm to/from the nominal value.

For the stem-to-tank welds and the welds between tank quarters, a careful analysis of the material shrinkage during the weld had to be done, since the shrinkage extent (ranging form 0.1 to ~ 0.6 mm depending on the beam features) can influence the resonator final frequency significantly.

The resonator assembly procedure and in particular the rough tuning sequence, which consists in a stepwise consistent reduction of both the stem length and the tank perimeter, were extensively disussed in Ref. (8).

EXPERIMENTAL TESTS ON THE STEEL MODEL OF SRFQ2

Bead-pull measurements

The electromagnetic characterization of the steel model of SRFQ2 is described in Ref. (9) and we only summarize here the results for the sake of completeness.

Bead-pull mesurements revealed the degree of accuracy of the resonator alignment. They were taken with a 6 mm diameter plexiglas sphere at 45° and a distance of 4 mm from the beam axis, showing that voltage drops and bumps along the structure are within ±2% (fig. 3). The value is acceptable for beam dynamics. We believe to have understood the reasons why we did not meet the goal value of $|\Delta V/V| < 1\%$: an unpredicted shrinkage of the structure during the last TIG-welds of the external stiffening cage not only resulted in a final average position of the four electrodes by ~0.2 mm closer to the beam axis, but also worsened the transverse relative positioning precision of the electrodes themselves (below 0.1 mm during assembly and only within 0.2 mm after that operation).

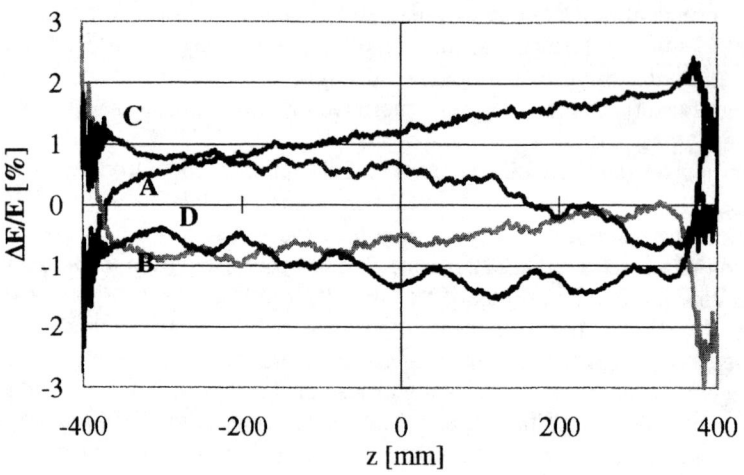

FIGURE 3. Result of the bead pull measurements taken along the structure in the four quadrants. Voltage drops are within ± 2% from −367 to + 367 mm, the vane length.

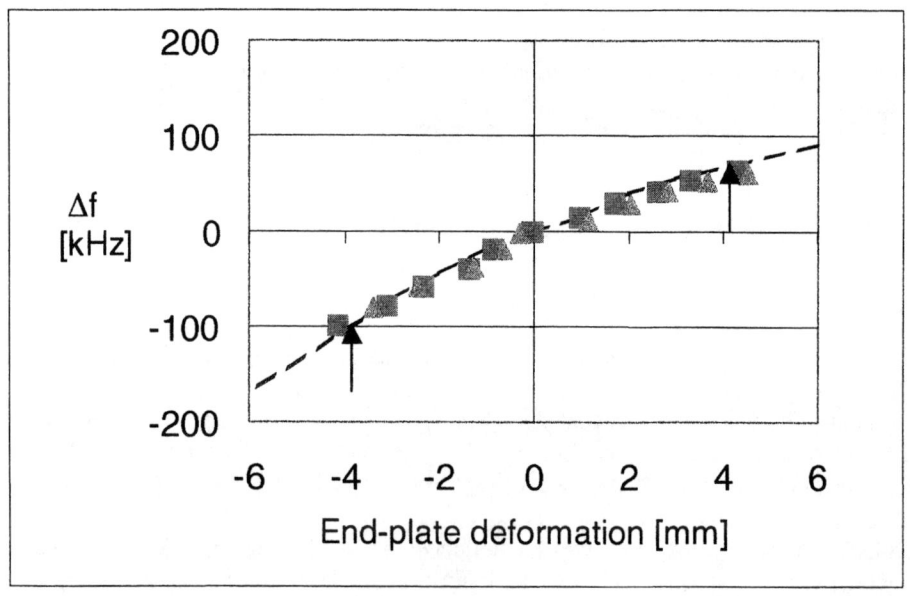

FIGURE 4. Measurement of the frequency range of one of the two fine tuners. The full fine tuning range is +100÷-200 kHz. The dotted line shows M.A.F.I.A. calculation result

Fine tuner

The whole fine tuning range, obtained by pushing/pulling both end-plates by ± 4 mm, fulfilled our expectation from the mechanics point of view: a +100÷-200 kHz range was measured, with a mechanical deformation of the Cu-based end-plates, which remains well within the elastic deformation regime of the material. The results are in excellent agreement with MAFIA predictions (9) (fig. 4).

Electro-mechanical stability

Large part of the structure design effort was devoted to find a good compromise between the electromagnetic features of the resonator and its mechanical stiffness. The latter was obtained primarily by the trapezoidal shape of the stems and the welding of a proper stiffening cage outside the resonator tank.

We assumed that, in order not to be endangered by the most common sources of mechanical excitation (e.g. vacuum pumps, cryogenic system), a safe lower threshold of mechanical eigenfrequency values for our s.c. resonator is 100 Hz (10).

The experimental spectrum of the mechanical eigenfrequencies obtained on the steel model of SRFQ2 shows that the lower eigenfrequency mode is found at $f_{1,exp}$=130 Hz (vs. $f_{1,th}$=151.2 Hz given by the numerical model obtained by means of the code I-DEAS (11)). This result is expected to translate into $f_{1,exp}$~120 Hz, by scaling Young modulus and density of the materials, for the being built Nb resonator stiffened with a Ti cage. Figure 5 shows the pattern of mechanical vibration of this mode.

Although the structure is reasonably stiff, it is however safe to foresee the employment of a fast tuning device, which proved extremely convenient in large-sized full Nb structures so far (12). In the case of SRFQ2 we shall adapt a spare VCX fast tuner, borrowed by Argonne National Laboratory (6), to our resonant frequency and to the frequency range to be controlled.

Slow drifts of the electromagnetic eigenfrequency with the liquid helium pressure (~10÷100 Hz/min) must be compensated by putting the fine tuner in a frequency feedback circuit, capable of compensating slow variations caused by drifts in the liquid He bath pressure. A system similar to the one we need was setup for ALPI full Nb QWRs (5). In the case of QWRs the frequency sensitivity to pressure variations is ~ 1 Hz/mbar, while 10 Hz/mbar is the value experimentally estimated for SRFQ2 within a maximum pressure fluctuation range $|\Delta P|$ < 50 mbar (9); on the other hand the frequency sensitivity of the SRFQ fine tuner is 10 times larger (37 Hz/μm vs. 4 Hz/μm). Hence, assuming that the mechanical features of the two tuners are comparable, they can cope with comparable pressure rates (the SRFQ tuner mechanics seems superior, but a response will only be given by the first test with the s.c. cavity). Recent results with QWRs show that their fine tuner feedback system can withstand pressure change rates as high as 5-10 mbar/min at values of electric peak field E_p and stored energy U which are consistent with those specified for SRFQ operation (13): we specified the cryogenics requirements of SRFQs consequently.

FIGURE 5. View of the oscillation pattern of the lowest mode of mechanical vibration as calculated by I-DEAS (amplitude is deliberately exaggerated).

Vacuum tests

After completion and before checking the resonant frequency change in liquid nitrogen, the SRFQ vacuum tightness was tested to be better than 1.2×10^{-9} mbar l/s. It was also experimentally verified (9) that the maximum deformation of longitudinal ribs in vacuum, amounting to -18 μm on average, is responsible for a $\Delta f_{res} = -15.7$ kHz, which is not very large compared to the fine tuner range, but must be taken into account in the rough tuning of the s.c. version.

Cold tests

Although the whole SRFQ2 model is made out of a single material (stainless steel), the structure $\Delta f/f$ during cooldown had to be tested for two basic reasons. First of all the steel parts of the SRFQ2 model were never annealed, and were consequently suspected of possible deformations at cryogenic temperatures. At the time when the cavity was conceived we were in fact not sure whether our vacuum furnace were capable of preserving the high RRR of Nb. Second, the large number of EBW joints could have also introduced stresses in the material which could result in mechanical deformation during cooldown.

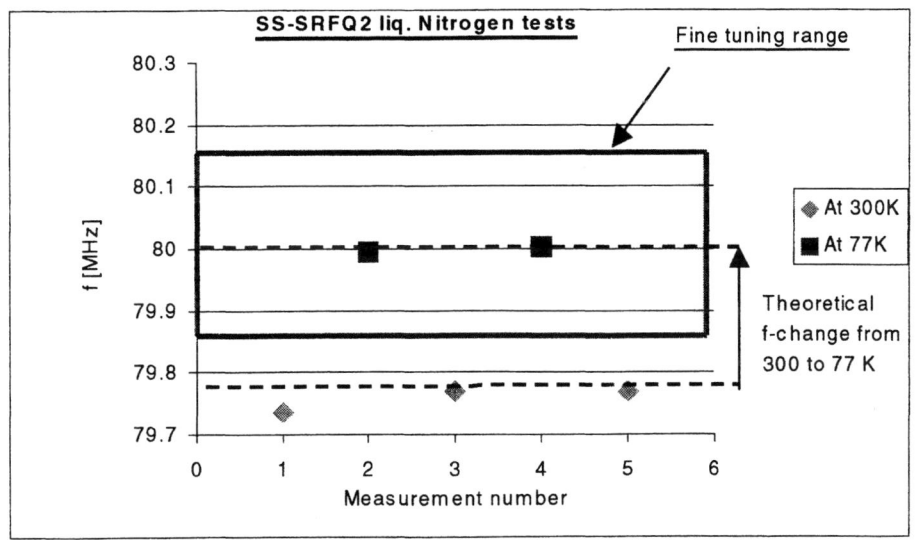

FIGURE 6. Result of two complete cooldown and warming up cycles of the steel model of SRFQ2: the expected frequency change is exceeded by 11% in the first cooldown, whereas it is respected within < 0.5 % in the following ones.

The cooldown test was arranged with a dummy basin, filled with liquid nitrogen, housing the resonator; the basin was located in the external tank of the being built SRFQ test cryostat and thermally insulated from it. The thermal contraction coefficient of stainless steel between 300 and 77 K is $\Delta L/L = 2.9 \times 10^{-3}$, only 5 % smaller than between 300 and 4 K: cooling down to liquid nitrogen temperature was hence judged to be sufficient as a test. The structure went through two complete cooldown cycles, the results of which are shown in fig. 6. It can be seen that, while the frequency increase during the first cooldown exceeds the theoretical prediction,

$$\frac{\Delta f}{f} = \alpha_{300-77K,SS}$$

($\alpha_{300-77K,SS}$ being the thermal contraction of stainless steel between 300 and 77 K)

by about 11%, subsequent warming up and cooling down cycles follow the theoretical prediction with an accuracy exceeding 0.5%.

The initial variation from the expected frequency change, which is anyhow much smaller than the fine tuning range of the structure, can be possibly ascribed to a stress relief of the material itself: this topic will be the subject of future investigations.

A collaboration with W. Singer of DESY (Hamburg,D) allowed us meanwhile to anneal Nb electrodes and stems in their furnace, which is fully devoted to thermal treatment of chemically etched high RRR niobium and hence extremely clean and reliable, whereas the components of the external stiffening cage were heat-trated in the INFN-LNL vacuum furnace. Hence, despite the difference between the contraction factors of Ti and Nb in the 300 - 4 K range (only 5 %), we are confident that the frequency change of the s.c. resonator will follow the theoretical prediction with more than sufficient accuracy.

Chemical etching

Chemical etching of the full Nb structure is foreseen to take place at CERN, in the frame of a collaboration contract between CERN and INFN-LNL.

Before structure assembly all joints are etched by 5-10 μm at Zanon spa before electron beam welding, but a deeper etching would substantially decrease the surface resistivity of the resonator and improve its performances: an optimum is expected to be reached at 100 μm material removal, but 40 μm are not too far from it (14).

At the end of the rough tuning procedure, before welding the stems to the tank, all electrode plus stems will be etched by 60 μm. Then the assembly will continue and the final frequency rough-tuned so that another 20-40 μm can be removed from the whole internal surface of the SRFQ after its completion. In such a way, 80÷100 μm will be eventually removed from the regions of higher E_p and H_p (electrodes and stems) and 40 μm elsewhere.

The steel model has already been used to test the etching procedure at CERN, where it has been seen in particular that:

- filling and evacuation time of the whole resonator takes 5÷10 s, appropriately negligible with respect to the treatment time (20÷40 min);
- beside the typical de-ionized (DI) water rinsing, high pressure rinsing (HPR) of the complex internal geometry of the resonator was optimised and the 100 bar DI water jets will be made directly impinge the whole resonator surface.

STATUS OF SRFQ1

SRFQ1, the first s.c. resonator on PIAVE beam line, was designed and will be built second, both because it is twice as long (and hence would have required a significantly larger construction time if built first) and because its stored energy U = 2.1 J is nearly twice smaller, which makes the resonator less sensitive to mechanical vibrations (15): SRFQ2 was hence considered more appropriate as a proof of principle than SRFQ1.

Despite the transverse resonator design was kept as similar as possible to SRFQ2 to take advantage of the technological steps made with that resonator (same stems and same cylinders on the rear of electrodes, only 1÷2 % smaller tank size), an exact estimation of the resonator size could only be determined through the development of a model (fig.7), since the size of the RFQ modulation was too small to be input into the MAFIA simulation.

FIGURE 7. Photo of the aluminum model of SRFQ1.

The development and the results of the model, built half-scale in Al, are thoroughly discussed in Ref. (16). Besides the determination of the correct size of the resonator, alignment tolerances of each electrode on the cartesian axes (± 0.1 mm) were shown to keep electric field unbalances along the structure within ± 1%.

No field drop was observed (consistently with MAFIA predictions) at the point where electrode halves are split and short-circuited. Splitting of the long electrodes is necessary, in order not to have to use a lerger size EBW chamber: this small complication translates into an additional, but certainly minor, step in the resonator assembly procedure.

CONCLUSION

The tests with the stainless steel model of SRFQ2 showed that:

- a ±0.2 mm alignment tolerances of electrode keeps voltage unbalances among the four quadrants along the structure within ± 2% (care will be taken to reduce the tolerances to ±0.1 mm in the s.c. structure so that this value reduces to ± 1%);
- the fine tuning range is large (-200 ÷ 100 kHz) and is in good agreement with the numerical estimation;
- the structure features a sufficient mechanical stiffness, with respect to both the vibration spectrum and the deformations under vacuum;

- cooldown of the structure to 77 K changed the electro-magnetic resonant frequency with deviations from the theoretical predictions which are much smaller than the resonator tuning range;
- the geometry was nicely adapted to the chemical etching apparatus of CERN and high pressure rinsing was optimised.

We are presently ready to start the rough tuning procedure with the Nb resonator, aiming at a room temperature frequency of 79.8 MHz, which takes into account the subsequent frequency increases given both by the final 20-40 μm material removal with the chemical etching (~ 60-120 kHz) and by the final cooldown to liquid helium temperature (~ 100 kHz). The s.c. resonator is expected to be ready for tests at 4 K by spring 1999.

ACKNOWLEDGMENTS

We acknowledge fruitful discussions with A. Facco on the frequency control versus mechanical vibrations and slow He pressure drifts and we thank L. Bertazzo, F. Stivanello and N. Dainese for their technical assistance.

REFERENCES

1. A. Pisent et al., Proc. of the 6th European Particle Accelerator Conference, 22-26 June 1998, Stockholm, 758
2. G. Bisoffi, G. Algise and A. Lombardi, Proc. of the 7th Workshop on RF Superconductivity, Gif-sur-Yvette, October 1995, 677
3. V.A. Andreev, G. Parisi, Proc. of the 1993 Particle Accelateron Conference Proc., Washington, DC., May 1993, 3124.
4. M. Bartsch et al., Computer Physics Communications 72 (1992) 22-39
5. A. Facco, V. Zviagintsev, S. Canella, A.M. Porcellato and F. Scarpa, Proc. of the Sixth European Particle Accelerator Conference, 22-26 June 1998, Stockholm, 1846
6. N. Added, B.E. Clifft, K.W. Shepard, Proc. of the 16th Int. Linear Accelerator Conference, 1992, Ottawa, Ontario, 181
7. C. Antoine and H. Safa, C.E.A. Saclay, DIST 91191 - Gif-sur-Yvette Cédex, private communication
8. G. Bisoffi et al., Proc. of the 1997 Particle Accelerator Conference, Vancouver, B.C., Canada, 12-16 May 1997, 1087.
9. G. Bisoffi et al., Proc. of the Sixth European Particle Accelerator Conference, 22-26 June 1998, Stockholm, 1840
10. D.L. Schrage, E. Swensen and B. Rusnak, Proc. of the 7th Workshop on RF Superconductivity, Gif-sur-Yvette, October 1995, 629
11. I-DEAS Finite Element Modelling, Structural Dynamics Research Corporation, 2000 Eastman Drive, Milford, OHIO 45150, USA
12. K.W. Shepard, Nuclear Instruments and Methods A382 (1996) 125
13. A. Facco, INFN-LNL, via Romea 4 I 35020 Legnaro – Italy, private communication
14. P. Kneisel and B. Lewis, Proc. of the 7th Workshop on RF Superconductivity, Gif-sur-Yvette, October 1995, 311
15. J.R. Delayen, G.J. Dick and J.E. Mercereau, IEE transactions on Nuclear Science, vol.35, 177
16. V. Andreev et al. "Scale room temperature model of the superconducting RFQ1 for the PIAVE linac", Proc. of the 1998 LINAC Conference, 24-28 August 1998, Chicago (IL, USA), in press

The Non-RFQ Resonators of the PIAVE Linac

A. Facco, F. Scarpa and V. Zviagintsev

INFN-Laboratori Nazionali di Legnaro, via Romea 4, I35020 Legnaro (PD), Italy

Abstract. The PIAVE superconducting linac is not only made of rfqs; 8 low β superconducting quarter wave resonators, working at 80 MHz, will give an equivalent acceleration voltage of 7.2 MV; a room temperature triple harmonic buncher working at 40 MHz will match the beam to the rfq input and 2 room temperature, 80 MHz rebunchers will match the beam produced by PIAVE into the linac ALPI. The construction and testing of the first superconducting cavity was more than successful: the resonator, at the design power of 7W, reached 6.3 MV/m corresponding to a total acceleration voltage of 1.13 MV at β = 0.047. This results allowed us to increase the linac design accelerating field from 3 MV/m to 5 MV/m. The 7 remaining superconducting cavities are presently under construction, and they will be completed within 1998. While the more conventional, room temperature 80 MHz rebunchers are at the design stage, the triple harmonic buncher, made of 2 quarter wave resonators, was constructed, tested and it is ready for installation.

INTRODUCTION

The goal of the PIAVE project [1,3] is the acceleration of ions produced with $m/q \leq 8.5$ by an ECR source located on a 320 KV platform. PIAVE will replace the LNL XTU Tandem as an injector for the superconducting linac ALPI [2], with significant improvements in terms of beam intensity and ion charge state. The input ion velocity is $\beta=0.0089$; the output velocity must be higher than to $\beta \geq 0.045$ for an efficient injection in ALPI. The first acceleration to $\beta=0.035$ is performed by two, 80 MHz superconducting rfq's, which must work with a fixed beam velocity profile. The second part of the acceleration is produced by 8, independently phased quarter wave resonators; in this case the beam velocity profile can be conveniently modified for different beams, the only limitations being the maximum acceptable rf defocusing [3]. The required accelerating field is $E_a=3$ MV/m; higher fields, however, can be used in the QWR section of PIAVE to further increase the final beam energy.

The longitudinal matching from the HV platform into the first rfq is performed by a room temperature, triple harmonic buncher working at 40, 80 and 120 MHz. The buncher is located in a transverse beam waist, 3.4 meters upstream the rfq input port; the calculated bunching efficiency is 68% and the longitudinal beam emittance is 3.75 MeV deg [4].

Two 80 MHz, room temperature quarter wave resonators will perform the longitudinal matching between PIAVE and ALPI. The wide range of possible output beam velocity prevents us from using mutigap structures, characterized by very high shunt impedance but narrow velocity acceptance; 2-gap structures with $\Delta W \geq 180$ kV acceleration voltage at $0.045 \geq \beta \geq 0.055$ are then required [1].

CP473, *Heavy Ion Accelerator Technology: Eighth International Conference,*
edited by Kenneth W. Shepard
© 1999 The American Institute of Physics 1-56396-806-1/99/$15.00

THE LOW BETA SUPERCONDUCTING QUARTER WAVE RESONATORS

The design of this bulk niobium cavity was obtained starting from the ALPI, β=0.055 resonators [5], the main difference being the shape of the drift tube region, where the gap-to-gap distance has been changed from 100 to 80 mm to bring the optimum ion velocity down to the value of β=0.047 (see fig.1). The main characteristics of the cavity are listed in tab.1 [6,7]. The 8 cavities will be installed in two cryostats, similar to the ALPI low beta ones except for the outer shields cooling, which will be obtained by means of liquid nitrogen instead of helium gas.

The resonators are made of full niobium; the inner surface of their outer shields, facing the outer conductors, are covered by 0.5 mm thick titanium sheets in order to simplify the operations in case of high temperature thermal treatment with the titanium sublimation technique. The tuning plates, like in the ALPI low β cavities, are made of lead plated OFHC copper. The resonators are equipped by the LNL mechanical dampers [7,8] in order to prevent high amplitude resonant oscillations of the 92 cm long inner conductors. The surface treatment consists of chemical polishing and high pressure water rinsing, followed by a final ethanol rinse: this relatively simple procedure was found to be, in the ALPI bulk niobium cavities, the best procedure to suppress electron loading at very high field.

TABLE 1. The PIAVE low β quarter wave resonator parameters.

	symbol	theory	experiment	units
effective length	l	0.18	-	m
Frequency	f	80	≈ 80	MHz
Optimum velocity	β	0.047	0.047	
Transit time factor	T_0	0.88	0.87	
Stored Energy	U/E_a^2	0.129	0.134	$J/(MV/m)^2$
Peak magnetic field	H_p/E_a	≈ 100		$G/(MV/m)$
Peak electric field	E_p/E_a	≈ 5		
Shunt resistance	R_{sh}'	17.7		$M\Omega/m$
Geometrical factor	Γ	15.4		Ω

The first β=0.047 cavity was constructed and successfully tested during 1997; no high pressure water rinse was done before the first rf test in superconductive regime. In comparison to what observed in the ALPI cavities, the response to external mechanical vibrations and to helium pressure variations did not change significantly; moreover, the changes in the rf geometry did not introduce any new multipactoring levels. The rf test results are shown in fig. 3: the low power quality factor exceeded 1.6×10^9 and the resonator could reach 6.3 MV/m accelerating field (equivalent to 1.13 MeV/q energy gain) at the design power dissipation of 7W. At the field of 3 MV/m required by PIAVE the power dissipation was below 1 W. The cavity, in spite of the less favourable geometry dictated by the necessity of accelerating low velocity ions, behaved as the ALPI type niobium cavities and exceeded the design requirements by far; further improvements could be expected after high pressure rinsing. After this test we started the construction of the remaining 7 cavities without any modifications in the design; the completion is expected within 1998.

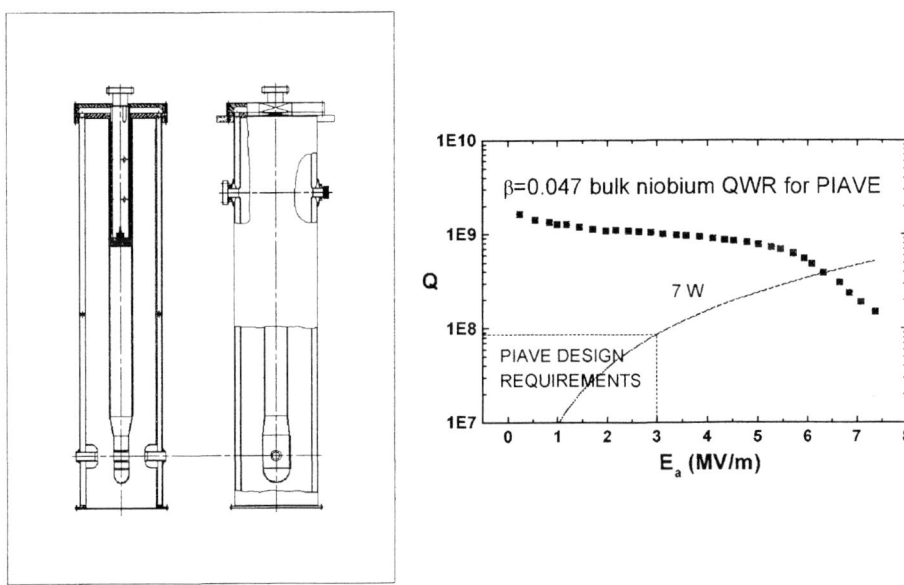

FIGURE 1. Schematic of the 80 MHz superconducting resonator of PIAVE.

FIGURE 2. Q-curve of the first $\beta=0.047$ cavity.

The test results, together with the data collected with the $\beta=0.055$ cavities, prompted us to increase from 3 to 5 MV/m the design accelerating field in the QWR section of PIAVE, changing from 0.045 to 0.052 the output velocity of the beam. Each of the 8 cavities will produce 0.9 MV acceleration voltage, giving a total of about 7 MV with two cryostats; moreover, working at more than 1 MV/cavity without exceeding the design power limits appears also feasible.

THE TRIPLE HARMONIC BUNCHER

The longitudinal beam matching at the input port of the superconducting rfq1 of the PIAVE project [1] requires a 40 MHz buncher with very high efficiency and very

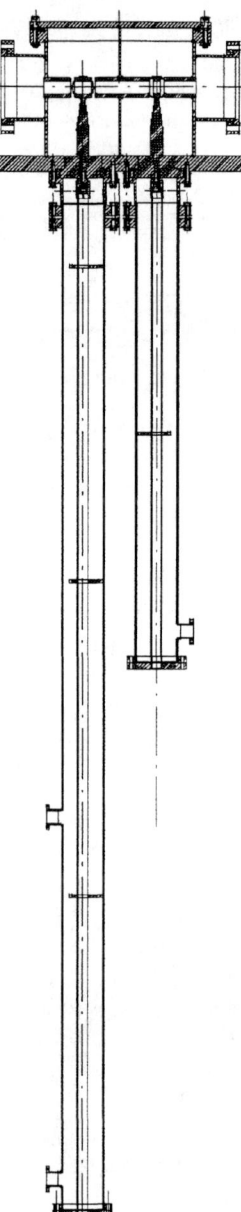

low emittance growth. These requirements were met in the design of a three harmonics buncher with two double gap (fig.3) [4,9].

This unusual structure has some features which cannot be obtained simultaneously by double-drift bunchers or by single gap, gridded ones: the short spacing and the location of the buncher in a transverse beam waist allow us to avoid non-linearities due to the radial dependance of the transit time factor in accelerating gaps; the high number of harmonics and the absence of grids allow us to obtain a high efficiency (about 68 %) and a good output emittance (3.75 *Mev deg*). The 40 MHz and the 120 MHz harmonics are applied to the same drift tube (they have practically the same transit time factor in a double gap structure) and they are generated by the same, 1.8 m long, coaxial resonator as $\lambda/4$ and $3\lambda/4$ modes. The 80 MHz harmonic needs a second, 0.9 m long quarter wave resonator and a second drift tube.

The velocity of the beam, produced by a ECR source and accelerated by a high voltage platform, will be $\beta=0.0089$ at the buncher, and the optimum lengths of the 40-120 MHz and 80 MHz drift tubes are 34 mm and 17 mm, respectively. The center-to-center spacing of the drift tubes is 120 mm, fitting well the length of the transverse waist at our disposal; the bore radius is 10 mm. The resonators are in air, easily removable for frequency adjustment, connected to the drift tube stem through vacuum tight teflon rf feedthroughs.

FIGURE 3. Schematic of the triple harmonic buncher.

TABLE 2. Triple harmonic buncher measured characteristics.

rf characteristics					
resonator length (cm)	f (MHz)	Qo	ΔW/q (kV)	P (W)	TTF
180.6	40±0.068	1330	3900	34	0.9@β=0.01
	120±0.126	2399	380	2.2	0.64@β=0.01
93.6	80±0.117	1545	1400	10.4	0.72@β=0.01
transmission coefficients between different ports					
signal source	Tpu 40 (dB)	Tpu 120 (dB)	Tc 40 (dB)	Tc 120 (dB)	
40 MHz	-28.5	-32.1	-	-13.1	
120 MHz	-42.9	-19.3	-41.4	-	

The frequency and voltage stabilization of two different harmonics in the same resonator are obtained by taking advantage of the different field distribution of the two modes, and properly locating tuners, couplers and pickups. The modes are powered independently by three solid state amplifiers, and the rf control is performed by means of standard ALPI rf controllers.

The buncher construction was completed during 1997, and then the buncher was tested off-line [10]. The results are shown in tab.2. Matching the couplers and the pickups in the two harmonics resonator required many steps. The final result is a sufficient decoupling of the 40 and 120 MHz modes, except for the transmission coefficient from the 40 MHz coupler to the 120 MHz one; this could not be kept below 13 dB. This will cause the dissipation of about 2W of the first harmonic through the third harmonic coupler; and the necessity of mounting on it a 10 dB attenuator to prevent cross-talk between the two rf amplifiers. The tuning is obtained by means of two capacitive tuners: the first one acts in opposite directions on the two frequencies and it is used to find the proper 1/3 ratio; the second one acts simultaneously on the two frequencies preserving the chosen ratio.

All other design requirements have been satisfied and the total power consumption of the buncher in operation is below 50 W. The buncher and related electronics is ready for installation in the beam line.

THE HIGH ENERGY REBUNCHERS

The two 80 MHz rebunchers, located at 3.5 m and 17.2 m downstream from PIAVE are presently being designed. Recent upgrading of the PIAVE beam dynamics [1] have put higher field requirements to these resonators with respect to the original design: the beam velocity is not anymore fixed to 0.045 but it can raised up to 0.055 and more; the cavities must provide a maximum field above 180 kV, to be increased according to the transit time factor at different velocities (see tab.3 and fig.4-5). The space available for each resonator is about 0.4 m. The preliminary calculations, performed by means of

TABLE 3. Energy parameters of the rebunchers.

Element	E_a (QWR)	$2 \cdot V_g \cdot TTF$	β	TTF	V_g	P
	MV/m	kV			kV	kW
Uranium 28/238	3	120	0.045	0.795	75	1.99
Uranium 28/238	5	180	0.052	0.902	100	3.50
Argon 14/39.9	5	180	0.074	0.910	99	3.43

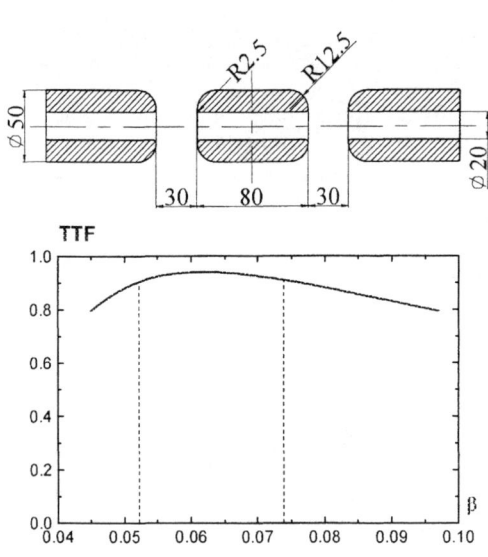

FIGURE 4. Geometry of the rebuncher drift tubes and TTF calculation.

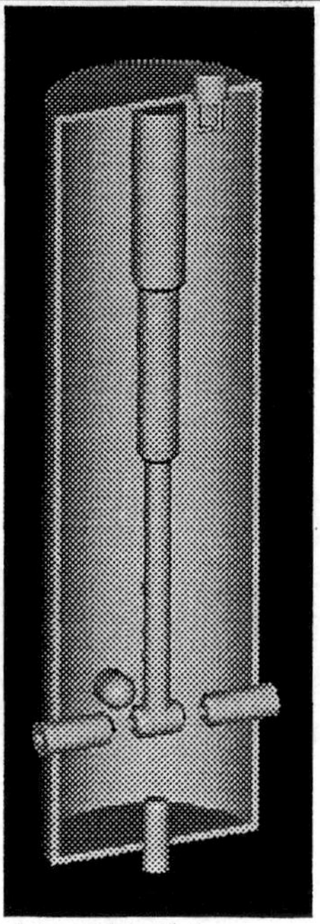

FIGURE 5. Schematic of the rebuncher

TABLE 4. Parameters of the rebunchers (SUPERFISH/POISSON calculations).

Operation frequency	f	MHz	80
Shunt impedance	$Rsh = V_g^2/(2 \cdot P)$	MΩ	1.43
Quality factor	Q_o		10500
Voltage on the gap	V_g	kV	100
RF power	P	kW	3.5
Peak and Kilpatrik limit ratio	E_p/E_k		0.58
Diameter of the cavity	D	m	0.3
Length of the cavity	L	m	1.1

the SUPERFISH and POISSON programs [11], have shown that room temperature, two-gap quarter wave resonators can fulfill the requirements with a power consumption of 3.5 kW (see fig.3 and tab.4).

The resonator construction should be completed within 1999.

CONCLUSION

The PIAVE triple harmonic buncher have been constructed and successfully tested off-line, and it is ready for installation. The design of the two PIAVE 80 MHz, room temperature rebunchers with $\Delta W \geq 180$ kV are now in an advanced stage. The rf test of the first $\beta=0.047$ superconducting quarter wave resonator have shown that this cavity can reach 6.3 MV/m at 7 W; further increase of performance can be expected after high pressure water rinse. Consequently, the original PIAVE design goal of 3 MV/m have been changed to 5 MV/m, with a significant upgrading of the linac output beam energy. The production of the 7 remaining cavities has started in summer 1998.

REFERENCES

1. A. Pisent et al., "The new LNL injector PIAVE, based on a superconducting RFQ," in Proceedings of the Sixth European Particle Accelerator Conference, Stockholm, Sweden, 22-26 June, 1998, pp. 758-760.
2. G. Fortuna et al., *Nucl. Instr. and Meth.* **A287**, 253 (1990).
3. A. Lombardi et al., "The new positive ion injector PIAVE at LNL," in Proceedings of the 1997 Particle Accelerator Conference, Vancouver, B.C., Canada, 12-16 May, 1997, pp. 1129-1131.
4. M. Comunian and A. Pisent, "LEBT and three harmonics buncher design for PIAVE", INFN- Laboratori Nazionali di Legnaro, Annual Report 1996, pp. 296-297.
5. A. Facco and J.S. Sokolowski, in Proceedings of the 4-th European Particle Accelerator Conference, Ottawa, Ontario, Canada, 1994, p. 2054.
6. A. Facco et al., "Rf testing of the first $\beta=0.047$ bulk niobium cavity for PIAVE", INFN- Laboratori Nazionali di Legnaro, Annual Report 1997, pp. 220-221.
7. A. Facco and V. Zviagintsev, "Mechanical stabilisation of superconducting quarter wave resonators," in Proceedings of the 1997 Particle Accelerator Conference, Vancouver, B.C., Canada, 12-16 May, 1997.
8. A. Facco et al., "New results of the LNL mechanical mode damper for low β cavities", INFN- Laboratori Nazionali di Legnaro, Annual Report 1997, pp. 222-223.
9. A. Facco and F. Scarpa, "The new triple harmonic buncher for the PIAVE project", INFN- Laboratori Nazionali di Legnaro, Annual Report 1996, pp. 298-299.
10. A. Facco et al., "Testing of the PIAVE triple harmonic buncher," INFN- Laboratori Nazionali di Legnaro, Annual Report 1997, pp. 218-219.
11. K. Halbach and R.F. Holsinger, "Superfish – A computer program," *Part. Accel.* **7(4)**, 213-222 (1976).

The Lead-Copper Option for Heavy Ion Accelerating Cavities: History, Status, and Future Prospects

John W. Noé

Department of Physics and Astronomy
University at Stony Brook
Stony Brook, New York 11794-3800 USA

Abstract. Superconducting radio-frequency (SRF) cavities have been used to accelerate heavy ion beams for the last quarter century. Over this period devices based on pure lead or a lead-tin alloy electrodeposited on to a copper structure have been a viable alternative to solid niobium cavities, with many comparative advantages such as lower unit cost, ease of re-coating and better thermal stability. We review the historical development, current status and future prospects of the lead-based SRF technology in the context of recent advances in solid and thin-film niobium.

I INTRODUCTION

There has been no general review of the lead-copper superconducting radio-frequency (srf) technology for heavy-ion accelerators since that by Delayen at the 1987 SRF Workshop [1]. While the information on srf properties of lead summarized then is still current, in the mean time there have been some significant advances in practical techniques and achievable performance for both lead-based and competing niobium technologies which justify a new discussion. The present review gives an overview of the historical development of the field and then offers some opinions about relevant factors, recent developments, and future prospects.

Of course the considerable success of heavy-ion booster linacs over the last two decades is due not only to there being practical and effective methods for creating the superconducting *surfaces*, both lead-based and niobium, but even more importantly, to the development of well-optimized accelerating *structures*. The history of this structure aspect has been thoroughly reviewed by Shepard at the last HIAT conference [2] and by many others. We will not emphasize it here other than to point out that the availability of the relatively simple and economical Pb-Cu electroplating technology greatly facilitated the development of the new structures.

Reviewing such a large field with such a long history and so many detailed variations in the techniques necessarily involves some difficult and subjective choices. In section II we have tried to mention every significant project and historical development, emphasizing what seem to be the most significant details. In section III we have attempted to draw some conclusions about the most relevant factors in srf electroplating from the various reports and personal experience.

It is of course not possible to mention every publication connected with each project, or even every relevant publication. Fortunately a limited number of conference proceedings contain most of the sequential status reports from the major projects of the last twenty years, as well as most of the various reviews of new developments. These include the proceedings of the current and previous HIAT conferences (NIM) and of the eight Workshops on RF Superconductivity, reports from the annual SNEAP meetings (World Scientific and RSI), and proceedings of Particle Accelerator Conferences (IEEE), from about 1975-85. The linac booster overview by Weisser and Lobanov [3] is an excellent source of information and references on the latest developments, particularly in niobium surfaces. Also, while it is outside the scope of this review, it should not be forgotten that lead-copper rf superconductivity has many non-accelerator applications. Some early examples are given in a review paper by Septier and Viet [4].

In the presentation of this paper at the HIAT conference a number of illustrations were shown. While none of these are included here most of them are available on the 'Stony Brook SRF Homepage' at http://resonator.physics.sunysb.edu.

II HISTORICAL OVERVIEW

It is convenient to describe the history of the lead-copper heavy-ion accelerator technology in terms of the following arbitrary 'technological stages.' All have involved the electrodeposition of micron thick lead or lead-alloy films on copper cavities, with many variations in the detailed techniques.:

A. **Basic research** on the rf superconducting properties of lead (various laboratories, and mostly from about 1960 to 1975);

B. **Applied research** on more practical and effective rf structures and lead surfaces (mostly at Caltech 1970-1975);

C. **Development and construction** of the first lead-copper superconducting booster linac (Stony Brook/Caltech collaboration, 1975-1983);

D. **Development and implementation** of the superconducting quarter-wave resonator (Stony Brook, Weizmann Institute, University of Washington, INFN-Legnaro, Tata Institute, 1981-1987);

E. **Further work** in novel structures and better surface preparation methods (Stony Brook, Oxford, Legnaro, Munich, 1980-).

A Basic research

There was a great revival of interest in superconductivity in the 1960's stimulated by the new fundamental theories [5,6]. The mood of that time was similar to the 'Woodstock' enthusiasm for high-Tc superconductivity that dates from March 1987, with hundreds of new experiments carried out in a few years. New ideas about efficiently producing high rf electric fields for accelerator and deflector applications soon evolved into feasibility studies at various research centers (Stanford, Karlsruhe, BNL, ORNL, Rutherford). Initially interest was concentrated on accelerating electrons or protons, and it was only after about 1970 that an equally strong enthusiasm for superconducting cavities for heavy ions took shape.

The various lead srf studies from this pioneering 1965–75 era have been tabulated by Delayen [1]. Typically the quality factor Q of some appropriate test cavity was measured at very low power as a function of temperature, frequency (rf mode), and possibly the ambient magnetic field. The average surface resistance R_s is then given by Γ/Q, where the geometrical factor Γ is measure of the magnetic field distribution that can be calculated or determined from the room temperature Q.

The key result is that R_s can invariably be expressed as the sum of the theoretically predictable component R_{BCS}, which decreases exponentially with falling temperature T, and a 'residual resistivity' R_{res} which is *practically independent of temperature*. The various experiments confirm that R_{BCS} scales with frequency as $\omega^{\simeq 1.75}$ as expected from the BCS theory and has about the expected magnitude. R_{res} generally increases with increasing frequency as well, but not in a predictable or consistent way from one experiment to another.

Several studies from 1965–75 and two from ~1985 are described next. When considering their results it should be kept in mind that some, but not all, test cavities are quite different (in size, mode, joints, field level, etc.) from practical accelerator cavities. In particular, the TE_{011} mode has zero surface electric field.

Some specific research programs

- John Pierce, a student in the Fairbanks group at Stanford, worked primarily at 10 GHz in a tiny (2×2 cm) cylindrical TE_{011} mode X-band cavity [7]. By carefully excluding ambient fields ($H < 0.1$ mG) he achieved $R_{res} \simeq 30$ nΩ – a remarkably low value for such a high frequency. Pierce noted that both thermoelectrically-induced H fields and poor vacuum can also cause residual losses.

- Similar experiments at S-band (2.9 GHz) were carried out by Hahn, Halama and Foster (HHF) at BNL in 1967–68, with the result $R_{res}= 50$ nΩ [8]. One interesting feature of their experiment was the use of relatively thin lead film (2.5 μm) to reduce the electronic mean free path l to a value ($l \simeq 0.59$ μm, estimated from the measured RRR value) believed to be optimum [18].

- Bruynseraede et al. at CERN also worked with an S-band TE_{011} cavity [9]. They tested Pb, In, and Pb-In alloy surfaces at both low and high rf magnetic fields. Their R_{res} for Pb is about double that reached by HHF.

- A very careful and well-documented study in a larger and more complex structure intended for actual acceleration (a 137 MHz helix) was carried out at Oak Ridge (ORNL) as part of an investigation into alternative designs for a new heavy-ion facility [10].[1] Procedures and some first results are described in a 1972 paper [11], while final measurements and analysis are reported in a dissertation by Judish [12] and a 1977 paper [13]. In this final data R_{res} of only 4.7 nΩ was recorded at very low field (energy content $U = 10^{-6}$ joule). This is the lowest R_{res} ever directly observed in lead.

- There were a number of important lead srf studies at the Karlsruhe Nuclear Research Center (KfK) between about 1965–75, both experimental (by thesis students Flécher and Szecsi and collaborators) and theoretical (by Halbritter):

  - Szecsi measured $R_{res}(\omega)$ from 375 MHz to 5 GHz in a novel 'half-wave' coaxial resonator that resembled two 20 cm long × 2.9 cm diameter QWR's placed mouth-to-mouth, with inner and outer conductors joined [14]. It was possible to make these current-carrying joints (which used Pb-to-Pb knife edges [15]) with low enough losses that differences between the even and odd harmonic modes were essentially negligible [14]. The same frequency dependence of $\omega^{1.73}$ was found for both R_{BCS} and R_{res}.

  - The work of Flécher et al. [16,17] broke new ground in a number of areas: surfaces were prepared by electroless deposition and evaporation as well as by electroplating; surfaces were over-coated with indium and made from lead-bismuth alloys; losses were studied at high field levels as well as low. Pillbox cavities operating in both TE and TM modes at 2–4 GHz were used, as well as two types of 750 MHz coaxial resonators. The first of these was a $\lambda/2$ design similar to Szecsi's, which could operate up to magnetic breakdown at $H_p = 660$ Gauss (at which point E_p was 20 MV/m) [17]. The second or λ type coaxial resonator had a gap in the inner conductor to produce enhanced electric fields. It ran at peak electric fields up to the onset of field emission at 30 MV/m. In both resonators Q declined by a factor of 3–5 between low-field and the highest fields.

    In the experiments on alternative coating methods good results were achieved with a fresh electroless solution and when evaporating in UHV of 10^{-9} Torr, but 10^{-5} Torr was not sufficient. It was also reported that R_s for lead alloyed with ~5 at.% bismuth or overcoated with 0.1–1.0 μm of indium was similar to that for pure lead [17].

---

[1] Jones's conclusion [10] that a superconducting (helix-based) linac would not be competitive with a room-temperature linac or cyclotron was reasonable in 1972, but could not of course forsee the new possibilities soon to be offered by the better-optimized split-loop structure.

- Halbritter's numerous papers and reports include a widely-used computer program for numerically evaluating R_{BCS} and various discussions of the lead material parameters and the residual surface resistance [18,19].

- Yogi's experiments at Caltech were designed to study the high field surface resistance (see II.B), but he also measured low-field R_{res} (= 20–50 nΩ) and R_{BCS} at 205 MHz for several 1.5–3.0 μm surfaces [1,20].

- In the mid 1980's Dietl and Trinks [21] measured the srf properties of the lead-tin alloy that was to be used on the Tritron cyclotron cavities (II.E). The re-entrant test cavity (f = 492 to 1685 MHz) was coated with a 5 μm surface of $\sim$2 at.% tin using a commercial methane-sulfonate (MSA) plating process from Schlötter GmbH [21,22]. Dietl used a "fluid substance like car polish" to mechanically polish the *plated* surface, which was then cleaned with acetone. Without this unusual treatment R_{res} was ten times higher [21], possibly due to surface roughness associated with the particular Schlötter formulation available at that time [22]. An preliminary report on this work [23] mentioned two other unusual methods for surface treatment of (pure lead) plated films that were said to give good results on small samples: electropolishing with acetic acid plus sodium acetate, and "vapor exposition."

- Some Stanford studies at about the same time by Campisi and others [24] provided information on the rf critical magnetic field (which in lead is somewhat higher than the bulk H_c) in the context of a possible pulsed superconducting accelerator. H_c is not directly important in a well-designed Pb-Cu resonator, which should be limited by overall dissipation rather than breakdown.

Procedures and techniques

Some common features of *most* of the studies mentioned above are: a careful attention to the preparation and cleanliness of the copper substrate, use of high purity chemicals, measures to avoid oxidation of the plated surface, good vacuum practice, and a milli-gauss test environment. Considerable foresight and ingenuity is evident in the specific precautions mentioned in the various papers. To cite just two examples: the 'voltage was already on when the cavity was lowered into the plating bath,' or 'cavity parts were machined using only ethyl alcohol as the lubricant.' Yet clearly just the fact of taking a particular precaution and then achieving a good result is not enough to prove that the precaution is essential. Thus perhaps the instances where a reasonably good result was achieved *without* extreme precautions are the most revealing and of the greatest practical significance.

There is certainly good cause-effect information about stray magnetic fields: the additional loss R_H has an $\omega^{\frac{2}{3}}$ frequency dependence and a roughly linear field dependence [1]. From Yogi's measurements $R_H \simeq 40$ nΩ for a 150 MHz cavity at 4.2 K in a 0.5 gauss external field (typical 'earth field'). Thus shielding to only 100 mG should be good enough in heavy-ion accelerators.

Another well-established determinant of R_{res} is vacuum quality, especially below room temperature. Halbritter has proposed a possible explanation of this in terms of surface fissures. Vacuum quality is an important practical issue, in part because it is much harder to control in accelerators, where the resonator and cryostat vacuum spaces are the same, than in the test cryostats of these studies, which generally have independent cavity pumping. (An exception is the lead-tin test cavity of Dietl and Trinks.)

Regarding lead coating methods, similar R_{res} results were obtained with many different types of plating baths based on fluoborates or MSA and bone glue, Shinol, or other commercial additives, and even by evaporation and electroless deposition under favorable conditions. Plating rates were typically *higher* than in more recent accelerator projects, 10–30 μm/hour. The various papers rarely mention plating anodes except for their purity, which varies. In the ORNL work, which achieved the lowest R_{res}, the anode was simply a rolled-up lead foil without any 'bag' [25].

Regarding the copper substrate, ORNL used a 'copper strike' (a routine industrial plating step) just before plating lead. This could effect not only the bonding and thermal conductivity of the substrate but also its microstructure, an important parameter. Some groups used electropolishing (HHF mention that this gave a lower R_{res}) and/or electrocleaning, but Pierce used only mechanical polishing with 600 grit paper.

Regarding post-plating treatments, most papers mention rinsing, drying with the aid of some volatile solvent, and transferring to vacuum or inert atmosphere as quickly as possible.[2] Several authors also mention a 'Varsene' or 'Varsene plus Duponol' dip to prevent oxidation; the Duponol may have acted as a type of wetting agent, but the Varsene is more of a mystery [26]. About an acceptable final appearance, all seem to agree that this should be smooth and 'silver-white.'

Surface resistance results

The basic research studies give the low-field surface resistance ($R_s = R_{\text{res}} + R_{\text{BCS}}$) as a function of frequency, temperature and material parameters. For accelerator applications one really wants to know the rf losses at high fields, but this is not considered a fundamental property and has only been systematically studied in Yogi's work. The apparent surface resistance $R_s(H)$ at high field is typically a factor of 2–4 greater than $R_s(0)$, and it is certainly no less, so the low-field data do give some idea what to expect.

Table 1 summarizes the 'best-reported' R_{res} values. In the Judish and Szecsi studies R_{res} was found to vary with frequency as ω^α with $\alpha = 1.23$ or 1.73, respectively. This scaling was used to predict R_{res} at 150 MHz, as shown in the last

---

[2] As an example of some of the many variations, Judish rinsed in two DI water baths (just 15 seconds each) and then immediately in three successive vats of electronic grade acetone [25]. He also used argon gas for drying and assembled his cavity in an argon glove box. Pierce, on the other hand, rinsed in flowing 'unaerated' solutions and used a nitrogen atmosphere.

TABLE 1. Residual resistivity measurements and extrapolated values

Study	Cavity type	Frequency	Best R_{res}	Alpha	$R_{res}(150)$
Judish / ORNL	Helix	137 MHz	4.7 nΩ	1.23	5.3 nΩ
Yogi / Caltech	Coaxial 'neck'	205 MHz	20 nΩ	~1.5	12.5 nΩ
Szecsi / KfK	$\lambda/2$ coaxial	376 MHz	35 nΩ	1.73	7.1 nΩ
Dietl / Munich	Re-entrant	1.04 GHz	342 nΩ	~2	7.1 nΩ
Flécher / KfK	TM pillbox	2.20 GHz	370 nΩ	~1.5	6.7 nΩ
Hahn / BNL	TE_{011} pillbox	2.83 GHz	51 nΩ	~1.5	0.6 nΩ
Pierce / Stanford	TE_{011} pillbox	10 GHz	30 nΩ	~1.5	0.1 nΩ

TABLE 2. R_{BCS} measurements and extrapolated values

Study	Surface	μm	Frequency	R_{BCS}	Alpha	$R_{BCS}(150)$
Judish / ORNL	Pb	~9	137 MHz	15 nΩ	1.90	20.3 nΩ
Yogi / Caltech	C.P. Pb	2–3	205 MHz	34.7 nΩ	~1.8	19.8 nΩ
Szecsi / KfK	Pb	~10	376 MHz	130 nΩ	1.73	26.5 nΩ
Dietl / Munich	Pb-Sn	~5	492 MHz	117 nΩ	1.90	12.2 nΩ
Flécher / KfK	Pb		2.20 GHz	2400 nΩ	~1.8	19.1 nΩ
Hahn / BNL	Pb	2.5	2.87 GHz	3200 nΩ	~1.8	15.8 nΩ

column. In the Pierce, Hahn, Flécher and Yogi studies frequency dependence was not measured, so $\alpha \simeq 1.5$ was assumed. Finally, in the Dietl lead-tin study a much steeper frequency dependence of $\alpha \simeq 2.9$ can be inferred from the R_{res} values for modes without joint currents. Since this scaling would predict $R_{res}(150)$ of less than 1.0 nΩ, which seems unlikely, we have assumed a more conservative $\alpha = 2.0$ instead. With these assumptions five $R_{res}(150)$ values are roughly 10 nΩ while the two values from TE_{011} mode cavities are less than 1 nΩ.

Table 2 summarizes 4.2 K R_{BCS} values extracted from the same experiments and the corresponding values at 150 MHz. R_{BCS} is expected to scale with frequency as ω^α with $\alpha = 1.75$–2.00 [1,12,20]. As in table 1 we used a measured or inferred α if possible and $\alpha=1.8$ otherwise. Three of the $R_{BCS}(150)$ values for pure lead are quite close to 20 nΩ while the Szecsi result is somewhat higher (26.5 nΩ) and the HHF result somewhat lower (15.8 nΩ). On the other hand R_{BCS} for lead-tin (Dietl) is significantly lower than 20 nΩ, presumably on account of the shortened electron mean-free path in the alloy.

B Applied Research at Caltech LTP

R & D work by the Low Temperature Physics (LTP) group at Caltech during the 1970's was of crucial importance for establishing a *practical* lead-copper heavy-ion accelerator technology.[3] LTP was founded and led by Prof. Jim Mercereau, who had previously developed the superconducting quantum interference device (SQUID) at the Ford Motor Company. The group included at various times several physicists still active in accelerator design (Ken Shepard, Jean Delayen), other researchers and students now in related fields (S. Sridhar, G.J. Dick, T. Yogi, H. Notarys), and some dedicated support people (S. Santantonio, E. Boud, D. Bell).

There were many other significant LTP contributions besides those described here. These include the direct rf feedback method of frequency control (Delayen thesis), various techniques for systematic cavity design and fabrication, and even some X-band R_s measurements on sputtered niobium films [27].

A new resonator concept

The LTP development that really moved the whole heavy-ion linac field ahead was the invention of a new type of low-velocity accelerating structure, the superconducting split-ring, by Shepard, Mercereau and Dick [2,28]. This device, now called the split-loop resonator (SLR), could provide 2–3 times more acceleration than the helix designs that had been under development in other laboratories for the previous decade, with much less vibration [2]. The prototype Pb-Cu SLR was tested in beam at Stony Brook in early 1976 [29]. By then the SLR concept had also been implemented in niobium by Shepard, Bollinger and others at ANL for use in the booster linac that soon evolved into ATLAS [2]. An important feature of the Caltech Pb-Cu SLR which differs from the earlier helix designs and from the ANL-type SLR is the cooling by pool-boiling helium. All subsequent Pb-Cu resonators have used this, except for the Tritron cavities described in II.E below.

The prototype SLR ran at 238 MHz; this was reduced to 150 MHz for the second generation SLR's designed for the Stony Brook linac. Lower frequency gives greater total acceleration and lower rf losses per unit surface area, but also of course larger size and greater stored energy. A even lower frequency of 97 MHz was adopted for the Nb SLR's at ANL, with the result that those cavities can easily produce over 1 MeV of acceleration per charge.

The original Caltech SLR design had four major OFHC-copper parts attached by indium joints: the loop assembly, a cylindrical housing, and the two circular 'tuning plates.' The components of the loop were joined by e-beam welding, a strong capability for which existed in the Caltech area. Based on experience gained with the low-beta SLR's for the Stony Brook linac, this basic design was later modified to have heavier loops welded into the housing.

---

[3] One could say that ANL and INFN-Legnaro have had a similarly dominant influence on the development of bulk-niobium and sputtered-niobium heavy-ion cavities, respectively.

A chemical polish for lead

An historically important Pb-Cu technique was the lead surface treatment developed primarily by Dick [30] and quite successfully demonstrated [4] on the prototype split-ring mentioned above and in a highly re-entrant 'post' resonator [30]. This lead 'chemical polish' was universally used in Pb-Cu linac cavities for the next decade. The concept was to create an optimal and stable lead surface *after* plating rather than *by* plating. Similar to the well-known CERN copper polish 'SUBU,' which is based on a hydrogen peroxide activated acid (sulfamic acid) combined with a viscous moderator (n-butanol), the lead chemical polish combines peroxide, a 2:1 nitric/glacial acetic acid mixture and chelate (sodium EDTA) in a water solution.[5] The essential idea of all these polishes is that through the action of the moderator high points on the surface become preferentially oxidized and then dissolved by the acid, leading to a polishing or leveling action. Fortunately the lead polish is used at room temperature unlike SUBU, which requires a strict 70±2 C.

The Caltech polish was typically used to remove about 5 μm from a 10 μm thick initial plating in about a minute. Proper transfer of the dissolved material away from the surface, which is important to avoid stains, was sensitive to the exact techniques used. Relatively small pieces could be agitated in a large tub of solution with good results, or solutions could be poured over them as they were spun on a turntable, but for larger pieces the only practical approach was quick filling and dumping. The final rinses involved large volumes of noxious, flammable and expensive chemicals (concentrated ammonia, reagent grade acetone).

While these procedures were involved and tricky to execute the result if all went well could be a *dramatically* metallic-looking and *very stable* lead surface with good high-field performance [28,30]. Such surfaces on installed linac cavities have performed with virtually no change for more than a decade. On the other hand, it must be appreciated that a very shiny final surface may still contain small defects detrimental to superconducting performance. Chemically polished low-beta SLR's removed from the Stony Brook linac clearly show traces of numerous microscopic nodules and whiskers in the original plating that were leveled, but not removed, by the polishing procedure.

In retrospect the main advantage of the lead chemical polishing procedure was in producing a visually attractive and very durable surface, not one with necessarily the highest performance. Some version of chemical polishing might still be appropriate for certainn small devices, but for regular accelerator cavities the technique has been superceded by direct plating of somewhat thinner lead or lead-tin layers.

---

[4] Note however that the prototype SLR was tested 1.8 K [28] and that the quoted 25 MV/m is a 'best result' which 'cannot be obtained reproducibly' [30]. On the other hand these test results were obtained without the now-customary pulsed rf and/or helium conditioning.

[5] According to a recent internet discussion [31] a good lead polish can also be made with glycerin as the moderator, which would more closely resemble the viscous butanol.

'Shinol'

While not a Caltech development *per se* the LTP group did introduce a new synthetic plating additive, Harstan Shinol LF-3, to the Pb-Cu community. A standard Harstan fluoborate bath[6] with 2% Shinol by volume was almost universally used by srf electroplaters for the next 15–20 years. Fresh Shinol baths were reported to have higher throwing power than bone glue baths and to consistently produce fine-grained deposits [20]. The Shinol LF-3 baths will long be remembered for their pungent odor and intense lavender/red color that left difficult to remove stains.

Thesis research of T. Yogi

T. Yogi's thesis research provided new quantitative information on lead srf properties at low frequency (205 MHz) and both low and high surface magnetic fields [1,20]. H_p of over 600 gauss (with E_p of only $\sim$3 MV/m) was obtained in a coaxial test resonator with a constricted region at the center of the inner conductor. After mechanical and electropolishing, lead was deposited in a Shinol bath at 27 μm per hour using ordinary lead foil anodes (purity 99.9%). Plated parts were rinsed in running DI water and then absolute ethanol, and blown dry with nitrogen gas. In some cases the plated parts were chemically polished after plating.

In Yogi's high field experiments $Q(H)$ was measured at 2.2 and 4.2 K. Q was found to gradually 'droop' in proportion to H^2 up to a 'transition field' H_{tr} above which it falls much more rapidly up to the rf critical field H_c^{rf} (which generally exceeds the thermodynamic critical field of 520 gauss at 4.2 K by 10–20 %). Of particular interest here is the way in which Q below H_{tr} depends on the 0.3–14 μm lead thickness: a thicker surface has a flatter Q curve (less droop) but also a lower transition field H_{tr}, and *vice versa*. Low-field Q on the other hand does not vary much with thicknesses. Yogi interprets the parabolic droop in Q in terms of localized heating at defect sites [20]. According to his argument the flatter Q curve for thicker surfaces is because 'smaller defects are covered as the lead layer builds up.' Regardless of the interpretation, Yogi's results can be used to pick an optimum thickness for any specific design-goal field level.

Another significant Yogi result concerns the effect of diminshed substrate thermal conductivity at a simulated braze joint [32]. When lead was plated directly over a $\sim$2 mm wide and deep 'joint' Q plunged by an order of magnitude as H_p was raised to just 100 gauss. In a second test, the braze fillet was covered with an electroplated copper film 100 μm thick before the lead plating. This time 700 gauss could be reached with only a modest fall-off in Q, similar to the Q behavior for solid copper with no joint. It is remarkable that the Q curve for lead over plated copper is actually a factor of two *better* than for the original copper substarte. This is reminiscent of Judish's result, where a record low R_{res} was achieved with a plated copper substrate.

---

[6] The 218 g/l of Pb in this standard recipe was far more than needed for resonator plating [37].

C   The Stony Brook linac

The Stony Brook heavy-ion linac [33] was the first full-scale application of the Pb-Cu srf technology. The project began with a collaboration agreement between Caltech LTP and the Stony Brook Nuclear Structure Laboratory in May 1975 and was completed in April 1983. The 16 'low-beta' (β_{opt}= 0.055) and 24 'high-beta' (β_{opt}= 0.10) SLR's were manufactured under the supervision of J. Delayen at and near Caltech and plated at Stony Brook by the chemical polishing method.

With this first complete machine the lead-copper srf technology entered the 'real world.' Plating procedures were directed towards producing and installing visually satisfactory resonators on a tight schedule; parts were replated if necessary, but the completed units were not cold-tested before installation. The linac environment also had many short-comings compared to the test cryostats used in basic research studies. For example, the resonators and super-insulated linac cryostats share the same vacuum space, and the linac working temperature is ~4.5 K. Also, although efforts were made to use nominally non-magnetic parts, there are two stepping motors attached to each SLR and the cryostats have no shielding.

The Stony Brook plating procedures have been described in detail by Burt [34]. The surface preparation included meticulous hand polishing down to the level of 1 μm rouge followed by cleaning with solvents, a soak cleaner, and sometimes a chemical etchant ('Chem-Polish'). Plating was at a very low rate of the order of 1 μm/hour in open tanks. The bath was routinely filtered before use and maintained by occasional charcoal filtering and replenishment of the Shinol additive. A complete plating cycle including set-up and preparation of solutions for the chemical polishing took about two days. No cryogenic tests were performed, but parts were routinely stripped and replated if visually unacceptable.

The six-watt on-line performance of the high-beta SLR's after conditioning improved with plating experience, from ~2 MV/m accelerating field initially to just over 3 MV/m for the best two resonators [33]. (E_a = 3.0 MV/m at 6 watts corresponds to a Q of $6.7 \cdot 10^7$ and an energy gain of 660 keV/charge.) While resonator performance was remarkably stable for many years, inevitably particular resonators and modules were damaged by vacuum accidents, coupler failures, beam deposits, etc. and needed to be replated. This was done at various times using several types of lead-tin plating baths [35,36]. In the most recent and also most successful replating program performance of over 3.0 MV/m was consistently achieved in two modules by a simplified 'environment-friendly' plating system (II.E) [37].

Practical operating fields in the installed low-beta SLR's turned out to be limited to ~2 MV/m by excessive vibrations of their thin loading arms driven by fluctuations in the cryogenic system similar to those recently studied at ANU [3]. This problem was eventually resolved by incrementally upgrading the four low-beta modules to quarter-wave resonators (II.D).

D The superconducting quarter-wave resonator

The superconducting quarter-wave resonator (QWR) has proven to be a very successful design and has been widely adopted. Like the SLR and SRFQ it was demonstrated first in lead-copper and later implemented in bulk niobium. The initial QWR development by Ben-Zvi and collaborators in 1981–83 was in the context of a creating a demonstration booster module for the Weizmann Institute tandem [38–40]. The wide velocity acceptance of the two-gap QWR soon led to its being utilized for the second full lead-copper booster linac at the University of Washington, which was completed in 1987. This was followed by a retrofit/upgrade project at Stony Brook, a second major linac project at INFN-Legnaro, and the development of a demonstration module at the Tata Institute. All of these Pb-Cu QWR's operate near 150 MHz and have β_{opt} between about 0.07 and 0.20.

The original QWR was fabricated from three major OFHC pieces joined by two e-beam welds at the thick 'shorting plate.' These welds in thick copper are difficult to execute well and have often caused problems. One weld was eliminated for the UW prototype, and this remaining weld was moved to the outer conductor in the retrofit design. Legnaro has gone even further, by replacing the weld with either a braze joint or no joint at all (one piece construction). Beam tubes are traditionally brazed in place, with fewer problems.

The QWR is a good structure for evaluating the intrinsic quality of various lead-copper surfaces on account of its modest peak fields (no 'weak link') and lack of significant joint losses. Its nearly-closed cylindrical geometry is convenient for copper polishing by tumbling and for plating using the resonator 'as its own container.' The down-side of this is that internal access can sometimes be rather limited, especially in QWR's with relatively low β_{opt} and hence small beam gaps.

The University of Washington linac

The 24 'high-beta' ($\beta_{\text{opt}} = 0.1$) and 12 'really-high-beta' ($\beta_{\text{opt}} = 0.20$) QWR's of the UW linac give it a mass range from protons to $A > 60$. The development and construction of this machine is very well documented in a series of papers, as is generally true of all of the major projects described here.

A prototype 'high-beta' was plated by Seattle and Stony Brook personnel at Stony Brook in 1984 using essentially the same chemical polishing method as used for the high-beta SLR's and the Brennan/BenZvi prototype. Variations included warming the bath to 35 C, filtering the bath while plating, and use of an anode bag. After He conditioning this QWR ran at 3 MV/m with only 2 W dissipation, and nearly 4 MV/m at 10 W. The Q of $3 \cdot 10^8$ at low field (also at 3 MV/m) is equivalent to $R_s \simeq 100$ nΩ. (This performance is similar to that of the original prototype.) The remainder of the UW resonators were plated in Seattle by essentially the same procedure, except that tumbling was used for surface finishing.

The average on-line field of the $\beta_{\mathrm{opt}} = 0.1$ and 0.2 units has been stable at 2.8 and 2.4 MV/m resp. for the last decade. The good performance is due in part to He conditioning with 1 kW pulses and also the ample refrigeration capacity of 10–15 watts per resonator. The average energy gain ($\Delta W = E_{\mathrm{acc}} \times L$) provided by the very large (36 cm dia.) $\beta_{\mathrm{opt}} = 0.2$ QWR is 0.86 MeV.

The Stony Brook retrofit/upgrade

After 1984 the Stony Brook group began development work on a retrofit/upgrade of the low-beta linac [41,42]. Even at the same nominal field level the new $\beta_{\mathrm{opt}} = 0.068$ QWR's could provide *double* the acceleration of the $\beta_{\mathrm{opt}} = 0.055$ SLR's for light ions ($\beta \sim 0.1$) with no loss of efficiency for the heaviest ones (Sneap 1990). This project was funded in 1987 but took until 1994 to complete because of somewhat limited resources (a staff of three) and various problems in obtaining satisfactory platings and test performance. Every resonator was cold-tested before use, and there were ~50 such tests. Problems overcome include porosity at the weld zone, which was treated by roller burnishing, severe multipactoring, and poor plating results due in part to the tight geometry and degraded Shinol baths.

A quite complete description of a successful plating technique and test performance for the second module was presented at HIAT-6 [42]. Rinsing and drying was done under nitrogen, following the method developed for the SRFQ, and the anode was made of lead-plated copper rods. The M2 QWR's all operated between 3.0 and 4.0 MV/m, both in tests and on-line, after high-power He conditioning. The best resonator R601 had $Q_0 = 4 \cdot 10^8$, or $R_s \sim 65$ nΩ in the off-line test. On-line performance was always essentially the same as in these tests, at least at high field, and the performance of the final two modules was similar to M2.

The Legnaro medium-beta linac

The 48 'medium-beta' QWR's of the INFN-Legnaro heavy-ion linac employ the lead-copper technology and have similar overall specifications to the Stony Brook/UW prototypes [2]. This is the largest set of cavities of the same type in any of the heavy-ion linacs, and considerable experience was gained in developing, fabricating and operating them. In the development stage the cavity geometry was changed from that of the classic QWR with a tapered inner conductor terminated by a drift tube into simply a uniform inner conductor with a beam hole terminated in a hemisphere. The simplified termination improves the E_p/E_a ratio by about 10% from the classic value (4.6 *versus* 5.0) while the straight inner conductor, which was introduced to reduce multipactoring, results in a 50% higher H_p ratio (104 *versus* 70 G/(MV/m)). (The higher H_p wouldn't matter with a Nb surface but might become relevant with Pb.) An important advantage of the simplified geometry is a much easier construction, with at most one current-carrying joint in the outer conductor made by vacuum brazing. The similar 'high-beta' QWR designed to be

Nb sputtered is even more simplified, with a continuously curving shorting plate and externally-mounted ground potential drift tubes. Legnaro has developed an impressive technique for making these resonator bodies from a single copper piece by deep machining (HIAT-6, HIAT-7).

As was the case in the Stony Brook QWR retrofit, but not at UW, plating methods evolved over the two years (1992–93) of production. Initially the Caltech chemical polish was used in a closed system under a nitrogen atmosphere, but this was later changed to direct plating of thin layers under nitrogen. In the final procedure EDTA, citric acid and hot (60 C) water were used for rinsing. As with QWR's elsewhere, tumbling methods are used to highly polish the copper surface. Resonators were tested in groups of four in an off-line module setup. The 7-watt operating fields of 1.5–3.2 MV/m generally improved with plating experience.

E Further Developments in the 80's and 90's

Most of the other new lead-copper developments in the 80's and 90's came out of Stony Brook, Oxford (ASI), Munich and Legnaro. Three new structures were demonstrated, two quite large, and there were various advances in fabricating resonators and preparing effective surfaces more simply and reliably. These developments overlapped in time and to some extent synergistically interacted.

'Thin-lead' and 'tin-lead'

While the Stony Brook and UW linacs were succesfully completed using the thick-lead chemical-polishing method, there was a strong feeling in the community by about 1984 that 'there has to be a better way.' At Stony Brook the change was to thin (1–2 μm) layers of lead-tin alloy, while thin pure lead was adopted by Oxford/ASI. After about 1990 the Legnaro group also embraced thin pure lead, and completed the ALPI middle-beta section with it. No one involved with these developments mentions that both alloy plating and relatively thin lead surfaces had already been tested with promising results during the basic research era (II.A).

Brennan [43] has described how R. Couglin, a member of the original Stony Brook plating team who had had prior experience with industrial plating, first suggested the use of a lead-tin bath for resonator plating. Lead-tin was well known in industry to 'throw'' better than lead and to form finer-grained surfaces more resistant to oxidation. Existing data showed that a lead-tin alloy with a modest tin content has a significantly higher T_c than pure lead [44]. Following some very encouraging test results in a QWR [45], lead-tin was adopted for all further work. Lead-tin has also been used on the Tritron cyclotron cavities and, in one non-accelerator application, for a kHz superconducting coil used for gravity wave detection.

The first Stony Brook lead-tin bath followed a standard Harstan/Shinol recipe for '93:7 bearing alloy.' The 93:7 ratio was thought to be lean enough to avoid the possible formation of 'islands of tin' from precipitation out of the solid solution.

Only some years later, during the SRFQ project, was it realized that 93:7 was a by-weight ratio corresponding to a higher than intended ∼11% by atoms. The tin was reduced to 2–4% by atoms for subsequent work, including the final plating of the SRFQ. In retrospect, the higher tin content originally used may have caused some hysteresis effects in measured Q_0 values, but this tin ratio could clearly give outstanding results [45]. One wild card here is that prolonged resonator baking could allow tin to migrate away from the surface. On the other hand, an unpublished RBS analysis of samples plated with the Harstan 93:7 bath did confirm that the surface tin ratio was similar to the bath ratio.

Delayen/ASI developments

Following the completion of the Stony Brook linac Delayen remained active in the Pb-Cu field primarily through Applied Superconductivity Incorporated (ASI). There were several important advances in structures and lead surfaces from ASI projects, which were often in collaboration with Oxford University [46]. In particular, ASI was responsible for the fabrication of ten Stony Brook type SLR's for Oxford. As is well known these were later transferred to Daresbury and then to ANU–Canberra. Delayen's development projects include a highly-optimized 'half-wave' resonator and two improved split-loop resonators with β_{opt}=0.045 and 0.055.

The half-wave resonator project [47] was historically very significant for at least two reasons. First, the half-wave was a beautifully optimized design which even took into account Yogi's $R_s(H)$ results. Clearly it was a practical design as well, with eg., well-designed welds. Second, the thin-lead plating techniques used were so well controlled that an excellent test result was obtained on the first test, and just the third plating attempt. It is remarkable that this device only 25 cm in diameter could provide an energy gain of >1.1 MeV ($E_{acc} \sim 4.5$ MV/m). Only brief 200-watt He conditioning was done, and it is plausible that 5 MV/m could have been achieved with acceptable dissipation through additional effort. The low-field R_s after conditioning was ∼100 nΩ ($Q_0 = 4 \cdot 10^8$), or roughly twice R_{BCS}.

Some details of the procedures may be significant. One is that the final finish of most of the copper surface before plating was relatively coarse (from abrasive pads); only the drift tubes had a high polish. Plating was done in a fresh Shinol bath at moderate speed (1.5 μm in 30 minutes). Post-plating treatment included soaking in chelate solutions and acetone, and argon gas was used for drying.

The Tritron cyclotron project

The physically largest Pb-Cu cavities built to date were developed for the 'Tritron' cyclotron project at TU-München [48,49]. The six identical 170 MHz cavities have a roughly conical shape 1.2 meters long with re-entrant sides that contain 20 accelerating gaps 6–13 cm wide. The cavities have a substantial energy content on account of their large size (∼2 Joules), but they operate at modest peak fields compared to typical linac resonators ($E_p \simeq 10$ MV/m, $H_p \simeq 200$ G). Frequency excursions are only ~ 10 Hz, allowing relatively easy stabilization.

The large size of the Tritron cavities required novel fabrication and cooling methods. Each cavity consists of two ~ 1 cm thick shells which were electroformed over fiberglass molds. (Electroformed copper has been used once before in a superconducting cavity, for a Harwell deflector.) The shells meet at a flat joint that carries no current. The cooling to 4.5–5.0 K is by flowing helium through attached pipes; this has been tested to 100 watts per cavity.

Cavity shells were plated by the Electrochemistry Group at the GSI using a commercial MSA-based lead-tin process (Slotolet KB, Schlötter GmbH) and methods (platinated titanium anodes, large open tanks, relatively fast plating) that are common in industry. Lead-tin (~ 5 μm with ~ 4 at.% tin) was choosen for its resistance to oxidation during the open-air handling of the large parts; tests also showed that lead-tin had significantly lower R_{BCS} than pure lead (see II.A). After plating the parts were first rinsed in plating acid to dilute the metal ions, then rinsed in a tank of DI water, and finally dried with a nitrogen stream [22]. Completed cavities have Q_0 of $\sim 3.5 \cdot 10^8$ ($R_s \sim 250$ nΩ).

A superconducting RFQ

Around 1990 a Stony Brook – Legnaro collaboration led by Ben-Zvi set out to develop a superconducting radio-frequency-quadrupole resonator (SRFQ) for the efficient acceleration of very slow ions [51]. The final design looks more like an SLR or Delayen's half-wave resonator than the usual room-temperature RFQ, with four massive and highly modulated vanes ($m = 4$) suspended by robust hanging tubes inside an *upright* cylindrical cavity. The prototype that was built and demonstrated in-beam in 1993 is optimized for ions with $\beta \simeq 0.03$ and $q/A \geq 1/6$. It operates at 57 MHz, which is the lowest frequency yet in a Pb-Cu cavity. No further work was done with this design past the demonstration stage, but Legnaro is actively pursuing a second generation niobium SRFQ for use with PIAVE (HIAT-98).

The SRFQ project provided valuable experience in the fabrication of large Pb-Cu devices [50]. The cavity was assembled by brazing in argon (hydrogen could not be used because not all copper was OFHC). A very simple procedure was used for the copper preparation before plating (sanding with 600 paper, scrubbing with abrasive pads wet with 'Micro' cleaning solution). Plating was accomplished without a separate tank, and rinsing and drying (which together took 30 minutes) was done in a nitrogen atmosphere. No solvents were used for the drying.

In the first two tests a good $Q \sim 10^8$ was achieved only at *very* low field, possibly on account of the poor thermal conductivity of some copper 'bubbles' that had formed during the brazing. The bubbles were peened down before the third and final plating (1.5 μm from a fresh Shinol bath with 2 at.% tin). The bath was heated to 35 C and filtered continuously during the plating. In the final test a fairly flat Q curve was found that turned down abruptly at ~ 1.25 MV/m due to a thermal/magnetic breakdown. Although this was short of the 2 MV/m design goal (at which field $H_p \simeq 400$ gauss, $E_p \simeq 15$ MV/m), the device did demonstrate 700 keV of energy gain with only 7 W of cooling. The flat portion of the Q curve corresponds to $R_s \sim 100$ nΩ.

A surface treatment for multipactoring

Lead-copper QWR's are especially prone to difficult multipactoring on account of certain geometrical factors [52] and their relatively 'soft' surface. Traditionally multipactoring has been treated by rf pulse processing, but this method is time consuming and not always effective. Building on some earlier surface-chemistry work at ANU, Noé developed a 'Freon-plasma' surface treatment that is invariably effective after only a few minutes of processing [53]. The technique proved invaluable during the QWR retrofit project and in tests of the Seattle '13/12' buncher resonator [54]. 'Freon processing' works on unplated copper QWR's and might be useful for niobium structures like the New Delhi QWR [2] as well.

Diagnostic thermometry

While diagnostic thermometry is a standard technique in $\beta=1$ niobium cavities, it has hardly been used with Pb-Cu devices. A 1993 study of the Stony Brook SLR shows the potential power of the method [55]. In a good resonator with low field emission > 90% of total losses were shown to come from the loop structure rather than anywhere on the outer housing (which includes two large demountable joints and the coupler hole). This heat loss distribution was nearly constant up to 3.8 MV/m, confirming the absence of significant field emission to that level. As it is now possible to model temperature distributions with standard codes [3], perhaps this promising diagnostic tool will soon come into greater use.

Environment-friendly plating

As described in II.B, a fluoborate plating bath with Harstan Shinol LF-3 was widely used after the early 1970's. However, *fresh* Shinol LF-3 was no longer available after 1987 [37] and the groups at Stony Brook, Legnaro and Bombay then actively using it were eventually forced to seek an alternative. At Stony Brook a generic gelatin-resorcinol additive was developed and used to replate a high-beta module with reasonably good results [56]; however the bath decomposed in a few months. The Tata Institute group tried both resorcinol and a different type of Shinol intended for solder plating, TLM-15, that was still available from a secondary supplier [57]. Finally, the Legnaro group successfully reverse-engineered Shinol LF-3, but have had difficulty obtaining the 'Ethomeen' ingredient [37].

The next step for Stony Brook was to abandon fluoborate chemistry completely in favor of an 'environment-friendly' commercial plating system[7] based on methane-sulfonic acid (MSA). This was very successfully used to replate two high-beta SLR modules, as described in considerable detail at Sneap-96 [37]. The MSA-based plating is superior to fluoborates in every regard except cost. Tata Institute and ANU are now using MSA plating, and the Schlötter 'FF' chemicals long used by Munich (but not specifically described in their reports) are also MSA-based.

The 'environment-friendly plating' scheme [37] included a less-friendly but very beautiful copper treatment with SUBU-10 (CERN) polish. Best performance was obtained however by 'spoiling' this with a mechanical polish with 1 μm rouge. The two SLR's thus treated were able to run at 4 MV/m with a dissipation of ten watts.

More recent Legnaro developments

After the completion of the medium-beta section of ALPI in 1993, the Legnaro srf group led by Palmieri put more development emphasis on sputtered niobium, which was then beginning to show great promise. Of course the sputtering-related work on surface preparation by tumbling, CERN-polish, electropolish, etc. indirectly benefits lead-copper as well, because surface quality is such a crucial factor in both technologies. Also of potential benefit to everyone are the new and very economical methods of copper fabrication by spinning, hydroforming, etc. that are being pioneered. Palmieri's group has developed a 'synthetic Shinol' plating additive and an acid passivating dip for stabilizing lead surfaces. It is their opinion that with use of this dip there is no reason to prefer lead-tin to pure lead [58,59].

III DISCUSSION

In the following we try to summarize what the cumulative lead-copper experience seems to be saying about various plating and other variables. These opinions should be tested by appropriate controlled experiments and diagnostic tests.

⋆ *Substrate condition.* A *sound* substrate is clearly crucial, but a high polish is probably not, and a 'too perfect' surface may be undesirable. How can 'Scotchbrite' surfaces work as well as they do?

⋆ *Activation.* Appropriate cleaning followed by thorough rinsing is essential. The safest activating dip is dilute plating acid (MSA).

⋆ *Plating chemistry.* Additive type and bath chemistry are not directly relevant to performance, but having a *fresh* bath seems important. MSA has become the preferred acid for environmental reasons, and the easier disposal of MSA baths will encourage bath replacement at appropriate intervals.

---

[7] The manufacturer LeaRonal was recently absorbed into Rohm and Hass and will be combined with another R & H company to become 'LeaRonal-Shipley.'

- ⋆ *Additivies* Modern baths such as LeaRonal Solderon have more ingredients and more organic content than historic bone-glue baths, but it is not clear that this is a problem; additive amounts can be reduced if necessary by short-pulse plating [60].

- ⋆ *Surface composition.* The lower BCS losses of lead-tin compared to pure lead seem to be not very relevant (since $R_{\rm res}$ is so far greater than $R_{\rm BCS}$), but lead-tin may offer practical advantages. More study is needed.

- ⋆ *Plating anode.* Lead-plated copper or passive (platinated) anodes are convenient, especially in awkward geometries, but well-cleaned lead sheets seem to be acceptable. Anodes should be removed from the bath after plating. Anode bags are only needed where debris can fall on a critical surface.

- ⋆ *Purity* Anode and bath purity is not a critical issue in general, but certain impurites (iron) could be very harmful.

- ⋆ *Plating rate* Fast plating has given good results, and may be desirable. With very large surfaces plating rate may be set by power supply considerations.

- ⋆ *Uniformity and coverage.* In a properly working bath reasonable uniformity can be expected. Complete coverage is crucial. Use flashing to get a good starting plate.

- ⋆ *Plating bath volume* It seems to be better to work with a generous volume of bath, rather than just enough to fill a small cavity. A surface-to-volume ratio issue?

- ⋆ *Agitation and filtration* It seems desirable to have some movement in the bath, from slow circulation through a filter or stirring, if only to counteract stratification. Filtering is important, but may not remove sub-micron tin precipitates without a 'flocculating agent' [37].

- ⋆ *Rinsing.* Quick rinses seem to be enough, although long ones under nitrogen also work. Excessive rinsing turns a hydrophobic surface hydrophilic. It may be helpful to bubble nitrogen through rinse water, or to 'spike' it with ammonia or acid.

- ⋆ *Drying.* Closed cavities can be dried slowly in an inert atmosphere. Immersion in a large tank of high-purity solvent works well with open structures, but is expensive and not environment friendly. Drying complex open structures with a gas stream is tricky because the droplets are so mobile; gentle blotting may help.

- ⋆ *Storage and handling.* Storage under nitrogen gas, even without a continuous purge, seems to be fine, as long as moisture is excluded. Even a one micron lead surface is surprisingly rugged, and can be wiped or even mechanically polished as needed.

- ⋆ *Joints* Best to have joints that carry little current if possible. Lead wire and tin-indium both give good results, as have Pb-Pb knife-edge joints (Szecsi). Thermometry can test joint quality.

IV CONCLUSIONS

The lead-copper srf technology has played an important role in the historical development of superconducting heavy-ion accelerators. It has offered quite competitive performance for far less cost and complexity than bulk niobium, thus accelerating the development and evaluation of new structures by a diverse community of potential users, a critical element of progress in the field.

The practical details of working in lead-copper have been greatly improved and simplified in the last decade, with the result that good results can now be more consistently obtained with less effort and environmental impact than ever before. Yet there remain many open questions as to why certain methods seem to be more effective than others, and why. Even in the best cases the surface resistance achieved so far is still several times higher than the BCS resistance, leaving considerable room for improvement. Usable accelerating fields of 4 MV/m in typical structures seem quite achievable, while the earlier ambitious design goal of 3 MV/m can be considered routine. Beyond that, the ultimate limit in continuous 4.2 K operation will be set by the practical critical field of $\sim$500 gauss [20], which in a well-designed structure like the QWR or half-wave should correspond to 5–6 MV/m accelerating field. Of course, improvements which reduce losses (raise Q) can reduce refrigeration costs even if significantly higher working fields cannot be attained.

At the same time, recent advances in sputtered-niobium technology, which shares with lead-copper many advantages such as a thermally stable substrate and ease of re-coating, already have pushed its anticipated performance to 6 MV/m or even more. If fields this high prove to be sustainable in prolonged regular operation, and there are no new impediments such as problems with frequency stabilization, then niobium-copper will clearly be the way to go for most future applications involving large numbers of modest-size cavities. It is likely though that there will always be some low-frequency applications where the device is so large, or so specialized, that only lead will do. Fortunately the performance of lead in very large devices has proven to be remarkably good, so that one need not be concerned about srf difficulties at the largest conceivable size.

By 'lead' or 'lead-copper' we of course do not mean to exclude lead-tin and possibly other alloys. Lead-tin clearly works quite well, but it will take further studies, both practical and fundamental, to decide it offers any decisive advantage over properly-handled pure lead.

ACKNOWLEDGEMENTS

I am indebted to the many friends and colleagues with whom I have collaborated on various srf projects for the last twenty years and/or who provided information and insights to incorporate in this review. I also wish to thank the Department of Physics and Astronomy at Stony Brook for its generous on-going support.

REFERENCES

1. Delayen, J.R., "Rf Properties of Superconducting Pb Electroplated onto Cu," in *Proceedings Third Workshop on Rf Superconductivity (SRF-3)*, ANL, September 1987 (ANL-PHY-88-1), pp. 469–489.
2. Shepard, K.W., *Nucl. Instr. and Meth.* **A382**, 125-131 (1996), and refs. cited therein.
3. Weisser, D.C. and Lobanov, N.R., this conference, and references cited therein.
4. Septier, A. and Viet, Nguyen Tuong, *Journal of Physics E: Scientific Instruments* **10** 1193 (1977).
5. Bardeen, J., Cooper, L.N., and Schrieffer, J.R., *Phys. Rev.* **108** 1175 (1957).
6. Mattis, D.C., and Bardeen, J., *Phys. Rev.* **111**, 412 (1958).
7. Pierce, John M., *J. Appl. Phys.* **44, No. 3** 1342-1347 (1973).
8. Hahn, H., Halama, H.J., Foster, E.H., *J. Applied Physics* **39** 2606-2609 (1968).
9. Bruynseraede, Y., *et al. Physica* **54** 137-159 (1971).
10. Jones, C.M., *Particle Accels.* **5** 45-60 (1973).
11. Jones, C.M., *et al., Particle Accels.* **3** 103-113 (1972).
12. Judish, John P., doctoral thesis, University of Tennessee, 1974.
13. Judish, J.P., *et al., Phys. Rev.* **B15** 4412-4424 (1977).
14. Szecsi, L., "Measurement of the Dependence on Frequency of the Residual Resistance of Superconducting Layers of Lead," in *Proc. VII Int. Conf. on High Energy Accelerators*, Yerevan 1969, pp. 691–695.
15. Kneisel, P., private communication, 1999.
16. Flécher, P., *et al.*, "The Preparation and Performance of Superconducting Cavities," in *Proc. Linear Accel. Conference, Los Alamos*, 1966, pp. 499–501.
17. Flécher, P., *et al.*, "Measurements of the Rf-Absorption of Superconducting Resonators," in *IEEE Trans. Nucl. Science NS-16* , 1969, pp. 1018–1022.
18. Halbritter, J., *Z. Physik* **238** 466–476 (1970).
19. Halbritter, J., *J. Applied Phys.* **42** 82 (1971).
20. Yogi, T., doctoral thesis, Caltech, 1977.
21. Dietl, L. and Trinks, U., *Nucl. Instr. and Meth.* **A284**, 293 (1989).
22. Trinks, U., private communication, 1998.
23. Dietl, L., *Proceedings SRF-2*, Geneva, 1984.
24. Campisi, I.E., SLAC reports and paper with G.J. Dick.
25. Judish, J., private communication, 1998.
26. Hahn, H., private communication, 1998.
27. Yogi, T. "Surface Resistance of Sputtered Niobium Film on Copper at 8.86 GHz," Caltech LTP internal report, August 1979.
28. Shepard, K.W., Mercereau, J.E., and Dick, G.J., "A New Superconducting Heavy In Accelerating Structure Using Chemically Polished Lead Surfaces," IEEE Trans. Nucl. Science, **NS-22, No. 3**, June 1975, p. 1179.
29. Dick, G.J. *et al., Nucl. Instr. and Methods* **138** 203-207 (1976).

30. Dick, G.J., Delayen, J.R., and Yen, H.C. "A Polishing Procedure for High Surface Electric Fields in Superconducting Lead Resonators," IEEE Trans. Nucl. Science, **NS-24, No. 3**, June 1977, p. 1130.
31. 'Ask the Experts Metallography Forum' at http://metallography.aasp.net/bboard/messages/684.html.
32. Yogi, T., *Cryogenics* **13**, 369-370 (1973).
33. Noé, J.W., *Rev. Sci. Instr.* **57 (5)**, 757-760 (1986), and references cited therein.
34. Burt, W.W., *Adv. Cryo. Eng. Mat.* **29**, 159 (1983).
35. Noé, J.W., Rico, J., Uto, H., *Proceedings SRF-6*, Cebaf, 1993, p. 160-166.
36. Ben-Zvi, I. et al., *Proceedings SRF-7*, Gif-sur-Yvette, 1995, p. 55-56.
37. Noé, J.W., "An Environment-Friendly Method for Resonator Surface Fabrication," *Proceedings 1996 SNEAP Conference*, (World Scientific Press), p. 76.
38. Ben-Zvi, I., and Brennan, J.M., "A Superconducting Quarter-Wave Resonator," BAPS, Spring 1982.
39. Ben-Zvi, I., and Brennan, J.M., *Nucl. Instr. and Methods* **212**, 73 (1983).
40. Ben-Zvi, I., et al., *Nucl. Instr. and Methods* **244**, 306 (1986).
41. Noé, J.W., et al., *Nucl. Instr. and Methods* **A287**, 240 (1990).
42. Noé, J.W., Rico, J., Uto, H., *Nucl. Instr. and Methods* **A328**, 285-290 (1993).
43. Brennan, J.M., "New Directions in Lead-Plating Technology," *Proceedings 1984 SNEAP Conference* p. 249-255.
44. Warren, W.H., Jr., and Bader, W.G., *Rev. Sci. Instr.* **40** 180 (1969).
45. Brennan, J.M. et al., IEEE Trans. Nucl. Science **NS-32** p. 3122 (1985).
46. Delayen, J.R., *Rev. Sci. Instr.* **57 (5)** 766–769 (1986).
47. Delayen, J.R. and Mercereau, J.E., *Nucl. Instr. and Methods* **A257** 71-76 (1987).
48. Trinks, U., *Nucl. Instr. and Methods* **A287**, 224-234 (1990).
49. Trinks, U. and Schütz, P., *Proceedings SRF-6*, Cebaf, 1993, 167-172.
50. Jain, A., et al., *Nucl. Instr. and Methods* **A328**, 251 (1993).
51. Ben-Zvi, I. et al., *Proceedings SRF-6*, Cebaf, 1993, p. 160-166.
52. Ben-Zvi, I., and Sokolowski, J.S., *Rev. Sci. Instr.* **57**, 776-779 (1986).
53. Noé, J.W., *Nucl. Instr. and Methods* **A328**, 291-292 (1993).
54. Storm, D.W., et al., *Nucl. Instr. and Methods* **A382**, 125 (1996).
55. Noé, J.W., et al., *Proceedings SRF-6*, CEBAF, 1993, pp. 1052-1064.
56. Lipski, A., et al., *Proceedings SRF-7*, Gif-sur-Yvette, 1995, p. 611-615.
57. Pillay, R., private communication.
58. Gustafsson, S., et al., *Proceedings SRF-7*, Gif-sur-Yvette, 1995 page 589-593.
59. Palmieri, V., *Proceedings SRF-8*, Abano Terme, 1997.
60. Sheppard, *Keith*, Stevens Institute of Technology, private communication.

Design of the Heavy-Ion Injector PIAVE

A.Pisent

INFN, Laboratori Nazionali di Legnaro, via Romea 4 I-35020 Legnaro, Padova (Italy)

Abstract A new injector for the superconducting linac ALPI is under construction at LNL. This linac (called PIAVE) will be able to accelerate ions up to uranium, for an equivalent voltage of about 8 MV, and will improve the performances of ALPI (presently injected by a 15 MV Tandem) both in intensity and ions species availability. We shall discuss the design choices adopted for PIAVE, with the use of two compact 80 MHz superconducting RFQ up to about 530 keV/u, followed by eight independently phased QWR (minor modification of those already installed in ALPI). This approach, made possible by a rather sophisticated beam dynamics design, allowed to concentrate the R&D effort on the development of the RFQ structure. Details like the optimization of the acceleration in the RFQ, the bunching system and the transition between the two RFQs will be discussed.

INTRODUCTION

The new injector under construction at LNL will produce beams of larger intensity and mass number than the ones that the TANDEM can produce nowadays. Moreover the third Experimental Hall with independent transfer lines allows additional flexibility, since the simultaneous use of New Injector+ALPI and Tandem beams by two different experiments will be possible.

Owing to the higher charge states delivered by the ECR source, an RF injector with an equivalent voltage of about 8 MV can substitute the 15 MV XTU Tandem. The Injector consists of Radio Frequency Quadrupole (RFQ) structures up to about 550 keV/u, plus independently phased Quarter Wave Resonators (QWR); this subdivision gives a good acceleration efficiency and allows the use of already developed QWR's. Fig. 1 shows a scheme of the linac, with the accelerating elements, the buncher, the superconducting RFQ's (SRFQ1 and SRFQ2) and the quarter wave resonators (QWR). A rather complete list of parameters is given for completeness in Appendix A.

The use of a sequence of short superconducting RFQ's for the acceleration of heavy ions was first proposed by I. Ben Zvi [1] and a first lead plated prototype was built and tested at SUNY with the participation of A. Lombardi from LNL (1988-91)[2].

CP473, *Heavy Ion Accelerator Technology: Eighth International Conference,*
edited by Kenneth W. Shepard
© 1999 The American Institute of Physics 1-56396-806-1/99/$15.00

After some years the construction of a new injector at LNL based on a SRFQ was proposed (1995)[3]. The following summer INFN funded a "progetto speciale" named PIAVE (Positive Ion Accelerator for Very low Energy). At present the production steps of the SRFQ have been tested with the completion of a SS model[4,5], and the construction in Nb will be ready in a few months [see G. Bisoffi, Friday]. The first QWR has been tested [6], the site is being prepared and the beam transfer lines are almost mounted.

In this paper we shall review the main design choices, with particular attention to the beam dynamics. These choices belong to a project under construction, and have therefore overcome all the verifications (review committees) but the experimental one.

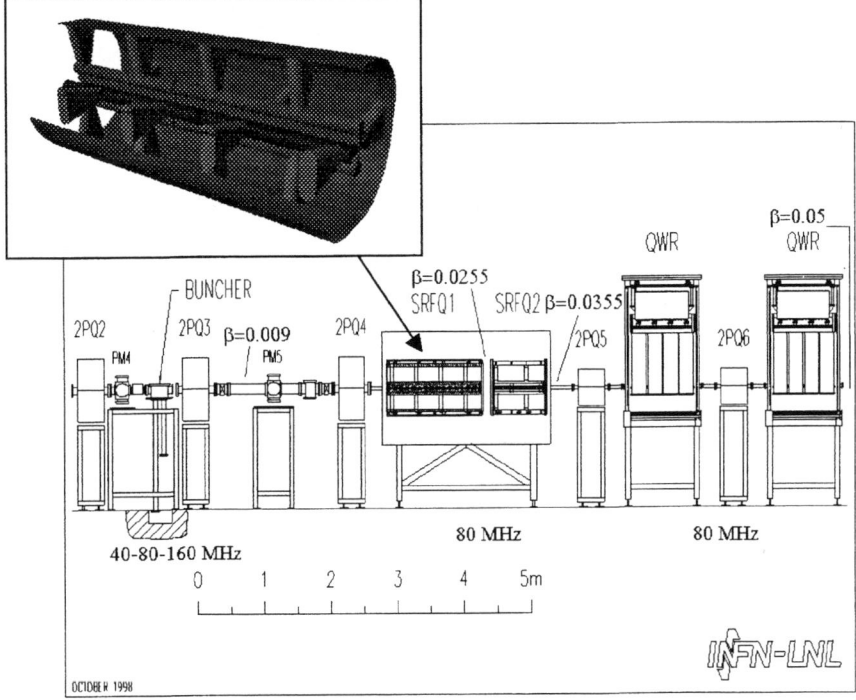

FIGURE 1. Lay-out of PIAVE (accelerating structures).

THE RFQ

Specifications

In Table I the RFQ specifications are listed. The table is divided between beam specifications and technological constrains.

Table I Specifications

q/M	28/238	
Energy range	38÷540	keV/u
β range	.009÷.035	
Acceptance (norm)	>0.8	mm mrad
Output long. emittance	<0.75	ns keV/u
Surface Field	<25	MV/m
Stored energy per tank	<5	J

The RFQ output energy has to be high enough to allow a convenient acceleration in an 80 MHz Quarter Wave Resonator (QWR). The bulk Nb QWR's that are being installed at LNL have β_0=.055 ($\beta_0\lambda/2$ is the effective distance between the two accelerating gaps), and with minor modifications β_0=.047 and a transit time factor equal to 0.85 after the RFQ was achieved. The main advantage coming from the the change of structure (from RFQ to QWR) lies in the better efficiency with which a set of independently phased resonators accelerate ions with different q/M ratios, since one passes from a definite velocity profile structure to a definite accelerating voltage accelerator.

The acceptance considers that the typical ion beam emittances from ECR sources, according to CERN, GSI and ANL experiences, are below *0.5 mmmrad*. The longitudinal emittance is kept as small as possible. The maximum electric field has been chosen according to LNL experience with bulk Nb QWR.

The maximum stored energy per cavity U has to be limited in a high Q resonator. Actually the required active power to have a Δf feedback control bandwidth is:

$$P_a = 2\pi U * \Delta f .$$

In order to have routinely a bandwidth of 20 Hz an active power of about 500 W is required. As a consequence the total power from the amplifier is 1 kW.

Design criteria

As in the design work done in Stony Brook (our starting point) big emphasis is given to the maximization of the accelerating field E_a [1]. Due to this an adiabatic bunching within the RFQ would have made the structure too long, and an external bunching has been chosen (at the expenses of a capture efficiency of about 60%).

Moreover, since we are limited by the maximum surface field E_S, we have to maximize the ratio E_a/E_S:

$$E_a = V\frac{kA(kR_0,m)}{4} \qquad E_s = \kappa(kR_0,m)\frac{V}{R_0}$$

$$\frac{E_a}{E_s} = \frac{kR_0 A(kR_0, m)}{4\kappa(kR_0, m)}$$

Here $k=2\pi/\beta\lambda$ is the wave number, m is the modulation factor, ratio between the maximum and the minimum vane-beam axis distance, A is the accelerating factor and κ the field enhancement factor. The ratio E_a/E_s is function of two geometrical parameters (average aperture in modulation wave length units and modulation amplitude) and has, for each m, a maximum as function of kR_0.

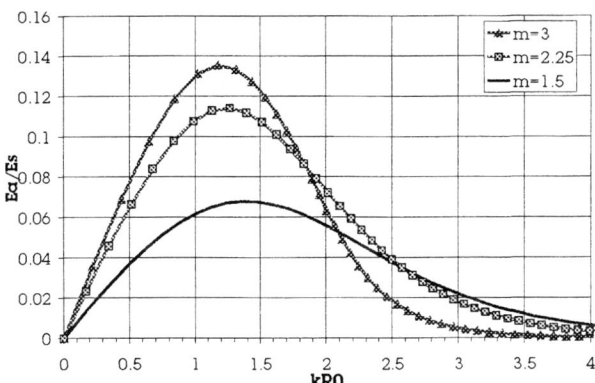

FIGURE 2. Ratio E_a/E_s as function of kR_0 for different modulation factor m.

Making use of the two terms potential for A, one gets the curves of fig. 2 with optimum values for the ratio E_a/E_s around $kR_0 \approx 1.4$ for a wide range of modulation values. Moreover the value of this maximum is a monotonic function of the modulation factor m, but with negative second derivative, i.e. the advantage of high modulations almost saturates, while all kind of exotic effects related to Bessel functions increase. For this reason we have chosen to remain below $m=3$.

The fields E_a and E_s are kept constant if V and R_0 are both proportional to β (stepwise from one tank to the next one). Increasing β the stored energy becomes a problem, since:

$$U = \frac{1}{2}(C/\ell)V^2 \ell$$

and the capacitance per unit length in a quadrupole is not very sensitive to the geometrical details (about 120 pF/m).

Also the transverse stability becomes a problem with increasing energy since the transverse phase advance

$$\sigma_T = \sqrt{\frac{B^2}{8\pi^2} - \frac{\sigma_L^2}{2}}$$

becomes imaginary. Indeed the focussing factor

$$B = \frac{eV}{mc^2}\left(\frac{\lambda}{R_0}\right)^2 = \frac{e}{mc^2}\frac{E_s}{\kappa}\frac{\lambda^2}{R_0}$$

scales as $1/\beta$, and the longitudinal phase advance:

$$\sigma_L = 2\pi\sqrt{\frac{eAV|\sin\phi_s|}{mc^2\beta^2}}$$

scales as $\beta^{-1/2}$ if V is proportional to β.

If we choose kR_0 and E_s at the beginning of the RFQ, both the average aperture R_0 and the voltage V (that we keep constant in the RFQ) are determined. Moreover stored energy per unit length (and therefore the maximum RFQ length fulfilling our specifications) and the transverse phase advance (for given accelerating factor) are determined. In fig.3 we show the stored energy per unit length and the transverse phase advance as function of RFQ input energy. Three cases are considered: 80 MHz kR_0=1.3, 80 MHz kR_0=.96, and 40 MHz kR_0=1.3. In the first and in the third case we apply almost rigorously our criterion, but at 80 MHz the transverse focusing is too weak, while at 40 MHz the stored energy is too large. As a consequence the dynamics of our second RFQ uses 80 MHz kR_0=.96.

But even if not completely applied the accelerating field optimization criterion is very useful for a pushed RFQ. Let us compare this design with a normal conducting RFQ design, in order to underline how a superconducting RFQ allows a high accelerating field.

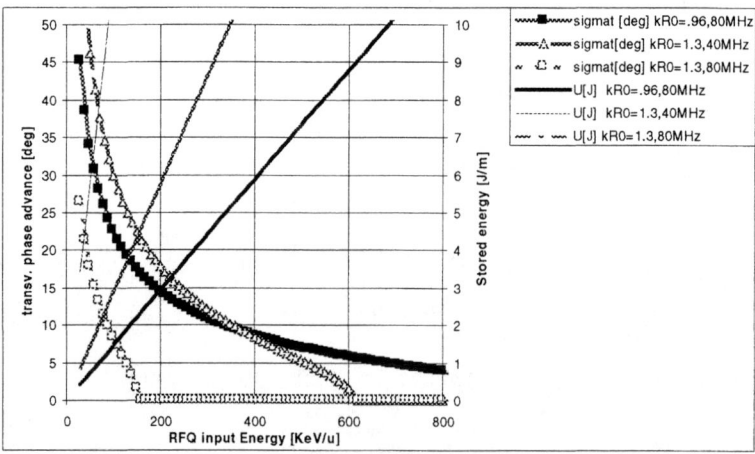

FIGURE 3. Transverse phase advance and stored energy per unit length as a function of the RFQ input energy in the range of the entire New Injector. A surface field 25 MV/m, an enhancement factor 1.3 and vanes capacitance 120 pF/m are assumed; moreover A=.6 and ϕ_s=-12 deg, as for the nominal SRFQ2 dynamics.

Our surface field limit (25 MV/m) is larger but comparable with

$$E_s = 2E_{Kilpatrick} = 21\,MV/m \quad @80MHz$$

that can be reached in a room temperature structure. Our stored energy limit (5 J/RFQ) has an equivalent dissipated power limit of 500 kW for each RFQ (if Q=5000). This limit would hardly be considered acceptable for a CW accelerator, and therefore a normal conducting RFQ cannot be optimized in this way. As a result room temperature RFQ's have an average accelerating field of the order of 1 MV/m, while our RFQ exceeds 2 MV/m.

The nominal case

An effective RFQ accelerator can be obtained by choosing $kR_0 \approx 1.4$ and an high modulation factor; our design value is $m=2.8$ for the linear RFQ, corresponding to $m=3$ taking into account the constant transverse radius geometry (multipole corrections). The voltage V is determined by the maximum surface field; the synchronous phase ϕ_S has to leave the required acceptance ($|\phi_S|$ and the RF defocusing must be small enough).

Short RFQ's have higher efficiency and smaller stored energy, but more inter tank drift spaces. These spaces have a physical length $d=200\,mm$ determined by the hardware and cause a drifting of the transverse ellipsis and beam mismatching in the following RFQ. At 80 MHz it is convenient to construct two RFQ's: the first relatively long, with small R_0 and V, and a second shorter, with higher R_0, V and E_a. The same optimization criterion brings to 3 RFQ's @40 MHz.

Indeed the sudden beginning of the acceleration spoils the longitudinal emittance, since a low $|\phi_S|$ has to allow transverse stability. In this way one is forced to work in the non linear region of the RF field, with the beam pulse relatively long produced by the external bunching system; as a consequence the longitudinal emittance is multiplied by 3 in the first ten cells.

To solve this problem we have introduced an adiabatic bunch compressor, i.e. a section where ϕ_S is linearly ramped from -40^0 to -18^0. At the same time the modulation is increased to avoid the strong parametric resonance (linear longitudinal-transverse coupling):

$$0.4 \leq \frac{\sigma_T}{\sigma_L} \leq 0.6$$

and to maintain a nominal acceptance* higher than 1.2 mm mrad. These two requirements determine the law for m. In fig.4 the main RFQ parameters as a function of cell number are shown.

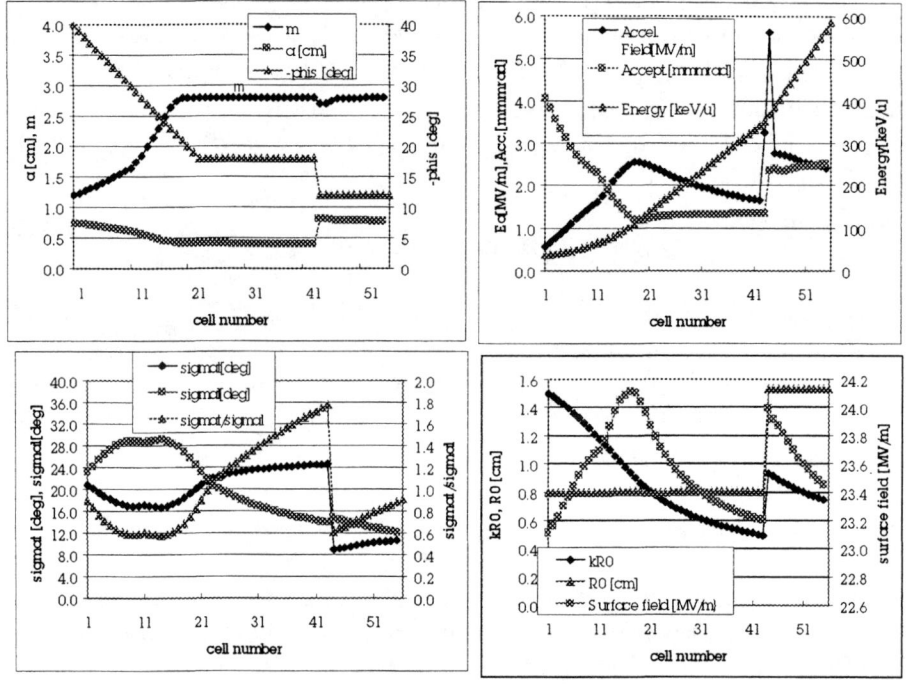

FIGURE 4. RFQ parameters as a function of the cell number.

At the entrance of the second RFQ the bunch is very short and $\phi_S = -12°$ can be used. There is a longitudinal emittance increase but the final value is within specifications. The nominal acceptance is more than *2.5 mm mrad* and can deal with the beam mismatch due to the drift space.

The superconducting RFQ's are being built in full Nb (3 mm thick everywhere besides the thicker modulated vanes) electron-beam welded. The resonator is a four road variant, with symmetrical supports (see fig. 1); SRFQ2 will be built first, so as to verify the phase locking with the larger stored energy.

---

* The nominal acceptance is the acceptance for a perfectly matched beam in a channel with adiabatically changing focusing functions. Unfortunately, due to our efficiency requirements, the focusing functions change rapidly respect to the betatron oscillations frequency. For this reason the effective acceptance has to be calculated "a posteriori" from simulations and is in our case around *0.9 mm mrad*.

Since SRFQ1 is roughly twice as long as SRFQ2, the RFQ cross section can be done almost identical, with two supports per electrode in SRFQ2 and four supports per electrode in SRFQ1. The construction of SRFQ1 will be carried out with two separate RFQ halves, very similar to SRFQ2, so that the prototyping experience will be reused. These two halves will be connected at the end by welding the two external tanks. The similarity of the two RFQs gives also advantages in the cryostat design.

The half cell.

As above stated the drift space of 200 mm between the two RFQs electrode terminations determines a certain beam mismatch in SRFQ2 and a consequent problem in keeping the specified acceptance. Cutting the electrodes where the beam envelopes have a waist, can minimize the problem [7]. This corresponds to a length

$$\frac{\beta\lambda}{2}(\frac{1}{2}-\frac{\phi_s}{\pi})$$

for the last cell of SRFQ1 and correspondingly for the first cell of SRFQ2 (almost *half-cell*). In Fig. 5 the transverse acceptance improvement due to the *half-cell* is shown by plotting the beam transmission of SRFQ1 and SRFQ2 as function of source emittance.

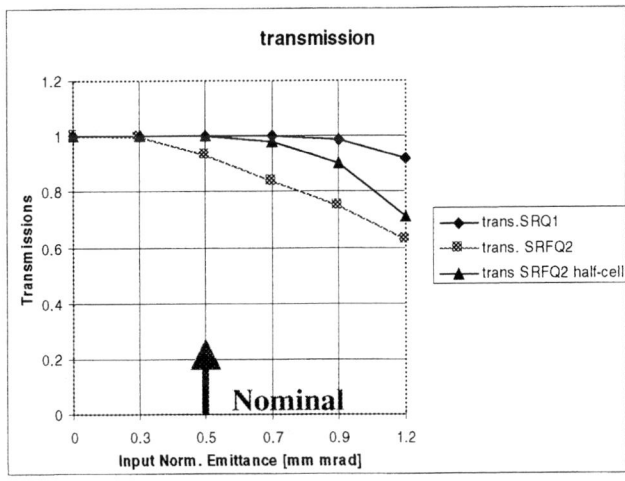

FIGURE 5. RFQ acceptance with and without half-cell.

This *half-cell* has been simulated using MAFIA electrostatic solver, so to have an accurate description of both the focusing and the accelerating field components. In doing this it is important to take into consideration the (local) electrode voltage related to ground, determined by the resonator supports geometry. The last support is indeed so close to the end wall that it behaves almost like a short circuit to mass; as a consequence the voltage of the quadrupole neutral axis (a part from modulation) is about $V/2$. This property has allowed to design the modulation in such a way that, in first approximation, this special cell has the focussing effect of half-cell (minimum mismatch) and the acceleration of a full cell (maximum acceleration).

In Fig.6 the accelerating field is plotted, as measured in the aluminum model of SRFQ1 [5]. The zoom of the last half-cell shows the comparison with MAFIA calculations. It is interesting to observe that this condition is possible only with an even number of regular cells.

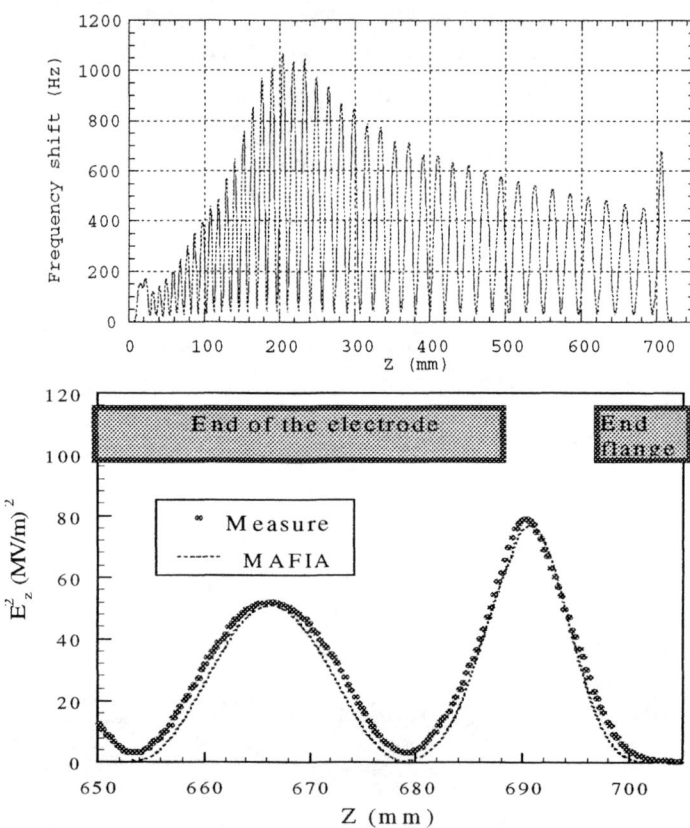

FIGURE 6. Accelerating field in SRFQ1, as measured in the aluminum model; comparison with the electrostatic MAFIA simulation of the half-cell.

BEAM TRANSPORT LINES

Two beam transport lines are necessary for PIAVE: the LEBT between the ECR source and the RFQ, and the MEBT between the superconducting RFQ cryostat and ALPI. In the first line the particles are bunched and matched transversally to the RFQ acceptance. In the second line they are accelerated by the 8 QWR and matched to ALPI.

The magnets, two 45^0 dipoles, two small 90^0 dipoles, and quadrupoles assembled in 4 triplets, 6 doublets and 2 singlets, have been delivered in the second week of 1998, together with their power supplies. An achromatic almost vertical line (18^0 inclination) guarantees the beam transfer from the accelerating column axes to the RFQ axes, with two dipoles, two triplets and a singlet. The whole line was mounted on a common girder and arrived preallined from the company; the beam is therefore reduced to a small spot in the buncher by means of two doublets, and matched to a smaller spot at the RFQ input by means of two similar lenses. The quadrupoles of the LEBT are characterized by a large bore aperture (120 mm diameter) and a short effective length. The main difficulties in these elements arose from the fringe field non linearities control (multipole components below .8% in 85% of quadrupole aperture have been obtained with a mechanical correction after preliminary measurements). In the MEBT two compact and strong doublets (0.8 T on pole tips) are placed between the cryostats, while the remaining elements are almost identical to the ones already in ALPI. This focusing structure, stronger than in ALPI, is the key point that allows the use of 5 MV/m in the QWR's.

We have selected a grid-less three harmonics buncher, based on two gaps coaxial resonators. In the first two gaps 40 and 160 MHz signals are applied, 80 MHz in the second two; the system is longitudinally compact enough to have a small spot in all the gaps and a good longitudinal and transverse emittance[8].

A prototype of the QWR bulk Nb cavities has been built and successfully tested, showing an accelerating field exceeding 5 MV/m [6]. This cavity is a modification of the low β cavities used in ALPI, with a new geometry in the beam port region in order to decrease $β_0$ from 0.055 to 0.047. Beam simulations show that for many beams it is indeed possible to use this additional acceleration. The last cavity of the first cryostat uses $\phi_S = 20^0$ (longitudinally defocusing) to have a regular phase envelope.

The new injector is completed by the line MEBT, that matches the beam into ALPI (in the three degrees of freedom). The line is composed by an achromatic L bend (identical to ALPI L bends), two (ALPI) triplets and two bunchers. The transverse focusing in the ALPI branch of MEBT is provided by the existing lenses. The two room temperature bunchers, operating at 80 MHz, are located in the beam waists before and after the L bend.

BEAM SIMULATION RESULTS.

Multi-particle simulations have been done using the LANL programs PARMTEQM, in the RFQs, and PARMILA, for the transfer lines and the QWRs section[9]. The same ensemble of 10000 particles is transported from the bunchers up to the end of the accelerator. The space charge can be completely neglected up to 5 µA, while we do not aspect more than 1 µA from the source. In the transition between the two RFQs particle trajectories are calculated with a dedicated code using the MAFIA fields.

In fig. 6 we summarize the results of simulations as function of the initial transverse emittance; namely the RFQ transmission, the total PIAVE transmission (including bunching efficiency), and the longitudinal emittance after the RFQs and after the QWRs are plotted. The RFQ gives an excellent longitudinal emittance, thanks to the continuos transverse and longitudinal focusing. In the QWR section there is an emittance growth due to the change of longitudinal and transverse focusing structure. Nevertheless correspondingly to the ECR nominal RMS emittance (0.1 mmmrad), 70% of the particles are transmitted with a final longitudinal emittance within the specifications.

In fig. 7 we show the transverse and longitudinal phase space in critical locations for the nominal beam. In particular it is possible to appreciate the longitudinal phase space evolution all over PIAVE and the residual mismatch at the transition between the two RFQs.

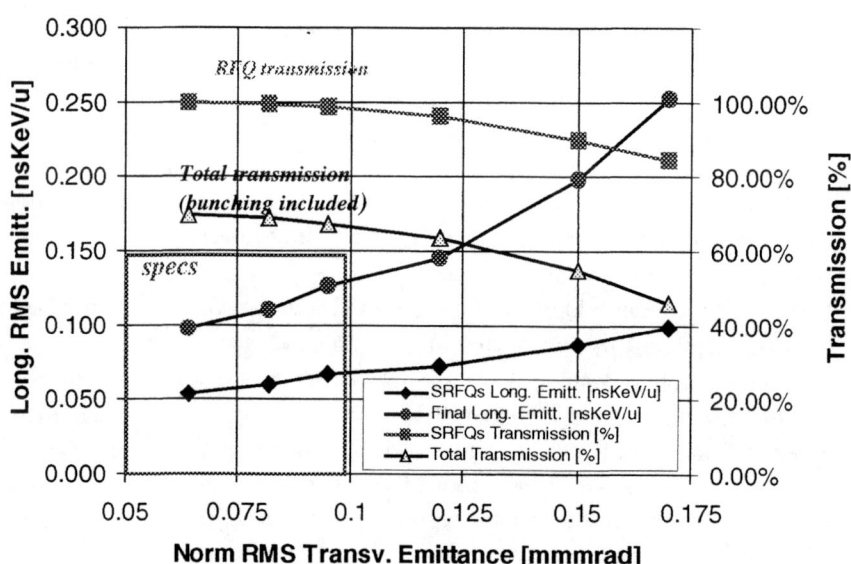

FIGURE 7. Long. Emittance and Transmission as function of Transverse Initial Emittance.

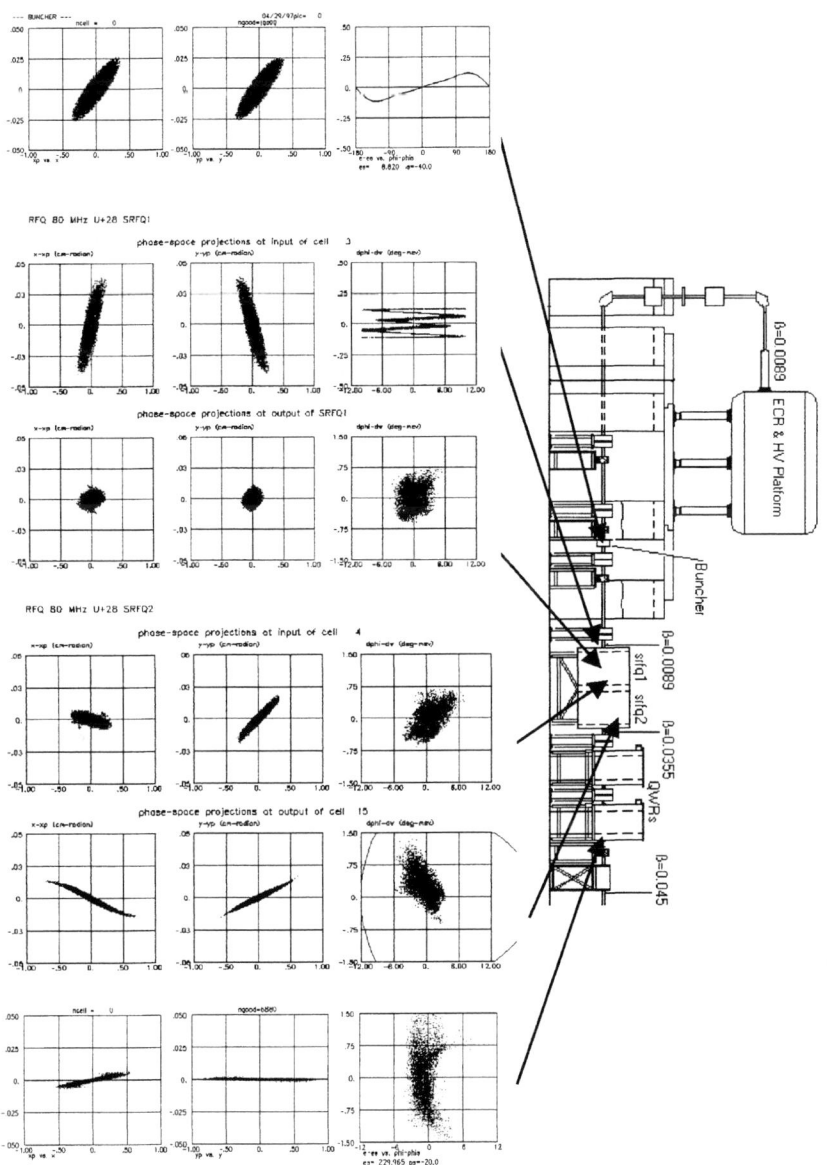

FIGURE 8. PIAVE layout and transverse and longitudinal phase space in various locations.

CONCLUSIONS

A nominal design for the New ALPI injector, based on two SRFQs and eight QWRs, has been determined; the input line, the SRFQs and the QWRs have been simulated with well proved codes, showing that all the specifications can be met. The amount of technological risk seems reasonable and well distributed between the various subsystems. In particular the use of compact 80 MHz SRFQs (possible with a sophisticated beam dynamics design) and of already developed QWRs are key points for the feasibility of the project. The construction is in progress.

ACKNOWLEDGMENTS

My thanks to A. Lombardi and G. Bisoffi, with whom in these years I had day by day discussions on the intersection areas of interest for the determination of this superconducting RFQ design. M. Comunian played a key role in the final parameter optimization. The collaboration with ITEP (Moscow) and the visits of A.Kolomiets,V. Andreev, S.Yaramishev and A. Balabin have given an indispensable contribution to the clarification of important points of this design.

REFERENCES

1. I.Ben-Zvi, A.Lombardi "Design of a Superconducting RFQ resonator" Particle Accelerators 35 (1991) 177.
2. A.Jain et al "Fabrication and test of a superconducting RFQ" Nucl. Instr. Meth. in Phys. Res. A 328 (1993) 251-254.
3. G.Bisoffi, P.Favaron, A.Lombardi, A.Pisent, R.Tovo "The Positive Ion Injector for ALPI", presented to the 7th International Conference on Heavy Ion Acceleration Technology, Canberra. Nucl. Instr. Meth. in Phys. Res. A 382 (1996) 245-251.
4. A. Pisent et al. "The New LNL Injector PIAVE, based on a Superconducting RFQ" to be published in EPAC 98 conference proceedings (1998).
5. G.Bisoffi this conference
6. A.Facco this conference
7. A.Kolomiets "Summary of beam dynamics simulation results for the new injector", LNL internal note (1995).
8. A.Facco "Testing of the PIAVE triple harmonic buncher" Laboratori Nazionali di Legnaro Annual Report 1997.
9. A.Pisent, M. Comunian "Complete simulation of the heavy ion Linac PIAVE." to be published in PAC Conference Proceedings (1997).

APPENDIX A:
MAIN PIAVE PARAMETERS

Source and LEBT

Ion source	ECR	14 GHz	
Mass to charge ratio	8.5÷2		
Platform voltage*	315	kV	
Energy	37.1	keV/u	(β=.0089)
Beam emittance	0.5	mm mrad	(5*ε_{rms} norm.)
Bunching system	3H	40÷80÷120	MHz

RFQ Accelerator

Radio Frequency	80	MHz	
Equivalent voltage	4.7	MV	(2.2 MV/m)
Output Energy	586	keV/u	(β=.0355)

	SRFQ1	SRFQ2	(bulk Nb)
Vane length	137.8	74.6	cm
Number of cells	42.6	12.4	
Voltage *	148	280	kV
Tank diameter (approx.)	65	65	cm
Max. surface B field*	280	295	G
Shunt impedance R_{sh}/Q	20	32	Ω–m
Max. Sur. E field*	24.1	24	MV/m
Max. stored energy*	1.8	3.5	J
Estimated Q	7E8	9E8	
Power dissipation (4K)*	<7	<7	W

QWR Section

Number of resonators	8	(bulk Nb)	
Output energy*	1.2	MeV/u	(β=.051)
Radio Frequency	80	MHz	
Optimum β	0.047		
Accelerating Field	3÷5	MV/m	
Shunt impedance R_{sh}/Q	580	Ω	
Quality factor Q	10^9		
Power per cavity (4K)	<7	W	

Matching Line to ALPI

Number of bunchers		2	(room temperature)
Buncher Eff. Voltage VT		<200	kV

---

* The values are referred to a mass to charge ratio 8.5, ($^{238}U^{+28}$).

Beam Test of Niobium Sputtered QWRs and Upgrading of ALPI Medium β Cavities

A.M. Porcellato, S. Y. Stark, V. Palmieri, G. Bisoffi, A. Dainelli, M. Poggi, L. Bertazzo, F. Stivanello, L. Badan, A. Beltramin, L. Boscagli, D. Carlucci, F. Chiurlotto, S. Contran, T. Contran, M. De Lazzari and L. Ziomi

Istituto Nazionale di Fisica Nucleare, Laboratori Nazionali di Legnaro, Via Romea 4, Legnaro (Padova),35020, Italy

Abstract Four new super-conducting Quarter Wave Resonators (QWR, β=0.13), obtained by niobium sputtered on an OFHC copper substrate, were recently installed and tested on the ALPI post-accelerator (1). The resonators substituted four similar ones that were operating at 4 MV/m at 7 W, because two of their couplers jammed. Now the cavities have a new coupler design and show an acceleration field, measured with different ion species, in excess of 6 MV/m at 7 W (about 1 MeV energy gain per cavity per charge unit). The resonators show excellent stability on long time runs even at such a high accelerating field. The high accelerating fields, together with the good stability and the low costs, made this technology almost ideal for heavy ion acceleration. Moreover the niobium sputtering process was applied to substitute the electroplated lead coating in four medium β (β =0.11) resonators after only minor changes in the resonator shape. Accelerating fields in excess of 4.5 MV/m a 7 W could be obtained in laboratory even though the cavity geometry was not optimised for the sputtering procedure. Beam tests are foreseen soon. The replacement of the installed resonators with sputtered units can either improve the ALPI performance or reduce the cryogenic power consumption.

INTRODUCTION

The technology of electroplated lead on copper was thought safer and less expensive for a laboratory without any experience in superconducting resonators when the construction of ALPI began in 1988. If fixing the procedure for QWR construction and plating took more effort than originally foreseen (2), nevertheless the choice allowed to have 44 accelerating resonators installed and operating at an average field of 2.7 MV/m, by the beginning of 1994 (3). The accelerating field is lower than in structures made of niobium, but the cavities were insensitive to mechanical vibrations or liquid helium bath pressure changes. This feature made the resonator control easier and allowed to operate the resonators reliably at the maximum field obtainable using the available 7 W cryogenic power per resonator.

Taking into consideration the better superconducting characteristic of niobium, an R&D program was in the meantime set to develop both the manufacturing technology of bulk niobium (4) and the sputtering procedure of niobium on copper (5). This last method of cavity production originally developed at CERN for 350 MHz β=1 structures (6), was particularly desirable because, even at the lower β range, allowed to combine, at reduced cost, the better superconducting characteristic of niobium to the mechanical stability of copper structures.

The sputtering technology was first successfully devoted to high β resonator development and can be now extended to upgrade the performance of medium β cavities. The results of beam test and the upgrading plan of ALPI are reported in the following paragraphs.

BEAM TEST OF SPUTTERED HIGH β RESONATORS

The first cryostat housing four sputtered resonators (CR20) was installed in ALPI in July '95 (7). The details of cavity electromagnetic characteristics, the procedure of cavity substrate production and sputtering, and off line performance are extensively described elsewhere (8).

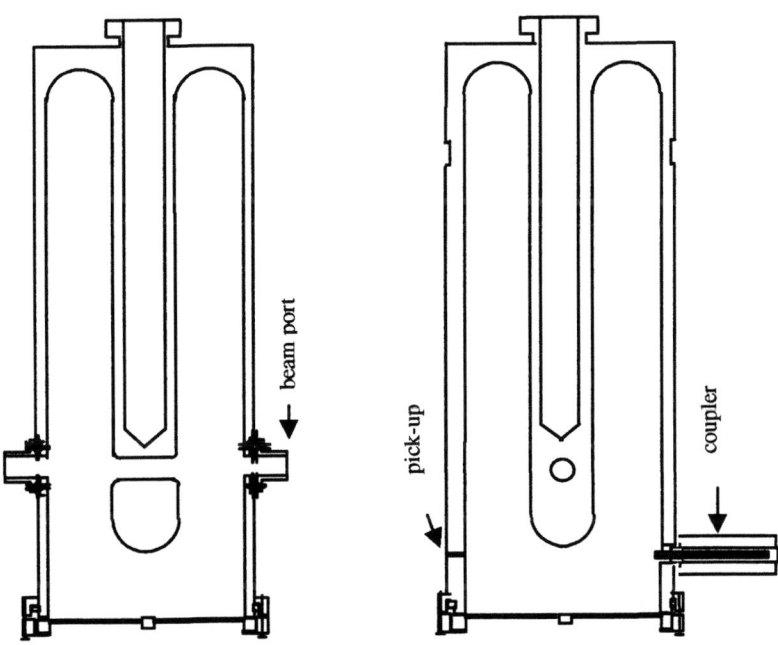

Figure 1. ALPI β=0.13 resonator views; to notice coupler and pick up position. The cavity has no brazed joint; beam ports are jointed to the resonator body by indium gaskets.

The performance of the resonators in the beam line was heavily affected by field emission that limited the obtainable fields at a level slightly higher than 4 MV/m. This problem was surely bounded to the procedure of repairing for a leak the cryostat without having the possibility, for lack of time, to dismount the resonators.

Moreover the jointed tuning and coupling mechanism mounted on the bottom plate that was originally designed for these resonators, showed in operation to be less friendly to be used than the standard separated systems installed in ALPI medium β resonators. The coupler movement drove a change in the position of the bottom plate and a related shift of the resonant frequency. Only after a couple of tuning adjustments to correct a slow drift of resonant frequency (few Hz) it was possible to obtain a steady condition. When two coupler lines jammed, not allowing resonators operation any longer, it was decided to remove the cryostat.

Four new resonators, without any brazing joint and with a lateral capacitive coupler, were then built. With the new coupler design the advantages of having the coupler hole located in low current area could be combined with the reliability of the same coupling mechanism as the one used for the other ALPI cavities.

The improved copper quality and the optimisation of sputtering parameters allowed to obtain resonators with an average field between 6 and 8 MV/m, as shown in fig. 2.

In spring '98 the cryostat CR20 was demounted, and the resonators substituted with the new units showing better performance.

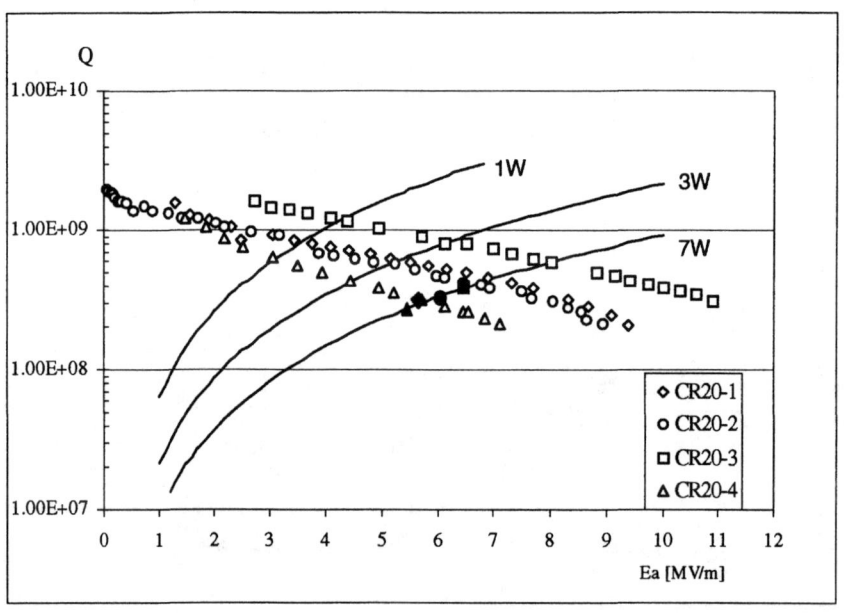

Figure 2. Q curves of high β resonators (empty bullets). The full bullets indicate the accelerating field of the four resonators measured by beam test.

The low field quality factors, measured on the beam line, did not change significantly with respect to the value measured in laboratory (1-2 x10^9). Before conditioning, the accelerating fields were anyway limited to 2.5-4 MV/m.

It is not possible, with the Alpi medium β cryostat, to align the resonators without opening the cavity beam holes: as a consequence the resonators are exposed to laboratory air for some hours during the alignment procedure and this explain the degradation in performance. A field level of 5.5-6.5 MV/m at 7 W could be reached after 3-6 hours of He conditioning, 1kW (pulsed) maximum RF power.

The accelerating fields obtained on the beam line are slightly lower than the ones measured in laboratory: it has to be verified if they can be improved by increasing the conditioning time.

The accelerating fields of the four resonators at 7 W dissipated power were confirmed by the energy gain obtained by a ^{28}Si^{9+} beam. The input beam energy was 221.3 MeV, that corresponds to a β =0.13 (T.T.F.$_n$ (0.13) > 0.999). The energy gain was determined by the current increase in a downstream dipole after turning on the resonator. The resulting accelerating fields for the four resonators were 5.8, 6.6, 6.1, 5.5 MV/m respectively.

The resonators were used to boost the energy of a 339.2 MeV ^{58}Ni^{14+} beam (β=0.112) by 54.1 MeV. The synchronous phase was set at −20°. An energy gain of 0.97 MeV per charge state and per cavity was measured. The resulting average accelerating field is 5.8 MV/m.

The electric field on beam line (fig. 2) and normalised TTF curve of this resonator (TTF$_n$ in fig. 3) were measured by bead perturbation technique.

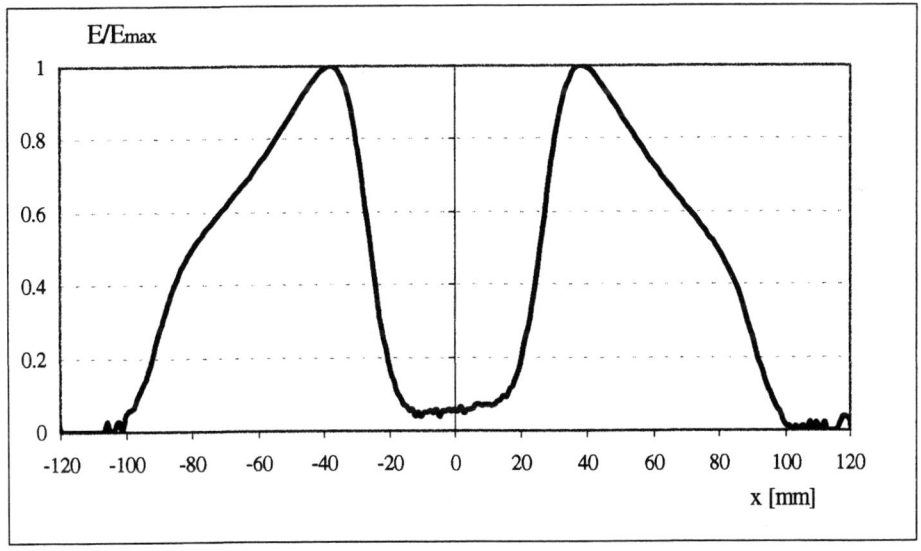

Figure 3. Electric field profile measured on the beam line of the high β resonator.

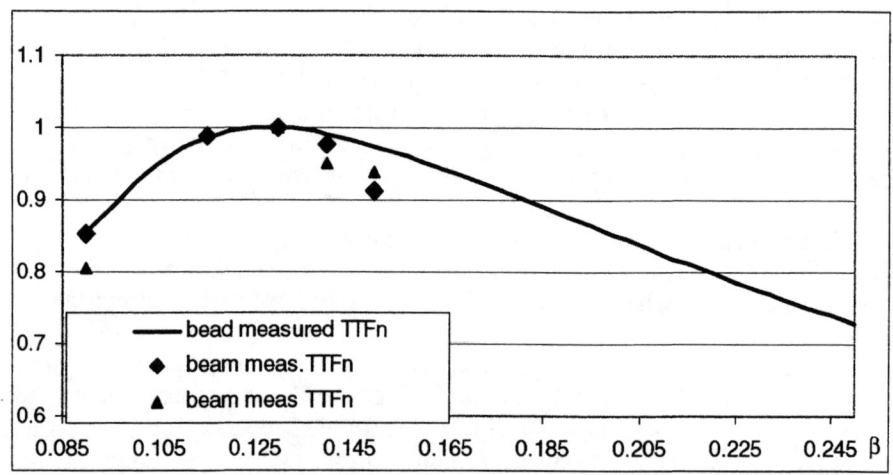

Figure 4. Normalised Transit Time Factor (TTF$_n$) of the high β resonator measured by bead perturbation technique. The bullets indicate TTFn values obtained in two different beam tests.

The optimum β (β$_{op}$) is 0.128, the related TTF (β$_{op}$) is 0.868. β$_{op}$ is lower than the one originally foreseen for the high β linac section. It was decided to reduce it because it gave the possibility to install 4 resonators instead of the foreseen three in each cryostat without necessity to develop a new cryostat and with the possibility to use standard ALPI components.

The cavities are very stable in frequency: The phase lock condition of the cavities could be maintained for more than 48 hours by the controllers using the 100 W amplifiers used for medium beta resonators (9).

UPGRADING OF MEDIUM β RESONATORS

In late spring '97 a leak opened in the valve that commutes between He pre-cooling and filling systems in the first installed cryostat (CR10) making the cryostat operation impossible. The cavities installed in the cryostat had an average accelerating field of 2.2 MV/m. We had four medium beta resonator bases available and it was decided to verify the possibility to obtain a good superconducting layer by Nb sputtering.

The medium β cavity (fig. 5) was designed before setting the most suitable geometry for the sputtering procedure and it presents some features that make it hard to obtain the same results reached in high β resonators. If good film quality could be obtained in the beam ports area [9] the corners that blend inner and outer conductor to the shorting plate (curvature radius of 10 mm) remain a critical issue. It was anyway thought that by means of low cost (250$) minor changes in the resonator copper base, an improvement in the performance with respect to lead plated resonators was possible.

It was anyway thought that by means of low cost (250$) minor changes in the resonator copper base, an improvement in the performance with respect to lead plated resonators was possible. The sharp inner edges of the resonator holes (pick-up, coupler, and shorting plate outlet for plating), that were critical for the sputtering were rounded to a curvature radius of 2 mm. Moreover an extension, sputtered together with the cavity, was mounted on the coupler hole in order to allow higher electromagnetic field to decay before meeting the coupler body normal conducting surface.

The performance obtained in laboratory by this sputtered resonator was lower than it would certainly be for suitable designed resonators but the fields obtained are in between 4 and 5 MV/m. They represent a substantial increase with respect to the average fields obtained by lead plated resonators (2.7 MV/m).

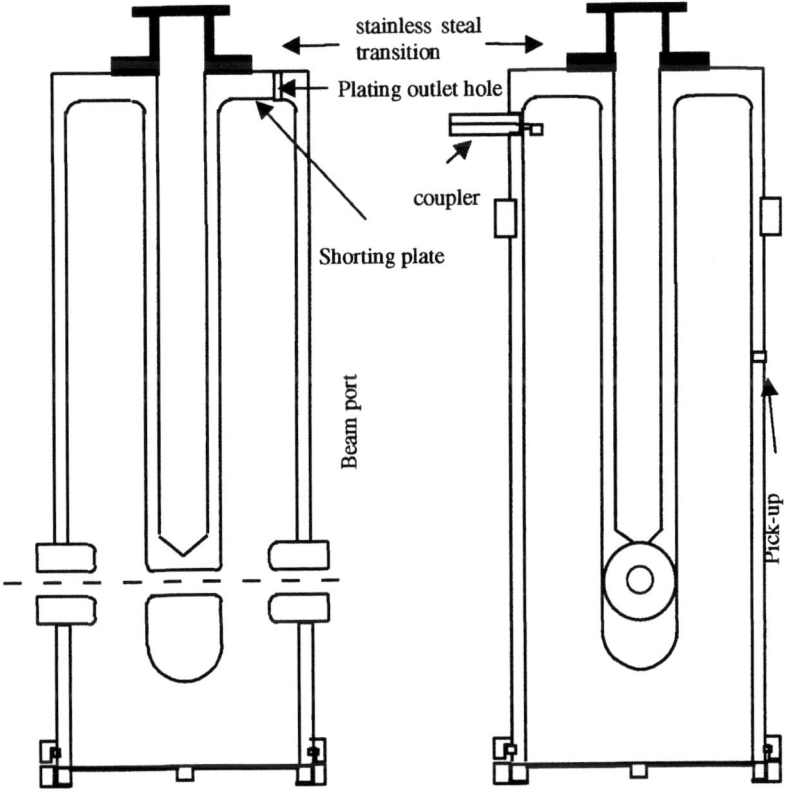

Figure 5. ALPI medium β resonator views. The stainless steal collar, the beam ports, the holding blocks are brazed to the resonator body. The resonator has a flat shorting plate in the edges of which it is difficult to obtain good quality superconducting layer.

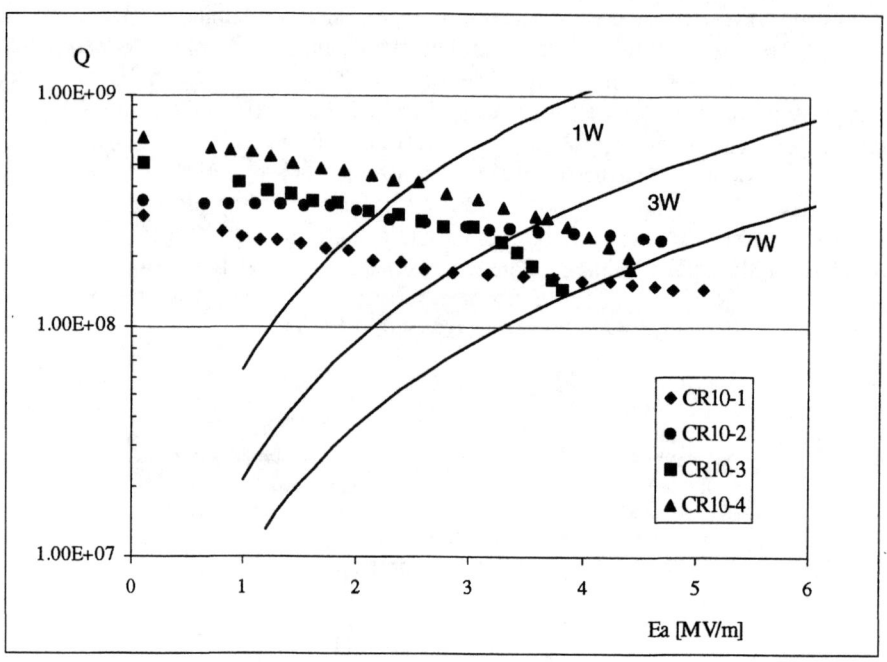

Figure 6. Performance of the four medium β ALPI cavities in which the lead layer was replaced by niobium sputtered coating.

The possibility to replace, at negligible cost, the lead plated with niobium sputtered resonators is very exciting. It opens the possibility to increase the beam energy without adding new cryostats or, if convenient, to halve the cryogenic power consumed by the resonator (11).

REFERENCES

1. G. Fortuna, R. Pengo,, G. Bassato, I. Ben-Zvi, J. D. Larson, J. S. Sokolowski, L. Badan, A. Battistella, G. Bisoffi, G. Buso, M. Cavenago, F. Cervellera, A. Dainelli, A. Facco, P. Favaron, A. Lombardi, S. Marigo, M. F. Moisio, V. Palmieri, A. M. Porcellato, K. Rudolph, R. Preciso and B. Tiveron., *The ALPI Project at the Laboratory Nazionali di Legnaro*, Nuclear Instruments and Methods **A287** (1990) pp. 253-256.
2. G. Fortuna., A. M. Porcellato, G. Bassato, A. Battistella, M. Bellato, L. Bertazzo, G. Bezzon, G. Bisoffi, G. Buso, S. Canella, M. Cavenago, F. Cervellera, F. Chiurlotto, A. Dainelli, N. Dainese, M. De Lazzari, A. Facco, P. Favaron, M. Lollo, A. Lombardi, S. Marigo, M. F. Moisio, V. Palmieri, R. Pengo, M. Poggi, J..S. Sokolowski, L. Badan, M. Barbadillo, R. Pegoraro, R. Preciso, and F. Scarpa, *Completion of the medium-β section of the ALPI SC-booster at LNL*, Nuclear Instruments and Methods **A328** (1993) pp. 236-241.

3. A. M. Porcellato, G. Bisoffi, S. Gustafsson, L. Boscagli, D. Carlucci, F. Chiurlotto, M. Morvillo, F. Stivanello, *Experience with the ALPI linac resonators*", Nuclear Instruments and Methods **A382** (1996) pp 121-124.

4. A. Facco and J.S. Sokolowski, The bulk niobium resonators program at LNL, Nuclear Instruments and Methods **A328** (1993) pp.275-278

5. V. Palmieri, *New developments in low beta superconducting structures*, Proc. of the 7th Workshop on RF Superconductivity, Gif-sur-Yvette, October 1995, pp 237- 258.

6. C. Benvenuti *Superconducting Coating for Accelerating RF cavities: Past, Present and Future*, Proc. of 5th Workshop on RF Superconductivity DESY, Hamburg, editor D. Proch, DESY M-92-01 (1992) pp 189-209.

7. V. Palmieri, A. M. Porcellato, V.L. Ruzinov, S.Y. Stark, L. Badan, A. Beltramin, L. Bertazzo, R. Preciso, F. Stivanello, G. Bisoffi, L. Boscagli, D. Carlucci, A. Dainelli, G. Fortuna, S. Gustafsson, A. Lombardi, M. Morvillo, *Installation in LNL ALPI linac of the first cryostat with four niobium quarter wave resonators*, Nuclear Instruments and Methods **A382** (1996) 112-117.

8. S. Stark, V. Palmieri, L. Badan, C. Durand, I.I. Kulik, R. Preciso, F. Stivanello, W. Venturini, S. Zandolin, *Niobium Sputter-coated QWRs*, Proc. of 8th Workshop on RF Superconductivity, Abano, October 1997, in press.

9. G. Bassato, A. Battistella, M. Bellato, and S. Canella, *Automation of procedures for locking ALPI resonators*, Nuclear Instruments and Methods **A329** (1993) pp. 195-198.

10. V. Palmieri, R. Preciso, V.L.Ruzinov, S.Y. Stark, L. Badan, A.M. Porcellato Sputtering of Niobium Film onto Copper Wave Resonators, Proc. of 5th Workshop on RF Superconductivity DESY, Hamburg, editor D. Proch, DESY M-92-01 (1992) pp 473-486

11. A. Dainelli et al., *Status and Operation of the XTU-ALPI Complex*, these proceedings

Status of the First Batch of Niobium Resonator Production for the New Delhi Booster Linac

Prakash N. Potukuchi[*] and Subhendu Ghosh[*]
Nuclear Science Centre, Aruna Asaf Ali Marg, New Delhi 110067, India
and
K.W.Shepard
Physics Division, Argonne National Laboratory, Argonne, IL 60439, USA

Abstract: This paper reports the status and details of the costs of construction of niobium superconducting resonant cavities for a linear accelerator, presently being built as a booster for the 15 UD tandem Pelletron accelerator at the Nuclear Science Centre, New Delhi. The linear accelerator will have three cryostat modules, each holding eight quarter-wave resonators. Construction of a batch of ten resonators for the linac started at Argonne National Laboratory in May 1997. For production, all fabrication and all electron beam welding is being done through commercial vendors. Details of construction and present status of the project are presented.

1. Introduction

A superconducting linear accelerator booster for the existing 15 UD tandem Pelletron accelerator [1], at the Nuclear Science Centre, is presently being constructed in collaboration with Argonne National Laboratory (ANL). A prototype quarter wave resonator (QWR) for the linac was designed, fabricated, and tested successfully at ANL [2]. The linac [3] will eventually consist of three cryostat modules each containing eight QWRs. Production of the first batch of ten resonators is presently nearing completion at ANL.

Figure 1 shows the schematic diagram of the quarter wave resonator. The resonator parameters (referenced to an accelerating gradient of 1 MV/m) are:

Resonant Frequency	97.0 MHz
Synchronous Velocity	0.08 c
Drift Tube Voltage	85 kV
Energy content	0.131 J
Peak Magnetic Field	106 G
Peak Electric Field	3.9 MV/m
Geometric Factor QRs	17.3
Active Length	15.9 cm

---

[*] Currently at the Physics Division, Argonne National Laboratory, Argonne, IL, 60439, USA

The cavity is formed entirely of niobium and is closely jacketed in an outer vessel of stainless steel, which contains the liquid helium required to cool the superconducting structure. Where the outer stainless steel jacket joins the niobium resonator (i.e. at the beam ports, coupling ports, and the slow tuner end of the niobium housing), a flange made of explosively-bonded niobium and stainless steel is used to provide a welding transition between the two materials. Details of the design have already been presented elsewhere [2,4].

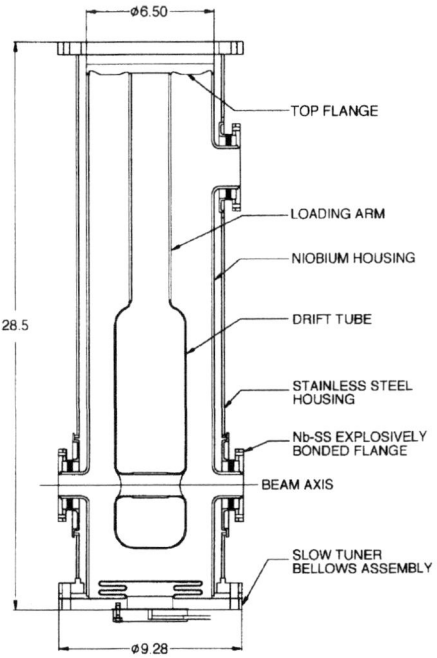

Figure 1. Schematic diagram of the 97 MHz Quarter Wave Resonator. Dimensions are in inches.

2. Resonator Fabrication

2.1 Mechanical Design of the Cavity

The niobium cavity parts are fabricated from both sheet metal and bar stock. Table 1 lists the various parts of the resonator, the material used, its thickness, and the various machining operations performed on them. Figure 1 shows the different parts of the cavity. There are thirty three electron beam welds per cavity, twenty one - 1/8" thick, eight - 1/16" thick, and four - 1/32" thick.

Item	Material	Rolling	Forming	Turning	Mill/Boring
Niobium Housing	Sheet, Gr-1 0.125" thk	X		X	X
Loading Arm	Sheet, RRR 0.125" thk		X	X	X
Drift Tube Cylinder	Sheet, RRR 0.062" thk	X		X	X
Coupling Port	Sheet, Gr-1 0.125" thk			X	X
Beam Port - Housing	Bar, Gr-1			X	X
Beam Port - Drift Tube	Bar, RRR			X	X
C.Port Extension Tube	Sheet, Gr-1 0.125" thk			X	X	X
Drift Tube Upper/End Cap	Sheet, RRR 0.062" thk			X	X
Seamless Tube (Drift Tube Port)	Tube, RRR			X	
Top Flange	Bar, RRR			X	
Nb-SS Transition Flange	Plate, ¼" thk Nb + ¼" thk SS			X	
Nb-SS Open End Flange	Plate, ¼" thk Nb + ¾" thk SS			X	
Nb-Cu Slow Tuner Flange	Plate, 0.062" thk Nb + ½" thk Cu			X	
Slow Tuner Bellows	Sheet, RRR 0.032" thk		X	X	

Table 1. List of the niobium resonator parts and the various machining operations performed on them (indicated by a cross). Gr-1 is Grade-1 material, RRR is high purity and high conductivity material, and thk denotes thickness.

2.2 Initial Development

The resonator production is being carried out in collaboration with commercial vendors in USA and India [5,6,7]. During the development of the prototype resonator most of the niobium machining, all the sheet metal work, and all the electron beam welding was performed in-house at ANL. Therefore before the production could start the technology of fabricating niobium resonators had to be transferred to the outside vendors. Considerable amount of time and effort went into training the vendor manpower to perfect the machining and sheet metal work with niobium. Several parts were formed, initially out of copper sheets, and later out of niobium, as practice pieces. Other machining operations, e.g. turning, milling and drilling were performed to identify suitable machining parameters, tools, coolants etc. Most of the dies made during the fabrication of the prototype resonator were used for forming parts. However, additional dies were made as felt necessary. Similarly, where appropriate, additional machining fixtures were made during the development process.

A major effort went into developing the electron beam welding parameters for welding niobium of different thickness [8]. The electron beam welder is a five-axis CNC machine with a movable gun and tilt, and a large vacuum chamber of size 138" x 108" x 107". The machine is equipped with a three head-stock rotary fixture with matching tail-stocks. In order to fully exploit the capabilities of the welder, and increase productivity, the weld fixturing was designed to perform multiple welds in a single pump down.

2.3 Fabrication Details

The contract for the first batch of production calls for fabricating ten resonators. Because of increased productivity from the electron beam welding machine, resulting in a cost reduction, we are aiming to produce twelve complete resonators and most of the parts for two more.

All the niobium material was received by November '97. Chemical analysis, RRR measurements and water dunk tests were performed on niobium samples to check for its purity. The fabrication work started in December '97.

The entire machining and welding effort has been done in four major groups of tasks in the following time sequence. Highlights of the major efforts are:

1. In the first task group the niobium housings and drift tube cylinders were rolled, and the extension tubes for the coupling port saddles were formed, and welded.

2. In the second task group the loading arm, drift tube beam port assemblies, and the coupling port saddles were completed. In all 98 welds were performed in a week.

3. In the third task group the bare niobium housings were completed by welding the beam and coupling ports to the housing cylinder. Additionally the drift tubes were also completed. In all 108 welds were performed in a week.

4. Following the fabrication of the drift tubes and loading arms, they were electropolished and heat treated. These two elements were then welded together to complete the central conductor assemblies.

2.4 Present Status

At the time of presenting this paper all niobium housings complete with beam and coupling ports, the central conductor assemblies, the top and open-end flanges, the stainless steel housings, and the beam and coupling port transition flange assemblies (niobium explosively bonded to stainless steel) are ready. This amounts to a total of about 80% of the machining and sheet metal work, and about 63% of the total electron beam welding work for the production.

Figure 2. Niobium Housings with the Beam & Coupling Ports

Figures 2 and 3 show all the niobium housings and the central conductor assemblies respectively. We plan to proceed with completing initially two bare niobium resonators (without the stainless steel outer jacket), which will be dunk tested in liquid helium. They will then be jacketed in the stainless steel outer vessel for vacuum tests. After the successful testing of the resonators in vacuum the remaining resonators will be completed. The initial two resonators have been tuned to frequency and welded to the top flange. They are being prepared for the final closure weld to the niobium housing. Work on the slow tuner system will be taken up after completing the fabrication of the first two cavities.

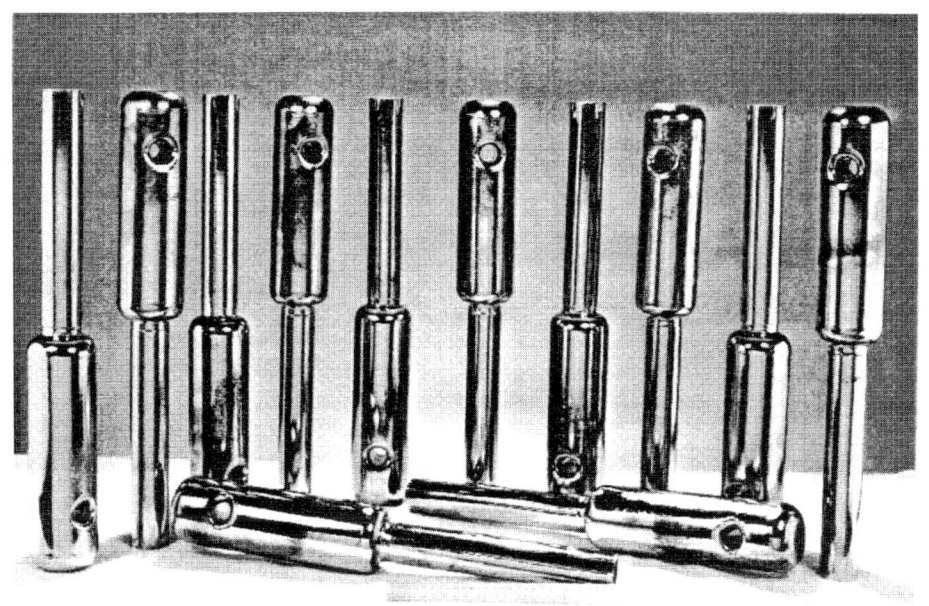

Figure 3. Central Conductor Assemblies

3. Costs and Effort

The man-hours effort for the project is shown in Table 2. We have chosen to present the cost of the project in terms of the man power effort since costs may vary from laboratory to laboratory, depending upon the overheads, cost of labor, etc. Several points should be noted:

1. The total niobium material ordered is about 1100 lbs., of which 750 lbs. is Grade-1 material and 350 lbs. is RRR grade material. This includes about twenty percent allowance for contingency.

2. We are building twelve complete resonators and most of the parts for two more. We consider the scope of the present work to be equivalent to building thirteen complete resonators.

3. The effort indicated in the category *"niobium machining"* includes the machining and forming of the niobium parts, and the machining of the niobium-stainless steel explosively bonded parts.

4. The effort indicated in the category *"engineering"* shows the engineering effort of the outside vendors only.

5. We choose to present the effort in the category *"electron beam welding"* separately, since the hourly rate for performing it is about three times more than the individual rates for *niobium machining, engineering,* and fabricating *fixtures* (which are all about the same).

Job	Man Hours Spent on Job	Estimated Man Hours to Complete Job
Niobium Machining	2250	450
Engineering	330	70
Electron Beam Welding	210	120
Machining & Welding Fixtures	650	70
Stainless Steel jobs	750	250
Electropolishing & Heat Treatment	500	700
Technical Coordination	2000	600
Testing	0	1200
Miscellaneous	1500	500

Table 2. Man hours effort.

6. The effort shown in the category *"stainless steel jobs"* includes the fabrication of the outer vessel, the stainless steel top flange, and flanges for the beam and coupling port transition flange assemblies.

7. The *"electropolishing & heat treatment"* work is performed in-house at ANL. The effort shown in the category also includes cleaning and etching of the niobium parts performed prior to electron beam welding them.

8. The effort indicated in the category *"technical coordination"* is the effort put in by the authors and includes resolution of all technical problems encountered during the fabrication process, fixture designing, material testing, directing the machining, forming and beam welding developments, participation in the beam welding of the resonator components, etc. A major portion of the effort in the categories *technical coordination* and *miscellaneous* was put during the initial development work. For constructing similar number of cavities in subsequent

productions we expect the total of this effort to reduce by about half. Similarly, all the effort in fabricating machining and welding fixtures, and some portion of the effort in engineering, will also reduce.

9. The effort indicated in the category *"miscellaneous"* includes all other administrative responsibilities, travel to vendor sites, correspondence, and overall coordination.

4. Future Plans

A liquid helium dunk test of the bare niobium portion of the first two resonators is scheduled for December '98. Vacuum tests will begin afterwards and are expected to be over by March '99. The remaining resonators are expected to be fabricated by April '99. Fabrication of the slow tuner bellows will begin after completing the initial two resonators. We estimate that all the twelve resonators along with the slow tuner bellows will be ready and tested by July 1999.

Acknowledgements

The authors would like to thank Mr. Edward Bonnema and Mr. Lester Dudek of Meyer Tool and Mfg. Inc., and Mr. Radwan Mourad and Mr. Tadeusz Hejna of Sciaky Inc., for their cooperation during this work. The authors thank Mr. Rajeev Mehta and Dr. Amit Roy for their help in getting the stainless steel jackets, and the beam & coupling port transition flanges fabricated at New Delhi. Special thanks are due to Mr. Mark Kedzie for his suggestions and help in electron beam welding of the niobium parts. The authors would like to thank Prof. G.K.Mehta, Dr. Walter Henning and Dr. Jerry Nolen for their continuing support.

References

[1] D. Kanjilal et. al., Nucl. Instr. and Meth. A328, (1993), 97-100
[2] K.W. Shepard, A.Roy and P.N.Potukuchi, Proc. of the 1997 Particle Accelerator Conference, May 12-16, 1997, Vancouver, BC, Canada, p3072
[3] A.Roy et al., in this Proceedings
[4] K.W. Shepard, A.Roy and P.N.Potukuchi, Proc. of the 1993 Particle Accelerator Conference, May 17-20, 1993, Washington, DC, p1045
[5] Meyer Tool & Mfg., Inc., 4601 Southwest Hwy., Oak Lawn, IL 60453, USA
[6] Sciaky Inc., 4915 W. 67th Street, Chicago, IL 60638, USA
[7] Don Bosco Technical Institute, Okhla Road, Jamia Nagar, New Delhi, 110025, India
[8] Mark Kedzie - ATLAS, ANL, Private communication

Study of Acceleration Across the TTF's Zero-Crossing Velocity in Independently Phased Linacs

SuehiroTakeuchi

Japan Atomic Energy Research Institute, Tokai Research Establishment
Tokai, Naka, Ibaraki 319-1195 Japan

Abstract. Quarter wave resonators used in independently phased linacs have a wide incident velocity acceptance above the velocity of about 0.5 β_{opt}, where β_{opt} is the optimum incident velocity for the resonators. Their transit time factor(TTF) crosses zero at the velocity of 0.5 β_{opt} and has a small negative peak at 0.35 β_{opt}. An acceleration of ions across the zero-crossing velocity starting from a velocity in the negative peak is discussed. For taking into account of velocity changes in the resonators, a numerical calculation was carried out and it was found that a very slow acceleration can occur if the rf phase is properly changed over the transitional region. In the calculation, Cl^{10+} ions and the shape of the quarter wave resonators of β_{opt} = 0.1 in the JAERI Tandem Superconducting Booster were chosen.

INTRODUCTION

Superconducting quarter wave resonators are used in many independently phased linacs to boost the energy of heavy ions from tandem accelerators. The resonators have two acceleration gaps and can accelerate heavy ions over a wide range of incident velocity above 0.5 β_{opt}. The transit time factor of the resonators of β_{opt} = 0.1 for the JAERI tandem booster is shown in Fig. 1, as an example[1]. The transit time factor(TTF) that characterizes the acceleration efficiency for different incident velocities. Normally, ions are accelerated where the factor is close to one. The TTF crosses zero at the velocity of 0.5 β_{opt} and has a small negative peak below the velocity. At the negative peak, an acceleration can take place if the synchronous phase is reversed. It is interesting to know whether it is possible in an independently phased linac to accelerate ions across the TTF's zero-crossing velocity starting from a lower incident velocity. In case of such a low incident velocity, non-linear effect is large[2]. This work showed that the acceleration is theoretically possible, by carrying out a numerical calculation.

METHOD OF CALCULATION

The energy gain from a resonator is usually given as

$$\Delta E = q\, E_{acc}\, L\, TTF(\beta)\, \cos\phi_s,$$

where q is the charge state of incident ions, E_{acc} the mean field gradient in the resonator, L the acceleration length, TTF the transit time factor, β the beam velocity and ϕ_s the synchronous phase. This formula is valid only for the case that velocity changes in the resonator are negligible. For the present case, it is quite necessary to take into account of the velocity changes, because acceleration and deceleration take place alternatively in the alternating electric field although the resultant velocity change after the resonator is very small.

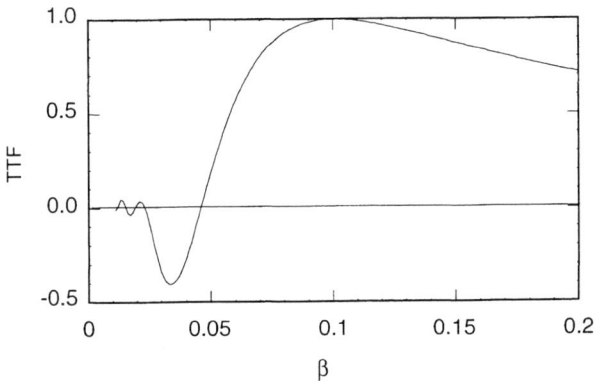

FIGURE 1. Transit time factor of a $\beta_{opt} = 0.1$ quarter wave resonator.

The calculation was carried out by numerically solving the one dimensional equation of motion;

$$m\, d^2 z/d\, t^2 = q\, e\, E(z)\, \sin(\omega t + \phi_s)$$

using the Runge-Kutta method, where m is the mass of the incident particle, z the position of the particle on the beam axis, e the unit charge, $E(z)$ the electric field amplitude along the beam axis and ω the angular frequency. In the calculation, Cl^{10+} ions and the shape of the quarter wave resonators of $\beta_{opt} = 0.1$ in the JAERI Tandem Superconducting Booster were chosen.

RESULTS

In Fig. 2, the field amplitude $E(z)$ for E_{acc} = 5 MV/m, the field acting on a particle of 50 MeV Cl^{10+} with a synchronous phase of -30° and the resultant velocity are shown as a function of particle position. The incident velocity in the calculation is appreciably higher than the TTF's zero-crossing velocity. It can be seen that there is a deceleration in the beginning half of the first gap and a substantial acceleration occurs in the second gap. Such calculations were repeated for different incident velocities and different phases.

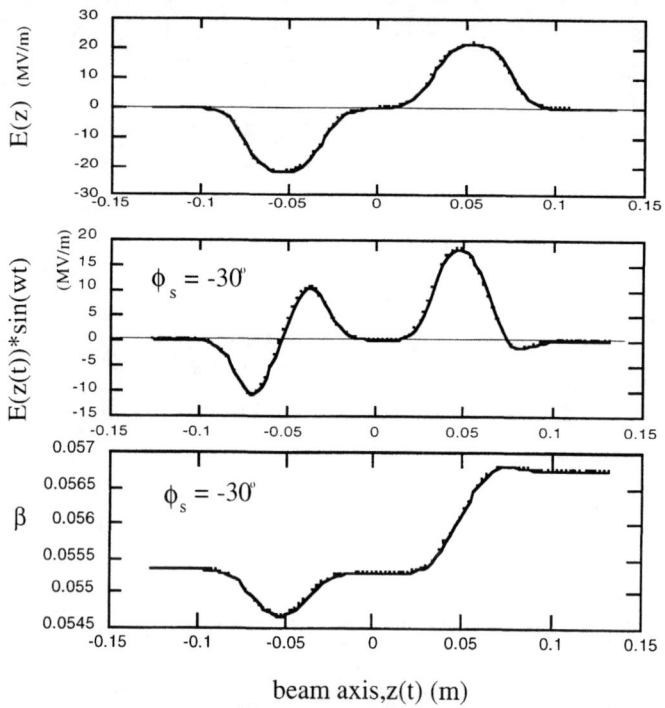

FIGURE 2. Field amplitude profile, electric field acting on an incident particle of 50 MeV Cl^{10+} ion and its velocity as a function of beam axis, calculated at a synchronous phase of -30°.

In Fig. 3, velocity increases/decreases from several incident velocities calculated at a field gradient of 5 MV/m are shown as a function of synchronous phase. The velocities that give no energy gain(thick solid curves) were found to move up in the negative side of synchronous phases and to move down in the positive side. One finds, for the incident velocity of 0.0459, two spans of phases giving small velocity increases not only in the normal acceleration phase span between -90° and +90° but also in the normal deceleration phase span between +90° and -90° through ±180°. And one can see that an acceleration over the transition zone is possible if one takes a path on velocity increasing phases as shown by arrows in the figure. For an actual beam with a finite emittance, however, it is presumably not easy to pass through the transition zone where the phase acceptance is half as much as that in the normal zone.

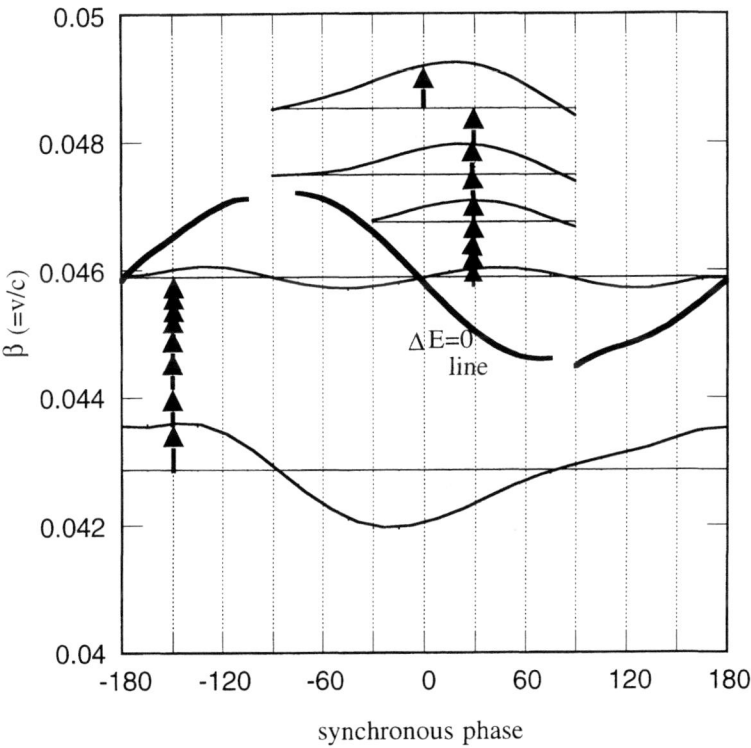

FIGURE 3. Velocity increases/decreases of Cl^{10+} ions as a function of synchronous phase from several incident velocities. Thick arrows indicate the accelerations on the path taken in the calculation of Fig. 3.

The movement of no-energy-gain incident velocity and the energy gain of ions injecting at the TTF's zero-crossing velocity depend on accelerating field gradient. The no-energy-gain incident velocity at $\phi_s = -150°$ is presented as a function of field gradient in Fig. 4. The higher the field gradient, the greater the displacement from the reference velocity of 0.0459 which refers to TTF = 0. The energy gain at the reference incident velocity and at $\phi_s = 30°$ is presented as a function of field gradient in Fig. 5. The gain changes as E_{acc}^2 or the acceleration efficiency(= gain/voltage) is proportional to E_{acc}.

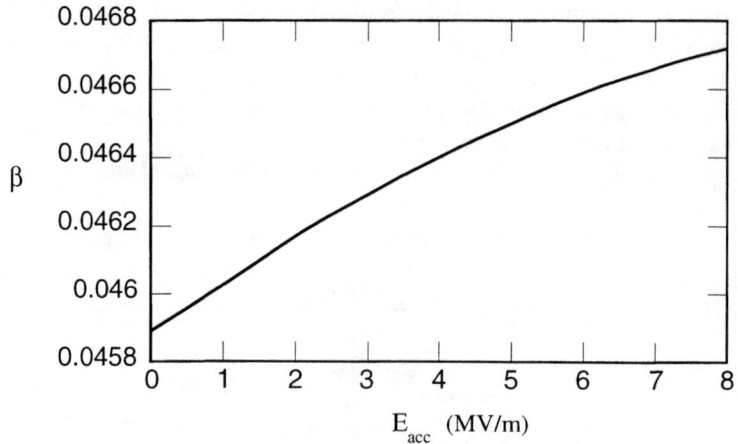

FIGURE 4. Field gradient dependence of the velocity which gives no energy gain $\Delta E = 0$ at a synchronous phase of $-150°$, calculated for 50 MeV Cl^{10+}.

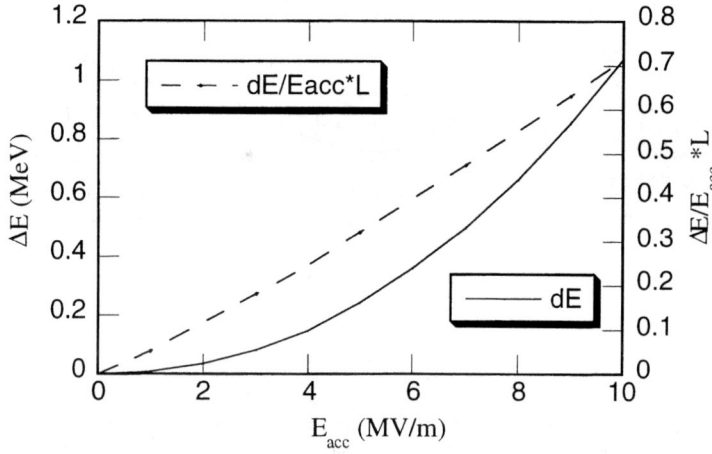

FIGURE 5. Field gradient dependence of the energy gain at a velocity of 0.0459 which gives TTF = 0 and at a synchronous phase of 30°, calculated for 50 MeV Cl^{10+}.

As an example of acceleration across the TTF's zero-crossing velocity, Fig. 6 shows the energy growth as well as energy gains calculated for an acceleration of Cl^{10+} with incident energy of 30MeV by 40 resonators with a field gradient of

5MV/m. The synchronous phase is changed from -150°, through +30° and then to -30° as is illustrated in the figure.

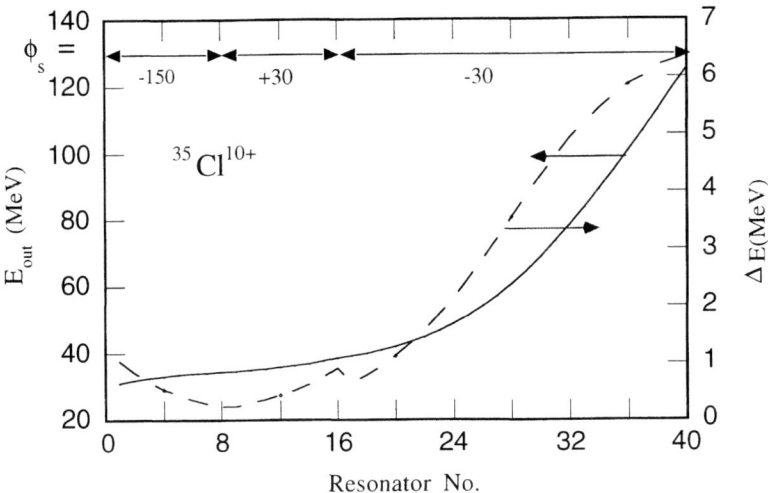

FIGURE 6. Energy growth(solid) and energy gain (broken) of Cl^{10+} as a function of resonator number.

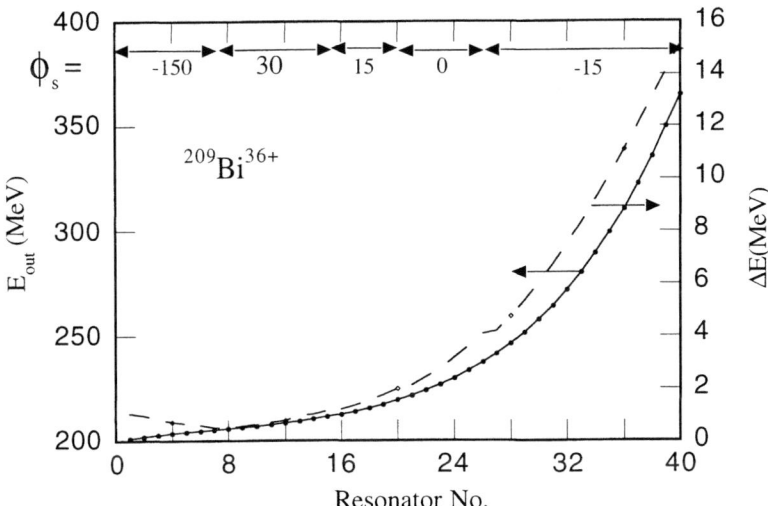

FIGURE 7. Energy growth(solid) and energy gain (broken) of Bi^{36+} as a function of resonator number.

For a practical application to a tandem booster linac like the one at JAERI, accelerations of very heavy ions are interesting. Another result calculated for $^{209}Bi^{36+}$ is seen in Fig. 7. Seen in Fig. 8 is the result of calculations for different ion species with respect to the energy gains and velocity increases at the reference TTF's zero-crossing velocities, at two different phases and at E_{acc} =5 MV/m.

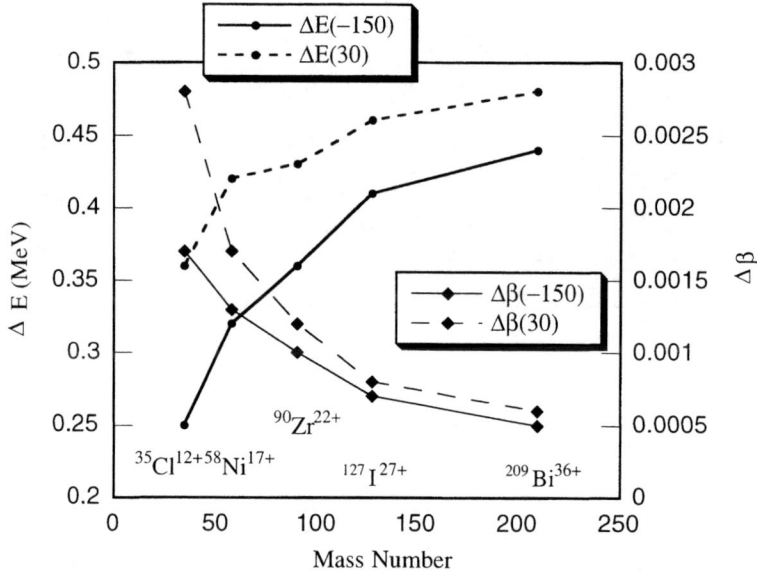

FIGURE 8. Energy gains and velocity increases calculated for different ions at the incident velocity of 0.0459(TTF=0) and with an accelerating field gradient of 5 MV/m.

CONCLUSIONS

Numerical calculations of heavy ion accelerations over the transition zone through quarter wave resonators were carried out taking velocity changes in the resonators into consideration. As a result, we found that the injection velocity where the energy gain becomes zero changes a lot depending on the phase as the field gradient goes high, that an acceleration phase span appears in the normal deceleration phase span at the transition zone, and so that it is possible to accelerate ions starting from a low injection velocity of TTF<0, crossing the velocity of TTF=0 and entering into the normal acceleration zone of TTF>0 with changing the synchronous phase. The higher the field gradients, the larger the acceleration effect. For an actual beam with a finite emittance, further calculations are wanted.

REFERENCES

1. S. Takeuchi, T. Ishii, H. Ikezoe, Nucl. Instr. and Methods A281(1989)426-432
2. J. R. Delayen, Nucl. Instr. and Methods A258(1987)15

A Distributed Control System Status Report of the Munich Accelerator Control

L. Rohrer and H. Schnitter

Beschleunigerlaboratorium der Universität und der Technischen Universität München,
85748 Garching, Germany

Abstract: A system of computers connected by a local area network (ARCNET®) controls the Munich accelerator facility. This includes ion sources, the tandem accelerator, the beam transport system, the gas handling plant, parts of experimental setup and also an ion source test bench. ARCNET is a deterministic multi-master network with arbitrary topology, using coax cables and optical fibers. Crates with single board computers and I/O-boards (analog, parallel or serial digital), dependent on the devices being controlled, are distributed all over the building. Personal computers serve as user interfaces. The LAN communication protocol is a client/server protocol. Communication language and programming language for the single board computers is Forth. The user mode drivers in the personal computers are also written in Forth. The tools for the operators are MS-Windows applications, programmed in Forth, C++ or Visual Basic. Links to MS-Office applications are available, too.

INTRODUCTION

The distributed computer control system for the Munich accelerator was put into operation for the first time 10 years ago, and it was developed further continuously since. It controls now the accelerator facility comprising ion sources, the 15 MV tandem accelerator, the beam transport system, and auxiliary systems (e.g., the gas handling plant), parts of experimental setup, and an ion source test bench.

HARDWARE

The architecture of our control system is decentralized. Its backbone is a local area network with crates containing single board computers and I/O boards on the machine side and personal computers used as control desks on the user side.

Local Area Net

The local area network is ARCNET®. The main reasons for this choice were the topology of the network and the bus arbitration. The net can be configured physically as a star or a bus or a combination of both, so it could be easily adapted to the topography of our laboratory. The transport medium is coaxial cable or fiber optics, the latter are used for devices at elevated potential or in areas with electrical interference due to high voltage sparks.

The bus arbitration is performed in a modified token passing scheme in which, after an automatic configuration run, all token passes are acknowledged by the node accepting the token. The maximum time for a round trip token pass is calculable and finite: in our network it is 20 ms at heavy traffic, but under normal conditions it is well below 5 ms and can be neglected compared to the rest of the data processing time.

Nodes for Devices

About 50 device nodes are spread throughout the building. These comprise a single board computer, an ARCNET interface board, and I/O-boards in an ECB-crate. When we started developing our control system ECB was a bus system with a wide distribution, mainly in Europe. Circuit boards of various kinds were available from many manufacturers. Our initial aim was to use commercially available boards, but in the course of the development more and more of them were made in house, partly for reliability and performance reasons but also to meet special requirements of existing devices. Some of the boards which we purchased in the past are not available any more, consequently we make the complete hardware in our laboratory.

The single board computer uses a Z280 CPU, which is a successor of the Z80. The other chips on the board are the ROM, non-volatile RAM, CIO timers, a real time clock, a watchdog circuit, and the ECB interface.

The ARCNET interface board consists of the chip set for the ARCNET (controller and line-driver), a Z80 CPU with RAM and ROM, and a dual ported RAM to communicate with the ECB bus. While the ARCNET controller handles the layer 1 to 2 of the ISO-OSI model the layer 3 to 4 is covered by the Z80 CPU.

The equipment of the crates with digital and analog I/O-boards depends on the application. Here follows a short description of our standard boards: For analog output two boards are available: one with 0V...10V output and 16 bit resolution, optically coupled, and the other one with 8 output channels -10V...+10V and 12 bit resolution. The output amplifiers of the latter are analog adders adding the DAC voltage to the ground potential of the devices to be controlled. This differential output is nearly as good as an optically isolated output if the ground potentials of the devices differ only by a few volts. For voltage monitoring we use a 12 bit ADC board with 16 multiplexed differential input channels (-10V...+10V). Parallel digital output is effected either by a unit with 16 relays or a 32 bit TTL output board. Matching to these boards a unit with 16 opto-coupled inputs and an other one with 32 TTL inputs are available. More and more commercial available devices are equipped with a serial input/output link. These are controlled by a serial I/O unit with cable or fiber optic. It has a microcontroller and a buffer memory on board, so the main CPU is relieved of time critical I/O events. An IEEE-488-bus interface is available, too.

In addition to these standard boards some special units were developed: For beam current measurement we use a picoampere-meter (30 pA to 1 mA full scale) with ECB interface, the input being well protected against high voltage surges, so that it can be used also on the injector platform. As mentioned above we made special boards to control existing equipment such as the pulsing system, the step motors of the slits, or the beam profile monitors.

Multiplexed serial data links control items in the terminal and in the dead sections of the tandem. The node on ground potential uses standard serial I/O-boards communicating via fiber and infra-red links through windows in the tank wall with relay stations inside the tank, where the signals are demultiplexed and forwarded by fiber optic links to the individual devices.

Control Desks

Since the control system is decentralized one can connect many control desks in arbitrary places to the ARCNET, and each of them can control any object.

The standard control desks are personal computers with Microsoft Windows 3.11. They are connected to an Ethernet LAN for data and program exchange with a file server. But the applications available to the operator for controlling and monitoring accelerator data can be executed without any connection to the file server. This increases the reliability of the system and eases the maintenance of accelerator components, because one can operate a control desk in any place near the object to be serviced.

For those who prefer control knobs to the mouse for the control of analogue values a mini control desk has been developed. It is a small box containing a Z8 microcontroller, two knobs with incremental shaft encoders, buttons to modify the shaft encoder speed, and two LCD units to indicate the names of devices under control and the actual values. The mini control desk is connected with a serial link to an ECB crate described above. To attach and detach devices a standard control desk is used.

Other Computers in the LAN

Two more computers are connected to the LAN. One of them, with the same hardware as the device nodes, is an address server, where the names of all items in the LAN and their ARCNET addresses are stored in a non volatile RAM. These data are retrieved whenever an object is attached to a control desk or to another node in the LAN. The other one, a personal computer, is an error logger. All nodes which do not have a standard output unit, e.g. a screen, send their error messages to the error logger. These messages are locally displayed and stored on a disk for future analysis, and they are forwarded to other computers in the LAN on request.

SOFTWARE

The programming language used in the single board computers and as the network communication language is Forth. This is a language optimally suitable for machine control because its easy access to the hardware. It can switch at any time to a compile mode and back to an execution mode. In the compile mode new procedures are created, which later can be used in the execution mode. Each single board computer contains an Open Network Forth (ONF), a Forth system with extensions for the network communication, resident in the ROM. During a download process the data stream coming from the net is interpreted as a source code to be compiled. So the computer learns its job. After that the incoming instructions are executed immediately. So the computer controls the accelerator hardware. The source code is also stored in a segment of the non volatile RAM, and the nodes can later self-boot using the code in the source code segment. However the program can be redefined at any time via ARCNET if any revision is necessary.

Network Philosophy

The communication between tasks on different nodes is based on the client/server scheme: if a client sends a request labeled with a message ID to another node the latter then acts as a server and responds with the same message ID. If the server performs an action correctly it sends 'ok' to the client, otherwise 'ko'. The client interprets the response and may perform an action routine or an error handling routine.

Different message types are available: Messages of the type REQUEST are used for instructions to be executed by the server and result in a RESPONSE message to the client. Other types are TRANSMIT for any other message to the interpreter/compiler (e.g., program code) and ERROR for messages to the error logger.

The main task for the processor on the ARCNET board is the routing of messages between tasks on different nodes and to prevent deadlock situations. It also controls the access to the node. Each node can be switched into a private or public state. An access violation to a node owned by an other client is responded to the requester itself and reported to the error logger.

Device Nodes

Objects are implemented as data structures comprising data fields and pointers to object specific procedures. Here some examples for the contents of the data structures:

current value, lower and upper limits, name of the scale unit, calibration constants, hardware addresses, pointers to I/O procedures and scaling routines. The approach to object oriented programming results in device-independent messages. A request can, for instance, set the field of the high energy analyzing magnet to a certain value, or it can start a gas transport procedure after closing the tandem tank: evacuate the tandem tank, let gas flow with controlled speed from the storage tank to the tandem, start the compressor if pressures are equal, stop the compressor if the tandem tank is full, close all gas and cooling water valves.

Addressing is performed by an EXPORT/IMPORT scheme. Calling EXPORT introduces the most recently defined object to the address server, and any other node can IMPORT it. The address server converts symbolic addresses to ARCNET ID-numbers. Onward the programmer can use REQUEST without caring where the object is located. IMPORT may also be used as a forward reference being resolved at the execution time.

Applications

The operator interface is based on Microsoft Windows with user mode drivers for the ARCNET communication written in Forth. Various tools[1], programmed in Forth, C++, and Visual Basic, are available to the operators. Links to MS-Office applications are available, too.

The standard control desk was designed for quick and easy configuration. A group of 12 arbitrary objects can be attached to a display with 12 fields. The current values are displayed as bar graphs with a refreshing rate of about 3 per second. One can set digital and analog values by mouse clicks to the objects. Switching from one group to another is done by menu picking. The program flow is controlled by a text file where groups, menus and other options are described.

A control desk with graphic presentation of the equipment and the processes is preferred to the standard control desk for getting a general view of the accelerator. The background of the desk top is a bit map (e. g., of an ion source or a beam extension). Data displays of variable size are arranged in appropriate positions. All other functions of both control desks are similar.

A sequencer has been developed for automatic control: In a script file one can lay down the flow of actions dependent on time and actual measurements. This program is regularly used for high voltage conditioning of the tandem, but occasionally also for other purposes, e. g. to start an ion source.

A program to calculate initial values for the accelerator is regularly executed before setting up the machine. This program uses data tables (e. g., an atomic mass table, K. Shima's stripper tables, data of our facility, and data from former runs of the program) in a Microsoft Access data base. The operator can choose from a menu which values he wants to calculate (negative ion injector, tandem, the pulsing system, beam transport system), he types the input data in a form, and the results are displayed immediately.

---

[1] in collaboration with FORTecH Software, Rostock

The program has access to the accelerator, one can set the calculated values by a mouse click.

The applications packet contains also a log book program to record the status of the machine on request. All data are stored in an Access data base, and a selection of them is printed on paper, too. If somebody wants to set up the machine with data logged earlier he can use a program with access to the machine to retrieve the old status.

A few application programs were made for special purposes, for instance to record a mass spectrum of an ion beam, and for special devices, for instance the beam profile monitors. Also a strip chart recorder was implemented.

Conclusions

Ten years experience in controlling the tandem accelerator including ion sources, beam line components, and subsystems have shown that the system serves its purpose excellently in spite of its low cost. The accuracy and the speed are adequate. A high flexibility is inherent to the nodes programmed in Forth. New devices can be added quickly, and also specific application programs can easily be implemented. Any hardware and software may be replaced or modified if necessary. The only condition is that it can be connected to the ARCNET and can be programmed to communicate in Forth. The architecture keeps the expenses for cable installations low, speeds up the operation because parallel processes can be executed, and facilitates modifications. At present we are in the process to change from Windows 3.11 to the Windows NT system.

Completion of the ATLAS Control System Upgrade

F. H. Munson

Physics Division, Argonne National Laboratory, Argonne, IL 60439

Abstract. In the fall of 1992 at the SNEAP (Symposium of North Eastern Accelerator Personnel) a project to upgrade the ATLAS (Argonne Tandem Linear Accelerator System) control system was first reported (1). Not unlike the accelerator it services the control system will continue to evolve. However, the first of this year has marked the completion of this most recent upgrade project. Since the control system upgrade took place during a period when ATLAS was operating at a record number of hours, special techniques were necessary to enable the development of the new control system "on line" while still serving the needs of normal operations. This paper reviews the techniques used for upgrading the ATLAS control system while the system was in use. In addition a summary of the upgrade project and final configuration, as well as some of the features of the new control system is provided.

INTRODUCTION

ATLAS is a heavy ion accelerator that has roots dating back to the early 1960's. ATLAS has been designated a National User Facility, and attracts experimentalists from all over the world. The facility operates 24 hours a day, seven days a week.

As early as the late 1970's it was recognized that ion accelerators and their associated peripheral equipment were becoming complex devices. Some of these accelerators were already becoming too complicated to be controlled and monitored with traditional methods. These methods often employed the use of several electronic racks consisting of a host of potentiometers, meters, dials, switches, and indicator lamps. Today there are conferences devoted to the subject of accelerator control systems.

The early 1980's saw the construction of the accelerator system that would later evolve to become today's heavy ion accelerator facility, ATLAS. This system consisted of an electrostatic Tandem accelerator injecting beams into a booster linear accelerator employing RF (Radio Frequency) technology. Those responsible for the design of the control system for the newly constructed linear accelerator realized that the complexity of this new accelerator called for a computer assisted control system. Consequently, the resulting accelerator system was controlled and monitored by a hybrid control system. The more traditional control system was retained for the

Tandem, while the computerized control system was implemented for the new linear accelerator.

By the end of 1991 two additional linear accelerator sections, an ion source, and a more elaborate beam transport system were added to the facility. Due to evolutionary and operational constraints, the resulting control system consisted of the older traditional methods and three separate isolated computerized control systems. In 1992 work began on upgrading the ATLAS control system to provide a more cohesive system based on a singular control and monitor design.

THE SYSTEM TO BE UPGRADED

Prior to the start of the upgrade, control and monitoring of the accelerator was accomplished through the use of three DEC (Digital Equipment Corporation) PDP-11 computer systems (2). Control and monitoring of the ATLAS PII (Positive Ion Injector) was performed by one of the above mentioned computer systems, while control and monitoring of the primary accelerator section was controlled by a second system. All beam transport components external to the accelerator were serviced by the third system. All three systems contained separate and isolated databases.

Two separate CAMAC (Computer Automated Measurement And Control) Serial Highway subsystems were used as the primary interface between the computer systems and the various accelerator components as shown in Figure 1.

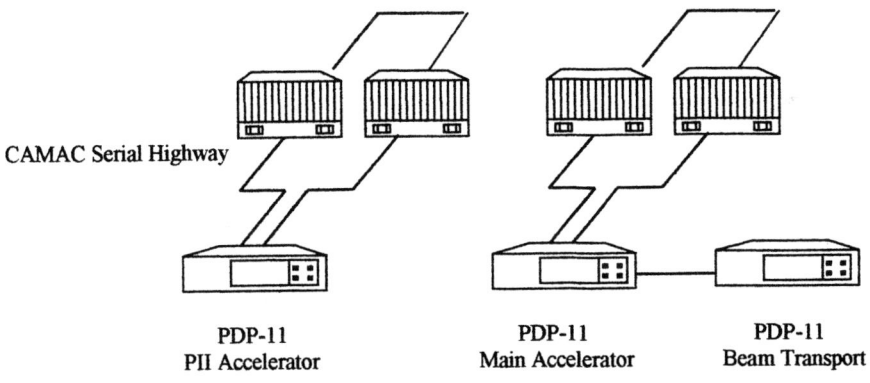

FIGURE 1. This figure depicts the system to be upgraded.

Control of the accelerator was accomplished using two separate control consoles. One console was remotely located, and was used to allow development of the new positive ion injector without interfering with normal day to day operation of the primary accelerator section. The second console was located in the main control room. Operator interfaces included a variety of touch screens, monitors, terminals, knobs, keypads, and switches.

UPGRADE STRATEGY AND IMPLEMENTATION

It was decided that the upgrade needed to provide a system that was modular, expandable with a minimum of effort, and conformant to current and emerging software and hardware standards to ensure long-term viability. Since the facility was entering a period where a record number of operational hours were planned, it was crucial that the new system be integrated into the overall control system without interfering with accelerator operations.

Specialized in-house written processes existed on the old control system that was the result of years of programming effort. The decision was made to provide a new system that was capable of accepting as much of the existing process code as possible. Due to the large investment in CAMAC components, the CAMAC I/O subsystem was retained, but the initial plan for the upgrade called for replacing all of the PDP-11 computers.

The staff responsible for the day to day operation of the control system, as well as achieving the upgrade goals, typically consisted of one full time system manager/programmer, one full time operator/programmer, and one part time co-op student/programmer. The size of this staff, given its responsibilities, provided a strong preference for acquiring an established control system software package that would form the real-time portion of the new control system.

Both commercial and non-commercial candidates were researched. Since the software package known as EPICS (Experimental Physics and Industrial Control System) was just being developed through a collaborative effort of participating national laboratories, there was a convincing motivation for adopting this control system software. However, only one package provided both CAMAC Serial Highway and CAMAC in-crate processing support as well as guaranteed technical support. Therefore, based on the size of the staff, and other considerations, the software package marketed under the name "Vsystem" by Vista Control Systems, Inc. was acquired (3).

In order to accomplish the previously described goals the initial upgrade strategy called for four overlapping phases. Since the initial goal of the upgrade was to centralize control of the entire accelerator in the main control room, the first step in phase one was to integrate control and monitoring of the positive ion injector into the main control system. The PII development computer system and the main computer system maintained two separate and isolated databases and two CAMAC subsystems. The two databases were combined to create one database, and the two separate CAMAC Serial Highways were combined to form one CAMAC Serial Highway.

Once this was accomplished the PII development computer system was no longer needed. Therefore, the first of the three PDP-11 computers was retired. The resulting system is depicted in Figure 2.

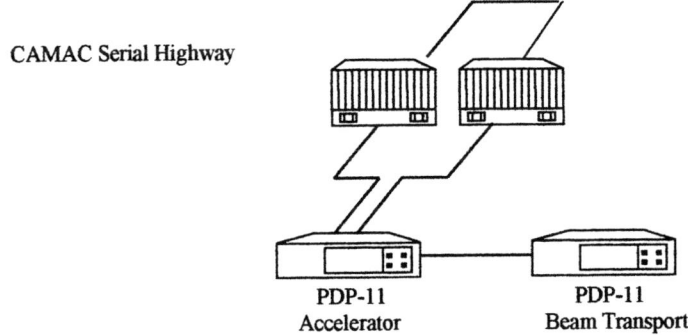

FIGURE 2. This figure shows the system configuration after the first upgrade phase.

The second phase called for porting all of the beam transport processes to the new control system. To help accomplish this a second I/O subsystem was installed. This was an Ethernet based LAN (Local Area Network), which is used to connect the various control system computers. A DEC MicroVAX was installed and connected to the newly installed LAN.

Two new database structures were installed on this new machine. The first database was the more dynamic "Vsystem" real-time database. This database provides the software or logical link to the CAMAC hardware. Changes made to this database causes CAMAC I/O to occur. The second database is a relational database, which uses the data management software product "Oracle Rdb" (4). This database contains static information about accelerator devices that would be inappropriate to store in the Vsystem real-time database. Information such as a device's name, the associated channel name in the real-time database, chassis rack location, and any other constant-like parameters associated with a device are stored in this database.

Cooperating communication processes were written for both the MicoVAX and the PDP-11. Since at this time the PDP-11 was still physically interfaced to the CAMAC subsystem, these two processes provided a means for the PDP-11 to issue CAMAC I/O requests on behalf of the newly configured computer system. This scheme provided the means for developing processes on the new system without affecting the configuration or operation of the old system.

Two additional cooperating processes were written for the MicroVAX and the PDP-11 that were responsible for maintaining database consistency between the two systems. This allowed an operator the flexibility of controlling a device from either the old system or the new system. If a device parameter was changed on the old system, the database of the new system would be updated. Likewise, if a device parameter was changed on the new system the database of the old system was updated.

At the end of the second phase of the upgrade all of the processes used for the control and monitoring of all beam transport devices were moved from the PDP-11 that was responsible for this task to the newly installed MicroVAX. This enabled the retirement of the second of the three PDP-11 computers. The configuration after phase two is shown in Figure 3.

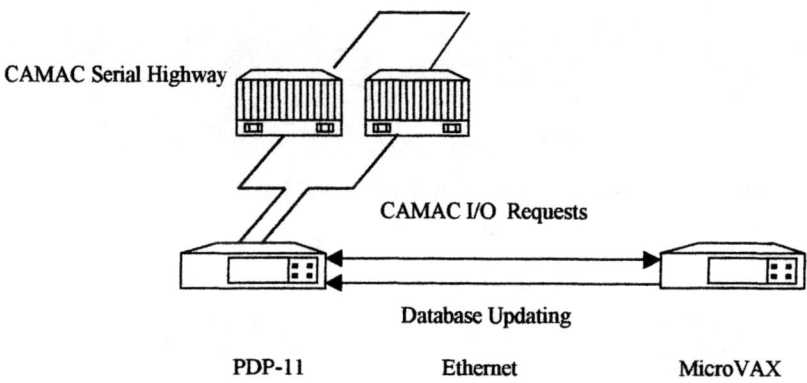

FIGURE 3. This figure is an indication of the system configuration after upgrade phase two.

During phase three, as more processes were moved to the new control system, the PDP-11 was issuing an increasing number of CAMAC I/O requests on behalf of the MicroVAX. This consequence lead to a degradation of overall system performance. It was at this time that the physical connection of the CAMAC Serial Highway was moved from the PDP-11 to the MicroVAX.

Also in phase three a method for storing complete accelerator tune configurations was put in place. The Oracle Rdb relational database plays a key role in the archiving of complete accelerator tune configurations. Using this archived information a tune configuration for a given experiment can be stored for future experiments. A Corel product called Paradox was installed to provide the operator with a graphical interface to the archived tune data (5). This was done to eliminate the need for an operator to learn, and use commands in SQL (Structured Query Language) to manipulate data in the relational database. The implementation of this interface provides the operator with a "point and click" windowing environment for retrieving data from the archive database.

When the final process that was executing on the remaining PDP-11 was moved from the PDP-11 to the MicroVAX, this remaining PDP-11 was retired. The configuration after phase three is shown in Figure 4.

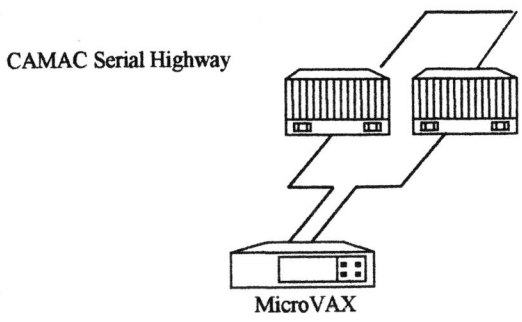

FIGURE 4. This figure shows the system configuration after the third upgrade phase.

Since from the beginning of the project it was planned that the interim MicroVAX computer system would be replaced when the DEC/Compaq Alpha technology became available, an inexpensive machine was purchased with a maximum memory limit of 64 MB (6). Once all control processes had been moved to the new system, and due to this memory limitation, system performance became less than desirable. Since the Alpha technology was now available, a new AlphaServer was purchased to replace the MicroVAX.

A new CAMAC Serial Highway interface and software driver was purchased from Kinetic Systems Inc., and both were tested offline (7). All control processes, databases, and operator display files were moved to the new AlphaServer. When all was ready the last step was to physically move the two CAMAC Serial Highway cables from the MicroVAX to the AlphaServer completing the fourth and final phase.

THE NEW CONTROL SYSTEM

The new control system provides the operator with all the features of the old control system as well as new features. The current hardware configuration of the new control system is depicted in Figure 5.

The new system consists of one AlphaServer, four AlphaStations, and seven PC workstations. The AlphaServer is at the core of the system, and provides the single link to the CAMAC Serial Highway. Two of the AlphaStations are physically located in the main control room, and provide the primary operator interface to the control system. The remaining two AlphaStations will be remotely located at the facilities' two ECR (Electron Cyclotron Resonance) ion source control consoles, and will provide an operator with a local interface to these devices. Six of the PCs are distributed throughout the facility and provide remote control and monitoring functions. The seventh PC is located in the main control room, and is used to provide a graphical interface to the archiving relational database system.

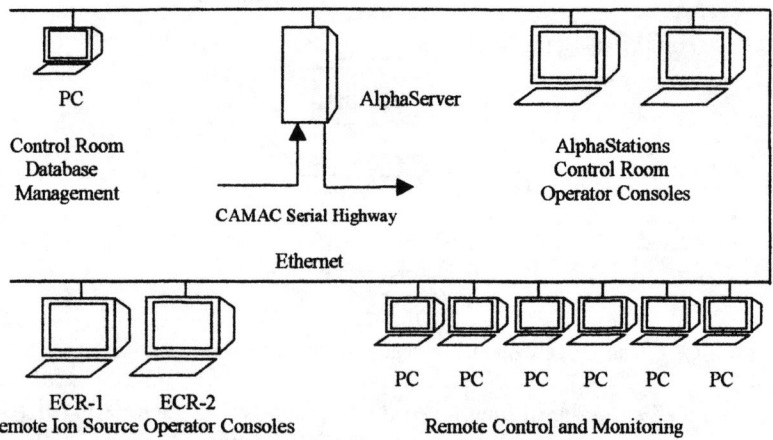

FIGURE 5. This figure depicts the current configuration of the control system.

Two I/O subsystems are utilized by the new control system. The first is a CAMAC Serial Highway, which currently links 18 crates, and operates at a clock speed of 2.5 MHz. The second subsystem is a 10 MB/s Ethernet LAN. This LAN is used to link all of the computers, and is the means by which data is exchanged.

Oracle Rdb and Paradox provide the database structures for archiving purposes. Vsystem provides the foundation for the real-time aspects of the control system. This system was chosen, as previously discussed, largely due to the relatively small size of the control system staff. Vsystem is a network distributed control system software that provides distributed database access and CAMAC I/O processing. While the ATLAS control system does not currently make use of the distributed CAMAC I/O processing feature, it does take advantage of the distributed operator interfacing capabilities. Through the use of the "X Windowing System", operator interface processing is distributed among the various workstations (8).

In-house written software comprises an estimated 75% of the control system processes. This software is written in "C", "FORTRAN", "SQL", or Paradox's native language. These processes were written specifically to address needs unique to the ATLAS facility. Some of these processes were available on the old system, and have been rewritten for the new system, while others were written specifically for the new system. A few of the processes that comprise the control system include the following:

- A process that automates the procedure of setting a resonator to its maximum energy gain amplitude and phase values is provided. Once initiated this process can typically set the entire LINAC (Linear Accelerator) to its full energy setting for a given ion species.

- Once the LINAC is in a full energy configuration, a process that automates the steps in setting the LINAC to a requested energy is made available.
- Integrated into the control system are processes that provide surface barrier detector measurements for both energy and time. Dynamic plotting of this data is an inherent feature of the system.
- A process is provided that continually measures "time of flight", and displays the result as an energy measurement.
- A process was written to periodically update the archiving database with values from the real-time database.
- An automated process is available to provide a complete mass scan of a given beam of ions. Once initiated this process ramps the magnetic field of a selected magnet plotting magnetic field versus faraday cup current.
- An automated process is provided that continually scans all operating resonators to ensure that each resonator continues to operate at the last requested phase setting.

FUTURE PLANS

ATLAS has two ECR ion sources. The newest ion source was brought online during the period when the control system was being upgraded. Consequently, control and monitoring of the new ion source was included in the new control system. The old control system, however, was retained for the original ECR ion source to ensure that the operating schedule of the accelerator would not be jeopardized. The new ECR source will now allow for the retrofitting of the original ion source's control system.

Demands on the system will no doubt grow in the future. It has been demonstrated that one of the better ways to deal with increased demand is distributed I/O processing. To ensure that the system will be prepared for future demands plans to implement distributed I/O are underway. To accommodate the increase in Ethernet traffic the control system LAN will be upgraded from 10 MB/s to 100 MB/s operation.

CONCLUSION

The control system resulting from the upgrade has met all of the design goals of the upgrade project. The development effort took place on line, and simultaneously with the operating schedule of the accelerator. Except for negligible periods of time the accelerator's operating schedule never suffered due to upgrade activities. This was accomplished during a period when the accelerator's operating schedule was designed to achieve a record number of operational hours.

REFERENCES

1. Munson F., Ferraretto M., and Rutherford B., "Status of the ATLAS Control System Upgrade," presented at the Symposium of North Eastern Accelerator Personnel, Chalk River, Canada, September 23, 1992.
2. Digital Equipment Corporation, Maynard, MA, USA.
3. Vista Control Systems, Inc., Los Alamos, NM, USA.
4. Oracle Corporation, Redwood Shores, CA, USA.
5. Corel Corporation, Jericho, NY, USA.
6. Compaq Computer Corporation, Houston, TX, USA.
7. Kinetic Systems Corporation, Lockport, IL. USA.
8. "X Windowing System", A windowing system that is the result of a collaborative effort of a consortium of vendors and the Massachusetts Institute of Technology.

* Work supported by the U. S. Dept. of Energy, Nuclear Physics Div., under contract W-31-109-ENG-38.

Status of the Linac Booster for NSC Pelletron

A.Roy, P.N.Prakash, B.P.Ajithkumar, S.Ghosh, T.Changrani, A.Sarkar,
R.Mehta, B.K.Sahu, A.Choudhury, J.Chacko, J.Anthony, M.V.Suresh
Babu, M.Kumar, S.A.Krishnan, A.Mandal, G.O.Rodrigues, R.Kumar,
R.K.Bhowmik and G.K.Mehta

*Nuclear Science Centre, Aruna Asaf Ali Marg, P.O.Box 10502,
New Delhi -110 067,India*

K.W.Shepard

*Physics Division, Argonne National Laboratory,
9700 South Cass Avenue,Argonne,IL 60439, U.S.A.*

Abstract. This paper reviews the progress made in the development of the linac booster for Nuclear Science Centre(NSC). The prototype resonator for the booster has surpassed design accelerating field. A cryostat has been fabricated to house this prototype for in beam tests. The cryogenic system has been installed and several rf modules have been constructed.

INTRODUCTION

The work for realization of the superconducting linac booster for the 15 UD Pelletron accelerator at NSC [1] is continuing in full swing to achieve the goal of accelerating heavy ions upto mass 80 above the Coulomb barrier. There has been progress in all the aspects of the Linac Booster project since the last HIAT conference. A brief review of the status is given below.

RESONATORS

The development of the optimized Quarter Wave Coaxial Line (QWCL) cavity resonator [2] for the linac booster has been completed. In recent cold tests the resonator has exceeded the design goal of 3 MV/m at an RF loss of 4 W. It has achieved a field of 4.2 MV/m @ 4 W and 5.0 MV/m @ 8 W of RF input. The resonator has now undergone long term stability tests and cycled through several thermal cycles.

CP473, *Heavy Ion Accelerator Technology: Eighth International Conference,*
edited by Kenneth W. Shepard
© 1999 The American Institute of Physics 1-56396-806-1/99/$15.00

It has arrived at NSC and is being readied for mounting in the buncher cryostat for tests in the beamline. The slow tuner bellows assembly has been tested at low temperatures and performed satisfactorily, providing a frequency tunability of about 70 kHz and causing no observable performance degradation at high field levels. Production of the resonators for the first linac module has started at ANL. The niobium housings, the central electrodes for the set of ten resonators and several parts have been already machined and welded. The first of this batch of resonators is expected to be tested shortly and the production is expected to be over by July next year. These resonators would be housed in the first linac module. Each module would have 8 such QWCL cavities and three such modules are now planned for the linac.

A project has been initiated to fabricate niobium superconducting resonators of the above design indigenously. A vendor at Thiruvananthapuram has been entrusted with this job with active involvement of NSC staff. Most of the machining of the parts have been completed and these are waiting for the e-beam welding.

CRYOGENICS

The cryogenic system is one of the vital components of the superconducting linac and consists of the helium liquefier, nitrogen liquefier, cryostats, helium purification and monitoring units.

A 600 W @ 4.5 K helium reliquefier plant has been installed and commissioned. In the first phase this system will provide 330 W cooling capacity without liquid nitrogen in the pool boiling heat exchanger. The machine can also deliver 1200 watts at 60 K with the addition of another expansion engine in parallel to the warm engine. A helium gas recovery system consisting of two 50 m^3 gas bags, a recovery compressor, a 15 cylinder manifold, has been commissioned and integrated with the existing helium refrigeration system. The recovery system has been designed to capture the outputs from all the major relief valves in the system and the vents in the case of power failure.

A helium gas purifier operating at 78K temperature has been designed and built indigenously for helium gas. This purifier has been used successfully to purify helium gas obtained from the vendors before using in the refrigerator. A helium impurity monitor using an arc-cell has been designed and tested. It is able to detect nitrogen impurities in helium at the level of 10 -100 ppm.

The closed loop liquid nitrogen reliquefier of capacity 5000 watts at 82 K, designed jointly by Nuclear Science Centre and M/s Stirling Cryogenics, the Netherlands, has been supplied and commissioned. It has delivered 5000 watts at 82 K with a spare capacity of 15%. The machine efficiently responds to the dynamic load variation in the range of 25% to 100% loads. When used as a liquefier, the machine has delivered 50 litres/hr of liquid nitrogen.

The resonators are to be mounted in specially designed vacuum cryostats. A versatile multipurpose test cryostat has been fabricated for testing resonators and solenoid magnets off-beam. The cryostat has a helium dewar of capacity 60 litres and two nitrogen dewars of capacities 40 litres each. It has several ports and feedthroughs that would also allow measurements of material properties at low temperatures. The prototype cavity resonator would be used as the superbuncher for the heavy ion beams before injection into the linac. A buncher cryostat has been designed and is under construction for this purpose. Preliminary design of the linac cryostat to house eight cavities and a superconducting solenoid has been made.

RF INSTRUMENTATION

Powering of superconducting resonators require special circuitry due to their high Q-factor. Several electronic modules are required to operate the linac for which the prototypes have been tested with resonators on line at Argonne Tandem-Linac Accelerator System and are being reproduced locally. These are, power amplifier, resonator controller and pin diode pulser module. Slow tuner assembly for the first linac module has been completed and tested. The VCX fast tuner mechanical assembly is under construction. A clock signal distribution system to provide phase reference signals at 97 MHz and its subharmonics has been designed.

Multiharmonic Buncher and Beam Sweeper

A multiharmonic buncher[3] has been jointly developed with ANL. The mechanical parts were made at NSC and the electronic circuits were wired at ANL. It has been installed at the low energy end of the Pelletron. In a test run with ^{28}Si beam the bunch width was found to be 1.5 ns with 50% bunching efficiency. This is lower than the theoretically predicted efficiency of about 65 % and was caused by a jitter in the third harmonic component during the run. A sweeper operating at 6.06 MHz has been designed to remove the dark current between the bunches.

A spiral cavity phase detector[4] installed in the beam line after the analyzing magnet of the Pelletron has been providing a stable reference pulse for timing experiments. This phase detector would be used for phase locking of the beam.

CONTROL SYSTEM OF LINAC

A control system running on a network of PC/AT586 computers has been developed for the Pelletron - Linac system. The design is based on a client server model with server directly connected to the accelerator by using a CAMAC serial highway which maintains an on-line database of all the accelerator parameters. All other computers are connected to the server by an ethernet link. The server would service the requests from all the client computers to access the machine parameters and keep the database periodically updated. Database integrity is maintained by allowing only one program to access the database and thus avoiding any conflicting requests. The client computers can run any program and they will have access to the runtime database through the server program. Several of them will act as operator consoles with some special hardware interfaced to it for supporting the assignable knobs and meters. The assignable control knob is an incremental input device connected to the PC ISA Bus. The concept of such a system with two consoles and a server has been tested by running the beam.

BEAM OPTICS

The energy gain, layouts of the beam line, ease of operation of the linac, depend very much on a thorough understanding of the beam optics and transport of the accelerated ions. The beam optics through the entire linac has been worked out using two computer codes, LINRAY [5] from ANL and NSCRAY [6], developed at NSC. The transverse focusing in the accelerating sections would be performed by superconducting solenoid magnets.

The energy gain, effect of misalignments of the resonators and the solenoids have been studied as well as the effect of randomly switching off of a few resonators and solenoids. Even with a solenoid in any cryostat switched off the beam can be transported to the target without any loss. The misalignment of the resonators does not pose a severe problem since misalignments of 1 mm can be tolerated, whereas the misalignments of solenoids by more that 0.1 mm produce a noticeable steering effect. A small magnetic steerer has been designed to go between the cryostats to correct for any such steering effects. Beam transport has been checked for the resonator fields varying over the range of quarter to one and a half times the design field as well as for all the resonators turned off. The placement of the buncher, linac modules and the rebuncher has been optimized from the consideration of close packing and the beam quality in both longitudinal and transverse phase space on target. Transverse emittance of the beam delivered by the Pelletron has been measured.

The quadrupoles and steering magnets for the beam lines have been designed in house and some of these have been fabricated indigenously.

CONCLUSION

The prototype resonator has been successfully tested and surpassed the design accelerating field. Production of the resonators for the first linac module has started at ANL. The civil construction for housing the linac and beam hall is complete. The cryogenic facilities comprising of liquid helium and liquid nitrogen systems are now ready. A test cryostat has been fabricated and linac cryostat is being designed. Beam optics studies have been completed and the beam line for the superbuncher installation has been laid. Several rf modules have been fabricated and tested. The development of the Linac booster is now in a critical stage with all the subsystems in place and work continuing for completing one linac module. This is expected to be done by the middle of next year.

REFERENCES

1. Reformulated Project Report for Phase II Accelerator Augmentation Programme, Nuclear Science Centre, Dec 1992 (Unpublished); P.N.Potukuchi et al., Proceedings of 6th Workshop on RF Superconductivity, CEBAF,1993, p 1184
2. K.W.Shepard and A.Roy, Proceedings of 1992 Linear Accelerator Conference, Ottawa,1992, p 425
3. F.J.Lynch, R.N.Lewis,L.M.Bollinger,W.Henning and O.D.Despe, Nucl. Instr. and Meth. Vol 159 (1979) p 245
4. S.Ghosh, R.Ahuja, S.Rao, A.Sarkar, D.K.Avasthi, D.Kanjilal, R.K.Bhowmik and A.Roy Nucl. Instr. Meth. in Phys. Res. A 356 (1995) 185
5. LINRAY, A computer programme for ion optics calculations, R.Pardo, ANL (Pvt. Communication).
6. NSCRAY, A computer programme to calculate the ion optics of a heavy ion linac, P.N.Prakash, A.Roy and A.P.Patro, NSC Technical Report, Ref: NSC/TR/PNP/95/110 (Unpublished)

Single-Gap Multi-Harmonic Buncher for NSC Pelletron

A.Sarkar[*], S.Ghosh[*], P.Barua[*], R.Joshi[*], R.Ahuja[*], S.Rao[*], S.A.Krishnan[*], A.J.Malyadri[*], R.Kumar[*], S.Gargari[*], S.Chopra[*], D.Kanjilal[*], S.K.Datta[*], A.Roy[*], R.K.Bhowmik[*], I.R.Tilbrook[+] and B.E.Clifft[+]

*Nuclear Science Center, Aruna Asaf Ali Marg, Post Box - 10502, New Delhi-110067, India
+Physics Division, Argonne National Laboratory, Argonne, 9700 South Cass Avenue, IL-60439, USA

Abstract. A single-gap multi-harmonic buncher, developed in collaboration with Argonne National Laboratory, has been installed in the pre-acceleration region of NSC Pelletron. This buncher is required for injecting bunched beam into the booster LINAC, presently under construction. A saw-tooth voltage generated across a single gap formed by a closely spaced pair of grids is used for bunching the dc ion beam produced by the Pelletron accelerator. This saw-tooth voltage is produced by adding a sinewave with its three higher harmonics in proper phase and amplitude. ^{28}Si beam has been bunched successfully using this buncher. The best FWHM of the bunched beam was 1.5ns and the maximum efficiency of bunching was 50%. The bunching voltage had no steering effect on the beam.

INTRODUCTION

The Argonne Tandem Linac Accelerator System (ATLAS) in the Physics Division at Argonne National Laboratory has been using single-gap multi-harmonic bunchers (1) for a long time. A similar buncher has been fabricated for the NSC Pelletron (2). This buncher is required for injecting bunched beam into the booster Linac (3), presently under construction. The buncher has been installed recently in the pre-acceleration region of the NSC Pelletron and tested with beam.

The theory of bunching of ion beams is well known. An ideal saw-tooth voltage can give the best energy modulation required for bunching a dc beam. This saw tooth voltage can be approximated with four sine waves, the fundamental and its three harmonics. The approximated saw-tooth can be written as:

$$V(t) = V_m (\sin \omega t + 0.40 \sin 2\omega t + 0.18 \sin 3\omega t + 0.06 \sin 4\omega t) \qquad (1)$$

A buncher using equation (1) as its energy modulation is called a multi-harmonic buncher. The frequencies used in the NSC multi-harmonic buncher are 12.125 MHz (fundamental), 24.25 MHz, 36.375 MHz and 48.5 MHz. All these frequencies are subharmonics of the LINAC frequency i.e. 97 MHz. The approximate saw-tooth voltage is applied across a single gap formed by a pair of grids placed very close to each other to bunch the dc beam. For a single gap buncher the condition for bunching is given by:

$$(dV/dt) = (2.8\ E_0^{1.5}.q^{0.5}) / (m^{0.5}.L) \qquad (2)$$

where V is the bunching voltage applied across the gap, qE_0 is the injection energy of the incoming ions, m is the mass of the ions and L is the effective length between the buncher and the point of time focus of the beam bunch.

BRIEF DESCRIPTION OF THE BUNCHER

The single gap multi-harmonic buncher developed for NSC consists of three major parts. These are: the mechanical assembly, the tank circuit and the electronics.

The Mechanical Assembly

The mechanical assembly consists of a 14 inch side cubical vacuum chamber made out of stainless steel. It has five ports of which two NEC type ports are used for beam inlet and outlet. The other three ports have conflat type flanges. One of these conflat ports is used for connecting the tank circuits to the vacuum chamber. The second one is used for the pick-up probes. The other conflat port is used for connecting an ion pump for additional pumping. The chamber also has two small view ports.
A pair of grids across which the bunching voltage is applied is etched out of 0.005 inch thick molybdenum having a separation of 1.5 mm. These grids are mounted on copper cones separated by machinable glass ceramic (MACOR) insulator. The cones are provided with cone extensions for proper field shaping. The whole grid assembly is housed in the vacuum chamber supported by two copper rods. Figure-1 gives a schematic diagram of the vacuum chamber showing the grid assembly. Co-axial water cooling arrangement is made to cool the copper rods holding the grid assembly. The grids are mainly cooled by conduction.

Tank Circuits

In order to excite the grids with modest power high Q resonant circuits are employed. The schematic of the tank circuits is shown in figure-2. The coils for 12.125 MHz and 24.25 MHz are cooled by co-axial water cooling arrangement. The 36.375 MHz coils are smaller and need no cooling. All the tuning capacitors used are 3 - 30 pF vacuum variable type. The 12.125 MHz tank circuit is tuned by changing the inductance of the coil. Special couplers for the different frequencies are used to feed in RF power to the tank circuits. The measured grid capacitance (including the assembly and extensions) is 26 pF.

Each tank circuit is tuned with coupler positions at critically coupled condition. The unloaded Q's of the different resonant circuits in air are 491, 677, 564 and 943 for 12.125 MHz, 24.25 MHz, 36.375 MHz and 48.5 MHz respectively.

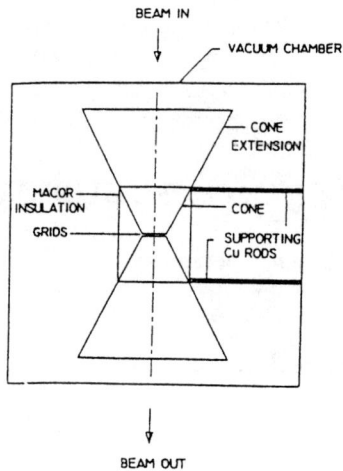

FIGURE I: SCHEMATIC DIAGRAM SHOWING
THE MECHANICAL ASSEMBLY OF
SINGLE GAP MULTI-HARMONIC BUNCHER

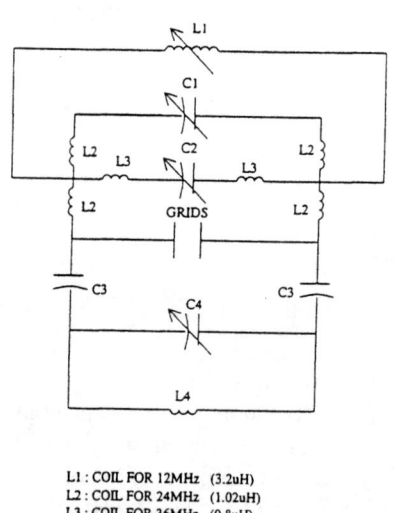

L1 : COIL FOR 12MHz (3.2uH)
L2 : COIL FOR 24MHz (1.02uH)
L3 : COIL FOR 36MHz (0.8uH)
L4 : COIL FOR 48MHz (1.02uH)

C1,C2,C4 = 3-30pF
C3 = 4.4pF

FIG:2   TANK CIRCUIT FOR MULTI-HARMONIC BUNCHER

Electronics

In order to generate the saw-tooth waveform, the amplitudes and phases of the four harmonics must be accurately set and maintained. This is accomplished by a set of 11 NIM modules housed in a specially wired NIM bin. A block diagram of the entire buncher electronics is shown in figure-3. The 12.125 MHz signal from the master oscillator is used to generate the harmonics. There is a local control box which provides signals to control the amplitude and phase of all the harmonics. A composite signal containing all the sinewaves added in proper amplitude and phase is produced. This composite signal is then amplified by a power amplifier built in-house and applied across the grids through the tank circuits. The power amplifier is a class AB push pull design and has a maximum output of 250 Watts. In order to measure the saw-tooth voltage generated across the grids there are two capacitive pick-up probes which pick up signals from the grids. Since the pick-up probes are capacitive, the pick-up is a differentiated form of the saw-tooth voltage. The saw-tooth voltage and the corresponding differentiated signal are shown in figure-4. The harmonics contained in the pick-up signal are amplified and filtered by the respective filters and applied to the feedback control loop for amplitude and phase locking.

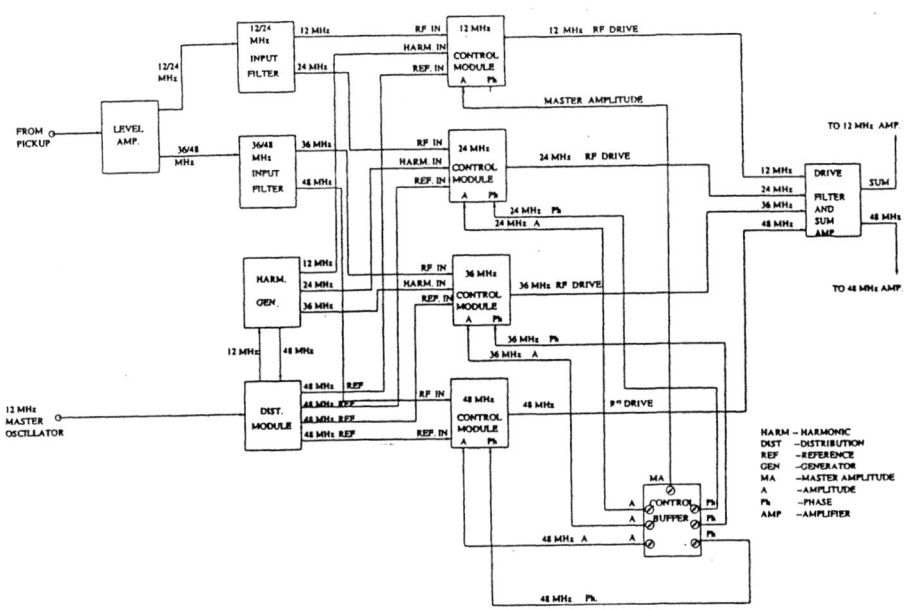

FIG. 3 BLOCK DIAGRAM OF BUNCHER ELECTRONICS

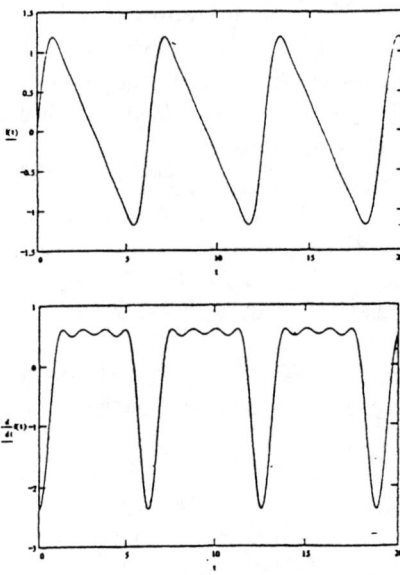

FIG. 4 SAW-TOOTH WAVEFORM AND ITS CORRESPONDING DIFFERENTIATED SIGNAL

TESTS AND RESULTS

Initially the buncher was tested with low and high RF power off-beam in the laboratory. It was then installed in the beam line in the pre-acceleration region of the Pelletron accelerator. As RF power was applied across the grids outgassing and multipactoring problems were encountered. These were solved by slowly raising the RF power on the grids over a long period of time.

^{28}Si beam with injection energies 200 KeV and 250 KeV was used for the buncher test. The bunched beam was intercepted by a thick Aluminium target placed in the post acceleration region of the Pelletron after the analyzer magnet. A Time to Amplitude Converter (TAC) was set with the gamma signal detected by a plastic scintillator coupled to a photomultiplier placed near the target as the start and the RF signal from the master oscillator as the stop. The TAC output was connected to a Multichannel Analyzer (MCA) to observe the bunched beam. The set-up for buncher operation is shown in figure-5.

Initially the buncher was calibrated using a single sinewave at 12.125 MHz. The higher harmonics viz. 24.25 MHz and 36.375 MHz signals were then added in proper amplitude and phase to generate the saw-tooth voltage and the improvements in bunching were observed. The fourth harmonic (48.5 MHz) signal was not used during the test. The FWHM of the bunched beam observed was 1.5 ns and the efficiency of bunching was found to be 50%. Figure-6 gives the spectrum showing the bunched beam as observed on the MCA.

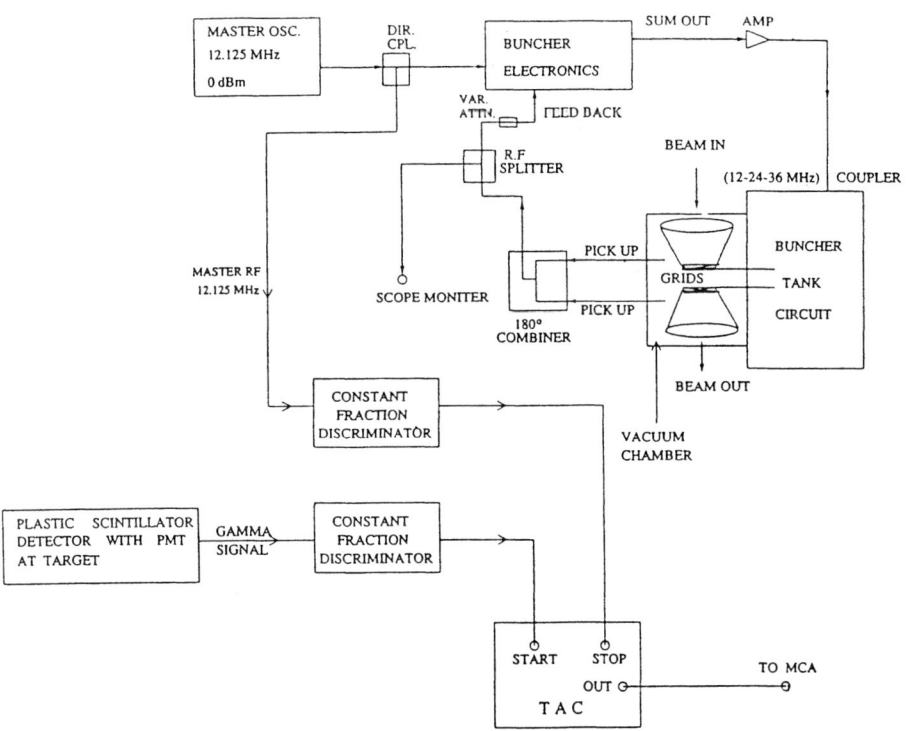

FIG-5 TEST SET UP FOR OPERATION OF HARMONIC BUNCHER

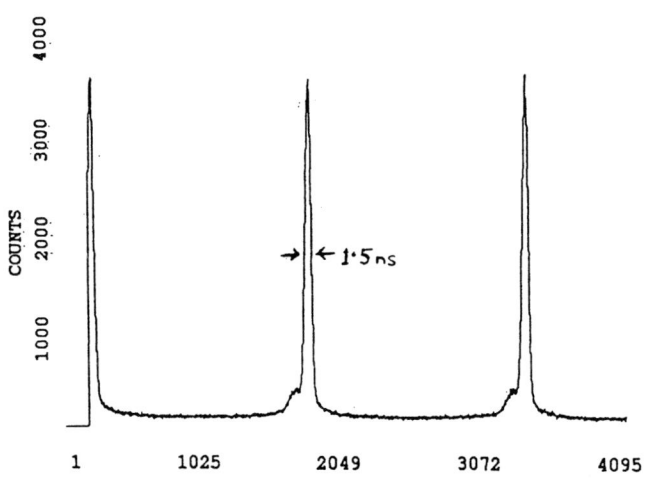

FIG. 6: SPECTRUM AS SEEN ON MCA SHOWING BUNCHED BEAM

ACKNOWLEDGEMENTS

The guidance and help received during the fabrication of the buncher electronics from Mr. W.E. Erdman and several other employees of the Electronics division at Argonne National Laboratory is highly acknowledged. The help and support by the employees of the machine shop in the Physics division at Argonne National Laboratory is also acknowledged. The authors would like to thank Prof. K.W.Shepard and Prof. J.A.Nolen for their support and guidance during the fabrication and testing of the buncher at Argonne National Laboratory. The co-operation from the entire Pelletron staff at NSC during the installation and testing of the buncher is acknowledged. The encouragement received from Prof. G.K.Mehta at different stages of the project is highly acknowledged.

REFERENCES

(1) F. J. Lynch et al, Nucl. Instr. and Meth. 159 (1979) 245 - 263
(2) D. Kanjilal et al, Nucl. Instr. and Meth. A 328, 97 - 100 (1993)
(3) P. N. Potukuchi et al, Proc. of sixth workshop on RF Superconductivity CEBAF, Newport News, VA, Oct 4 - 8, 1993

A Very Low Intensity Ion Beam Detector System

G.P. Zinkann, B. E. Clifft, J. A. Nolen, R. C. Pardo, C. E. Rehm, W. Q. Shen

Argonne National Laboratory, Argonne, Illinois

Abstract. A new time-of-arrival (TOA) control system has been developed at the Argonne Tandem Linear Accelerator System (ATLAS) for very low intensity radioactive ion beams.. This new system utilizes a plastic scintillator which intercepts a small fraction of the beam edge particles and measures the time-of-arrival of a bunched beam at the entrance to the ATLAS superconducting linac after acceleration through the ATLAS 9MV tandem injector A radioactive ^{56}Ni beam has been accelerated through a 9MV tandem and then injected into a superconducting linear accelerator for further acceleration The detector has been successful at stabilizing the pre-tandem bunching system with beam currents as low as 2×10^4 particles per second.

This paper will describe the electronic circuitry developed for this system and details about the control range and limits.

INTRODUCTION

The acceleration of beams of unstable nuclei has opened up new research frontiers. Critical cross sections for astrophysical processes previously impossible to obtain, qualitatively new and unexpected nuclear structure effects in nuclei far from stability, completely new approaches to studies of nuclear decays, reactions and structure have triggered much excitement for this new dimension in nuclear research.

In response to this demand, a variety of radioactive beams are in various stages of development and eight different exotic beams have been used for research at ATLAS. Examples of beams available for research at ATLAS are, ^{17}F, ^{18}F, ^{8}B, ^{21}Na, ^{25}Al ^{56}Co and ^{56}Ni. The intensities of these beams delivered on target are up to 5 X 10^6 particles per second(pps). The ^{56}Ni beam required the development of the new low-beam-current detector system in order to accelerate this very low intensity beam in the ATLAS linac.

^{56}Ni was produced at the Intense Pulsed Neutron Source (IPNS) at Argonne utilizing the ^{58}Ni(p,p2n)^{56} reaction. A 50 MeV proton linac irradiated a ^{58}Ni sample with a proton beam current of approximately 10μA for 16 hours. The ^{58}Ni sample was already mounted in a holder necessary for installation into the ATLAS tandem's SNICS[1] negative-ion cesium sputter source.

The ATLAS facility is a heavy ion, superconducting linear accelerator, with the capability of switching between a positive ion injector (PII), and a negative ion injector (Tandem). The Tandem injector is a 9-MV tandem electrostatic accelerator with a negative ion sputter source. The two injectors couple into the 'Booster' linear accelerator. The Booster consists of 46 independently phased, split-ring niobium superconducting resonant structures.

A variety of effects can cause transit time variations during acceleration through the tandem. Some examples of these are stripper foil thickening, source platform voltage variations, tandem voltage profile variations, or poor tandem voltage regulation due to low beam currents on the analyzing slits.. Such variations, if not corrected, would cause the ion bunch to arrive at the linac at the wrong time. Proper acceleration through the linac requires a stable time-of-arrival (TOA) at the entrance to the linac. The transit time through the tandem is typically 10μsec, it would be preferable to match the arrival time with an accuracy of 0.1 nsec, clearly some form of phase stabilization is need.[1]

The detector used for normal intensity beams from the tandem is a room temperature helix resonant structure which measures the TOA of the beam near the entrance of the linac. The beam bunch excites a the helix resonator and the induced RF phase with respect to the master oscillator is a measure of the time of arrival. The lowest beam sensitivity for this helix structure is around 1 enA. With the need to accelerate and stabilize the time of arrival of a low intensity beam, a new method of stabilization was necessary.[1]

APPARATUS

The new low-beam-current detection system is located about 0.25 meters downstream from the Tandem energy analyzing slits. This position, shown in figure 1, is desirable because the charge state ,mass, and energy of the ion of interested have been selected by the 90 degree analyzing magnet. The device is on a precision adjustable bellows feed-through, Model number PMZ-275-2, purchased from Huntington Mechanical[2]. This enables precision insertion, approximately 0.4mm position repeatability, of the scintillator tip into the outside fringes of the beam. The assembly is mounted vertically on the beam line A vertical insertion, rather than a horizontal insertion, is desirable, because in the horizontal position it is possible to experience a tandem voltage fluctuation that may cause the 90 degree analyzing magnet to sweep the full beam current onto the scintillator and thereby damaging the scintillator.

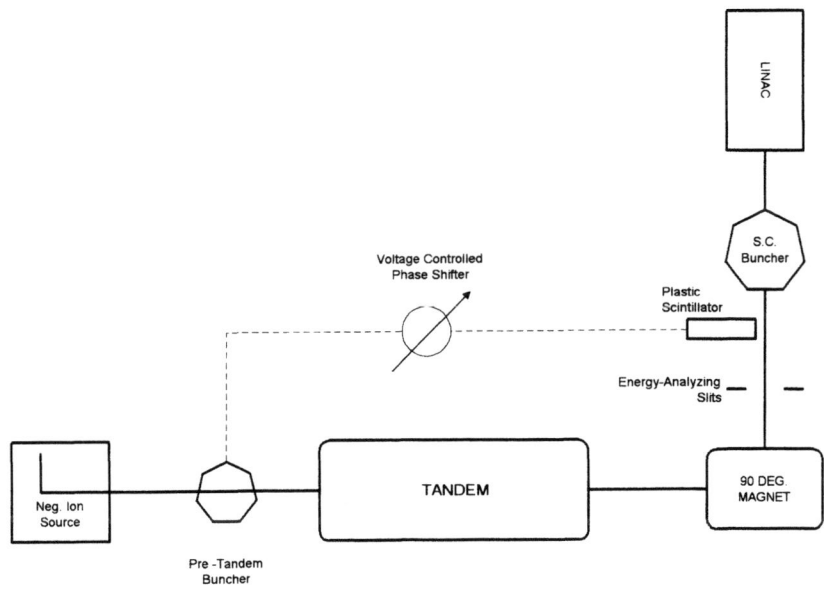

FIGURE 1. Diagram of the tandem accelerating system showing the scintillator slit position.

The scintillator material is supplied by Bicron Corporation[3], number BC408, and is coupled to a photomultiplier tube, R1635, model WB1432 purchased from Hamamatsu Corp[4]. The scintillator tip is the only piece of the assembly that is in the vacuum. An O-ring seal on the shoulder of the scintillator material provides the vacuum / atmosphere interface between the scintillator and the tube. This construction minimized the amount of material in the vacuum space, and allowed the tube and socket to be replaced without the need for venting the vacuum. Figure 2 illustrates the vacuum assembly of the apparatus.

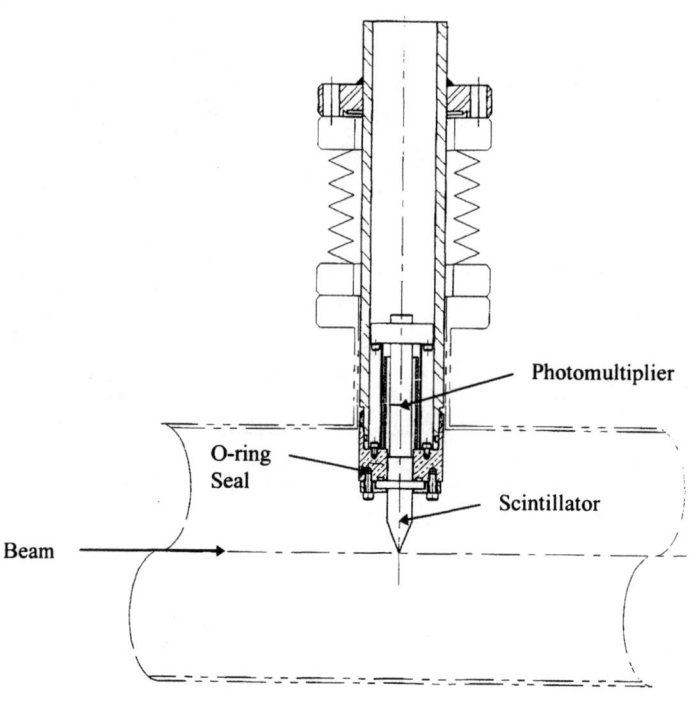

FIGURE 2. Drawing of the scintillator on the bellows feed-through

A phase-lock control circuit was constructed to use the pulses generated by intercepting a small fraction, approximately 1%, of the beam current on the scintillator and convert it into an input control signal for the existing voltage-controlled phase shifter of the pre-tandem buncher. The pulses from the photomultiplier tube of the scintillator are fed into an octal discriminator. The discriminator level is set to reject any equivalent low energy noise out of the scintillator thereby filtering out any signals that are not within 10% of the beam energy. The TAC output is split into a single channel analyzer (SCA) and an Evens Electronics[5] Gated Integrator Module (GIM), model # 4138A. The SCA upper and lower discriminators are set to bracket the desired signal level from the TAC. The GIM only samples the time-to-amplitude converter (TAC) output level at a valid gate signal and it's output holds the TAC signal as a DC voltage. This DC voltage is amplified and offset using an internal DC offset adjustment. This is then fed into a voltage controlled phase shifter that controls the phase of the pre-tandem buncher. The DC offset sets the "correct" value so as to produce no shift at the desired beam TOA. Any error in the beam TOA forces the phase shifter to the correct TOA. Figure 3 is a block diagram of this circuit.

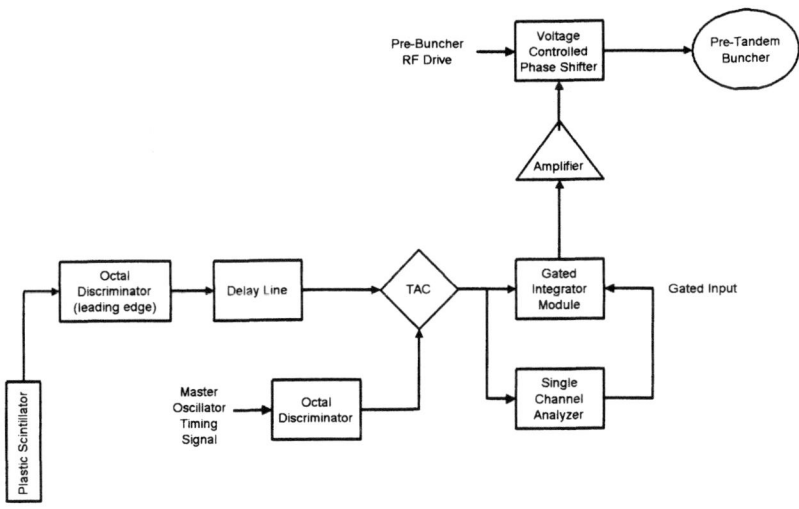

FIGURE 3. Block diagram of the electronics for the phase lock loop of the scintillator detector.

TEST

To test the entire system a guide beam of $^{28}Si^{+5}$ was accelerated through the Tandem. The $^{28}Si^{+5}$ has the same charge to mass ratio of the radioactive beam $^{56}Ni^{+10}$. By using the stable Si beam, the accelerator can be tuned with a measurable amount of beam and then by dramatically attenuating the beam test the new low beam current detector system.

The normal configuration of the spiral detector and the 90 degree magnet energy analyzing slits were used to set up the accelerator. A silicon barrier detector, located at the exit of the booster accelerator, measured the timing centroid of the guide beam. The timing centroid was observed for a period of time to determine that the terminal voltage was stable and not contributing to any shifts in the centroid. Once the timing centroid was determined, and a calibration of the silicon barrier detector's count rate vs beam current was performed, the beam was attenuated to a current of 1.68 $\times 10^4$ pps. At this low beam current the scintillator was inserted to yield a count rate of 200 Hz, which was approximately 1.2% of the beam current, and the phase lock loop control of the pre-tandem buncher was switched to scintillator control system. This rate was sufficient to allow active feedback for TOA variations slower than approximately 60 Hz.

To introduce a timing shift into the beam, and to test the control capabilities of the new detector system, the platform voltage of the negative ion source was varied. The normal operating voltage of the platform is 150kV. At 150kV the timing centroid on the surface barrier detector was measured. The control loop was switched off and the centriod was measured again, it remained the same. The platform voltage was then dropped to 149kV. The timing centroid shifted 12.61ns. The deck voltage was returned to 150kV, the loop was switched on. To check the effectiveness of the phase lock loop the deck voltage was changed to 148kV, with the control loop on, and the resulting change in the timing centroid was 0.67ns. This 2kV test shift is well in excess of the maximum normally expected shifts of 100 -200 volts. We infer from these measurements that the TOA of the beam is stabilized to approximately 0.1nsec for reasonable, short term, fluctuations in the tandem accelerator.

Table 1 Summary of the Open and Closed Loop Results

	ΔV Platform	Terminal Voltage	ΔT
Open Loop	1kV	8.450MV	12.6ns
Closed Loop	2kV	8.450MV	0.67ns

Stability tests were conducted for two days. The timing centroid was monitored on the silicon barrier detector using 1.68×10^4 pps of $^{28}Si^{+5}$ tuned through the accelerator. The tests showed a maximum long term drift in the timing centroid of 0.6ns.

This work was supported by the U.S. Department of Energy, Nuclear Physics Division, under Contract W-31-109-Eng-38

REFERENCES

1. F. J. Lynch, R. N. Lewis, L. M. Bollinger, and O.P. Despe, Necl.Instr. and Meth. 159, 245 (1979)
2. Huntington Mechanical Laboratories Inc. Mountainview, California
3. Bicron Newbury, Ohio
4. Hamamatsu Corporation Bridgewater, New Jersey
5. Evans Electronics Berkely, California

IV
ION SOURCES

Development of ECR Plasmas for Radioactive Ion Beams

R. Geller, J.L. Bouly, J.F. Bruandet, N. Chauvin, J.C. Curdy, T. Lamy,
H. Nifenecker, P. Sole, P. Sortais, J.L. Vieux-Rochaz

Institut des Sciences Nucléaires 53, avenue des Martyrs 38026 Grenoble Cedex France

Abstract. ECR plasmas are utilized for : 1) Charge breeding of 1+ RIB into N+ RIB in continuous regime with an efficiency for one given charge of 10 % for noble gases and about 5 % for solid elements. 2) Charge breeding with beam bunching (bunch duration 20 ms, 5 Hz) was obtained for Rb^{15+} ions with an efficiency of 2.2 %. These results are very reproducible and need only about 200 W of RF power. The number of ions contained in one bunch exceeds 1000 times those achieved with EBIS systems. The ECR trap is better suited for pulsed post acceleration. 3) Ion accumulation in the ECR plasma trap may become a method for realizing a radioactive target.

ECR ION TRAPS AND ECR ION SOURCES

For ion trap issues, one wonders if it is possible to improve the devices, knowing that they already operate at the limits of Coulomb repulsion between ions in an optimum vacuum. Therefore, we propose straight forwardly a radically different approach : let us utilize a dense support gas plasma as a storage medium, since in a plasma the ion/ion repulsion is screened by the electrons. The radioactive ions can for instance be accumulated in an ECR plasma if their accumulated density remains small compared to the global plasma density and their fate would be similar as the fate of the other ions. In other words they are confined, multicharged, recombined, diffused to the walls... or extracted in continuous or pulsed regimes as in the classical ECR ion sources (ECRIS). Now we recall that in routine ECRIS the global ion density N^+ is about 10^{11} cm^{-3} so we can accept an accumulated radioactive ion density of 10^{10} cm^{-3} and we emphasize that such densities are orders of magnitude larger than those of the usual traps.

However ECRIS normally work with neutral gas feeding which involves many atom/wall stickings with long release times and these delays must be avoided when radioactive ion beams (RIB) containing short life isotopes are considered (1), (2). Therefore we developed a special charge breeder ECRIS where the gas feeding is replaced by a 1+ RIB which is focussed and does not interfere with walls before being trapped in the plasma. This ECRIS worked with 1+ ions of Ar, Kr, Xe, Zn, Rb,

CP473, *Heavy Ion Accelerator Technology: Eighth International Conference,*
edited by Kenneth W. Shepard
© 1999 The American Institute of Physics 1-56396-806-1/99/$15.00

Pb and support gases like O_2, N_2, etc. (see Table. 1). The 1+ ions are introduced either from the extraction or injection side. The breeder efficiency for one given charge is defined as

$$\eta = \frac{\text{number of n+ charged ions extracted /sec}}{\text{number of 1+ ion injected /sec}}$$

In continuous regime we found for instance $\eta = 10\%$ for Ar^{10+} Kr^{12+}, Xe^{14+} and $\sim 5\%$ for Rb^{15+}. All the results were quite reproducible in routine DC operation with 100 watts of RF power in the ECR.

However a pulsed charge breeder trap, tuned as an ion buncher, could be of great interest when only pulsed post accelerators are available.

Hence our next task was to realise an ECR charge breeder/buncher which would transform a DC 1+ RIB (general case) into a n+ ion bunch for pulsed accelerators, in other words to build an ECR ion trap (ECRIT) with high efficiency.

TABLE 1. Breeder efficiencies for various ions. Different 1+ ion sources (ECRIS, hollow cathode, thermoionisation) have been used to produce rare gas, metallic and alkali ions, the n+ charge states given are those which have been optimised.

1+ ion	N+ ion	η ECRIS
Kr	Kr^{11+}	10 %
Ar	Ar^{9+}	9 %
Zn	Zn^{9+}	3.5 %
Pb	Pb^{24+}	4.2 %
Rb	Rb^{15+}	5 %

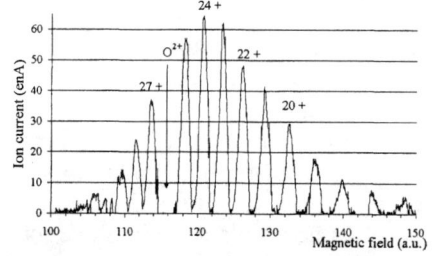

FIGURE 1. $Pb^{1+} \rightarrow Pb^{n+}$ charge spectrum

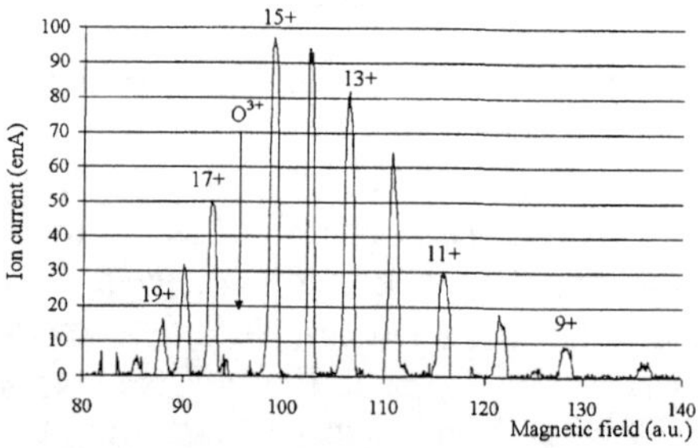

FIGURE 2. Rb^{n+} spectrum for a 135 nA Rb^{1+} injected beam

Charge spectra obtained by the 1+ → n+ method have been measured in the case of Pb and Rb (for example). In the case of an incident beam intensity of 150 nA of Pb^{1+} the n+ beam was peaked on the 24+ charge state (Fig. 1).

Depending on the tuning of the n+ ECRIS the charge spectra can easily be tuned on a specific charge state, for example in the case of a 135 nA Rb^{1+} ion beam, we optimised the 15+ charge state (Fig. 2).

ECRIS and ECRIT principles and timings

The main processes in ion traps are : 1. Ion confinement, 2. Ion expulsion, 3. Ion injection. In ECRIT systems none of these processes can rely on electrostatic gates since electrostatic fields cannot penetrate into a neutral plasma. Thusly:

1. The ECRIT confinement is based on magnetic structures (called min B done with coils + hexapoles) which trap essentially electrons having large magnetic moments $\mu = \dfrac{W_\perp}{B}$ (acquired during the ECR) : the so obtained negative space charge then traps the ions through ambipolar forces.
2. The ion expulsion process is less trivial : instead of cancelling the magnetic system which would take a rather long time, we create a sudden plasma instability which expulses the plasma in a short bunch of a few milliseconds, (this instability is simply obtained by cancelling suddenly the RF power of the ECR). The principle called afterglow process, developed in 1988, is currently utilised with ECRIS coupled to synchrotrons (1), (3).
3. The ion injection process was developed more recently in 1995. It is based on deceleration of the injected 1+ RIB inside the support gas plasma through ion/plasma collisions. The most efficient slowing down occurs when the RIB particles penetrating inside the plasma have velocities of the same order as the average velocity of the support gas plasma ions i.e. energies of a few eV. Then one single ion plasma collision yields a 90° deflection and the injected ions starts a squeezed helical path leading to more collisions… thermalisation, trapping and charge breeding (2)

ECRIS and ECRIT are built with the same components but their conceptual principles are different. An ECRIS cannot operate with a maximum ion confinement since the plasma leaks towards the extraction zone would be very small and the intensity of the extracted n+ beam would consequently be very small too. Therefore in order to obtain decent n+ beams, the magnetic confinement requires always a compromise between ion trapping and ion extraction (Fig. 3). On the opposite, the decrease of the confinement yields always more plasma leaks and subsequently more available ions for the extraction (but with lower charge states).

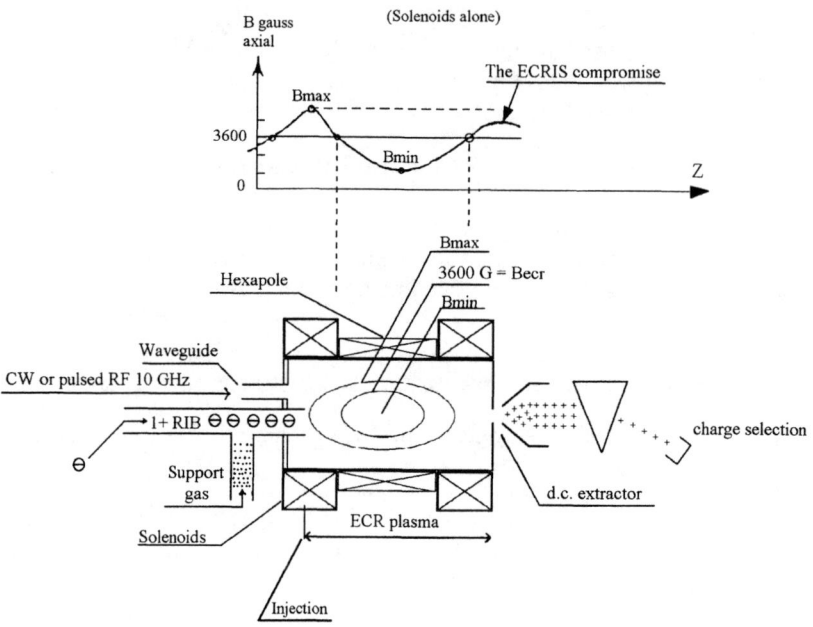

FIGURE 3. ECRIS charge breeder for RIB : gas feed replaced by continuous 1+ RIB injection and support gas feeding. CW or pulsed RF power ; asymmetric magnetic trap for ion confinement/extraction compromise

Such a compromise is not requested in the ECRIT. The ECRIT mode consists of optimising the trapping of the 1+ ions independently of the extracted n+ ion current. During the trapping time, however, n+ ions are also created (and are trapped too). They are therefore ready to be extracted with the help of the afterglow process which suddenly opens an ion expulsion gate. Thus it becomes possible to independently accumulate ions of given species in the ECRIT and extract them in a pulsed mode, exactly like in the 1+ injection systems presently used in the slow injection EBIS (4). However we insist upon the fact that unlike the EBIS the charge breeding time is not the trapping time. In an EBIS the recombinations are negligible. In an ECRIT after the process of the successive ionisations, the average charge states reaches a maximum (which results from an equilibrium between step by step ionisations and recombinations). We can call this breeder time also "useful exposure time of the injected ions to electron bombardment" ; but beyond this useful exposure time, the ions are not immediately lost. They remain trapped by the ECRIT confinement system and accumulated until they are either really lost on the walls through diffusion or suddenly lost by the expulsion process (Fig. 4).

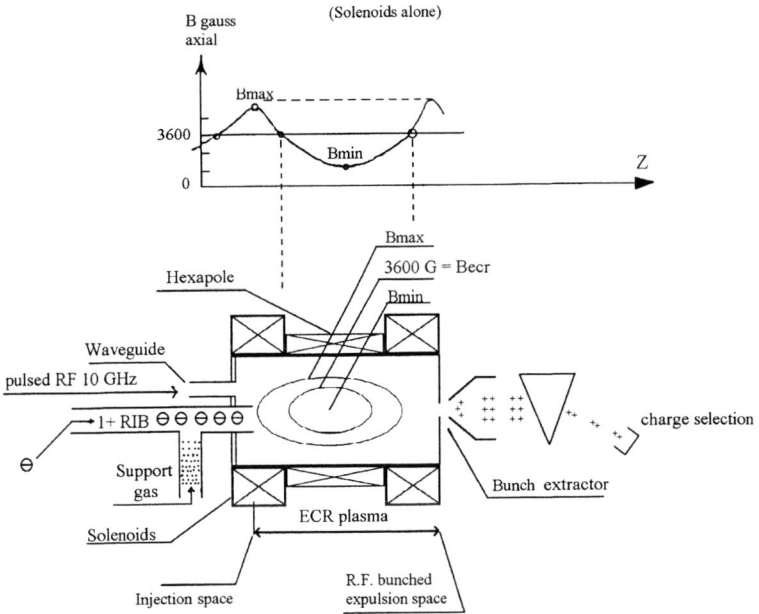

FIGURE 4. ECRIT breeder buncher : continuous support gas feeding, continuous or pulsed 1 + RIB ions ; symmetric magnetic trap for maximum ion accumulation, RF power cancellation (bunch extraction).

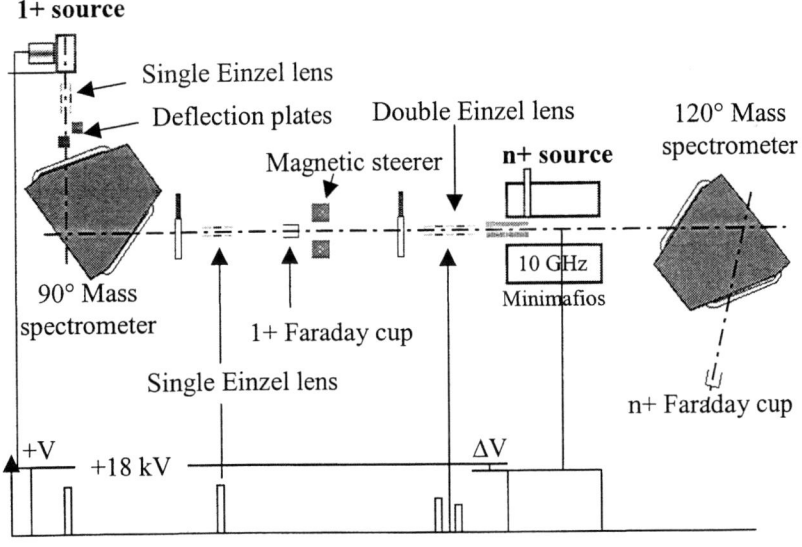

FIGURE 5. Experimental setup for the 1+/n+ breeding and the ECRIT working.

The potential applied to the 1+ source can be varied independently of the one applied to the n+ source, therefore the energy of the 1+ ions can be varied in order to optimise their capture by the plasma of the n+ ECRIS ; the potential difference between the two sources is called ΔV. We used three different 1+ ion sources (a thermoionisation producing Rb^{1+} ions, an ECRIS or a hollow cathode ion source producing Zn^{1+}). When varying the ΔV one can see, despite the difference of energy spread of the three ion sources, that the width of the curves is almost the same ≈ 10 eV (Fig. 6).

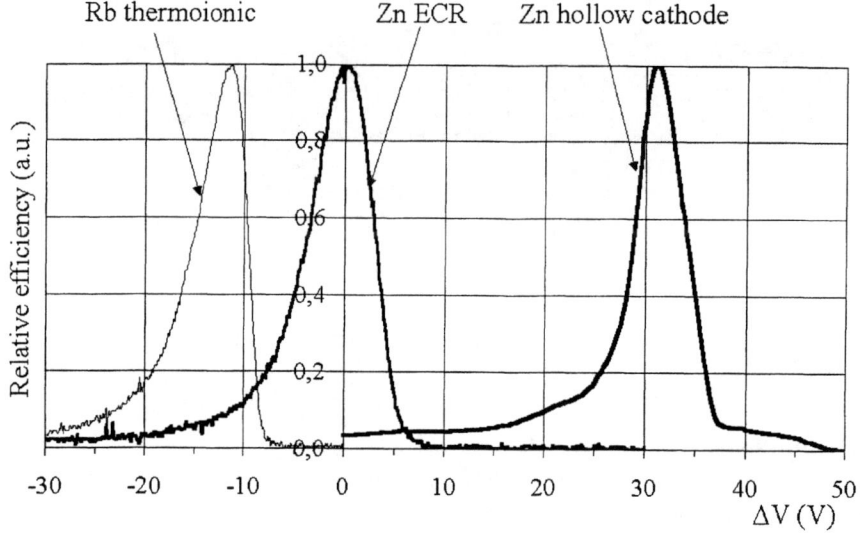

FIGURE 6. Variation of the relative efficiency versus ΔV.

Experimental facts about the ECRIT parameter (5), (6)

As described in a recent letter (5), in our first experimental demonstration of the ECRIT principles we worked with Rb^{1+} ions. A Rb^{1+} ion beam of constant intensity I_{Rb}^{1+} is injected in the ECRIT in continuous or in a pulsed mode during a time t_{inj}. The RF power which in this case is also pulsed is cancelled after a Δt time following the end of the Rb^{1+} pulse. During Δt, the Rb^{1+} ions are both multi-ionized and trapped in the plasma (6) (Fig. 7).

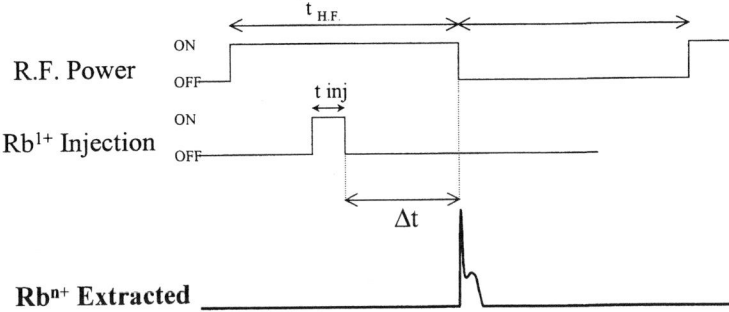

FIGURE 7. ECRIT breeder buncher, timing of RF power; 1+ RIB injection and bunches. During Δt the Rb^{1+} ions are multi-ionized and trapped in the plasma and they are brutally extracted by the cancellation of the RF.

In a typical example of accumulation trapping and pulsed extraction where the Rb^{1+} beam had an intensity $I(Rb^{1+})$ of 600 nA, t_{inj} = 50 ms, t_{HF} = 1500 ms and Δt = 800 ms, we found that 800 ms after the end of the Rb^{1+} injection Rb^{15+} ions still remain in the plasma as evidenced by their presence in the afterglow pulse (Fig. 8).

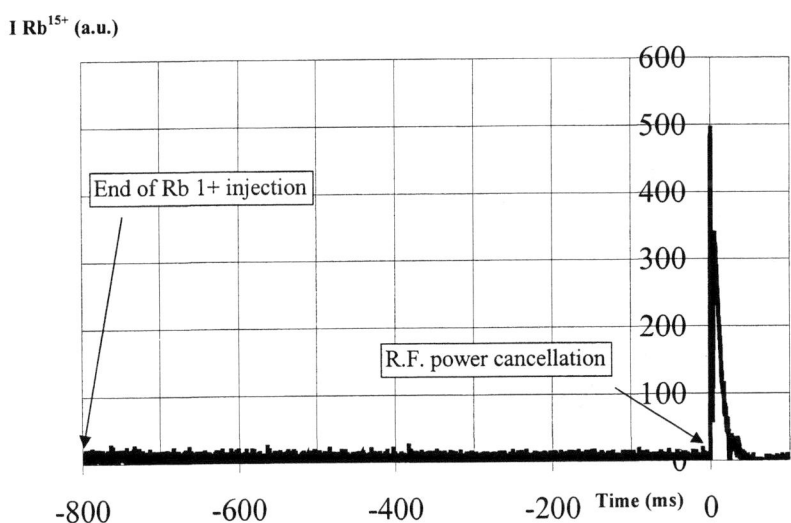

FIGURE 8. ECRIT breeder buncher. The Rb^{1+} beam has an intensity $I(Rb^{1+})$ of 600 nA t_{inj} = 50 ms, I_{HF} = 1500 ms and Δt = 800 ms. Notice that 800 ms after the end of the Rb^{1+} injection, Rb^{15+} ions still remain in the plasma as evidenced by their presence in the afterglow pulse.

This gave a first idea of the confinement time : > 500 ms. Next we looked for the charge breeder times. The method was : a) superposing the end of the t_{inj} and the end of the RF power pulse (i.e. $\Delta t = 0$) : b) gradually varying the duration of t_{inj} and observing the time dependence of the Rb^{n+} ions extracted.

In the case of Rb^{13+}, Rb^{14+}, Rb^{15+}, we observed a linear increase of the $Rb^{n+}(t)$ extracted for $t_{inj} \geq 30$ ms; therefore, beyond 30 ms an equilibrium between charge breeding and recombination is reached. According to these results we concluded that the charge breeder times are small with respect to the possible confinement times. Under these conditions, we calculated that the Rb^{15+} ion confinement lasts in average 520 ms. In the case of Rb^{15+}, all the ions contained in the extracted bunch give a breeder efficiency $\eta \approx 2.2$ % when injecting a 200 ms pulse of Rb^{1+}. Fig. 9 shows the fine structure of such an extracted Rb^{15+} bunch (which is quite reproducible). In the bunch one sees the successive non linear expulsions of the ions after the RF power is cancelled : the total bunch has a duration of 20 ms but roughly 10 % of the ions are expulsed during the first millisecond. The Rb^{1+} beam injected has an intensity of 400 nA. The Rb^{15+} afterglow measured has a maximum peak intensity of 11.5 µA. In 20 ms following the RF power cancellation 2.4×10^{10} particles of Rb^{15+} are extracted corresponding to a total number of Rb ions extracted greater than 10^{11} for the whole spectrum. These results show that the charge trapping capacity of the ECRIT appears to be important. Note that in this configuration 2.7×10^9 Rb^{15+} ions are already extracted in the millisecond following the RF cancellation. The integration of the afterglow signal during 20 ms after RF cancellation gives a $Rb^{1+} \rightarrow Rb^{15+}$ transformation efficiency of 2.2 % (Fig. 10).

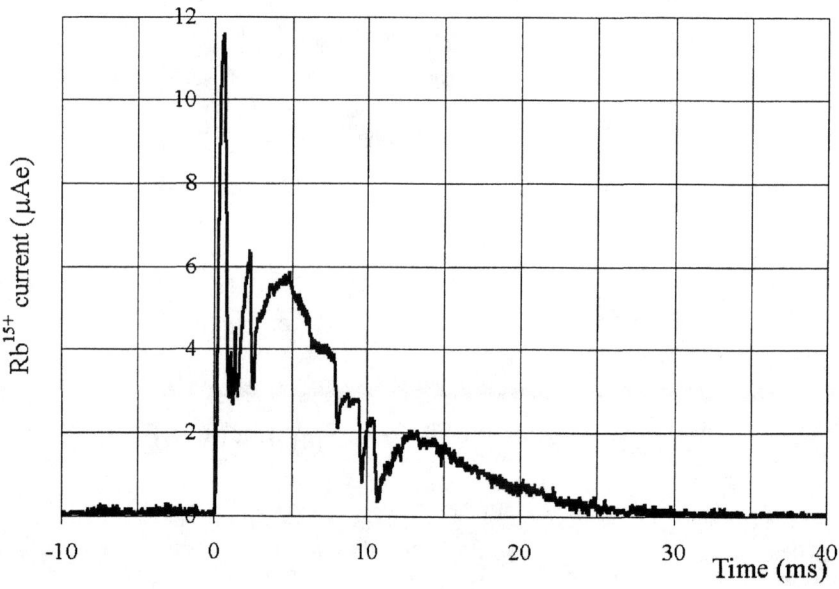

FIGURE 9. Typical Rb^{15+} bunch : the fine structure exhibits as expected successive non-linearities, total pulse duration is 20 ms but 10 % of the ions are expulsed in the first millisecond.

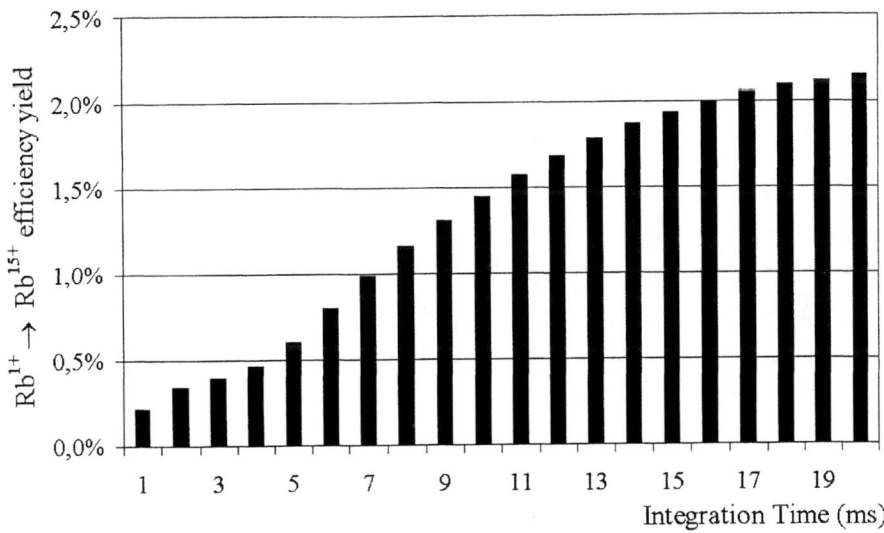

FIGURE 10. Cumulated efficiency of the ECRIT mode.

Thus during a time $\tau_{conf} = 520$ ms the injected 1+ RIB particles are captured and charge breeded. When for instance the 1+ beam is a constant amplitude DC beam of 0.2 µA the total number of all expulsed particles (contained in the sum of all ion bursts) gives the total operation efficiency T which is estimated to 15 % and remains valid up to an injected RIB flow of ~ 3.12^{11} ions.

Let us now stop the afterglow triggering. The real accumulation efficiency without extraction is even better than T since the particle losses of the afterglow process are now avoided. We then find that at least 3.10^{11} RIB particles are permanently accumulated in the ECR plasma and if they occupy a volume of 30 cm^3, the radioactive ion density is well about 10^{10} cm^{-3} as stated in the introduction. Hence the ECRIT plasma can become a rather dense target of radioactive ions… (and could be utilized in ion/ion colliders) (Fig. 11). Further we can explore the potentialities of this ECRIT as an ion accumulator with or without charge breeding.

On the other hand we proposed the use of ECRIS (Fig. 3) for continuous RIB accelerators such as GANIL SPIRAL or PIAFE. Taking into account some cyclotron losses the total efficiency for Rb^{15+} would be around 3 %. However this same device utilised with a pulsed linac with 10 % duty cycle would loose 90 % of the RIB and the efficiency would drop to 0.3 %. In this case the use of the ECRIT buncher would be very advantageous even if $\eta = 2.2$ % since then nearly all the Rb^{15+} ions contained in the bunch are accepted by the linac (see Fig. 12, 13) (7).

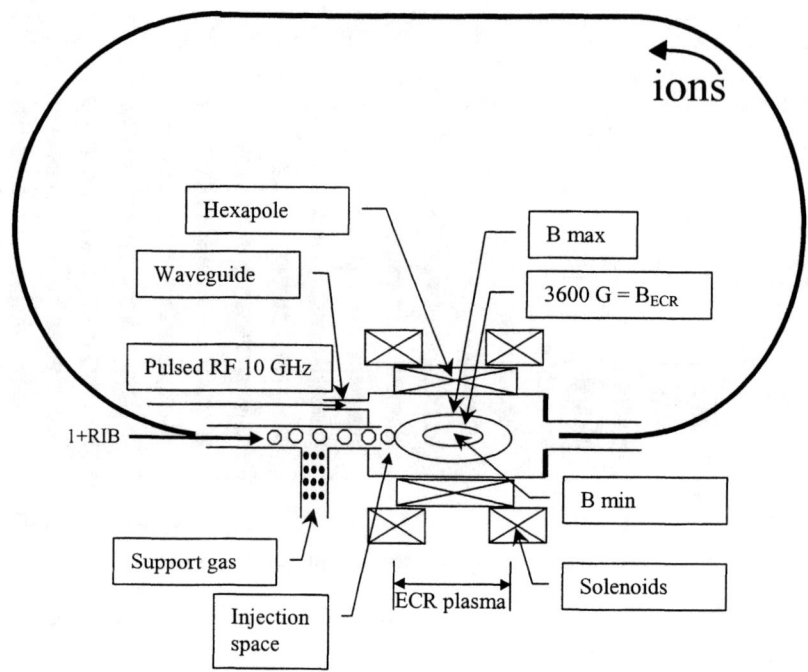

FIGURE 11. ECRIT accumulator target : same as fig. 5 without ion bunch extraction.

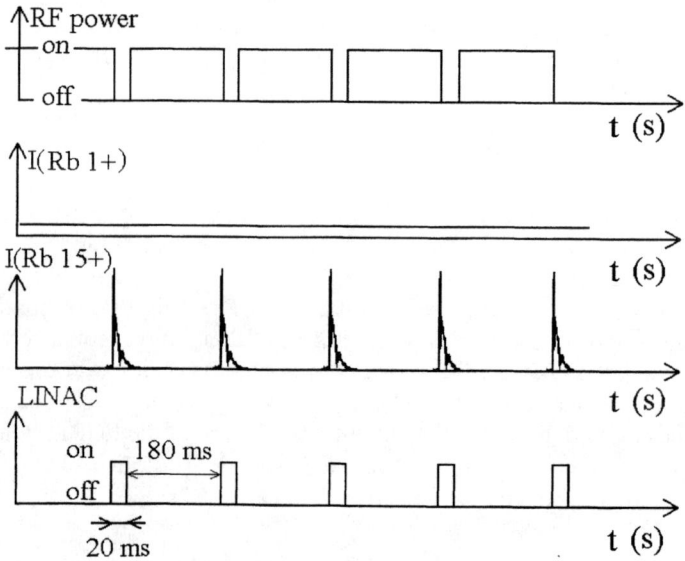

FIGURE 12. Example of timings of an ECRIT breeder buncher for a pulsed linac with 10 % duty cycle.

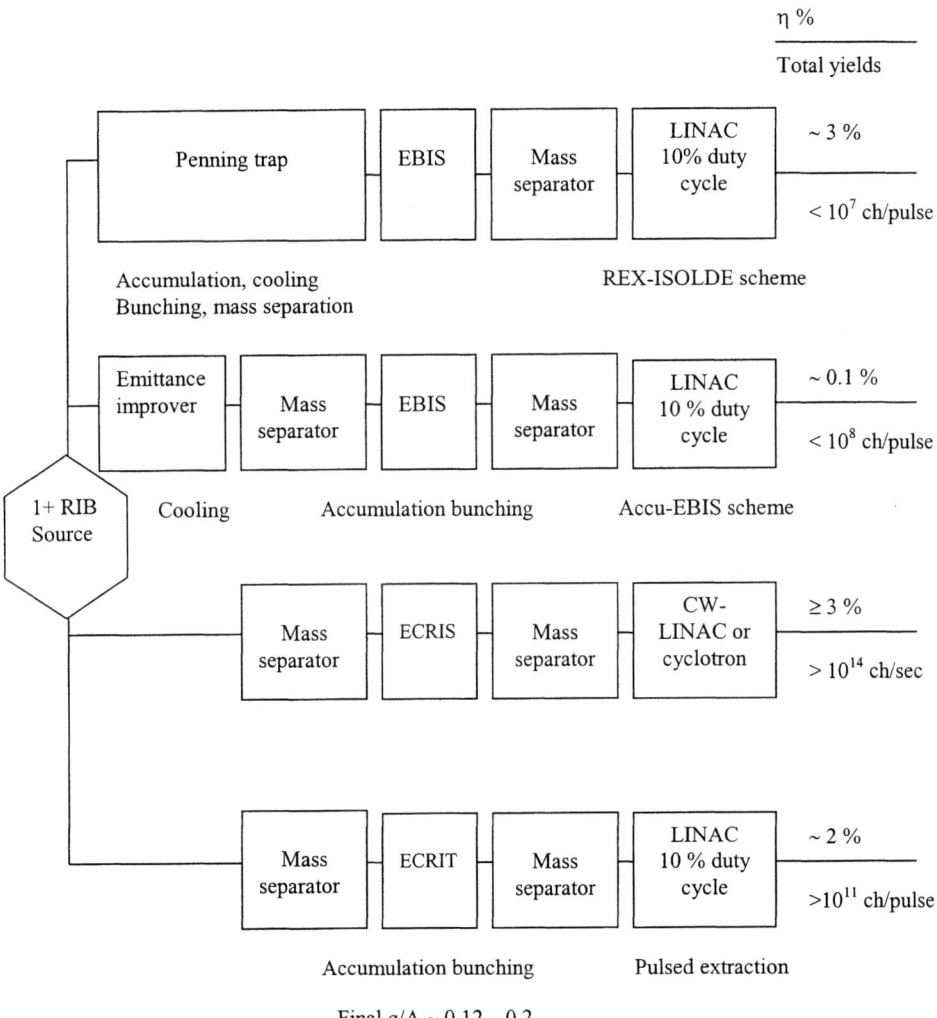

FIGURE 13. Schemes of different charge breeding modes with EBIS and ECRIS and realistic data of global efficiencies η and particle yields expressed in charges per pulse (7). Note that for ECR plasma devices the efficiencies are better, the systems are simpler and cheaper and in addition they can deliver at least thousand times more particles per pulse.

ECRIT AS INTERNAL TARGET IN ION RINGS ?

The ECR trap being able to store up to 3.10^{11} radioactive ions during 0.5 second in a volume of 30 cm x 1 cm^2, it gives the possibility to fill a storage ring. It can also be used as a target of radioactive nuclei. Some prospective applications of these possibilities are considered. In particular we have investigated the interest of associating a storage ring with an ECR trap. Figure 14 is a schematic drawing of the arrangement.

It is, first, instructive to compare the luminosity of the normal situation where an accelerated radioactive beam is sent on a stable target, to that where a stable high energy beam is stored in a ring and crosses an ECR trap at each turn. In the first case (case 1), the initial 1+ radioactive beam has intensity i. The acceleration efficiency is ε while the useful thickness of the target is n_T. The luminosity in the standard situation is, therefore, $\pounds = \varepsilon i n_T$. A typical useful thickness is around 500 µg/cm^2, namely $n_T = 1.4 \times 10^{18}$. In the second case (case 2) the intensity of the beam is I. The ECR trap is filled with a radioactive 1+ beam of intensity i. The confinement time being 0.5 second the number of trapped ions is 0.5i. The surface of the plasma is S and its length L.

The luminosity is $\pounds = \dfrac{0.5 i I}{S}$.

In order for case 2 geometry to be more advantageous than case 1, the condition $\dfrac{I}{S} > 2\varepsilon n_T$ must be realized.

To be specific let $\varepsilon = 0.05$, then $\dfrac{I}{S} > 1.4 \times 10^{17}$. For S = 1 cm^2 one gets I > 2.3 x 10^{-2} particle ampere.

Among other possible geometries, the most promising is the case 3 where radioactive species are injected both in the trap and in the ring. The number of trapped ions is $0.5 i_1$ while the number of ions injected in the ring is 0.5 εi_2 leading to a current $I_2 = 0.5 \varepsilon v i_2$.

The luminosity is, then $\pounds = 0.25 \, \varepsilon v \dfrac{i_1 i_2}{S}$.

It is important to note that, in this case, the luminosity is proportional to the square of the activity of the radioactive source.

A preliminary analysis shows that the 0.1 pA ring current should be within present technical possibilities, given the rather large cross-section of the beam (about 1 cm^2). For stable beams high intensity injection should also be possible. However, for case 3 the injection of a high enough ion number in the ring requires injection on a large number of turns, since the instantaneous RIB current is limited to 2 10^{13} particles/sec, due to the present 10 ms extraction time of the ECRIT. (The time available for one turn injection is about 1 µs, thus, at each turn the number of injected charges is only 2 10^7. it follows that, in order to reach the quoted 2.10^{11} stored ions one would require injection on 10^4 turns. This will certainly require beam cooling, and, thus put a limit on the life-time of the ions which could be stored. Further work is needed on that topic ; in particular any gain on the ECRIT extraction time will ease the injection problem. Extraction times in the range of a fraction of ms seem to be within reach.

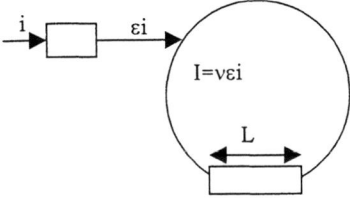

FIGURE 14. Ring/ECRIT system – Ring current $\nu\varepsilon i$ – Rotation frequency ν - Number of charges εi.

REFERENCES

1. Geller, R., Electron Cyclotron Resonance Ion Sources and ECR plasmas., edited by I.O.P. publishing, Bristol & Philadelphia, 1996, p. 256-265.

2. Belmont, J.L., et al, Proc. $13^{\text{ème}}$ Internal Workshop ECR Ion Source, edited D.P. May Cyclotron Institute Texas A.M., 1997, p. 1.

3. Sortais, P., Rev. Sci. Inst. Vol. 63, n° 4, 1992, 2087.

4. Visentin, B., et al., NIM B, 101, 1995.

5. Chauvin, N., et al., NIM A, 1998, to be published.

6. Lamy, T., et al., Rev. Sci. Inst. 69, 3, March 1998, p. 1322.

7. Kester, O., Belmont, J.L., 1998, ISN report.

8. Geller, R., Nifenecker, H., Nucl. Phys. News., to be published.

High Charge State ECR Ion Sources: Status and Developments

S. Gammino, G. Ciavola

INFN-Laboratori Nazionali del Sud, Via S. Sofia 44, 95123 Catania, Italy

Abstract. Electron Cyclotron Resonance (ECR) ion sources are used for many applications, ranging from the production of high intensity proton beams to the production of high charge state ion beams. This review is focused on the production of ions for heavy ion accelerators, which take advantage of the high reliability and stability of the beams produced by ECR ion sources as well as of their capability to produce currents in the order of some eμA of fully stripped light ions and of high charge states heavy ions. The improvement of performance through the years has been steady and an energy increase of the accelerated beams has followed. A brief summary of the main breakthroughs will be given, as the gas mixing and the electron donors.
The role of the magnetic field and frequency will be described with a particular attention to the High B mode concept and to the method of multiple frequency injection.
Some future developments that will increase the currents of the highest charge states and the design of some "3rd generation ECR ion sources" will be considered in the following.

ECR ION SOURCES: MAIN FEATURES

Since the beginning of '80s the Electron Cyclotron Resonance ion sources (ECRIS) have had an increasing number of applications in the accelerator facilities, boosting beam energies and intensities. The operating principle of ECRIS [1,2] is based on the production of a plasma in a chamber where the pressure is in the range of 10^{-6} to 10^{-7} mbar, when microwaves of frequency f are injected in presence of a magnetic field which fulfills the condition $B=B_{ECR}=2\pi f$ m/e over a closed surface inside the chamber; the electrons are accelerated by the microwave field and if the ion lifetime τ_i is sufficiently high, sequential ionization occurs and highly charged ions (HCI) are obtained, depending on the available electron energy. Computer codes are now quite adherent to the experimental results [3,4] but there is not any model that is able to predict the charge state distribution (CSD) exactly. The experience suggests that some parameters are to be optimized, particularly the electron density in the plasma n_e (to which the beam intensity is related), τ_i, to which the average charge state $<q>$ is linked and the electron temperature T_e. The so-called "quality factor" $n_e \tau_i$ [5] and T_e give an estimate of the achievable $<q>$, as shown in fig. 1. It may be observed that the production of 1 eμA of Ar^{16+} is as difficult as the production of 1 eμA of Ne^{10+} and the experiments have confirmed this estimate, at least for gases, whilst for metallic ions the "production difficulty" as vapor often depresses the CSD. Different methods (sputtering, oven, MIVOC, laser ablation, etc.) are used to optimize the production of metallic ions, as described in [2,6].

In spite of the success of fig. 1, the treatment in terms of $n_e \tau_i$ is not exhaustive and ECRIS modeling is unsatisfactory. Modeling often did not take into account a basic parameter as the neutral pressure (the charge exchange process makes lower the probability to get HCI) or the coupling efficiency of microwave power P_{rf} to the ECR plasma and a reasonable approximation of the plasma behavior is not available. We will not enter in details of the different theoretical approach, described in [1,7] but we will underline in the following some basic facts, with clues for the future.

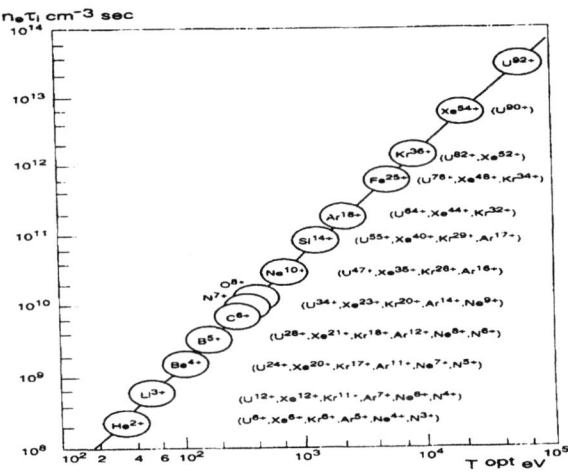

FIGURE 1. Optimum charge states versus $n_e \tau_i$ and electron temperature [7].

SCALING LAWS AND HIGH B MODE

In 1986 some qualitative "scaling laws" relating experimental results and macroscopical parameters were proposed by Geller [8]:

$$q_{opt} \propto \log B^{3/2} \quad (1)$$

$$q_{opt} \propto \log f^{7/2} \quad (2)$$

$$q_{opt} \propto P_{rf}^{1/3} \quad (3)$$

$$P_{rf} \propto V f^{1/2} q_{opt}^3 \quad (4)$$

$$I^{q+} \propto f^2 M_i^{-\alpha} \quad (5)$$

301

where q_{opt} is the optimum charge state, B is the magnetic field, V is the plasma volume, I^{q+} is the intensity of the charge state q, M_i is the ions mass and α is an adjustable parameter. These laws worked fairly good for many ECRIS, but some other effect (gas mixing, electron injection, wall coating, biased disk), observed in the following years, had even a higher impact on the HCI buildup.

A particular attention has been paid to the laws (1) and (2), i.e. to the magnetic field scaling and to the frequency scaling. Some experiments have put in evidence the dependence of the HCI production rate not only on the magnetic field strength, as suggested by the formula (1), but also on its shape. In fact the confining trap of an ECR source is given by the superposition of mirror field and hexapolar field, then the losses from the plasma are not isotropic.

The role of the field strength is simplified by the High B mode (HBM) concept [9], based upon the magnetohydrodynamical condition for quiet plasma:

$$\beta = p_{particle}/p_{magnetic} \approx (n_e k T_e) / (B^2/2\mu_0) < 0.01 \qquad (6)$$

where $p_{particle}$ is the pressure of charged particles in the plasma, $p_{magnetic}$ is the magnetic pressure, $T_e \gg T_i$. If $n_e \approx n_{cutoff}$ and $T_e \approx 10$ keV, (6) can be written:

$$B/B_{ECR} > 2 \qquad (7)$$

	SC-ECRIS	CAPRICE	SERSE	AECR	RIKEN
Frequency (GHz)	6.4	14	14	10+14	18
Radial field (Tesla)	0.57	1.05	1.54	0.85	1.2
Axial field (Tesla)	2.1	1.4	2.7	1.7	1.4
B(rad)/B(ECR)	2.47	2.08	3.05	2.3+1.7	1.85
Argon 11^+ (eμA)	200	200	257	270	280
Argon 16^+ (eμA)	2.5	1	17	21	18
Krypton 25^+ (eμA)	10	2	20	19.4	-
Xenon 30^+ (eμA)	5	-	12	10	10.5

TABLE 1. The main features of some of the best ECR ion sources.

According to this concept, a higher B/B_{ECR} increases the electron temperature and the ion lifetime so that fully stripped heavy ions could be obtained, but T_e depends also on the power density inside the plasma and microwave coupling to the plasma is often limited by other instabilities rather than magnetohydrodynamical. Moreover, the charge exchange process becomes crucial as soon as electrons of inner shells are to be removed and even in presence of high production rates of HCI inside the plasma for HCI, the amount of HCI in the extracted beam may be low.

Since its formulation in 1990, the HBM concept has been proven by many experiments carried out with different sources, operating at different ECR frequencies. The results have been remarkable, not only for SC-ECRIS, which used it for the first time [11] and for the TAMU source [12] which immediately after used a similar field configuration, but even other sources [13,14] are now operating with quite high B_{max}/B_{ECR} and outstanding results. Tab. 1 compares some results from the best ECRIS with their operational B_{max}/B_{ECR}. It can be observed that only the RIKEN source [15] works with a ratio lower than two, but it benefits of a higher cutoff density because of the higher frequency. The lists given in tab. 1 is limited for brevity to few sources which are particularly suited for HCI production.

THE "TRICKS"

During last fifteen years, many methods, often empirically found, have given more stable and efficient HCI production. For some of them there is not yet a clear explanation, because of the intrinsic difficulty of ECRIS modeling, which involves not only atomic and plasma physics, but also material science, microwave engineering, etc. In the following we will mention the most relevant breakthroughs, to which experts refer to as "tricks" with some modesty.

Gas mixing

A phenomenon which has not been completely understood, but that has taken deep roots in the ECRIS field is the so-called gas mixing effect [16], discovered by A.G. Drentje et al. in 1983. If we mix a consistent amount of Oxygen to any species to be ionized (or Helium in the case of oxygen and lighter elements), the CSD improves. Gas mixing stabilizes the plasma, the amount of RF power injected into the plasma chamber (and then the electron temperature) can be increased and the CSD is shifted towards higher $<q>$. Different models have been proposed [17,18] and ion cooling seems to be a reasonable explanation: the lightest ions decrease the ion temperature of the heaviest ions and the ion escape rate from the plasma decreases. Finally gas mixing has the effect to rise both T_e and τ_i.

The electron donors

ECR plasmas have an electron deficit, because the electrons escape along the loss cone with a higher rate than the escape rate of the slow ions; positive plasma potential appears, which shorten the ion lifetime in the plasma and then the ionization rate. The use of tricks to provide electrons increases the plasma density and lowers the plasma potential, making longer the ion lifetime and higher the ionization probability. Different systems have been used through the years, starting with the use of a first stage where plasma is created in a relatively high-pressure region. Later on, the wall coating was successfully applied, followed by other "tricks", some of them being used for a short time (like the e-gun) and some others (wall coating, biased disk, Al chamber) being now used worldwide.

Wall coating, e-gun and Al chambers

In 1986 it was observed that SiO_2 or Th coating increased the intensities of the extracted beams [8,19]. In fig. 2 the CSD with and without coating are shown; the shape of the CSD does not change and the current increase can be explained in terms of a higher n_e. These materials act as electron donors, the plasma density rises and the amount of ionizing interactions increases. Not many potential materials have been checked and Al_2O_3 is the most used because of the high duration of the coating and the good secondary emission. The injection of cold electrons into the plasma, by using an electron gun, was also tested [20]. The performances of AECR with e-gun were about three times better but some difficulties in the tuning of the source were reported. Even in this case the improvements are probably connected with n_e increase, whereas τ_i is unchanged.

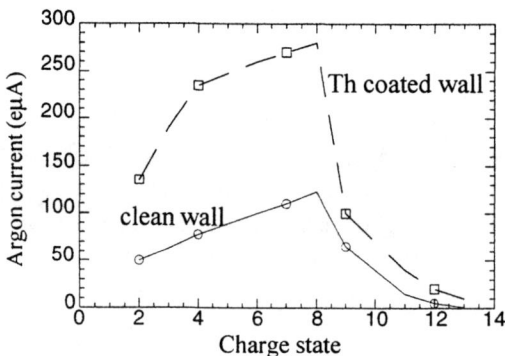

FIGURE 2. The effect of the wall coating [11].

Now these methods have been replaced by the use of Aluminum plasma chambers [15,21,22]. The Al chamber permits to operate with lower pressure and rf power, thus limiting plasma instabilities and charge exchange process. It is remarkable that the RIKEN source [15] gave better results with Al chamber and without gas mixing, showing that quieter plasma is generated. The improvement of base pressure is very important to minimize charge exchange processes and recombinations in the plasma or at the extraction. It has been widely demonstrated that in order to produce He-like or H-like light ions it is necessary an operational pressure of few 10^{-7} mbar or better. Therefore very "clean" Al chambers must be used, as well as the radial pumping [14] and large chamber volumes [11,12,13].

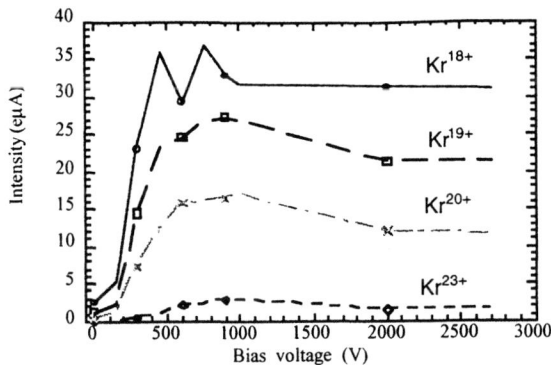

FIGURE 3. Current of medium and high charge states for different value of bias.

Biased disk

In 1990 it was observed in Grenoble that the presence of a 180 V negatively biased disk inside the plasma chamber improves the source performance [23]. Shortly after a similar experiment was carried out at KVI in Groningen [24], demonstrating that a negative bias up to 600 V is even more effective to rise electron density.
More measurements were carried out at MSU-NSCL [23] and fig. 3 shows that the increase of HCI currents saturates at higher voltage than for lower charge states. Unfortunately, it is not useful to increase the bias above 1 kV because the plasma would not trap more energetic electrons supplied by the disk.

The afterglow mode

It was shown on the source MINIMAFIOS [23] that during the rf pulse the plasma attains a quiescent state, which is abruptly spoiled when the pulse ends. The plasma density goes to zero and there is an enhancement in the extracted beam, because all the ions are lost in some ms. The enhancement factor can be three to ten, depending on the features of the source. This mode of operation (afterglow mode) is a very effective way to operate the ECRIS for pulsed accelerator and it is now successfully applied for CERN accelerator complex [27], at GSI, etc.

THE MAGNETIC FIELD SCALING

At MSU some experiments have been carried out with SC-ECRIS to test the field scaling [25,27] owing to its tunable hexapole and to the high axial field.

These unique features allowed making comparative tests for increasing values of magnetic field at 6.4 GHz and at 2.45 GHz, answering to some basic questions:
1) How important is the absolute value of the confining field?
2) How much are CSD and overall intensities affected by the magnetic topology?
3) What is the role of the plasma volume? Larger volumes [28,29] are useful?
The experiments have shown that there is an optimal value for the axial field B_2 on the extraction side, because confinement process is competitive with the extraction process and even though more HCI are produced inside the plasma, the ions cannot be efficiently extracted ($I^{q+} \propto 1/\tau_q$). In fig. 4 the O^{6+} currents are reported for different values of B_1/B_{ECR} (the axial field on the injection side) and B_{rad}/B_{ECR} (the radial field). Neither gas mixing nor biased disk were used, so that the magnetic field was the only variable parameter. The best currents at 2.45 GHz were obtained for $B_{rad}/B_{ECR} \approx 4$ and at 6.4 GHz for $B_{rad}/B_{ECR} \approx 2.5$, probably because the higher cutoff density at 6.4 GHz makes the plasma more stable and then the benefits of a higher radial field are observed at lower B_{rad}/B_{ECR} ratio. These tests have also shown that the optimum value of the injection end axial field is about 5 B_{ECR} [30] for both the frequencies and that low frequency ECRIS can produce HCI, provided that the magnetic field of the source is high enough [25,30].
This suggestion has been successfully applied at LBNL [31] and UCL [32].
Other tests were carried out in order to check the presence of volume effects with SC-ECRIS but the results were negative [25]. Recently tests of "volume effects" carried out with the ORNL Caprice [30] have given positive indications, showing that this argument deserves more investigations.

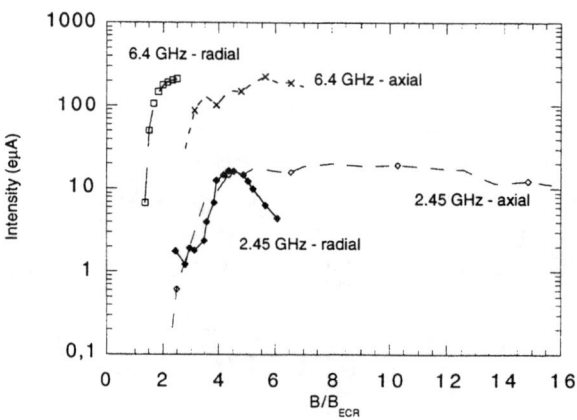

FIGURE 4. - O^{6+} intensities at 2.45 GHz and 6.4 GHz versus B/B_{ECR}.

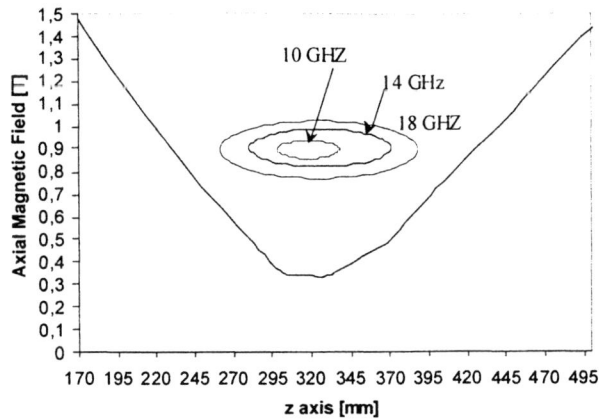

FIGURE 5. Multiple ECR surfaces inside the SERSE chamber.

FREQUENCY SCALING AND MULTIPLE RESONANCES

A particular relevance among the different scaling laws has been always given to the frequency scaling law (2). Electron density has a cutoff which depends on the square of the frequency, but the interplay of frequency, magnetic field and T_e is not so simple and it depends on the scaling law (3) which links the increase of the electron temperature to the power, in turn limited by the instabilities.

It was shown that higher power can be coupled to the plasma by using multiple frequencies. The outstanding results obtained by AECR [14,22] after the increase of magnetic field and with the two frequencies heating can be explained in terms of β. In fact the B_{max}/B_{ECR} ratio is now 1.7 for the 14 GHz resonance and 2.3 for the 10 GHz resonance (tab. 1). With respect to the formula (7), the source AECR is distant from the typical condition for HBM operations at 14 GHz, but it is close to the optimum for 10 GHz resonance, which increases the electron temperature.

Moreover the ionization probability is sharply increased by the presence of two regions of ion-electron interactions. The increase is not linearly dependent on the number of resonant surfaces, but it is expected that the use of a third well-separated frequency should increase more the ionization probability (fig. 5).

SERSE (Superconducting EcR ion SourcE)

The source was designed following the High B mode concept, which leads to the use of superconducting magnets if the frequency of operation is 14 GHz or higher.

The details of the construction of SERSE at INFN-LNS can be found in [10,13,33]. A schematic drawing of SERSE is shown in Fig. 6: the axial field is given by a set of three coils, the central one working with a reverse current; the hexapole is made of six flat race-track coils encased in an aluminum alloy supporting cylinder [33]. The maximum axial field is 2.7 T and 1.6 T on the two sides and the maximum radial field is 1.55 T. The SERSE external envelope, which delimits the cryostat vacuum chamber, is made of iron thick enough to serve as a magnetic shield. The chamber diameter is 13 cm and its volume is 5.6 liters.

The superconducting magnets make possible to operate routinely this ECR ion source not only at 14 GHz with a very high level of confinement, but also to 18 GHz and in the gyrotron domain (28-35 GHz), but in this case HBM operations are not allowed. The small volume high field configuration have given the best CSD because of its high magnetic pressure and its high power density, and SERSE operating at 14 GHz is able to produce 1 eµA of Ar^{17+} and 0.6 eµA of Kr^{30+}, if the base pressure is in the high 10^{-8} mbar.

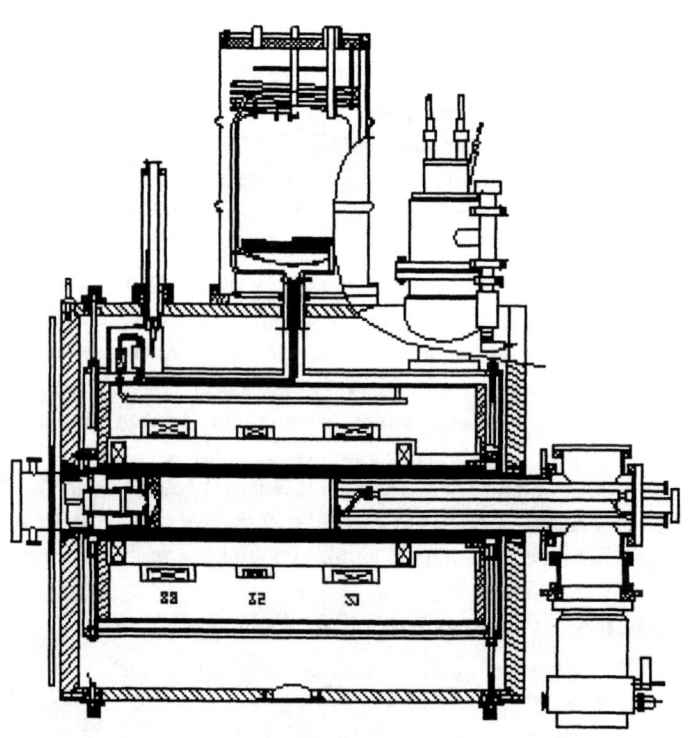

FIGURE 6. The design of the source SERSE.

HBM operations will be soon systematically verified both at 14 GHz and at 18 GHz. After that, the simultaneous use of the two frequencies will be studied, according to the experience of LBNL [22] and ANL [34] with the total available power of 4 kW (the injection section of SERSE has two separate waveguide input, each one allowing to inject about 2 kW). A new injection flange, with modified waveguide input, will allow injecting a third frequency in the range 8.5-12.5 GHz.

3rd GENERATION ECRIS : FUTURE PERSPECTIVES

New ECR sources, some of them gyrotron-based, are under study [35,36,37]. In order to work in High B mode ($\beta \approx 0.01$ or 0.02) at 30 GHz (i.e. $n_e \approx 10^{13}$ cm^{-3}) with $T_e > 10$ keV, they needs an axial field above 4 T and a radial field up to 3 T. In order to operate such a source in an effective way it is not sufficient to have safely operating coils, but also to reach a vacuum in the order of few 10^{-8} mbar.

The coupling of high power high frequency microwave to the plasma needs a careful study about the effects of high levels of power deposition per unit volume; some experience has been obtained in a pulsed mirror trap [38] with a quite low τ_i and $n_e \approx 10^{13}$ cm^{-3}, where it has been observed that the average charge state for Argon is 12^+. These results are very promising especially for sources working in the afterglow mode.

Another attractive design of 3rd generation ECRIS can be originated by the ECR ion trap [39], proposed by Sortais, to be built with a hybrid magnet (superconducting solenoids and permanent magnet hexapole).

CONCLUSIONS

Magnetic field scaling and frequency scaling have sharply increased the production of high charge states ion beams. In the next future the availability of SERSE as well as of other ECRIS with high magnetic field will open new field of investigation and the experiments performed with these sources will perhaps disclose new interesting domains. The optimum use of the "tricks" and the complete understanding of these phenomena will allow further increase of electron density and temperature, and finally of high charge state beam currents. The use of multiple and/or variable frequency injection is expected to boost the electron temperature, allowing to produce quite intense beams of highly charged heavy ions. In fig. 7 the highest charge states obtained with currents above 1 eμA in 1988 and in 1998 are shown for some ions. We have no doubt that the gain will be so nice even in next ten years.

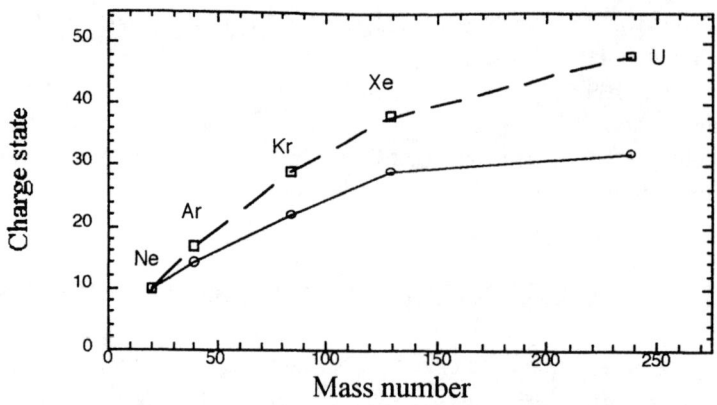

FIGURE 7. Charge states obtainable (I>1 eμA) in 1988 (solid line) and 1998.

ACKNOWLEDGMENTS

The suggestions of G. Melin, D. May, A. G. Drentje are gratefully acknowledged.

REFERENCES

1. R. Geller, Electron Cyclotron Resonance Ion Sources and ECR plasmas, Institute of Physics Publishing, Bristol, 1996
2. B. Wolf, Handbook of Ion Sources, CRS Press, Boca Raton (1995)
3. G. Shirkov, Rev. Sci. Instr. 67(3), 1996, 1158
4. P. Sortais Rev. Sci. Instr. 67(3), 1996, 867
5. R. Geller, Rev. Sci. Instr. 69(3), 1998, 1302
6. D. May, "ECR ion sources for accelerators", Proc. 15th Int. Conf. On Cyclotron, Caen (1998) to be published
7. A. Girard, G. Melin, Nucl. Instr. & Meth. A382 (1996) 252
8. R. Geller et al., " The Grenoble ECRIS status 1987 and proposal for ECRIS scaling", Proc. 8th Workshop on ECRIS, East Lansing, (1987) 1
9. S. Gammino, G. Ciavola, LNS internal report 90/4
10. G. Ciavola, S. Gammino, Rev. Sci. Instr. 63(4), 1992, 2881
11. T. A. Antaya, S. Gammino, Rev. Sci. Instr. 65(5), 1994, 1723
12. D. May et al., Rev. Sci. Instr. 69(2), 1998, 688
13. P. Ludwig et al., Rev. Sci. Instr. 69(2), 1998, 653
14. D. Xie and C.M. Lyneis, Rev. Sci. Instr. 69(2), 1998, 625

15. T. Nakagawa, "Recent progress of RIKEN 18 GHz ECRIS", Proc. 13th Workshop on ECRIS, College Station, (1997) 10
16. A.G. Drentje, Nucl. Instr. & Meth. B9 (1985) 526
17. G. Shirkov, "Interpretation and numerical simulation of the gas mixing effect in ECR ion sources", Proc. 13th Workshop on ECRIS, College Station, (1997) 78
18. G. Melin, A.G. Drentje, A. Girard, D. Hitz, subm. to Phys. of Plasmas (AIP)
19. C.M. Lyneis, "Operating experience with the LBL ECR source", Proc. 8th Workshop on ECRIS, East Lansing, (1987) 42
20. C. Lyneis, Z.Q. Xie, "Preliminary performance of the AECR", Proc. 10th Int. Workshop on ECR ion source, Oak Ridge (1990) 47
21. D. Hitz et al., "The new 1.2 T CAPRICE source: presentation and results", Proc. 12th Workshop on ECRIS, RIKEN (1995), 126
22. Z.Q. Xie and C.M. Lyneis, "Improvements on the LBL AECR source", Proc. 12th Workshop on ECRIS, RIKEN (1995), 24
23. G. Melin et al., "Recent developments and future projects on ECR ion sources at Grenoble", Proc. 10th Workshop on ECRIS, Oak Ridge (1990) 1
24. S. Gammino et al., Rev. Sci. Instr. 63(4), 1992, 2872
25. G. Ciavola et al., "Tests of volume and magnetic field scaling on SC-ECRIS", Proc. 12th Workshop on ECRIS, RIKEN (1995), 156
26. C. E. Hill and K. Langbein, Rev. Sci. Instr. 69(2), 1998, 643
27. S. Gammino et al, Rev. Sci. Instr. 67(12), (1996) 4109 and refs therein
28. A. Girard, "New data and comments on the ECR behavior", Proc. 11th Workshop on ECRIS, Groningen (1993) 86
29. G.D. Alton et al., Rev. Sci. Instr. 69(6), 1998, 2305
30. S. Gammino, G. Ciavola, "ECR ion sources and scaling laws", Proc. 14th Int. Conf. on Cyclotrons (1995) 377
31. D. Wutte et al., Rev. Sci. Instr. 69(2), 1998, 712
32. C. Barué et al., Rev. Sci. Instr. 69(2), 1998, 764
33. P. Ludwig et al., "First operations of the SERSE superconducting source", Proc. 13th Workshop on ECRIS, College Station, (1997) 119
34. M. Schlapp et al., Rev. Sci. Instr. 69(2), 1998, 631
35. S. Gammino, "A study for future upgrading of ECR ion sources", 28th European Cyclotron Progress Meeting, Dubna, (1994) unpublished
36. C.M. Lyneis et al., Rev. Sci. Instr. 69 (2), 1998, 682
37. T. Nakagawa, private communication (1997)
38. S. Golubev et al., Rev. Sci. Instr. 69 (2), 1998, 634
39. P. Sortais, Rev. Sci. Instr. 69 (2), 1998, 656

Preliminary Ionization Efficiencies of ^{11}C and ^{14}O with the LBNL ECR Ion Sources

Z.Q. Xie, J. Cerny, F.Q. Guo, R. Joosten, R.M. Larimer, C.M. Lyneis,
P. McMahan, E.B. Norman, J.P. O'Neil, J. Powell, M.W. Rowe,
H.F. VanBrocklin, D. Wutte and X.J. Xu

Lawrence Berkeley National Laboratory, Berkeley, CA 94720

P. Haustein

Brookhaven National Laboratory, Upton, NY 11973

Abstract. High charge states, up to fully stripped ^{11}C and ^{14}O ion, beams have been produced with the electron cyclotron resonance ion sources (LBNL ECR and AECR-U) at Lawrence Berkeley National Laboratory. The radioactive atoms of ^{11}C and ^{14}O were collected in batch mode with an LN_2 trap and then bled into the ECR ion sources. Ionization efficiency as high as 11% for ^{11}C^{4+} was achieved.

INTRODUCTION

BEARS (Berkeley Experiments with Accelerated Radioactive Species) is an initiative to develop a radioactive ion-beam capability at the 88-Inch Cyclotron Facility of Lawrence Berkeley National Laboratory. It involves the production of radioactive isotopes at an existing medical cyclotron on site and a gas-jet transport of the isotopes through a 300 meter capillary to the 88-Inch Cyclotron Facility. The radioactive isotopes are then injected into the electron cyclotron resonance ion sources to be ionized to high charge states and accelerated by the cyclotron for nuclear science experiments. Because of the lower yield of the radioactive isotopes relative to the stable species, efficiencies of target production, transport, ionization and acceleration all play critical roles in the production of radioactive ion beams. In addition to the tests on the target production and transport, developments have been carried out with the ECR ion sources to study and to optimize the ionization efficiencies. The first two radioactive isotopes to be developed are ^{11}C ($t_{1/2}$ = 20.3 min) and ^{14}O ($t_{1/2}$ = 70.6 sec). This article presents and discusses the preliminary results.

ECR ION SOURCES

An electron cyclotron resonance ion source (ECRIS) reliably produces singly to highly charged ion beams with high intensities and high ionization efficiencies. It is the dominant heavy-ion source and has dramatically enhanced the capabilities of many heavy-ion accelerators worldwide. With its two high charge state ECRISs, the LBNL ECR and the LBNL AECR-U, the 88-Inch Cyclotron has evolved from a light-ion cyclotron into a very versatile accelerator capable of accelerating ions from hydrogen to the heaviest natural element--uranium (1). Other applications of ECRIS are found in atomic physics research, production of radioactive ion beams and industry ion implantation.

The LBNL ECR Ion Source

The LBNL ECR was completed in 1984 and has been reliably operating since then. Its maximum peak magnetic fields on axis are 0.4 and 0.3 Tesla at the injection and extraction regions, respectively. The maximum radial field is 0.3 Tesla at the inner surface of a copper plasma chamber of 45 cm length with a 9 cm inner diameter. There are six large slots between the sextupole magnet bars in the plasma chamber for oven access and radial pumping, as shown in Fig. 1. The conductance of these six slots are calculated to be about 2000 l/s. This ion source operates at 6.4 GHz with a microwave-driven first stage to provide additional cold electrons to the main ECR plasma for the production of highly charged ions. Although the LBNL ECR does not produce ion beams with charge states as high nor intensities as great as the LBNL AECR-U, it continues to provide most of the intermediate charge state ion beams for the cyclotron. The detailed performance of this ion source has been reported in many previous publications (2).

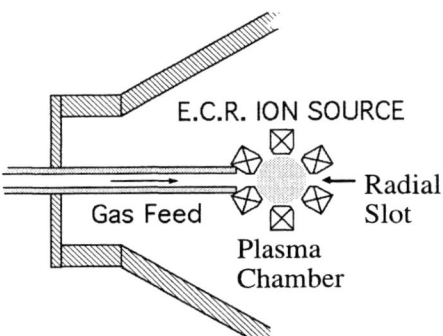

FIGURE 1. The cross section view of the LBNL ECR ion source. The plasma chamber has six large radial slots for pumping and oven access. In the ionization efficiency measurements of ^{11}C and ^{14}O, the activities were fed through one of the radial slots to the plasma chamber.

The LBNL AECR-U Ion Source

The LBNL AECR-U is a higher performance ion source built in 1990 and it was upgraded in 1996 to further enhance its performance. Its maximum peak fields on axis are 1.7 and 1.1 Tesla at the injection and extraction regions, respectively, almost a factor of 4 higher than the LBNL ECR. The maximum radial field at the inner surface of the plasma chamber is 0.85 Tesla. The plasma chamber, made from aluminum for higher yield of secondary cold electrons, is 30 cm in length with an inner diameter of 7.6 cm. Like the LBNL ECR, there are also six slots for pumping and easy oven access in the plasma chamber although the slots are much smaller. The conductance of these six slots are calculated to be about 200 l/s. Shown in Fig. 2 is an elevation view of the LBNL AECR-U ion source. The plasma of the AECR-U is driven by microwaves of two-frequency (14 and 10 GHz) launched through two of the three off-axis waveguides terminated at a bias plate in the injection region. The working gases are bled into the source through one of the waveguides. This ion source is well optimized for the production of highly charged ions (3). It is one of the few ECRISs that have combined all of the recent ECRIS techniques, such as multiple-frequency plasma heating, good plasma chamber surface coating with high yield of secondary cold electrons and high magnetic mirror fields (4). It has produced many record charge states and beam intensities.

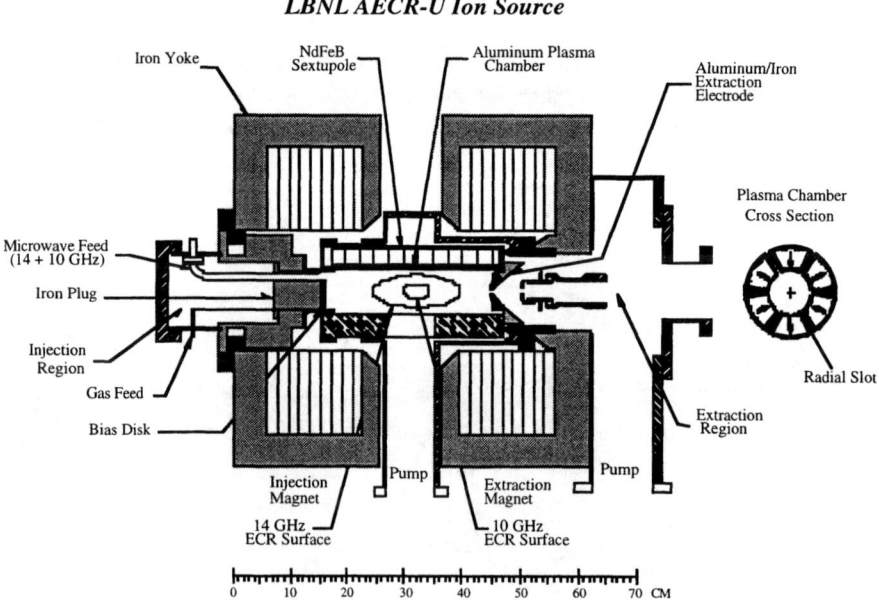

FIGURE 2. An elevation view of the LBNL AECR-U ion source. Only one of the waveguides is shown. The cross section of the plasma chamber is shown on the far right side.

TESTS AND RESULTS

Tests of ionizing the ^{11}C and ^{14}O isotopes were first done with the LBNL ECR ion source. These radioactive isotopes were produced either in a N_2 gas target or a boron nitride (BN) target. A 10 MeV proton beam with intensities up to 5 μA from the 88-Inch Cyclotron was used to irradiate the targets. The initial transport method involved attaching the activity to small aerosol clusters suspended in the carrier gas which is a standard technique which was expected to achieve high transport efficiencies. Unfortunately, it was found that this method failed to transport significant amounts of ^{11}C or ^{14}O. The problem was that most of the activity was in the form of gaseous compounds and did not attach to the aerosol clusters. The fraction of ^{11}C in a chemical form that could be transported was only on the order of 0.1-0.5%. However, this small amount of activity was successfully injected into the LBNL ECR, and an extracted beam of ^{11}C^{1+} was identified, although at very low intensities.

The second method, a cryogenic trapping technique, was found to be much more effective for transporting ^{11}C and ^{14}O. Shown in Fig. 3 is the cryogenic trap set-up. The N_2 target/carrier gas was passed through a coil of 3 mm o.d. stainless steel tubing, about 1.5 m long, submerged in liquid nitrogen. About 50% of the total produced activity is successfully trapped. After stopping the gas flow and allowing the remaining nitrogen to be pumped away, the trap was warmed up quickly to 195 °K by an alcohol bath containing dry ice. This temperature increase releases 90% of the trapped ^{11}C (believed to be ^{11}CO$_2$) and about 40% of the trapped ^{14}O (believed to be N^{14}O or N_2^{14}O). The released gas was then radially bled into the LBNL ECR plasma through an adjustable needle valve. Efficient production of high charge state ions from ECRIS requires low pressures, typically of the order of a few x10^{-7} to 10^{-6} Torr, which limit the gas load into the ion source. So care has to be taken to minimize the amount of non-radioactive gas that was trapped and introduced to the ECRIS along with the radioactivity. Otherwise too high a gas flow will greatly reduce the ionization efficiencies of the high charge state ions. By using the dry-ice/alcohol bath, rather than warming the trap to room temperature, the resulting ECR gas load was more controllable and reduced by 90%.

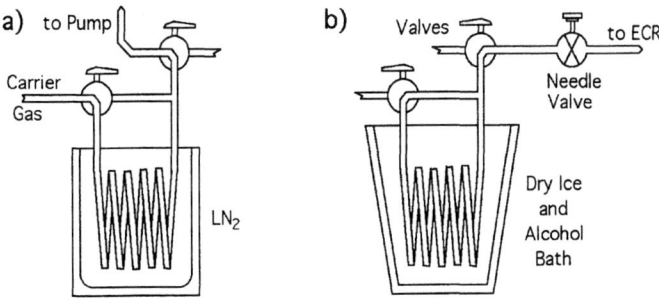

FIGURE 3. Cryogenic trapping system: (a) trapping and (b) release at dry ice temperatures into the ECR ion sources.

With this batch mode trapping, the ^{11}C activity was measured at the trap before it was released to the ion source. Due to its shorter lifetime, the trapped activity of the ^{14}O in each test run was not measured before releasing to the ion source but the previous measurement of about 50% trapping efficiency of the target yield of ^{14}O was assumed. This assumption of the ^{14}O trapped activity leads to a large uncertainty in determining the ionization efficiency. The ionization efficiency of each ion species is determined by normalizing the measured activity of the particular charge state at a time t back to the trap at t = 0. The release times of the activity from the trap were typically about one minute for the ^{14}O (about one half-life) and a few minutes for ^{11}C (about 0.2 to 0.3 half-life) with trapped radioactivity of about 0.5 to 1 mCi. Although the dry-ice/alcohol bath greatly reduces the gas load, there is still significant amount of stable gases released relative to the ECR ion source mass-handling capability. The short release time and the amount of non-radioactive gases make it unlikely that the ECR source tunings were fully optimized for each charge state as would be the case for constant mass flow in stable beam operation. Nevertheless, with this cryogenic trapping method and the ion source operated at microwave power (6.4 GHz) of 200 to 300 Watts, beams of both ^{11}C and ^{14}O with charge states of up to 5+ were successfully extracted from the LBNL ECR ion source. Listed in Table 1 are the best efficiencies achieved. The ionization efficiency for $^{11}C^{4+}$ was about 0.9% with the LBNL ECR ion source. Similar results were found for 1+, 2+ and 3+ charge states, while the efficiency for $^{11}C^{5+}$ was only about 0.1%, due to the difficulty in stripping an S-shell electron in the LBNL ECR ion source. Initial measurements of ^{14}O in 3+, 4+ and 5+ charge states indicate efficiencies in the range of about 0.4 - 0.5%.

Following the tests with the LBNL ECR, the new tests were carried out with the AECR-U to achieve better ionization efficiencies of the high charge state ions. The same cryogenic trapping technique was used to deliver the ^{11}C and ^{14}O activities to the ion source. A different gas injection method was used with the AECR-U. The ^{11}C and ^{14}O isotopes were fed off-axially into the plasma chamber through a 6.3 mm o.d. stainless steel tube terminated at the injection region. Axial injection could achieve a higher trapping of the atoms in the ECR plasma, since the path length of the atoms through the plasma is three times more longer than the path length with the radial injection. In the tests, the AECR-U typically operated with two-frequency plasma heating with total power up to 2.2 kW available from the two klystrons (1.6 kW of 14 GHz and 0.6 kW of 10 GHz) to optimize each charge state with the stable species. As expected, the ionization efficiency for the high charge state ^{11}C and ^{14}O ions are much higher than from the LBNL ECR. High charge states up to fully stripped ^{11}C (2%) and ^{14}O (0.4%) ion beams were extracted from the AECR-U. The AECR-U results are listed in Table 1 in comparison to the results from the LBNL ECR along with the AECR-U ionization efficiencies for stable isotopes of ^{12}C and ^{16}O measured with a calibrated CO leak. The highest efficiency for a single radioactive ion species is 11% for $^{11}C^{4+}$ which is about a factor of 2 lower than the efficiency for $^{12}C^{4+}$. This indicates that there may be still room for further improvement for the ionization of the ^{11}C isotopes.

TABLE 1. Preliminary ionization efficiencies with the ECR and AECR-U ion sources

ION	ECR (%)	AECR-U (%)	AECR-U[a] Stable Species (%)
$^{11}C^{1+}$	1.1		
$^{11}C^{2+}$	0.7		
$^{11}C^{3+}$	0.4	4	
$^{11}C^{4+}$	0.9	11	24
$^{11}C^{5+}$	0.1	4	14
$^{11}C^{6+}$		2	
$^{14}O^{3+}$	0.4		
$^{14}O^{4+}$	0.4		
$^{14}O^{5+}$	0.45		
$^{14}O^{6+}$		3.6	26.7
$^{14}O^{7+}$		1.2	5.6
$^{14}O^{8+}$		0.4	

[a] Stable species are ^{12}C and ^{16}O measured with a calibrated CO leak of flow rate of 9.5 pµA.

DISCUSSIONS AND FUTURE DEVELOPMENT

As shown in Table 1, the ionization efficiencies of the high charge state ions with the AECR-U are much higher than with the LBNL ECR. This is mainly because the AECR-U has a much stronger magnetic field configuration which can support a hotter plasma that is essential to the production of highly charged ions. The other contributions are the off-axial gas introduction, lower chamber conductance and aluminum oxide chamber surface which has a low sticking coefficient. In the LBNL ECR, which has a cooler plasma with presumably lower density, a good portion of the radially injected atoms may just make one radial pass through the plasma and then get pumped away at the opposite large slot or stuck on the copper plasma chamber surface.

In the production of the radioactive ion beams, the source hold-up time can dramatically reduce the ionization efficiency, especially for the short-lived isotopes. The source hold-up time is a function of the source geometry (conductance), plasma chamber surface condition (sticking) and plasma density that determines the ionization length, etc. Figure 4 shows the evolution of the measured activities and the deconvoluted beam rate for $^{14}O^{7+}$ from the AECR-U ion source. The trap valve was opened at time t = 0 and the vertical solid line indicates when the trap valve was closed. The system (transfer line and source) hold-up time for the $^{14}O^{7+}$ is about 20 to 30 seconds. For ^{11}C the system hold-up time is about 8 to 10 minutes which much longer than the ^{12}C hold-up time measured with CO_2 gas on the AECR-U source. We plan to make further tests to resolve the difference.

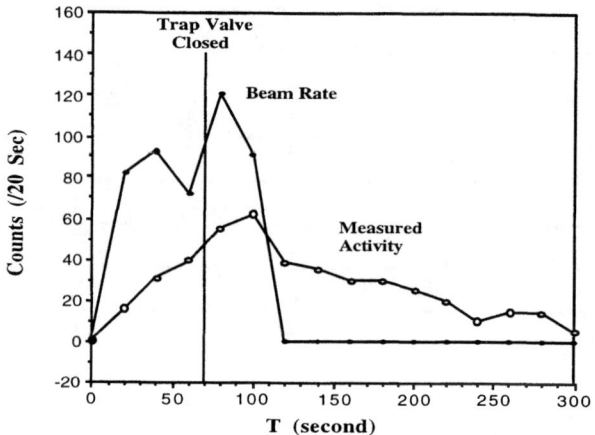

FIGURE 4. Evolution of the measured activity of 2.31 MeV γ-ray and the deconvoluted beam rate of $^{14}O^{7+}$ beam produced with the AECR-U ion source. The solid vertical line indicates the time when the trap was closed. The source hold-up time for the $^{14}O^{7+}$ is about 20 to 30 seconds.

The ionization efficiencies achieved for the high charge state of ^{11}C and ^{14}O ion beams with the AECR-U are comparable to the overall efficiency of the 1+/n+ method using a high efficiency ion source to produce the 1+ radioactive ion beams and then either by accelerating and stripping or injecting the 1+ ions into an ECRIS to produce the desired high charge states (5). If the radiation damage to the permanent magnets is minimized by careful engineering, using an ECRIS to directly and efficiently produce the highly charged radioactive ion beams will be an alternative technique for ISOL type facilities. With this target-ECRIS scheme, the transfer line with various traps can be used to reduce the mass load to the ECRIS. In addition the transfer line can also carry out the selection of certain isobar elements by careful control of the temperatures of the transfer line and traps. The operation of this target-trap-ECRIS system can be cw or batch mode depending on the nature of the desired radioactive species. The detailed of the isobar selection will be addressed in more detail in the future studies.

The first accelerated radioactive ion beam from the 88-Inch Cyclotron was successful carried out recently by transporting the ^{11}C activity in a batch mode from the medical cyclotron in Building 56 to the 88-Inch Cyclotron. With a trap activity of 165 mCi, a $^{11}C^{4+}$ beam of 100 MeV with intensity of 3×10^7 ions/second was extracted from the cyclotron. Figure 5 shows the evolution of the measured activities and the deconvoluted beam rate for the extracted $^{11}C^{4+}$ ions. The system hold-up time is about 10 minutes which is about the same as the measured transfer line and source hold-up time. The total system efficiency of this test is about 0.6% from the trap to the exit of the cyclotron. With further improvements, a total system efficiency of 1 to 1.5% is feasible. So with the ^{11}C production at the medical cyclotron of approximately 1×10^{11} atoms/sec (6), a ^{11}C beam of intensity up to a few x 10^8 ions/sec can be expected for the completed BEARS system. Similar projections for ^{14}O lead to initial beams of a few x 10^6 ions/sec.

Further tests will be conducted to maximize trap efficiencies and minimize the non-radioactive mass load and transport times. In addition, tests to develop other light-mass proton-rich ion beams, such as ^{13}N, ^{15}O, ^{17}F, ^{18}F and ^{10}C are being considered.

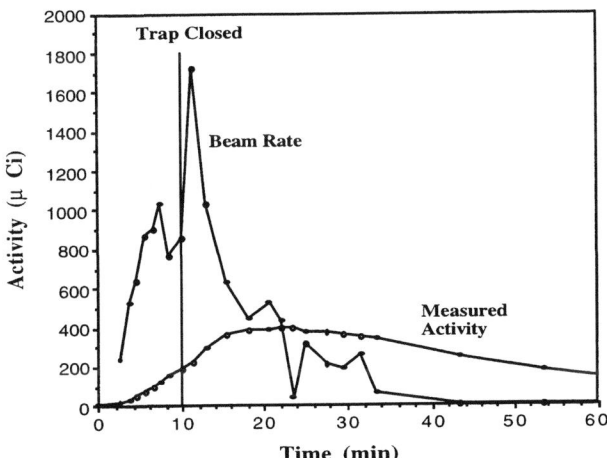

FIGURE 5. Evolution of the measured activity and the deconvoluted beam rate of a 100 MeV $^{11}C^{4+}$ beam extracted from the 88-Inch Cyclotron. The solid vertical line indicates the time when the trap at the AECR-U ion source was closed. The system hold-up time is about 10 minutes.

ACKNOWLEDGMENTS

This work was supported by USDOE, Division of Nuclear Physics, under contracts DE-AC03-76SF00098 at LBNL and DE-AC02-98CH10886 at BNL.

REFERENCES

1. Lyneis, C. M., Xie, Z. Q. and Clark, D. J., Proceedings of the 14th Conf. on Cyclotrons and Their Applications, Cape Town, South Africa, Oct. 8 -13, 1995, pp. 173-176.
2. Lyneis, C. M., Proceedings of the Int'l Conf. on ECR Ion Sources and their applications, E. Lansing, Michigan, NSCL Report #MSUCP-47, 1987, p. 42.
3. Xie, Z. Q. and Lyneis, C. M., Proceedings of the 13th Int'l Workshop on ECR Ion Sources, College Station, Texas, Feb. 1997, p. 16.
4. Xie, Z. Q., Rev. Sci. Instrum. **69**, 625 (1998).
5. Geller, R., Tamburella, C. and Belmont, J. L., Rev. Sci. Instrum. **67**, 1281 (1996).
6. Powell, J., *et al.*, Proceedings of the 2nd Int'l Conf. on Exotic Nuclei and Atomic Masses, Shanty Creek Resort, Michigan, June 1998, to be published.

ECR Ion Source Developments at the Oak Ridge National Laboratory

Y. Liu, G. D. Alton, and F. W. Meyer

Physics Division, Oak Ridge National Laboratory, P. O. Box 2008, Oak Ridge, Tennessee 37831-6368

New techniques for enhancing the performances of electron cyclotron resonance (ECR) ion sources are being investigated at the Oak Ridge National Laboratory. We have utilized the multiple discrete frequency technique to improve the charge state distributions extracted from conventional magnetic field geometry ECR source by injecting three frequencies into the source. A new flat central magnetic field concept, has been incorporated in the designs of a compact all-permanent-magnet source for high charge-state ion beam generation and a compact electromagnetic source for singly ionized radioactive ion beam generation for use in the Holifield Radioactive Ion Beam Facility (HRIBF) research program. A review of the three frequency injection experiments and descriptions of the design aspects of the "volume-type" ECR ion sources will be given in this report.

INTRODUCTION

In recent years, considerable progress has been made in electron cyclotron resonance (ECR) ion source technology in terms of their capabilities for generating high-charge-state ion beams as well as total beam intensities. ECR ion sources are being widely used for the production of highly charged ion beams for heavy ion accelerator based fundamental and applied research. Efforts are under way at the Holifield Radioactive Ion Beam Facility (HRIBF) to develop high performance ECR ion sources for the generation of radioactive ion beams (RIBs). In conventional minimum-B ECR ion sources, narrow bandwidth, single frequency microwave radiation produces thin annular, ellipsoidal-shaped ECR surfaces which constitute a small percentage of the plasma volume and consequently, the efficiency of RF power coupling is limited by the sizes of the ECR surfaces in these sources. It has been suggested that the performances of ECR ion sources can be significantly improved by injecting multiple-discrete or broadband microwave radiation into conventional minimum-B ECR ion sources (1, 2) or by tailoring the central region of the magnetic field so that it is resonant with single frequency microwave radiation (3-5). With multiple discrete frequency microwave radiation simultaneously launched into a minimum-B ECR ion source, one can generate multiple, separated and nested ECR heating surfaces. Consequently, more RF power can be coupled into the plasma, thus, heating a much larger electron population to higher energies, the effect of which is to produce higher charge-state distributions and high intensities within a particular charge state than

CP473, *Heavy Ion Accelerator Technology: Eighth International Conference,*
edited by Kenneth W. Shepard
© 1999 The American Institute of Physics 1-56396-806-1/99/$15.00

charge-state distributions and high intensities within a particular charge state than possible with single-frequency heating. The multiple-frequency method was first demonstrated by Xie and Lyneis (2) by injecting two frequency microwave radiation, well separated in frequency, into the LBNL AECR ion source to enhance the production of highly charged Bi and U ions. They have shown that multiple discrete frequency heating also improves the plasma stability, which benefits the production of highly charged ions (6) as predicted in Ref. (5). We have successfully utilized two- and three-frequency plasma heating in the Oak Ridge National Laboratory (ORNL) Caprice ECR ion source (7) to improve the high charge state distributions extracted from the source (8).

An alternative to the multiple-frequency technique has been proposed (3-5) that employs a new magnetic field configuration which has an extended central flat region tuned to be in resonance with single-frequency microwave radiation. Because of the extended resonant plasma volume, significantly more RF power can be coupled into the plasma, resulting in heating of electrons over a much larger volume than is possible in conventional ECR ion sources. The ability to ionize a larger fraction of the particles in the plasma volume effectively reduces the probability of resonant and non-resonant charge exchange, thereby increasing the residence time of an ion in a given charge state and the probability for subsequent and further ionization. All other parameters being equal, the "volume" ECR source should result in higher charge-state distributions, higher beam intensities, and improved operational stability (1). We are presently fabricating a compact, all-permanent-magnet volume-type ECR ion source that incorporates the flat-field concept for high charge-state ion beam generation. A compact electromagnetic volume-type ECR ion source has also been designed for singly charged radioactive ion beam generation for the HRIBF research program. The results derived from the multi-frequency plasma heating experiments and the design details of an all permanent-magnet volume-type ECR ion source for multiply charged ion beam generation as well as those for a volume-type ECR ion source for low charge-state RIB generation will be described in this report.

MULTI-FREQUENCY MICROWAVE PLASMA HEATING

We have conducted comparative studies to assess the relative performance of the ORNL Caprice ECR ion source for the production of multiply charged ion beams when excited with one, two and three-frequency microwave radiation. The ORNL Caprice ECR ion source (7) is equipped with a solenoidal minimum-B axial magnetic field and a hexapole cusp radial magnetic field for confining the plasma in a multi-mode cavity. The normal operating frequency for the source is 10.6 GHz, supplied by a klystron, coupled into the cavity through a coaxial waveguide injection system. In order to simultaneously inject three frequencies into the plasma chamber of the source, it was found necessary to design and fabricate an appropriate waveguide/injection system to avoid cross coupling of

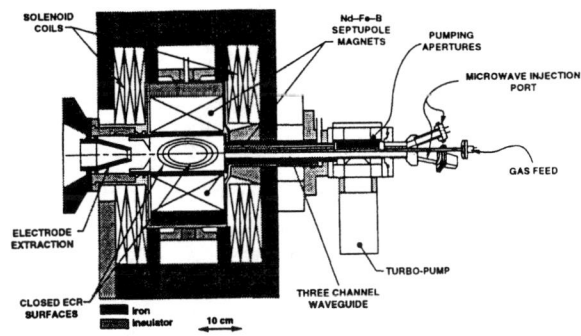

Fig. 1. ORNL-DWG 97-100794. Schematic Drawing of the ORNL Caprice ECR ion source equipped with a three-frequency microwave injection system.

the radiation in the waveguide system. The ORNL Caprice source, equipped with the new three-frequency injection module, is shown in Fig. 1. Microwave radiation between 10 and 14 GHz can be injected into the source. The studies were conducted with the existing 10.6 GHz, 1 kW, klystron power supply and two TWT-based microwave power supplies with rated powers of 80 and 200 W, respectively.

The charge state distributions and ion beam intensities within a particular charge state for Ar and Xe feed gases were measured by injecting multiple discrete-frequency microwave radiation into the source. For all the measurements, the 1 kW, 10.6 GHz microwave power supply was used to saturate the charge-state distributions for Ar^{q+} (132 W) and Xe^{q+} (290 W) before adding the other frequencies. Fig. 2 shows intensity versus charge state for Ar^{q+} obtained individually with 10.6 GHz (132 W) and 11.57 GHz (50 W) microwave radiation, and with both frequencies in combination [10.6 GHz (132 W) + 11.57 GHz (50 W)]. The corresponding charge-state distribution is displayed in Fig. 3. We note that the charge-state distributions resulting from 10.6 GHz or 11.57 GHz single-frequency heating were almost identical, while the charge-state distribution from two-frequency heating is clearly different; the high-charge states are seen to increase while the low-charge states decrease, even through the most probable charge state remains the same. The intensities of Ar^{q+} with charge states higher than the most probable (q=8) were increased by ~2 over those

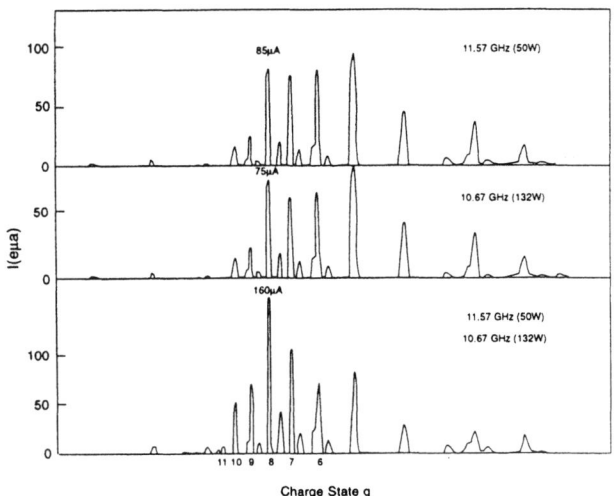

Fig. 2. ORNL-DWG 97-100716. Intensity vs. charge state for Ar^q ion beams extracted from the ORNL Caprice ECR ion Source. Upper curve: 50 W of 11.57 GHz TWT power (nonoptimized); middle curve: 132W of 10.6 GHz klystron power (optimized); and lower curve: 132 W of 10.6 GHz klystron power (optimized) plus 50 W of 11.57 GHz TWT power (nonoptimized).

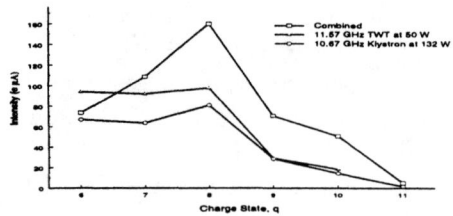

Fig. 3. ORNL-DWG 97-123882. Charge-state distributions for Ar^{q+}; lower curve: generated with 132 W of 10.67 GHz klystron power (optimized); middle curve: generated with 50 W of 11.59 GHz TWT power (nonoptimized); and upper curve: generated with a combination of 132 W of 10.67 GHz klystron power (optimized) and 50 W of 11.57 GHz TWT power (nonoptimized).

when only single-frequency klystron microwave power was used.

One, two and three-frequency heating experiments were also conducted with Xe feed gas with 10.6 GHz (290 W), 10.6 GHz (290 W) + 11.57 GHz (40 W) and 10.6 GHz (290 W) + 11.57 GHz (40 W) + 12 GHz (52 W) microwave power. Because of power supply limitations, we were unable to optimize the ion source performance for the cases of two- and three-frequency heating. Fig. 4 shows the Xe charge-state distribution produced when the source was operated with one, two and three frequency microwave radiation, respectively. It is clear that with the addition of the second and third frequencies, the most probable Xe^{q+} charge state moves to higher values by one unit and the intensities for the high-charge states are increased by ~3 over those for the saturated, single-frequency 10.6 GHz (290 W) case. Although maximum intensities from the source could not be obtained because of problems with the TWT power supplies, our results clearly illustrate that the performance of conventional geometry ECR ion sources can be significantly improved by use of multiple-discrete frequency plasma heating.

DESIGN FEATURES OF AN ALL-PERMANENT MAGNET VOLUME ECR ION SOURCE

A compact, all-permanent magnet, single-frequency ECR ion based on a novel magnetic field configuration has been designed and is presently under construction (9). The source is designed to achieve a large, on-axis ECR "volume," which allows ECR power to be efficiently coupled about and along the axis of symmetry. A

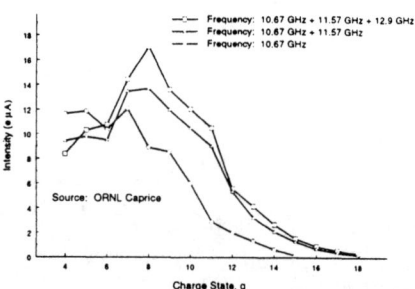

Fig. 4. ORNL-DWG 97-123883. Charge-state distributions for Xe ion beams extracted from the ORNL Caprice ECR ion source equipped with a three-frequency microwave injection system; lower curve: 290 W of 10.6 GHz kylstron power (optimized); middle curve: 290 W of 10.6 GHz klystron power (optimized) and 40 W of 11.59 GHz TWT power (nonoptimized); and upper curve: combination of 290 W of 10.6 GHz klystron power (optimized), 40 W of 11.57 GHz TWT power (nonoptimized), and 52 W of 12.9 GHz TWT power (nonoptimized).

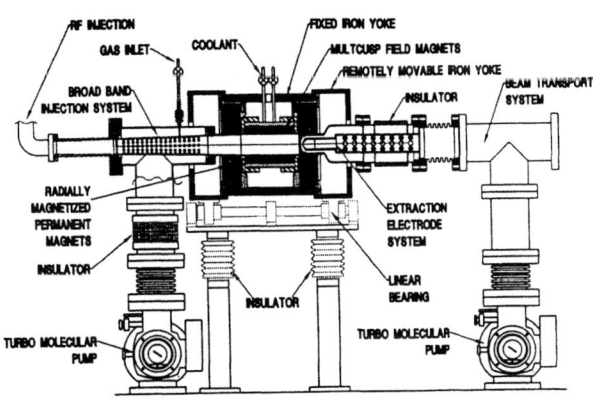

Fig. 5. Schematic view of the flat-*B* ECR ion source.

schematic representation of the source is illustrated in Fig. 5 and the axial magnetic field is displayed in Fig. 6. As noted, the axial magnetic field profile is flat (constant mod-B) in the center which extends over the length of the central field region along the axis of symmetry and radially outward to form a uniformly distributed ECR plasma "volume". This magnetic field design strongly contrasts with those used in conventional ECR ion sources where the central field regions are approximately parabolic and the consequent ECR zones are "surfaces". According to computational studies (3) the new configuration will result in dramatic increases in the absorption of RF power, thus enabling the heating of electrons over a much larger volume, thereby increasing the electron temperature and "hot" electron population in the plasma. The axial mirror field is produced by two, 50-mm thick, annular NdFeB permanent magnets radially magnetized in opposite directions and separated by ~ 150 mm. Specially designed and positioned iron cylinders are used to create the flat central field region between the mirror magnets. The source is designed to operate at a central frequency of ~ 6 GHz and the flat magnetic field region can be adjusted by mechanical means to tune the source to the resonance condition within the limits of 5.6 to 6.9 GHz. The plasma confinement magnetic field mirror has a ratio B_{max}/B_{ECR} of slightly greater than two.

Since the radial magnetic field distribution is proportional to $B = B_0 \, r^{N/2-1}$ where N is the number of cusps, r is the radial distance from the center of the device to the tip of the magnet, a high-order multicusp field for confining the plasma in the radial direction can increase the

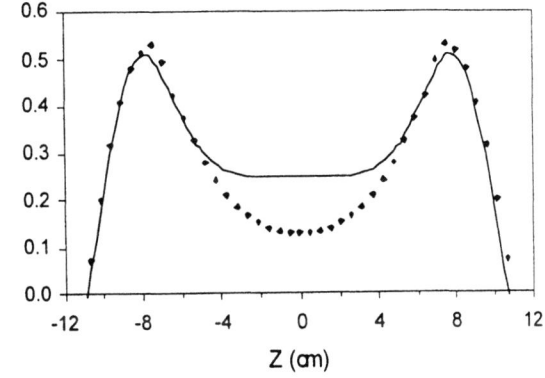

Fig. 6. Axial magnetic field profiles of the "volume" (solid line) and conventional minimum-*B* configuration "surface" (dotted line) ECR ion sources.

resonant volume in the radial direction. Therefore, instead of a sextupole field, commonly used in conventional minimum-B ECR ion sources, a 12-pole multicusp field was designed for the new source. Twelve NdFeB bar magnets, equally spaced in an alternating polarity arrangement around the circumference of a 57.2-mm diameter, water cooled Cu magnet holder, are used to produce the desired field for radial confinement, as shown in Fig. 7. In combination with the axial mirror field, a magnetic field strength of 5.1 kG, approximately equal to that of the axial mirror field, is generated at the inner wall of the plasma chamber. For comparison, the radial field profile for a sextupole configuration (N = 6) is also plotted in Fig. 7. As noted, the region over which the field region is uniform is much greater for the N = 12 multicusp field.

The cylindrical cavity plasma chamber is made of Al and is 15.6 cm in length and 5.4 cm in diameter. Computational design studies were performed for several different RF injection schemes, using the finite element code ANSYS (10). A broadband RF injection system was then selected for the new ion source. It is a long, precisely tapered rectangular-to-circular transition section, starting from a rectangular WR137 waveguide and ending with a circular diameter that matches the dimension of the plasma chamber. The transition from rectangular to circular is very smooth so that it has excellent voltage standing wave ratio (VSWR) while converting the rectangular waveguide TE_{10} dominant mode to the dominant circular waveguide TE_{1np} eigenmodes with the RF power concentrated in the resonant plasma volume and the E-vector oriented perpendicular to the magnetic field direction for efficient electron heating. The mechanical design of the source is very flexible in that it can be converted from a "volume" source to a "surface" source and vice-versa by simply adding/removing a Fe ring to/from the central region between the mirror magnets. The resulting minimum-B axial magnetic field profile, after adding the Fe ring, is also shown in Fig. 6. The multicusp field can also be changed to an N = 6 field distribution when the source is configured as a conventional "surface" source. Comparisons will be made of the performances of the "volume" and conventional single-frequency "surface" ECR sources in terms of the charge-state distributions and intensities within a particular charge-state for each configuration.

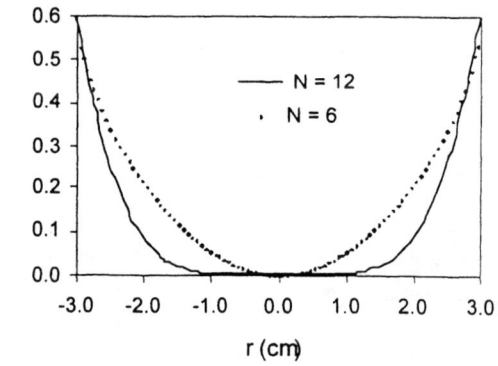

Fig. 7. Comparison of $N = 6$ and $N = 12$ cusp radial magnetic field profiles. The effect of increasing the number of cusps from $N = 6$ to $N = 12$ is apparent: the lower order multiple field results in a much smaller resonant plasma volume.

A COMPACT VOLUME ECR ION SOURCE FOR RIB APPLICATIONS

The HRIBF at the Oak Ridge National Laboratory is an on-line isotope separator (ISOL) facility for the production, generation, and post acceleration of radioactive ion beams for nuclear physics and nuclear astrophysics research. Radioactive nuclei produced in the target by the ISOL technique are transferred to the appropriate ion source where they are ionized and injected into the accelerator. Since many chemically active radioactive species are released from the target in molecular form, an ion source that can simultaneously dissociate molecules and efficiently ionize their atomic constituents is highly desirable. Compared to other ion sources, the ECR ion source is the best choice for the efficient formation of ions from gaseous and volatile compounds. The source has proved to be a very efficient means for dissociating and ionizing members of traditionally difficult gaseous molecular species such as H_2 and CO, etc. (11,12).

A compact volume ECR ion source operating at 2.45 GHz has been designed and will be developed for use at the HRIBF for the generation of RIBs. We believe that the volume-type ECR ion source offers the prospect of high yields of low charge-state radioactive ion beams with good emittance and low energy spread because of the fact that the ions are formed in close proximity to the optical axis of the source, thus can be more efficiently extracted without having to pass through an extended region of strong radially directed magnetic field. Since negative ions beams are required for injection into the 25 MV tandem accelerator that is used for post acceleration of RIBs for the HRIBF research program at ORNL, the source will be optimized for the generation of low charge-state positive RIBs as required for efficient charge exchange conversion to negative ion beams. The source should also be simple, reliable, and inexpensive with long operating lifetime. These guidelines have led us to the design of the simple electromagnetic ECR source illustrated in Fig. 8. The axial magnetic mirror field is produced with mirror coils and trim coils in combination with properly arranged ion yokes. The trim coils are driven in an opposite direction to those of the mirror coils. When the currents in the trim coils are adjusted properly, the resultant field is very uniform with magnitude of 875 G, corresponding to the ECR condition for 2.45 GHz microwave radiation. The axial field profile for the source is shown in Fig. 9. The mirror field is designed to have a small mirror ratio of ~1.2 at the ion extraction end and a high mirror ratio of ~ 2 on the RF injection end of the source. The cylindrical plasma chamber is made of Ta and is about 18 cm in length and 7.6 cm in diameter. Ta is used because of the high temperatures required to effect fast diffusion from target materials. Proton beams from the ORIC will pass through the plasma chamber at a right angle with respect to the ion extraction axis, as shown, and interact with target materials diametrically opposed. This arrangement provides direct coupling of the radioactive species to the plasma chamber following their diffusion release from the target material.

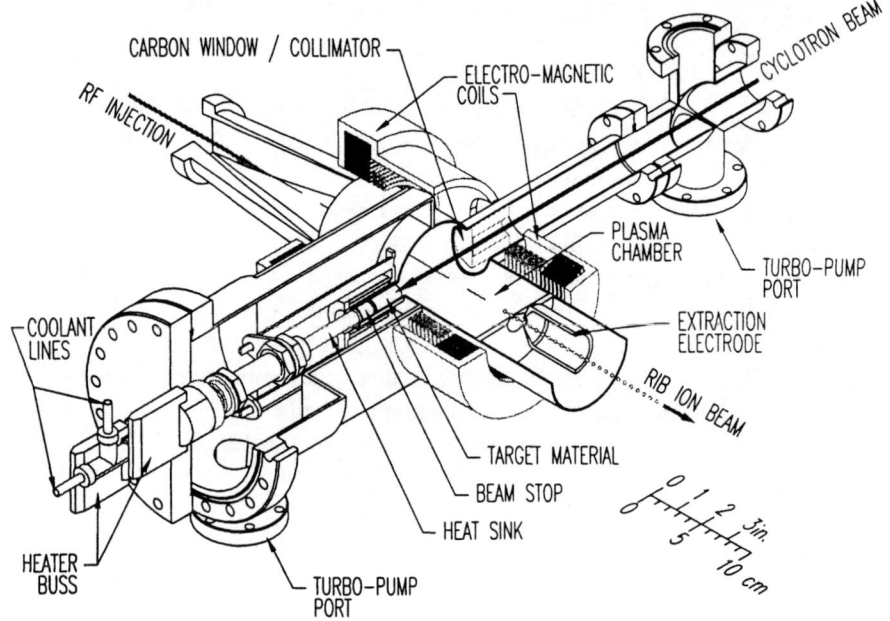

Fig. 8. Isometric drawing of the 2.45 GHz ECR ion source for the generation of RIBs.

No multicusp magnetic field will be used for plasma confinement in the radial direction. This makes the source much simpler, less expensive and durable for RIB applications by avoiding the use of permanent magnets that are subject to neutron degradation. An ECR plasma source has been developed that does not have radial plasma confinement and exhibits high degrees of molecular dissociation and high intensities for the production of singly charged ion beams (12, 13).

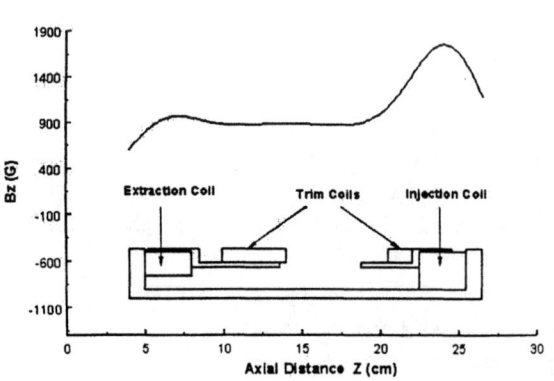

Fig. 9. Schematic layout of mirror and trim coils and the axial magnetic field profile for the 2.45 GHz volume ECR ion source.

ACKNOWLEDGMENTS

Research at the Oak Ridge National Laboratory is supported by the U.S. Department of Energy under contract DE-AC05-96OR22464 with Lockheed Martin Energy Research Corp.

REFERENCES

1. Alton, G. D., Proc. of the 14th Int. Conference on Cyclotrons and their Applications, ed. J. C. Cornell (Capetown, South Africa, October 8-13, 1995), World Scientific, Singapore, p. 362.
2. Xie, Z. Q. and Lyneis, C. M. in Proc. of the 12th Int. Workshop on ECR Ion Sources (Wakoshi, Japan, April 25-27, 1995), eds. M. Sekiguchi and T. Nakagawa, INS-J-182 (1995) 24.
3. Alton, G. D., and Smithe, D. N., *Rev. Sci. Instrum.* **65** (1994) 775.
4. Alton, G. D., *Nucl. Instr. and Meth.* A **382** (1996) 276.
5. Alton, G. D., and Smithe, D. N., *Physica Scripta* **T71** (1996) 66.
6. Xie, Z. Q., *Rev. Sci. Instrum.* **69** (1998) 625.
7. Jacquot, B. and Pontonnier, M., *Nucl. Instr. and Meth.* A **287** (1990) 341.
8. Alton, G. D., Meyer, F. W., Liu, Y., Beene, J. R., and Tucker, D., *Rev. Sci. Instrum.* **69** (1998) 2305.
9. Liu, Y., Alton, G. D., Mills, G. D., Reed, C. A., and Haynes, D. L., *Rev. Sci. Instrum.* **69** (1998) 1311.
10. ANSY is a finite element code marketed by ANSYS, Inc., Cannonsburg, PA 15317, USA.
11. Xie, Z. Q., Cerny, J., Guo, F. Q., Joosten, R., Larimer, R. M., Lyneis, C. M., McMahan, P., Norman, E. B., O'Neil, J. P., Powell, J., Rowe, M. W., VanBrocklin, H. F., Wutte, D. and Xu, X. J., The 8th International Conference on Heavy-Ion Accelerator Technology, October 5-9, 1998, Argonne, IL, USA (these proceedings).
12. Schneider, R. J., Von Reden, K. F., Wills, J. S. C., Diamond, W. T., Lewis, R., Savard, G., and Schmeing, H., *Nucl. Instr. and Meth.* B **123** (1997) 546.
13. Wills, J. S. C., Lewis, R. A., Diserens, J., Schmeing, H., and Taylor, T., *Rev. Sci. Instrum.* **69** (1998) 65.

A High Efficiency, Kinetic-Ejection Negative Ion Source for RIB Generation

G. D. Alton, Y. Liu, C. Williams and S. N. Murray

Physics Division, Oak Ridge National Laboratory, P. O. Box 2008, Oak Ridge, TN 37831-6368 USA

Chemically active radioactive species, diffused from RIB target materials, often arrive at the ionization chamber of the source in a variety of molecular forms. Because of the low probability for simultaneously dissociating and efficiently ionizing the individual atomic constituents of molecules with conventional hot-cathode electron-impact ion sources, the species of interest are often distributed in several mass channels in the form of molecular side-band beams and consequently, their intensities are diluted. The sputter negative ion beam generation technique offers an efficient means for simultaneously dissociating and ionizing highly electronegative atomic species present in molecular carriers. We have incorporated these principles in the design and fabrication of a kinetic ejection negative ion source and evaluated its potential for generating $^{17,18}F^-$ beams for the Holifield Radioactive Ion Beam Facility astrophysics research program. The source utilizes Cs^+ beams to bombard condensable fluorine compounds that emanate from a target material, such as Al_2O_3, and are transported to the cooled inner surface of a conical-geometry cathode where they are adsorbed. The energetic Cs^+ beams efficiently dissociate these molecules and sputter their constituents. Since the work functions of cesiated surfaces are low, highly electronegative species such as fluorine are efficiently ionized in the sputter-injection process. Measured efficiencies for ionizing atomic fluorine, dissociated from condensable compounds that are formed by reactions of SF_6 with fibrous Al_2O_3 material, exceed 6.5%. In this report, we describe the mechanical design features and principles of operation, and present emittance, F^- yield and ionization efficiency data derived from off-line, experimental evaluation of the source.

INTRODUCTION

Production of radioactive species at the HRIBF [1] is effected by directing beams of either 10-60 MeV protons/deuterons, or 10-100 MeV $^3He/^4He$ from the Oak Ridge Isochronous Cyclotron (ORIC) onto refractory target material maintained at high temperature and close-coupled to an ion source. Because the HRIBF relies on the 25 MV tandem electrostatic accelerator for propelling beams to research energies, negative ion beams must be injected into the device. With the exception of the noble-gas elements (Ne, Ar, Kr and Xe), most chemically active RIB species can be formed as atomic or molecular ions by utilization of either the sputter or charge-exchange negative ion generation processes. The charge-exchange process relies on sequential electron transfer interactions between an initially positive-ion beam and a suitably chosen exchange vapor to form the negative-ion beam. As a consequence, the over-all efficiency is often low due to the product of probabilities for first forming the positive ion beam that then must be converted to negative polarity. In addition, the charge-exchange process degrades the quality of beams (emittances) because of its collisional nature.

^{17,18}F beams are of considerable interest for astrophysics research because they play important roles in the hot CNO cycles and rp-processes responsible for heavy element synthesis in the universe. While it is easy to generate intense beams of stable F⁻ in sputter-type negative ion sources conventionally used at tandem based facilities, the generation of adequate intensities of ^{17}F⁻ for RIB applications has proved to be especially difficult because of the sequential times required for diffusion of the isotope from the interior of the target material and effusive flow to the ion source in relation to its lifetime ($\tau_{1/2}$:65s). The times required for each of these processes strongly depend on the chemical reactivity of the species with the target material and with the materials of construction of the vapor transport system as well as the geometry and operational temperature of the target/ion source system. In general, these delay times must be reduced to values commensurate with the lifetime of the particular RIB species in order to provide adequate intensities for RIB research.

Chemically active radioactive species, such as F, more often than not, arrive at the ionization chamber of the source in a variety of molecular forms. Because of the low probability for simultaneously dissociating and efficiently ionizing atomic constituents of molecules containing the element of interest by use of conventional hot-cathode electron-impact ion sources, the desired species are often distributed in several mass channels in the form of molecular side-band beams and consequently, the RIB intensity is diluted. This effect is illustrated in Fig. 1 which shows the mass distribution of fluorine-rich ion beams extracted from the electron beam plasma ion source (EBPIS) (2) used for positive ion beam generation at the HRIBF. The beams were formed from compounds synthesized in chemical reactions between SF$_6$ and fibrous Al$_2$O$_3$ target material under identical conditions to those used to test the source described in this article. As noted, only 13% of the ^{19}F species appear in the atomic mass-channel. Since the F⁺ must be subjected to charge-exchange conversion to form F⁻, the over-all efficiency of the process is < 0.06% at 10% charge-exchange efficiency.

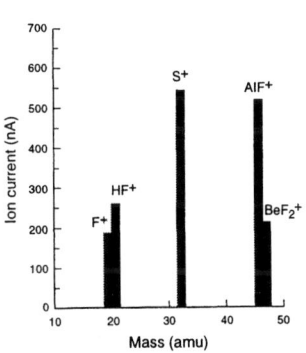

Fig. 1. Magnetic analysis mass spectrum for beams extracted from the EBPIS (2) illustrating that the source is ineffective for simultaneously dissociating and ionizing the constituents of fluorine-rich compounds. Note that only 13% of the atomic ^{19}F appears in the 19 mass-channel.

The sputter negative ion beam generation technique is a particularly effective means for simultaneously dissociating molecular carriers and efficiently ionizing highly electronegative atomic constituents present in the molecule. For these reasons, we have conceived, designed, fabricated and extensively evaluated an efficient kinetic-ejection negative-ion source for the direct generation of $^{17}F^-$ and $^{18}F^-$ beams for potential use in the astrophysics research program at the HRIBF. The sputter generation technique also overcomes the rather severe and chronic poisoning effects incumbent with sources equipped with conventional thermal-evaporation-type LaB_6 surface ionizers (3). In this article, we describe the mechanical design features, principles of operation and optics of the source, and provide operational parameter, emittance, and efficiency data for the generation of F^-.

DESCRIPTION OF THE SOURCE

Design features

The design features of the kinetic ejection source are illustrated schematically in Figs. 2 and 3. The source is designed to mount in the same vacuum housing used for positive sources such as the electron-beam-plasma source (EPBIS), routinely used at the HRIBF for RIB generation (2). The source consists of a Re-lined Ta target-material reservoir transversely attached to a Re-lined, Ta vapor-transport tube that directs particles emanating from the target material reservoir toward the ionization chamber of the source. (Re is used as a liner, where practicable, to reduce the effusive flow times from the target to the ionization region of the source because of its low enthalpy of adsorption properties for many elements.) The temperature of the target material is controlled by varying the current through a triple-pass, cylindrical Ta heater that co-axially surrounds the target material reservoir. The transport tube is independently heated by passing a current along the tube. Although the target material reservoir and the transport tube are mechanically attached, their temperatures can be independently controlled with little coupling. SF_6 is injected at a precisely controlled rate into the target reservoir where it reacts with fibrous Al_2O_3 to form fluorine-rich compounds that are transported to the ionization region of the source. Cs vapor is

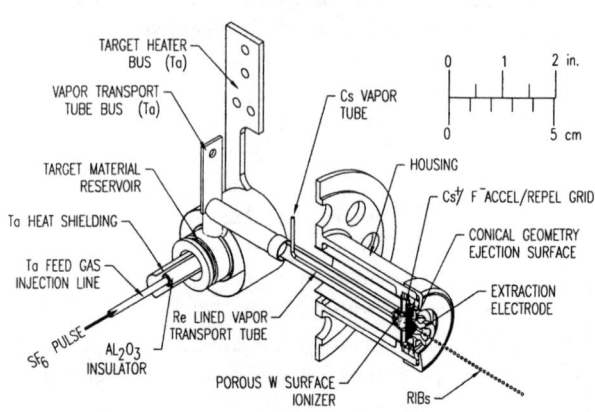

Fig. 2. Isometric cut-away drawing of the kinetic ejection negative ion source designed for potential use at the HRIBF for radioactive ion beam generation.

transported from an externally controlled Cs-oven through an independent transport tube coaxially located within the vapor transport tube and weld attached to a 50% porosity W ionizer. Since the ionizer is welded to the exit end of the transport tube, changes in transport-tube heating power result in changes in the temperature of the ionizer. As is well known, the W-ionizer must be operated at temperatures above the critical temperature (~ 1100 °C) for efficient surface ionization of Cs^0 (4). Therefore, it was necessary to design the integrated system so that the optimum ionizer and transport-tube temperatures could be simultaneously achieved by adjusting the current through the transport tube.

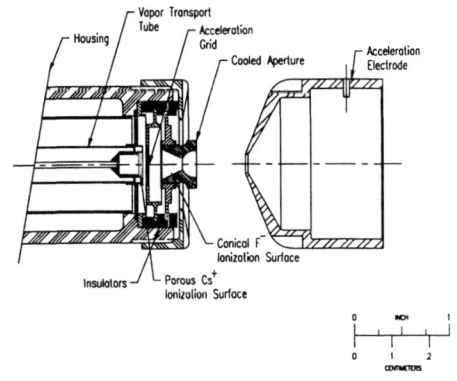

Fig. 3. Side-view of the negative ion formation region of the kinetic ejection negative source showing the Cs-ionizer, the acceleration grid, and the conical-geometry cathode.

Principles of operation

The species of interest, emanating from the target material in vapor form, exit the vapor transport tube through an annular aperture surrounding the W-ionizer where they strike the surface of the conical-geometry cathode equipped with provisions for gas/H_2O or liquid nitrogen cooling for condensing or perhaps, trapping molecules. However, the effects of cooling the cathode on the ionization efficiencies were not measured during these investigations. Such studies will be emphasized in a future paper. Cs^+ ions, formed by surface ionization are accelerated through a negatively-biased, ~ 90%-transparency, Mo-grid, maintained at voltages up to −300 V where they impact the inner surface of the cathode and kinetically eject particles from the surface. The Cs^+ beam, in combination with neutral Cs^0 vapor, lowers the work function (5), dissociates carrier molecules and kinetically ejects their constituents. The mechanism responsible for negative ion formation is well understood (6) and is rather efficient for many atoms/molecules with intermediate to high electron affinities. Moreover, it does not rely on thermal evaporation of the species of interest to effect ionization as does conventional surface ionization. Thus, the adsorption surface can be operated at low temperatures for volatile molecules, thereby increasing their residence times and consequently the probability for dissociation and ionizing their constituents. In cases where the species of interest has a high electron affinity, but is an constituent of a carrier molecule that has a low or even negative electron affinity, as is the case for F in AlF, and cannot themselves be efficiently formed as negatively ions, it is quintessential to be able to dissociate the carrier molecule in order to ionize the species of interest and concentrate them into a single mass-channel.

Cs^+ and F^- beam optics

The electrode systems for the acceleration of Cs^+ and extraction of F^- were computationally designed by the use of the simulation code PBGuns (7). The Cs^+ electrode system was designed to have a high perveance for Cs^+ beams while efficiently extracting negative ions formed on the conical surface during the sputter process. The space charge limited current flow between two parallel electrodes is given by the familiar Childs-Langmuir relation

$$I(mA) = 1x10^3 \{(4/9)\varepsilon_o(2e/M)^{1/2}A/d^2\}V^{3/2} \quad (1)$$

where ε_o is the permitivity of free space, e is the charge on the electron, M is the mass of Cs, A is the area of the ionizer and d the spacing between the grid and the ionizer surface. The calculated perveance for the planar geometry ionizer grid electrode system with spacing 1 mm is: $P = 1.5 \times 10^{-7}$ [A/V $^{3/2}$]. This correlates to space charge limited Cs^+ beams, respectively, of 424 µA at a grid voltage of –200 V and 779 µA at a grid voltage of –300 V.

The optics of the system were designed to assure uniform irradiation of the surface of the conical geometry cathode with Cs^+ beam intensities sufficiently high to ensure optimum probability for dissociating adsorbed fluorine-rich molecules and negatively ionizing their atomic fluorine constituents. Cs^+ beam trajectories accelerated through the gridded electrode and onto the conical geometry cathode are illustrated in Fig. 4.

Analogously, the negative ion optics were designed to optimize extraction of sputter ejected F^-.

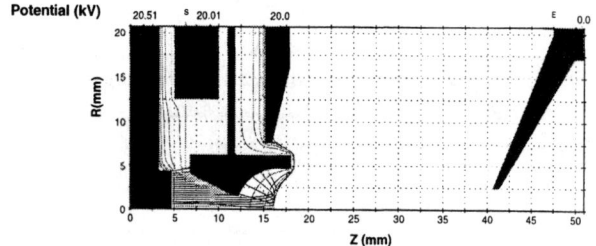

Fig. 4. Simulation of the Cs^+ beam optics from the W-ionizer surface through the negatively biased acceleration grid and onto the conical-geometry cathode where negative ions are kinetically ejected. The simulations were performed with PBGuns (7).

The results of these simulation studies are shown in Fig. 5. A grid voltage of at least –10 V to –20 V, more negative than the potential of the conical geometry cathode was found to be necessary in order to optimize negative-ion extraction. The action of the low gradient field is to reflect negative ions that otherwise would drift toward the grid and be lost. Particles are extracted from the source through a 2-mm diameter aperture in the apex of the cone by the penetrating electrostatic field produced between the conical geometry cathode, maintained at ~ –20.3 kV potential relative to the ground potential extraction electrode.

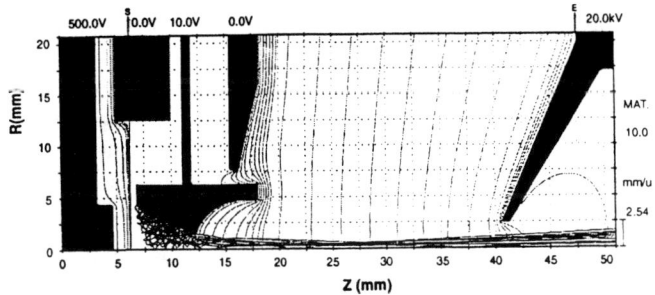

Fig. 5. Simulation of the F⁻ beam optics during ejection and extraction from the conical-geometry cathode surface. The grid-electrode is biased slightly more negative than the conical-geometry cathode to accelerate the F⁻ beam toward the exit aperture. The simulations were performed with PBGuns (7).

The SF_6 feed system

Fluorine-rich gases such as CF_4, CCl_2F_2 and SF_6 are viable candidates for use as fluorinating agents in the system because they are chemically inert and thermally stable at low temperatures but decompose at elevated temperatures, releasing known amounts of F into the high-temperature ion-source. Of course, the thermal dissociation fractions of the molecule in relation to the temperature distribution within the source must be known. The thermal decomposition characteristics of molecules such as these can be calculated by use of thermo-chemistry codes such as HSC (8), ThermoCalc (9) and ChemSage (10). SF_6 was chosen because of its high concentration of F and the ease at which it can be dissociated to release F and F_2 gas for reaction with the Al_2O_3 fibrous target material used in testing the ionization efficiency of the source (11). In order to make a viable estimation of the efficiency of the ion source for on-line F⁻ generation, it is quintessential to evaluate the source with compounds identical to those released during on-line operation. The computed equilibrium distributions of fluorine-rich compounds that result from the reaction of SF_6 with Al_2O_3 at various temperatures (10) are shown in Fig. 6. At temperatures above ~ 1400 °C, the principal components are AlFO and AlF both of which are gaseous at temperatures above ~ 253 °C.

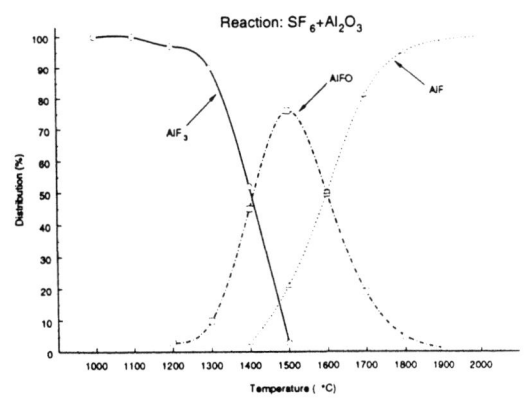

Fig. 6. Equilibrium distributions of principal fluorine-rich compounds resulting from the interaction of SF_6 with Al_2O_3, as calculated with ChemSage (10).

(AlF has been confirmed as the principal carrier of fluorine during on-line operation of the EBPIS.)

The feed system for gases consists of a small stainless-steel chamber of known volume that is equipped with both capacitance manometer and Bourdon gauge pressure monitors. The chamber is filled with the chosen feed gas (SF_6) where it is pressure fed from the chamber into the source through a precision, calibrated-leak, also made of stainless steel, designed to introduce $\sim 3.6 \times 10^{13}$ atoms/s of Xe into the source at 1 bar pressure across the leak. Feed gas from the calibrated leak is introduced into the source in pulsed mode by opening and closing a mechanical gate-valve in series with the calibrated leak and the high-temperature ion source. The time dependence of the rise and fall of the negative ion beam intensity at opening/closing provides information on the delay-times for transport of the fluorine-rich compounds to the ionization region of the source. The transport tube, located between the mechanical valve and target-material reservoir, is made of Ta; the feed gas injection end of the tube is equipped with a heat-sink avoid dissociation temperatures for the SF_6 feed gas at entrance to the target material reservoir. (SF_6 begins to thermally dissociate at $\sim 650\ ^\circ$C.) The interface connection between the gas feed tube and the target material reservoir is sealed with a close-fitting Al_2O_3 insulator to prevent leaks at the interface while providing thermal isolation of the tube.

OPERATIONAL PARAMETERS

In order to optimize the ionization efficiency of the source, it is necessary to know the dependencies of both Cs^+ and F^- beam intensities on the important operational parameters of the source. The most important operational parameters include: negative ion beam intensity versus Cs° oven temperature; negative ion-beam intensity versus ionizer temperature; and negative ion-beam intensity versus kinetic energy of the Cs^+ beam. The Cs^+ beam intensity can be estimated from Eq. 1 or from power supply drain currents. Other parameters such as negative-ion beam intensity versus target material reservoir temperature, negative-ion beam intensity versus cone temperature and negative-ion beam intensity versus cone voltage are also very important but will not be presented in this article because of brevity considerations. A more detailed account of the source with all parameters given will be the subject of a forthcoming article. All of these parameters were carefully measured for the source. Once established, the source is extremely easy to operate with the power supply inputs for control of the Cs oven, ionizer and target material reservoir temperatures and voltages for the grid electrode and conical geometry cathode voltages usually fixed. The source operates very stably and reliably without the need for constant operator attention and, to date, has demonstrated a continuous operational lifetime exceeding one month.

F^- beam intensity versus Cs oven temperature

The probability for negative ion formation during sputtering depends exponentially on the difference between the electron affinity E_A of the atom/molecule and the value of the work function ϕ of the surface (7); the work function, ϕ, in turn, depends on the Cs coverage (5). The flow of Cs through the ionizer provide Cs^+ beams for bombarding the conical-geometry surface from which the atoms/molecules are sputtered as well as a neutral Cs^o flux for lowering the work function. Therefore, knowledge of the dependence of negative ion yield on Cs oven temperature is quintessential for optimum performance of the source. Figure 7 displays F^- yield versus Cs oven temperature for the source. As noted the optimum temperature is ~ 200 °C.

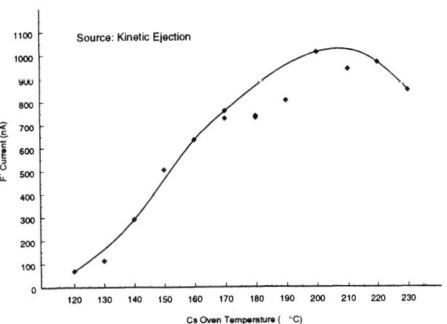

Fig. 7. F^- beam intensity versus Cs-oven temperature for the kinetic ejection negative ion source.

F^- beam intensity versus ionizer temperature

The W-ionizer temperature must be operated above the characteristic onset for positive ion formation, commonly referred to as the critical temperature (4). For Cs^o vapor incident on or passing through a hot W ionizer, this temperature is ~1100 °C. As noted in Fig. 8, the transport tube/Cs-ionizer can be operated over a rather wide range of temperatures without compromising the F^- beam intensity, thus simplifying source operation.

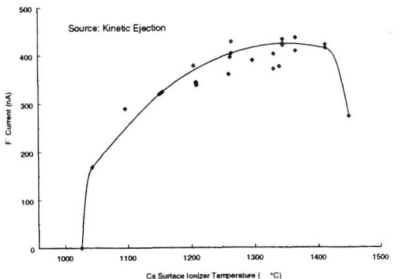

Fig. 8. F^- beam intensity versus W-surface ionizer temperature for the kinetic ejection negative ion source.

Negative ion beam intensity versus Cs^+ beam energy

Since the sputtering process is sensitively dependent on projectile energy in the low energy regime, it is crucially important to supply a Cs^+ beam with sufficient intensity and energy so that the F^- beams intensity versus ion energy reaches a constant value. As noted in Fig. 9, the negative ion beam intensity reaches saturation at

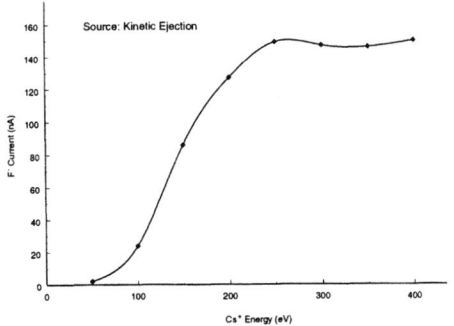

Fig. 9. F^- beam intensity versus Cs^+ beam energy for the kinetic ejection negative ion source.

~ 200 eV Cs^+ beam energy signifying that the intensity and kinetic energy of the Cs^+ beam is adequate for dissociating and sputtering the constituents of fluorine compounds.

SOURCE PERFORMANCE AND BEAM QUALITY CHARACTERISTICS

F^0 ionization efficiency measurements

In on-line experiments at the HRIBF, $^{17,18}F$ are produced within the volumes of refractory oxide targets through fusion-evaporation reactions such as $^{16}O(d,n)^{17}F$ or $^{18}O(p,n)^{18}F$. Because of the relatively short half-life of ^{17}F ($\tau_{1/2}$: 65s), the diffusion and effusive flow path lengths must be kept as short as practical. Fortunately, several oxides (Al_2O_3, Y_2O_3, ZrO_2, SiO_2, Ta_2O_5, Hf_2O_3), can be formed into highly permeable woven mats made of long, thin fibers (2-10 μm diameter) which combine the short diffusion length/high permeability (low density) characteristics required of ISOL targets for efficient diffusion release and effusive flow transport to the ion source (11). The most likely release products will be in the form of oxy-fluorides or metal mono-fluorides. In on-line tests, conducted at the HRIBF, fibrous Al_2O_3 has been used for the production and efficient release of $^{17,18}F$ (12) and, therefore, was used as the target material during testing of the source. Other more refractory oxides (such as ZrO_2 and HfO_2) are presently under evaluation for $^{17,18}F^-$ generation, as well. As noted in Fig. 6, the principal carrier of F from Al_2O_3 will be AlOF and AlF whenever the target is operated within the temperature range of 1500 °C - 1700 °C. Moreover, they are gaseous at temperatures above -253 °C.

Furthermore, once formed, AlF does not react appreciably with Cs^0 over a broad range of temperatures used for routine operation of the source nor does it react with H_2 used as a reagent for picking up elemental F from metal surfaces and efficiently transporting the product, HF, to the ionization region of the source. However, both F and F_2 readily react with H_2 or Cs^0 to form, respectively, HF or CsF, both of which are thermally stable up to ~1650 °C.

Figure 10 displays efficiency versus time for forming F^- during the fluorination of Al_2O_3 and transport of fluorine-rich compounds to the conical geometry sputter surface. The data are

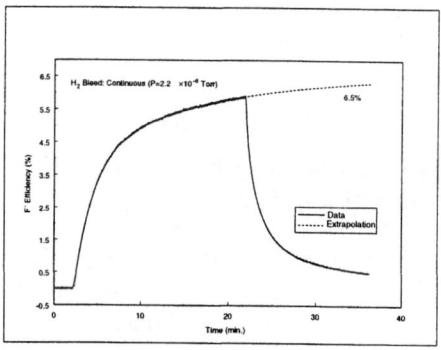

Fig. 10. Efficiency for forming F^- versus target reservoir temperature for the kinetic ejection negative ion source. The data were taken without coolant on the conical geometry cathode.

typical of those derived from the source after outgassing and reaching operational temperatures. The data were taken by injecting a long pulse of SF_6 into the target matcrial reservoir where it reacted with fibrous Al_2O_3 target material with a continuous bleed of H_2 into the source. While H_2 was found to reduce effusive flow times by ~ 5, the result of these studies show that it does not affect the ionization efficiency of the source. As noted, the efficiency reaches 6.5%, a respectably high value considering the fact that the signal is derived from molecular dissociation of a variety of fluorine-rich compounds. The value is even more impressive when one considers that no coolant was used on the cathode. By its addition, the efficiency of the source is expected to increase to even higher values. This study will be reserved for a future and more comprehensive article on the source.

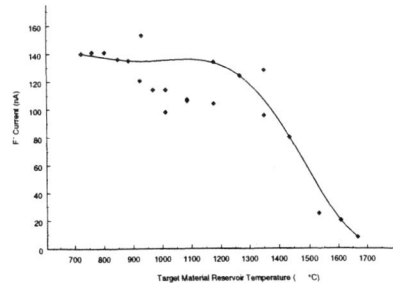

The efficiency for F⁻ formation is found to decrease rather abruptly at target temperatures above 1400 °C, approaching those for on-line operation. An example of this behavior is shown in Fig. 11. Although the origin of the

Fig. 11. Efficiency for F⁻ formation versus target material reservoir temperature for the kinetic ejection negative ion source. Equivalent flow rate of SF_6 into the source: 5.1 μA.

decrease in efficiency is still not known, equilibrium distributions for SF_6 interacting with Al_2O_3 fibrous target material, as shown in Fig. 6, suggest that the decrease may be attributable to changes in composition of the compounds with temperature or possibly to high pressures in the extraction region of the source.

Mass Spectra

The mass spectrum obtained by magnetic analysis of beams from the source is displayed in Fig. 12. As noted, the spectrum is simple with essentially 100% of the F⁻ in the mass-19 channel, in contrast to the spectra (Fig. 1) extracted from its hot-cathode positive counterpart, the EBPIS, where, only 13% of the F⁺ is found to reside in the mass-19 channel. In both instances, the beams were formed, under identical conditions, from compounds, synthesized in chemical reactions between SF_6 and fibrous Al_2O_3 target material.

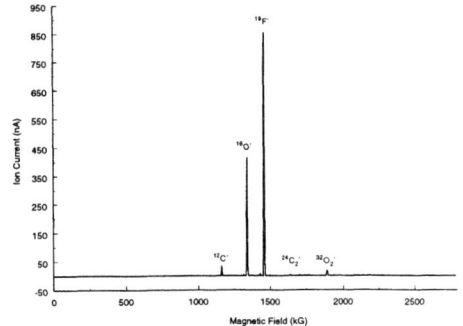

Fig. 12. Mass spectrum for the kinetic ejection source. Note that 100% of the ¹⁹F appears in the mass 19 channel This spectrum can be contrasted with that from the positive EBPIS, displayed in Fig. 1, where only 13% of the ¹⁹F appears in the 19 mass channel.

Emittance data

Isobaric contamination problems are expected to be a serious problem during on-line $^{17,18}F^-$ beam generation because of the high electron affinity of oxygen (1.46 eV) and its pervasiveness in the mass spectra of beams formed by the sputter process. This problem will be especially true when the targets are made of fibrous metal oxides. Therefore, the beam quality (emittance) of the beam is important because it affects mass resolution of these contaminants with existing magnetic analysis systems. The normalized emittance of the source is shown in Fig. 13 in comparison to that of the EBPIS. As noted, it is ~ 2 times larger than that of the EBPIS at the 80% contour level. The emittances of sputter generated negative ion beams are typically higher than positive produced by electron bombardment due to the inherently larger energy spread associated with the sputter process.

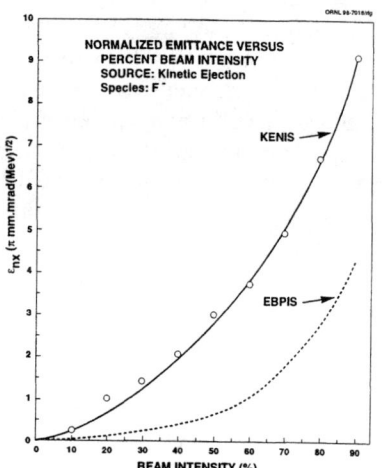

Fig. 13. Normalized emittance ε_n versus percentage of total beams for the kinetic ejection negative ion source.

ACKNOWLEDGEMENTS

Research at the Oak Ridge National Laboratory is supported by the U.S. Department of Energy under contract DE-AC05-96OR22464 with Lockheed Martin Energy Research Corp.

REFERENCES

1. G. D. Alton and J. R. Beene, *J. Phys. G: Nucl. Part. Phys.* **24** (1998) 1347.
2. G. D. Alton, *Rev. Sci. Instrum.* **65** (1994) 1141
3. G. D. Alton, M. T. Johnson, *Nucl. Instrum. and Meth. A* **328** (1993) 154.
4. G. D. Alton, *Rev. Sci. Instrum.* **59** (1988) 1039.
5. G. D. Alton, *Surf. Sci.* **175** (1986) 226.
6. J. K. Nørskov and B. I. Lundqvist, *Phys. Rev. B* **19** (1979) 5661.
7. PBGuns is a product of Thunderbird Simulations, Garland, Texas.
8. HSC is a chemical reaction and chemical equilibrium composition computer code and is a product of Outokumpu Research Oy, Pori, Finland
9. ThermoCalc is a chemical equilibrium composition and phase-diagram code developed by the Royal Institute of Technology, Stockholm, Sweden.
10. ChemSage is a chemical reaction and chemical equilibrium composition computer code and is a product of GTT Technologies, Herzogenrath, Germany
11. G. D. Alton, *Application of Accelerators in Research and Industry*, edited by, J. L. Duggan and I. L. Morgan, CP392, AIP Press, New York, 1997, 429.
12. D. Stracener, H. K. Carter, J. B. Breitenbach, J. Kormicki, J. C. Blackburn, M. S. Smith, and D. W. Bardayan, *Application of Accelerators in Research and Industry*, edited by, J. L. Duggan and I. L. Morgan, CP392, AIP Press, New York, 1997, 393.

A High-Intensity, RF Plasma-Sputter Negative Ion Source

G. D. Alton, R. Lohwasser, B Cui,* Y. Bao,* T. Zhang,*
C. A. Reed

Oak Ridge National Laboratory, P. O. Box 2008, Oak Ridge, TN 37831-6368 USA
*China Institute of Atomic Energy, Beijing, China

A high-intensity, plasma-sputter negative-ion source based on the use of RF power for plasma generation has been developed that can be operated in either pulsed or dc modes. The source utilizes a high-Q, self-igniting, inductively coupled antenna system, operating at 80 MHz that has been optimized to generate Cs-seeded plasmas at low pressures (typically, < 1 mTorr for Xe). The source is equipped with a 19-mm diameter spherical-sector cathode machined from the desired material. To date, the source has been utilized to generate dc negative-ion beams from a variety of species, including: C$^-$(610 µA); F$^-$(100 µA); Si$^-$(500 µA); S$^-$(500 µA); P$^-$(125 µA); Cl$^-$(200 µA); Ni$^-$(150 µA); Cu$^-$(230 µA): Ge$^-$(125 µA); As$^-$(100 µA); Se$^-$(200 µA); Ag$^-$(70 µA); Pt$^-$(125 µA); Au$^-$(250 µA). The normalized emittance ε_n of the source at the 80% contour is: ε_n = 7.5 mm.mrad.(MeV)$^{1/2}$. The design principles of the source, operational parameters, ion optics, emittance and intensities for a number of negative-ion species will be presented in this report.

INTRODUCTION

Negative-ion source technology has steadily advanced over the years, in keeping with the continual demand for higher intensity negative ion beams with improved beam qualities for a variety of tandem electrostatic accelerator-based fundamental and applied research as well as low-energy atomic physics research applications. The advancements in this technology have centered about the development of versatile cesium- and plasma-sputter negative-ion sources, including those described in Ref. 1. The tandem accelerator has also either been used [2] or considered for use as an injector for synchrotron heavy-ion accelerators [3]. The plasma-sputter negative-ion source is well suited for this application in that pulsed negative-ion beam intensities exceeding the practical value of ~200 µA (peak intensity) can be delivered to the synchrotron from the tandem electrostatic accelerator for a wide variety of heavy-ion species. It also offers the prospect of use for batch-mode generation of radioactive ion beams for injection into tandem electrostatic accelerators for post acceleration because of the perfect over-lap of the plasma particles that sputter the sample and the area of the sample irradiated by the production beam. In general, the plasma-sputter negative-ion source generates higher beam intensities with improved emittances than their Cs-sputter counterparts. Therefore, several plasma-sputter negative-ion sources have been developed over the past few years for a wide variety of applications, including those described in Refs. [4-9]. These sources all utilize hot cathodes for plasma ignition with the exception of the source described in Ref. 9, which uses RF power. Because of the erosional nature of the hot-cathode plasma discharge, cathodic wear limits the lifetimes of sources based on this principle. The use of RF power for plasma ignition, in principle, overcomes this handicap, reduces the complexity of operation and lowers overall source maintenance. For these reasons, we have chosen RF as the means for powering the discharge in the plasma-sputter negative-ion source described in the present article. (The studies and developments associated with antenna design and optimization for the source were the central objective of a thesis by one of the present authors [10].)

DESCRIPTION AND DESIGN ASPECTS OF THE SOURCE

The negative-ion source, displayed schematically in Fig. 1, is based on the use of RF power for plasma-sputter generation of high-intensity negative-ion beams. The source utilizes a high-Q, self-igniting, inductively-coupled antenna system, operating at 80 MHz, that has been optimized to generate Cs-seeded plasmas at low pressures (typically, < 1 mTorr for Xe). Figure 2 schematically illustrates the power supply arrangement of the source. The source can be operated in either pulsed or dc modes for the generation of negative ion beams; by simply reversing the polarity high voltage power supply, the source can also be use to generate positive-ion beams in either of the two modes of operation. The source is modular and can be quickly installed-into or removed-from the stainless steel, re-entrant, vacuum chamber, attached to the metal-to-

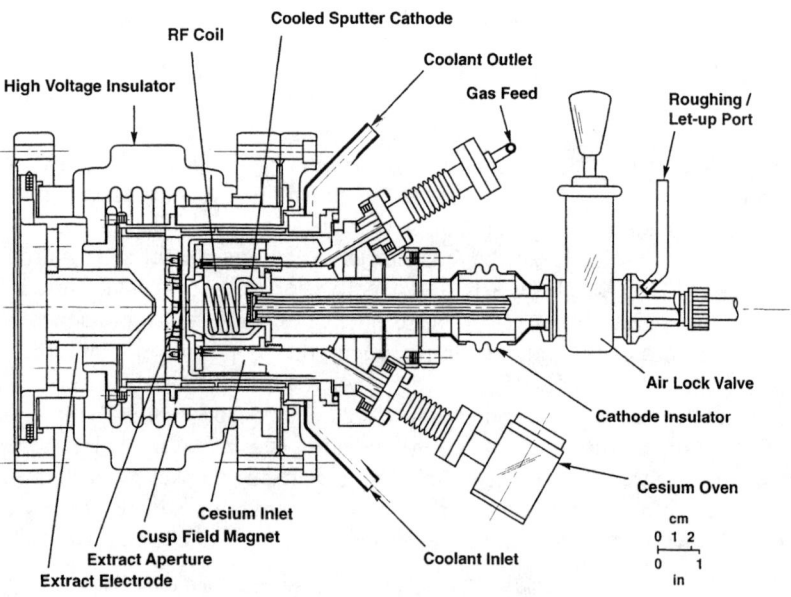

Fig. 1. Schematic Drawing of the RF plasma-sputter negative-ion source.

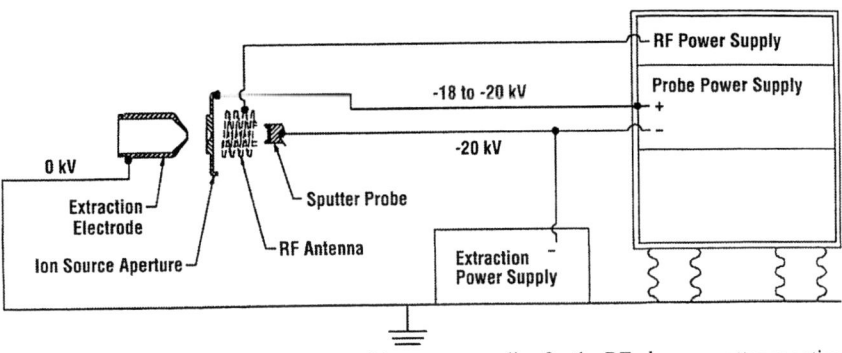

Fig. 2. Schematic drawing of the arrangement of the power supplies for the RF plasma-sputter negative ion source.

ceramic (Al_2O_3) high voltage insulator. The extraction-electrode system is attached to the ground plane of the high voltage insulator. The main vacuum chamber is equipped with permanent magnets positioned around the periphery of the chamber to form a longitudinal magnetic field (maximum field strength: 380 G) for confinement of the plasma in the radial direction. The axial magnetic field distribution, shown in Fig. 3, is designed to reach a maximum value at the aperture of the source and to drop off quickly toward the sputter sample position. This magnetic field geometry was conceived to eliminate magnetic field induced non-uniformity effects at the sample surface and to guide the accelerated particles along field lines toward the extraction aperture. This concept is highly successful as evidenced by the uniformity of the sputter patterns on samples that have undergone extensive bombardment in the plasma discharge. The sputter-probe sample can be withdrawn into the vacuum air lock for quick change without having to interrupt the main vacuum in a time period of ≤ 5 minutes.

The following materials are used for various components of the source: (1) modular re-entrant ion source housing: stainless steel; (2) high voltage insulator: stainless steel flanged, metal-to-ceramic (Al_2O_3) bonded with metal-to-metal seals; (3) sputter sample feed-through: stainless-steel flanged, metal-to-ceramic (Al_2O_3) bonded with metal-to-metal seals; (4) filament and RF feed-through: stainless steel flanged, metal-to-ceramic (Al_2O_3) bonded with metal-to-metal seals; (5) internal sputter probe insulator: BN; (6) vacuum/air-lock and isolation valve: stainless steel with elastomer-seals; and (7) H_2O-cooled, sputter sample holder: stainless steel and copper with an elastomer seal.

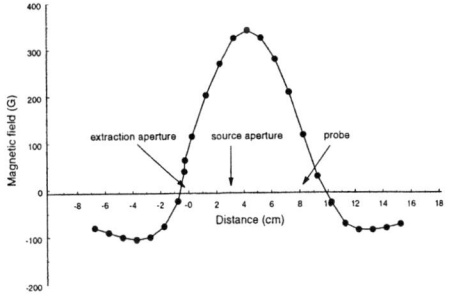

Fig. 3. Axial magnetic field distribution used to confine the plasma in the radial direction. The field is formed with permanent magnets placed around the periphery of the reentrant vacuum housing.

Principles of operation

The sputter negative-ion formation process depends exponentially on the difference between the electron affinity E_A of the species of interest and the value of the work function ϕ of the surface and, therefore, low work functions are desirable [11]. Cs vapor is very effective in lowering the work functions of surfaces either statically or dynamically [12] and is therefore fed into the Xe supported discharge at a controlled rate. Since the sputtering process continually removes adsorbed materials, the dynamic method continually supplies Cs^o to the surface and thus offers a means for providing close to optimum surface work functions. The sample is H_2O cooled to enhance Cs^o condensation on the surface for this purpose. The sputtering process is effected by accelerating Xe^+ particles extracted from the plasma up to 2 keV in energy where they impinge and uniformly sputter the spherical-sector voltage sputter sample. The interface between the spherical-sector geometry sample and the plasma forms a spherical-geometry lens system that focuses convergent negative-ion beams through a 4.5-mm diameter extraction aperture, located at a distance equal to the radius of curvature (50 mm) of the sample. This geometry minimizes aberrations, reduces Cs^o vapor and Xe gas flow from the source, and ensures efficient ion extraction from the source.

DESCRIPTION OF THE RF PLASMA GENERATION SYSTEM

Stable operation of the source at minimal neutral density within the plasma chamber is quintessential for negative-ion-beam generation applications since prohibitive losses can occur during transit through the plasma due to the rather large cross sections for electron collision-detachment ($\sigma \cong 10^{-14} - 10^{-16}$ cm^2) at pressures in excess of 10 mTorr [13]. Thus, one of the principal objectives of the project was to design and develop a high-Q antenna system capable of self-ignition and maintenance of low-density plasmas at low operating pressures. These objectives were accomplished (1) by choosing a rather high operating frequency (80 MHz) to effect a higher degree of ionization within a given plasma volume: (2) by designing an inductively coupled, high-Q antenna system and matching circuit for efficient coupling of RF power to the plasma-discharge; and (3) by evaluating the performance of a given antenna in terms of the parameters of self-ignited plasmas. A schematic representation of the RF coupling system is shown in Fig. 4.

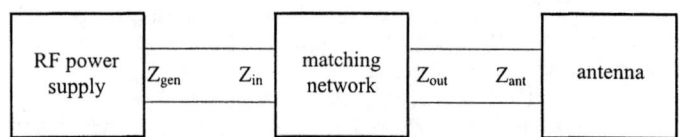

Fig. 4. Schematic representation of the RF coupling system.

The impedance matching network

The power P adsorbed by the load or antenna, that in turn, provides the energy to generate and sustain the plasma, can be expressed through the following relation:

$$P = U^2 Z_{ant} / (Z_{gen} + Z_{ant})^2 \qquad (1)$$

where U is the voltage of the RF power supply and Z_{gen} and Z_{ant} are complex representatives of their respective resistances. The impedances are optimally matched whenever $Z_{gen} = Z_{ant}$.

Since the plasma density may vary over several orders of magnitude during start-up and operation of the source, this causes an impedance mismatch, thus necessitating the insertion of a tunable impedance matching network between the RF generator and antenna. The function of an impedance matching network is to minimize power transmission losses between two discrete circuits with different impedances. Several schemes have been developed for impedance matching, including the π-network, illustrated schematically in Fig. 5, chosen for optimizing the coupling of power into the plasma for our applications.

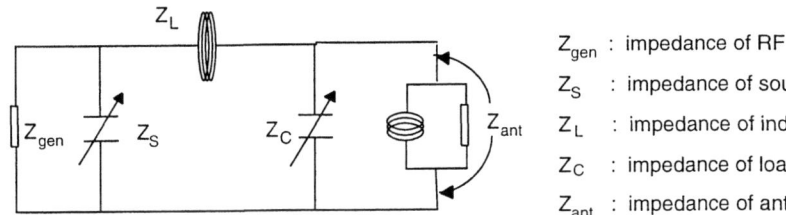

Z_{gen} : impedance of RF power supply
Z_S : impedance of source capacitor
Z_L : impedance of inductive coil
Z_C : impedance of load capacitor
Z_{ant} : impedance of antenna

Fig. 5. Schematic diagram of the π-network used to match the impedance of the RF generator to that of the antenna and thereby optimize coupling of the RF power to the plasma.

The antenna design

The principal objective of this aspect of the project was to develop antennae for efficient coupling of power into plasmas for ion source applications for which spiral geometry antennae are desirable. Spiral antennae were wound from 3-mm diameter Cu tubing with variable outer diameters and numbers of turns to approximate the specified inductance value for the antenna. The matching network, shown schematically in Fig. 5, was then constructed and close coupled to the ion source. Variable capacitors were used for both C_S (10 to 77 pF) and C_c (100 to 1000 pF) so that the circuit could be tuned to optimal plasma generation conditions to evaluate the effectiveness of a particular antenna design for coupling power into the plasma, as well as to compensate for variations in total capacitance and inductance in the circuit attributable to connecting wires and induced capacitance in a particular antenna. The circuit provided easy tuning over a wide range of plasma densities for evaluating a variety of antenna designs. The physical shape, geometry, size, and materials of construction of the antennae and RF feed-through affect the electrical characteristics of the circuit that must be optimized so that it is resonant with the operating frequency of the RF generator. The resonance frequency for an antenna with inductance L and capacitance C is given by the familiar relation

$$f_{res} = 1/2\pi\sqrt{LC} \ . \tag{2}$$

Thus, the resonant frequency f_{res} of the antenna can be adjusted to agree with the operating frequency f_{gen} by changing the LC product.

The antennae considered for the plasma sputter negative ion source applications were formed from 3-mm diameter Cu tubing by varying the diameter, pitch, and number of turns. The use of Cu tubing permits H_2O or air cooling of the antennae during operation. The ability to alter the capacitance or inductance of a particular antenna and thereby bring the circuit into resonance with the drive frequency of the generator provides an easy method for optimizing the coupling between the antenna and plasma. For example, the inductance can be varied by changing the number of turns or the diameter of the coils while the capacitance can be varied by changing the pitch of the coils (spacing between coils). The inductances of prospective antennae were calculated by use of the following formula

$$L(n,r,l) \approx \frac{r^2 \cdot n^2}{16.4 \cdot 10^6 (9r + 10l)} \tag{3}$$

where n is the number of turns and r is the radius of the coil of length l.

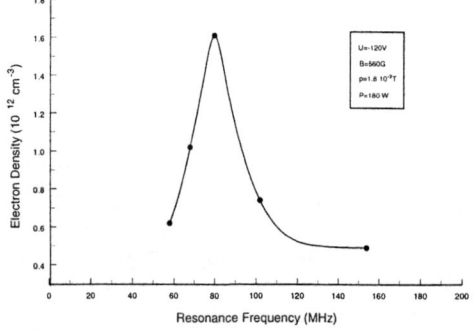

Fig. 6. Electron density versus RF power for five antennae wound with the same radii r but different number of turns n. The central antenna is in resonance with the 80 MHz RF generator.

The various antennae were each evaluated by measuring the densities of their self ignited plasmas by use of a Langmuir probe placed within a vacuum chamber identical in size to that of the application source. Figure 6 displays electron density versus RF power for five antennae wound for different resonant frequencies. The antennae all have the same coil diameter and pitch but different number of turns. As noted, the central antenna is wound so that it is in resonance with the 80 MHz RF generator frequency, thus illustrating the importance of making the resonance frequency, f_{res}, of the combination antenna/feed-through system identical to that of the RF generator, f_{gen}.

From these studies, an easy an effective method for designing antennae/feed-through systems was developed. Equations 2 and 3 can be combined to arrive at an expression for the radius r of the antenna in terms of its length *l*, number of turns *n*, operating frequency f_{res} and capacitance *C* of the feed-through system given by

$$r(n,l) \approx \frac{1.87 \cdot 10^6 + 2.04 \cdot 10^3 \sqrt{1.68 \cdot 10^5 + f_{res}^2 C n^2 l}}{f_{res}^2 n^2 C} \quad . \tag{4}$$

Final tuning of the coil can be effected by adjusting length *l* and turn-to-turn spacing *d*. The presence of a plasma will cause slight changes in the resonance frequency of the circuit due to changes in the capacitance of the system. This can be compensated for by changing the resonant frequency of the coil to slightly lower values. For the antennae used in these studies, we found that the optimum resonance frequency was 78 MHz.

For plasma-sputter negative-ion source applications, low plasma densities are desirable due to the fact that the negative-ion beam will be greatly attenuated by passage through regions of high plasma density. Also the inner diameter of the coil must be greater than the diameter of the sputter sample. For the source described in this report, the parameters of the optimum coil wound from 3-mm diameter oxygen-free Cu tubing are: *n*: 4; r_c: 21 mm; $ø_t$ (mm): 3; d_c: 7.8 mm; l_c: 29 mm; S_T: 25.4 mm. The antenna is coated with a thin layer of porcelain to prevent sputtering of the copper coil during operation; the porcelain does not change the characteristics of the antenna and is flexible enough to allow changes in turn-to-turn spacing required for fine tuning to the desired resonant frequency (80 MHz). An isometric drawing of the coil is shown in Fig. 7.

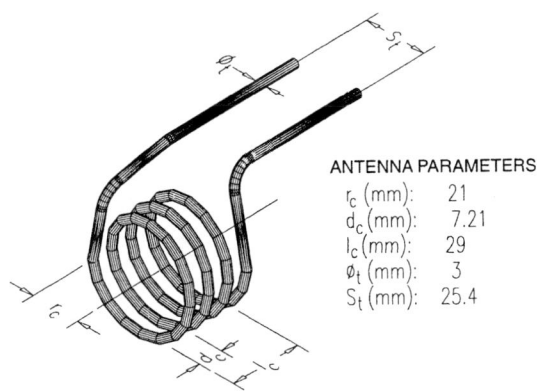

Fig. 7. Isometric of the optimized antenna for the plasma-sputter negative ion source, the parameters for the antenna are: r_c (mm) = 21; *n* = 4; d_c (mm) = 7.25; l_c (mm) = 29; S_T (mm) = 25.4; $ø_t$ (mm):3 .

OPERATIONAL PARAMETERS

In order to realize optimum performance of the plasma-sputter negative-ion source, it is necessary to know the dependence of relative or absolute negative ion beam intensity on the important operational parameters which include: Cs oven temperature; discharge support gas pressure; *RF* discharge power; and sputter-probe voltage.

Negative-ion beam intensity versus cesium oven temperature

The flow of Cs into the source offers a general and dynamic means of lowering the work functions of surfaces and at the same time, a means for overcoming deleterious poisoning effects that may occur due to the presence of certain impurities in the discharge. Since the effective Cs coverage depends on the difference between the rate of arrival and departure of Cs vapor to and from the sputter probe surface, the value of the work function of the surface and consequently the negative-ion beam intensity is sensitively dependent on $Cs°$-oven temperature [12]. The knowledge of the negative-ion yield or relative negative-ion yield on this parameter is quintessential for optimizing negative-ion yields from the source. Relative negative-ion yield versus $Cs°$-oven temperature, obtained at fixed RF power, Xe support-gas discharge pressure, and sputter sample voltage are displayed in Fig. 8. Once known, the $Cs°$-oven temperature, as a general rule, can be set at a fixed value of $\lesssim T_{max}$ where $T_{max} \sim$ 240-245 °C. It also important to note that the $Cs°$ oven temperature cannot be used as a real-time control parameter because of the characteristically slow response times required for thermal equilibrium processes. Values higher than T_{max} should be avoided since higher temperatures can be deleterious to the intensity and lead to operational instability due to sparking.

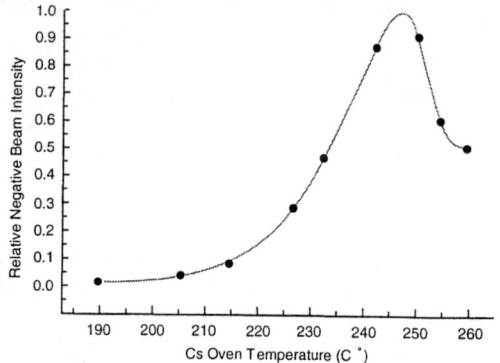

Fig. 8. Relative negative ion beam intensity versus $Cs°$ oven temperature for the RF plasma-sputter negative ion source. Sputter sample: Pt; support gas: Xe; pressure: 1×10^{-5} Torr.

Negative-ion beam intensity versus Xe support-gas pressure

Stable operation at minimal neutral density within the plasma chamber is quintessential for negative-ion beam generation applications since prohibitive losses can occur during transit through the plasma due to the rather large cross sections ($\sigma \sim 10^{-14} - 10^{-16}$ cm^2) for electron collision-detachment at pressures in excess of 10 mTorr [13]. Thus, one of the principal objectives of the project was to design and develop a high-Q antenna system capable of self-ignition and maintenance of low-density plasmas at low operating pressures as discussed previously. Figure 9 displays negative-ion beam intensity versus externally measured Xe pressure for fixed Cs oven temperature, RF power, and Cs sputter-probe voltage. The pressure in the plasma chamber is ~10 times these values as measured with a capacitance manometer. For a given support-gas species, the collision-detachment cross section depends on the electron affinity of the species as well as the velocity of the particle and thus, the magnitude of detachment loss will, in general, vary from species to species at a given pressure.

Negative-ion beam intensity versus RF power

Figure 10 illustrates the dependence of negative-ion beam versus RF power for C⁻. The RF power required to reach saturation of the negative-ion beam intensity depends on the sample material and source operational parameters used during the experiments. In general, the RF power required to reach maximum negative-ion beam intensities for the materials used in this study ranged between 40 and 100 W.

Negative-ion beam intensity versus sputter-probe voltage

Since the sputter ratio varies from material to material, the probe voltage required to reach steady-state negative-ion beam intensities also varies from species to species. Easily sputtered materials such as Cu and Au reach steady state conditions at much lower voltages than does a lower sputtering material such as C. An example of the dependence of negative-ion beam intensity versus sputter-sample voltage for C is displayed in Fig. 11.

NEGATIVE-ION BEAM INTENSITY AND EMITTANCE DATA

Negative-ion beam intensity data

The mass spectra from an Au sample are displayed in Fig. 11. As noted, the spectra are very clean from this species

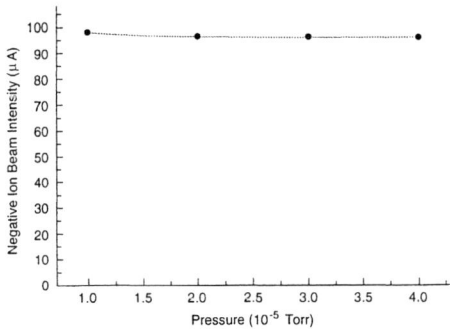

Fig. 9. Negative-ion beam intensity versus Xe external pressure for the RF plasma-sputter negative ion source. Sputter sample: Au; Cs°-oven temperature: 260°C.

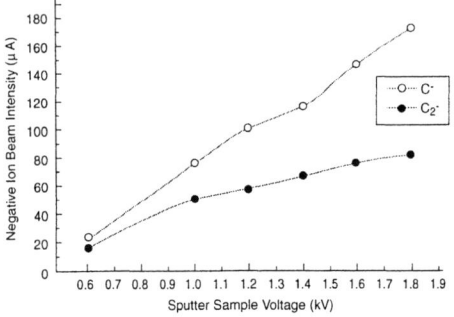

Fig. 10. Relative negative ion beam intensity versus sputter-sample voltage for the RF plasma-sputter negative ion source. Sputter sample: C; support gas: Xe; pressure: 1.4×10^{-5} Torr; Cs°-oven temperature: 200 °C.

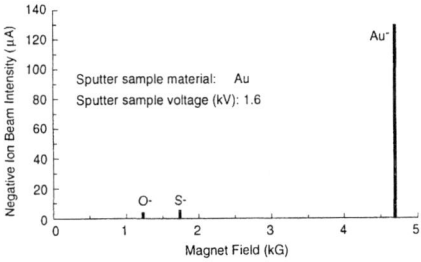

Fig. 11. Mass spectra taken from an Au sputter-sample. Sputter-sample voltage: 1.6 kV.

with only low intensities of O⁻ and S⁻ present. To date, the source has been utilized to generate *dc* negative-ion beams for a variety of species from a 19-mm diameter spherical-sector cathode, including those shown in Table I.

Table I: Negative-ion beam intensity data for the RF plasma sputter negative-ion source.

Species Sputter	Probe Material	Beam Intensity (μA)
C⁻	C	610
F⁻	LiF	100
Si⁻	Si	500
S⁻	ZnS	500
P⁻	GaP	125
Cl⁻	NaCl	200
Ni⁻	Ni	150
Cu⁻	Cu	230
Ge⁻	Ge	125
Se⁻	CdSe	40
As⁻	GaAs	100
Se⁻	CdSe	200
Ag⁻	Ag	70
Au⁻	Au	250
Pt⁻	Pt	125

Emittance data

The emittance of the source was measured using the equipment and procedures described in Refs. 14 and 15. Normalized emittance versus percentage of total negative-ion beam for 200 μA Cu⁻ beam is shown in Fig. 12. As noted, the normalized emittance ε_n for 80% of the total beam is $\varepsilon_n \sim 7.5$ π mm.mrad $(MeV)^{1/2}$. In general, the emittances of the source are slightly lower than those measured for Cs-sputter sources even though the intensities from the present source are in general higher for a given species.

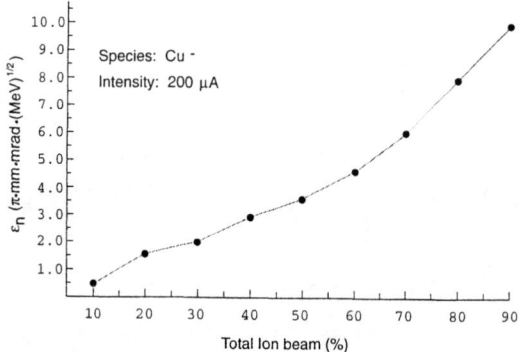

Fig. 12. Normalized emittance ε_n versus percentage of total negative ion beam for the RF plasma-sputter negative ion source.

ACKNOWLEDGMENTS

Research at the Oak Ridge National Laboratory is supported by the U.S. Department of Energy under contract DE-AC05-96OR22464 with Lockheed Martin Energy Research Corp.

REFERENCES

1. G. D. Alton, *Nucl. Instr. and Meth. B* **73** (1993) 221.
2. P. Thieberger, M. McKeown and H. E. Wegner, *IEEE Trans. Nucl. Sci. NS-30 (4)* (1983) 2746.
3. G. D. Alton and C. M. Jones, *Nucl. Instr. and Meth. A* **244** (1986) 170.
4. H. H. Andersen, IEEE Trans. *Nucl. Sci. NS-22* (1975) 1632; P. Tykesson, H. H. Andersen and H. J. Heinemeier, *IEEE Trans. Nucl. Sci. NS-23* (1976) 1104.
5. G. D. Alton and G. K. Blazey, *Nucl. Instr. and Meth.* **166** (1979) 105.
6. G. D. Alton, Y. Mori, A. Takagi, A. Ueno and S. Fukumoto, *Nucl. Instr. and Meth. A* **270** (1988) 194; Y. Mori, G. D. Alton, A. Takagi, A. Ueno and S. Fukumoto, *Nucl. Instr. and Meth. A* **273** (1988) 5.
7. Y. Mori, *Nucl. Instr. and Meth. A* **328** (1993) 146.
8. G. D. Alton, *Rev. Sci. Instr.* **63** (1992) 2455.
9. J. Ishikawa, H. Tsuji, Y. Okada, Y. Toyota and Y. Gotoh, *Vacuum* **44** (1993) 203.
10. "High Efficiency RF Plasma Generation Systems for Ion Source Applications," R. Lohwasser, Diplom Thesis, Dept. of Phys. Ludwig Maxmilians Univeristät, München, Germany, June 1997 (unpublished).
11. J. K. Nørskov and B. I. Lundqvist, *Phys. Rev. B* **19** (1979) 5661.
12. G. D. Alton, *Surface Sci.* **175** (1986) 226.
13. J. B. Hasted, *Physics of Atomic Collisions,* 2nd Edition (Elsevier, New York, 1972) Ch. 8.
14. G. D. Alton and J. W. McConnell, *Nucl. Instr. and Meth. A* **268** (1989) 445.
15. G. D. Alton, and R. W. Sayer, *J. Phys. D: Appl. Phys.* **22** (1989) 557.

A Multi-Sample Cs-Sputter Negative-Ion Source

G. D. Alton, B. Cui,[*] Y. Bao,[*] C. A. Reed, J. A. Ball, C. Williams

Oak Ridge National Laboratory, P. O. Box 2008, Oak Ridge, TN 37831-6368 USA

A multi-sample Cs sputter negative-ion source, equipped with a conical-geometry, W-surface-ionizer has been designed and fabricated that permits sample changes without disruption of on-line accelerator operation. Sample changing is effected by actuating an electro-pneumatic control system located at ground potential that drives an air-motor-driven sample-indexing-system mounted at high voltage; this arrangement avoids complications associated with indexing mechanisms that rely on electronic power-supplies located at high potential. In-beam targets are identified by LED indicator lights derived from a fiber-optic, Gray-code target-position sensor. Aspects of the overall source design and details of the indexing mechanism along with operational parameters, ion optics, intensities, and typical emittances for a variety of negative-ion species will be presented in this report.

INTRODUCTION

Negative-ion sources based on the sputter principle have been used for many years for injection into tandem electrostatic accelerators for fundamental nuclear and astrophysics research applications, and for applied research such as Accelerator Mass Spectrometry (AMS), high-energy ion implantation and modification of materials. In addition, these sources are being used in a variety of low-energy, atomic and molecular physics research applications. As a consequence of these and other demands for sources with improved intensities and beam qualities, this technology has reached a relatively high degree of maturity. Several sources, designed to accommodate single samples, have been developed over the years, including those described in Refs. 1 and 2. For applications that require short-duration beam-on-target, including AMS, it is desirable to be able to quickly and remotely change samples without vacuum disruption to save time and to preserve similar surface conditions between samples. A few sources have been reported that have been designed with this capability, including sources described in Refs. 3-5. In this report, we describe a multi-sample Cs sputter negative-ion source, equipped with a conical-geometry, surface ionizer that permits sample changes without disruption of on-line accelerator operation. Aspects of the overall source design and details of the indexing mechanism along with operational parameters, ion optics, intensities and emittances for a wide variety of negative ion species are provided in this report.

---

[*] Visiting scientist from the China Institute of Atomic Energy, Beijing, China.

CP473, *Heavy Ion Accelerator Technology: Eighth International Conference,*
edited by Kenneth W. Shepard
© 1999 The American Institute of Physics 1-56396-806-1/99/$15.00

DESIGN FEATURES

As noted in Figs. 1 and 2, the source consists of five main assemblies: the high-voltage acceleration electrode assembly; the ion source module; the vacuum-air-lock assembly; the H_2O-cooled, rotary-motion target-holder assembly; and the indexing mechanism/target position read-out assembly. The ion source is designed to fit into the same ionization-chamber and vacuum chamber assemblies that accommodate single-sample sources routinely utilized for the generation of stable ion beams for the nuclear and astrophysics research programs at the Holifield Radioactive Ion Beam Facility (HRIBF) of the Oak Ridge National Laboratory (ORNL). The vacuum chamber of the source is re-entrantly attached to the high-voltage side of a metal-to-Al_2O_3 bonded high-

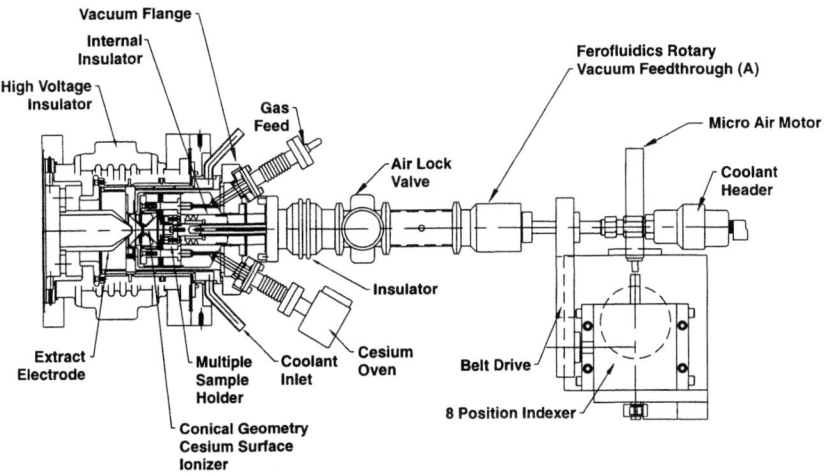

Fig. 1. Multi-sample negative-ion source (top view).

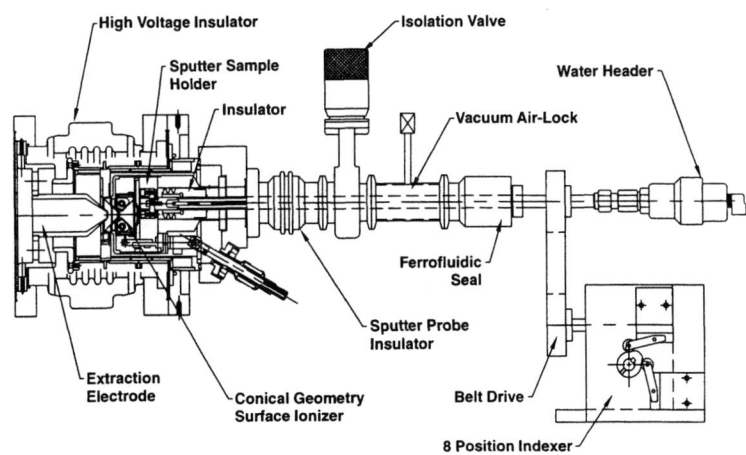

Fig. 2. Multi-sample negative-ion source (side view).

voltage insulator assembly while the acceleration electrode system is attached to the ground-side of the assembly. The source module can be quickly removed/installed from the vacuum housing by simply loosening/tightening thumbscrews. Power for the conical geometry Cs surface ionizer is fed through the stainless-steel ionization chamber support flange by means of a specially designed metal-to-ceramic bonded insulator, sealed with metal-to-metal gaskets. Cs vapor, from an external oven, is introduced into the ionization region of the source through the main vacuum flange by means of a metal-to-metal gasketed, transport tube, designed so that the transport tube is spring loaded against the body of the ionization chamber to eliminate vapor losses at the interface while allowing the oven assembly to be removed from the source module for servicing. Provisions are also made to feed gases through an analogous feed-through/transport system for in-situ formation of specific electro-negative compounds that serve as carriers for atomic species with low-negative electron affinities.

The sample indexing mechanism

The samples are screw attached to a Cu or Al holder that, in turn, is screw attached to a H_2O-cooled, Cu heat-sink assembly. The sample coolant is fed down and back through a coaxial conduit that allows H_2O to flow directly onto the Cu heat-sink assembly to which is attached the sample holder; the assembly is mounted at sample-plus-source potential. The sample holder is offset relative to the optic axis of the source (as illustrated in Fig. 3) so that a simple rotary motion can be executed to move individual samples into beam position. Because of physical restrictions in radius, in combination with the type of indexing mechanism chosen for moving samples into beam position, the sample holder is limited to eight, 7.9 mm-diameter samples. The belt-driven, coolant shaft is vacuum-sealed with a low friction, ferro-fluidic rotary-motion feed-through. Sample indexing is effected by applying 5 – 7 Pa air-pressure through plastic tubing to an electro-pneumatic

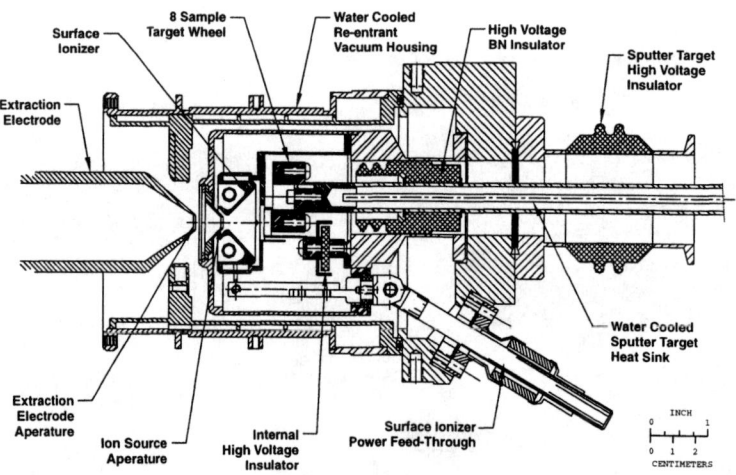

Fig. 3. Cross-sectional top-view of the multiple-sample, Cs-sputter negative-ion source showing details of the W-ionizer/sputter-sample and ion extraction electrode systems.

switch system, located at ground potential, that actuates an eight-position, Geneva-type mechanism mounted at source potential. In-beam-position samples are identified by means of an LED display derived from a fiber-optic link in combination with a Gray binary encoder wheel mounted on the indexing mechanism. This system eliminates the necessity of having to mount power supplies at source potential or source-plus-sample potential that would otherwise be required to identify and sense sample position.

Ion optics

Since the rate of sputtering depends on the physical properties of the particular projectile/target combination as well as the energy and beam intensity of the Cs^+ projectile, the negative beam intensity depends on the Cs^+ ion beam intensity and the efficiency of transport through the electrode system. Therefore, the ion optics of both the Cs^+ and negative-ion beams are very important. The space-charge limited flow for Cs^+ ion current, I_{Cs}^+, extracted through a potential difference V, is given by the familiar Childs-Langmuir equation

$$I_{Cs}^+ (A) = PV^{3/2} \qquad (1)$$

where P is the perveance for the conical-geometry/sputter-sample electrode system with value $P = 4.9 \times 10^{-9}\ [A/V^{3/2}]$. The computer code described in Ref. 6 was utilized to assist in the design of the ionizer/sputter sample electrode system. The computed optics for space charge limited extraction of Cs^+ at 2 kV are shown in Fig. 4 while those for the corresponding negative-ion beam, generated in the sputter process, are shown in Fig. 5. Simulation of the negative-ion beam optics are made under the assumption that the negative ions leave the surface in a typical sputter-particle energy-angular distribution and mimic the Cs^+ ion beam distribution on the surface in terms of initial intensity profile. As noted, negative-ion beams are efficiently transported back through the electrode system. One of the excellent features of sources based on this principle is that negative-ion beam intensities can be easily controlled between zero and saturation by simply changing the potential difference between the sputter sample and ionizer.

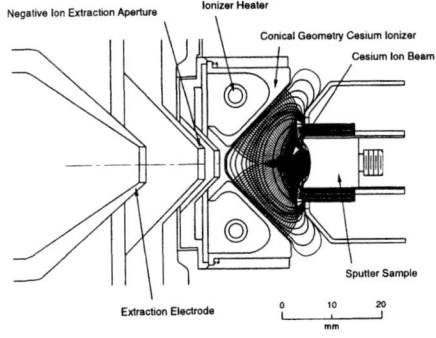

Fig. 4. Cs^+ beam optics as simulated by use of the computer code described in Ref. 6.

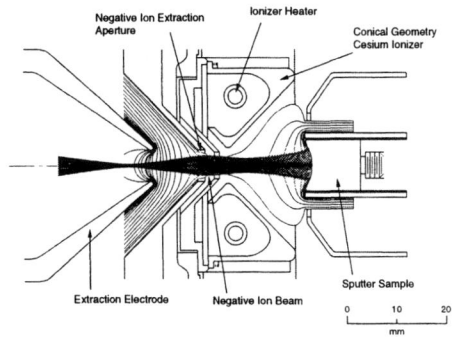

Fig. 5. Negative-ion beam optics as simulated by use of the computer code described in Ref. 6.

OPERATIONAL PARAMETERS

In order to realize optimum performance of the source, it is quintessential to know the dependence of the negative-ion beam intensity on important operational parameters of the source, including Cs oven temperature; ionizer temperature; and sputter-probe voltage.

Negative-ion beam intensity versus cesium oven temperature

Since the probability for negative ion formation depends exponentially on the difference between the electron affinity E_A and the work function ϕ of the surface that the particle leaves, low work function surfaces are prerequisite for reasonable negative ion formation efficiencies [7]. The flow of Cs into the source offers a general and dynamic means for lowering the work functions of surfaces and a means for overcoming deleterious poisoning effects that may occur due to the presence of certain impurities in the ionization chamber of the source. Since the effective Cs coverage depends on the difference between the rate of arrival and departure of Cs vapor to and from the sputter probe surface, the value of the work function of the surface and consequently the negative-ion beam intensity is sensitively dependent on Cs-oven temperature [8]. Knowledge of the negative-ion yield or relative negative-ion yield on this parameter is quintessential for optimizing negative-ion yields from the source. Relative negative-ion yield versus Cs-oven temperature, obtained at fixed ionizer power, and sputter-sample voltage are displayed in Fig. 6. Once known, the Cs-oven temperature, as a general rule, can be set at a fixed value up to T_{max} where T_{max} is $\sim$ 200 °C. It also important to note that the Cs-oven temperature can not be used as a real-time control parameter because of the characteristically slow response times of thermal equilibrium processes. Values higher than T_{max} should be avoided since higher temperatures can be deleterious to the intensity and lead to operational instabilities due to sparking.

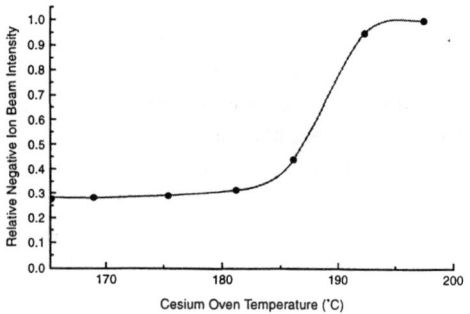

Fig. 6. Relative negative-ion beam intensity versus Cs-oven temperature for the multi-sample, Cs-sputter negative-ion source equipped with a conical geometry ionizer. Species P$^-$; Sputter-sample voltage: 6 kV.

Negative-ion beam intensity versus ionizer current

The surface ionization process is characterized by an abrupt initiation of ionization that occurs at a specific temperature, often referred to as the critical temperature [9]. This abrupt onset is associated with the fact that Cs$^+$ formed on the surface is bonded to the surface by the negative image charge, induced into the surface. The critical temperature for Cs occurs at ~ 1100 °C. This value correlates to an ionizer current of ~ 26 A, as noted in Fig. 7. Once the relationship between negative-ion yield and ionizer temperature or ionizer current/power is known, then the ionizer can be operated at fixed value. Typically the present source operates at a fixed ionizer current of ~26-27 A.

Negative-ion beam intensity versus sputter-probe voltage

Since the sputter ratio varies from material-to-material for a given projectile at fixed energy, the probe voltage required to reach steady-state negative-ion beam intensities also varies from species to species. An example of this dependence is displayed in Fig. 8 for a GaP sputter sample. Higher sputter-ratio materials such as Au reach saturation at lower sputter-probe voltages than do lower sputter-ratio materials such as C, and therefore, every compound or elemental material has a different intensity vs. sputter-sample voltage dependence.

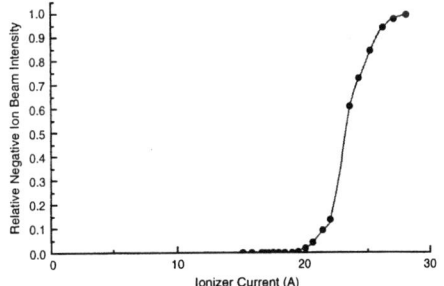

Fig. 7. Relative negative-ion beam intensity versus ionizer current for the multi-sample, Cs-sputter negative-ion source equipped with a conical-geometry ionizer. Species: C⁻ Sputter-sample voltage: −2 kV; Cs-over temperature: 195 °C.

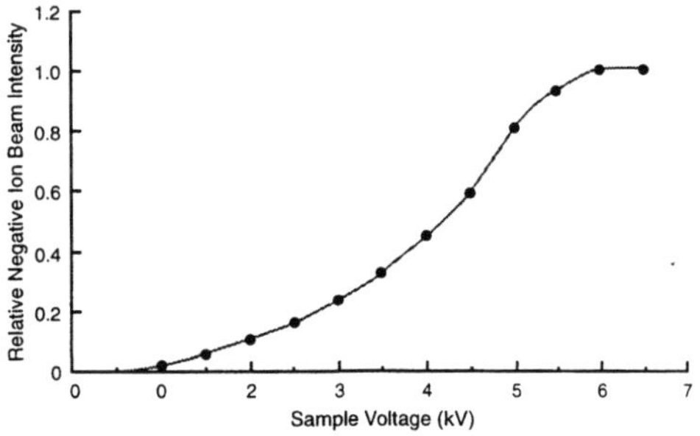

Fig. 8. Negative-ion beam intensity versus sputter-sample voltage for the multi-sample Cs-sputter negative-ion source equipped with a conical geometry ionizer. Species: P⁻; Cs-oven temperature: 195 °C.

NEGATIVE-ION BEAM INTENSITY AND EMITTANCE DATA

Negative-ion beam intensity data

Single-sample sources, equipped with this ionizer geometry, have been utilized for several years at the HRIBF for stable beam generation. Intensity data for several negative ion species derived from operation of these sources are shown in Table I.

Table 1: Negative-ion beam intensity data

Species	Sputter-Probe Material	Beam Intensity (µA)
C^-	C	160-275
Li^-	Li	3-12
B^-	B(90%) + Ag(10%)	30-60
O^-	Metal Oxides	250-310
F^-	LiF, CaF	50-110
Si^-	Silicon	300-440
P^-	GaP, InP	90-125
S^-	ZnS	60-110
Cl^-	NaCl	100-150
Ni^-	Ni	80-120
Cu^-	Cu	100-150
As^-	GaAs	50-60
Se^-	CdSe	30-100
Ag^-	Ag	20-50
Pt^-	Pt	40-80
Au^-	Au	50-90

Emittance Data

Emittances have measured for the single-sample equivalent of the present source for a number of species using the equipment and procedures described in Refs. 10 and 11. An example of the normalized emittance versus percentage of total negative-ion beam for a Ni-beam is shown in Fig. 9. In general, the emittances for the source are very close to those measured for Cs sputter sources equipped with other ionizer geometries. The emittances vary with species (see, e.g., Ref. 12) and typically have values ranging between 6 and 8.5 πmm.mrad.$(MeV)^{1/2}$ depending on the species.

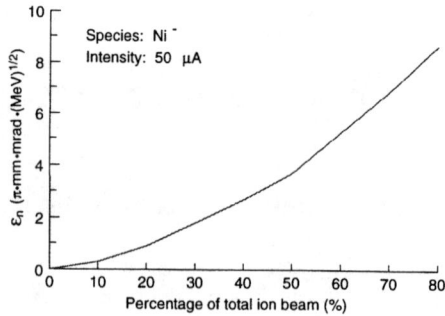

Fig. 9. Normalized emittance ε_n versus percentage of total negative-ion beam. (Taken from a single-sample source with identical ionizer and extraction optics.).

ACKNOWLEDGMENTS

Research at the Oak Ridge National Laboratory is supported by the U.S. Department of Energy under contract DE-AC05-96OR22464 with Lockheed Martin Energy Research Corp.

REFERENCES

1. G. D. Alton, *Nucl. Instr. and Meth.* B **73** (1993) 221.
2. G. D. Alton, *Rev. Sci. Instr.* **65** (1994) 1141.
3. R. Balzer, G. Bonani, M. Nessi, Ch. Stoller, M. Suter and W. Wölfli, *Nucl. Instr. and Meth.* B **5** (1986) 226.
4. I. D. Proctor, *Nucl. Instr. and Meth.* B **40/41** (1989) 727.
5. G. D. Alton, *Rev. Sci. Instr.* **63** (1992) 2450.
6. PBGuns, Thunderbird Simulations, Garland, TX.
7. J. K. Nørskov and B. I. Lundqvist, *Phys. Rev. B*, **19** (1979) 5661.
8. G. D. Alton, *Surf. Sci.* **175** (1986) 226.
9. G. D. Alton, *Rev. Sci. Instr.* **59** (1988) 1039.
10. G. D. Alton, J. W. McConnell, *Nucl. Instr. and Meth.* B **24/25** (1987) 826.
11. G. D. Alton and J. W. McConnell, *Nucl. Instr. and Meth.* A **268** (1988) 445.
12. G. D. Alton and R. W. Sayer, *J. Phys. D: Appl. Phys.* **22** (1989) 557.

Design and Development of the ECR Ion Source Control System at LNL

S. Canella*, M. Cavenago*, G. Delfitto†, G. Abrioni*

* I.N.F.N.- Laboratori Nazionali di Legnaro; Via Romea, 4 ; 35020 Legnaro (PD) Italy
† Universita di Padova, Dip. di Fisica and I.N.F.M, via Marzolo, 8, I-35131, Padova

Abstract. The design and development of the supervisory and control system for an Electron Cyclotron Resonance (ECR) Ion Source at Laboratori Nazionali di Legnaro (LNL) are described. Both the hardware and the software architectures of the system are presented, together with its links with the control systems of the superconducting linear accelerator ALPI and of the new injector PIAVE. A number of different control devices and instruments for measurements are used in the ECR, and this feature forced the integration in the same system of both traditional home-made available hardware/software solutions and of commercial products for standard industrial control systems. The new control system for the ECR Ion Source is planned to be put into operation in some months.

1. INTRODUCTION

The Electron Cyclotron Resonance ion source (ECR) Alice is installed and operated on a 350 kV platform since 1996 with manual control and no platform voltage [1]; operation with platform voltage and extraction of beam is expected for late 1998. Scope of this paper is the 'computer control' both of the ECR and of the platform, intended to control and monitor most major functions (see section 2 for a list), both from a local operator on the platform and from a remote operator in the accelerator console. The distinctive features and requirement of the control system of any ECR ion source installed on a high voltage platform [2] are evident considering that: a) no copper wire can pass from the operator console to the ECR; b) ion source requires flexible operation modes, since performance of an ion source depends on detailed adjustment and ECR is intended to give any generic ion; c) an ECR ion source consists of a number of different subsystems, namely high voltage power supply (ps), high current ps, turbopump controllers, vacuum gauges readers, stepping motor controllers, and so on; this calls for a plenty of different software drivers. Note also that: d) the high voltage platform environment can be equated to an industrial environment for electrical noise, both inherent and accidental, motivating short cable connections.

Consideration b) encourages local manual control of the ion source during tests and characterization of any specific ion, while a) excludes the use of any remote control panels easily supplied by ps manufacturers, so that only a computer network (with fiber optics links) is an adequate solution for remote operation. Both from the hardware and from the software point of view this new control system emphasizes integration with the control system for the Tandem-ALPI accelerators produced at LNL [5], but some special arrangements had to be planned for two reasons: since most of ECR hardware components are installed on the high voltage (HV) platform, user access must be provided both locally (on the HV platform) and remotely (in a control room at ground voltage); 2) the ECR Ion Source has the special needs of the monitoring the extracted beam current, which requires strict coordination of some hardware and made unusable the standard tools of the Tandem-ALPI accelerators.

As general backbone for communication we choose to rely only on one traditional Ethernet, ported on a fiber optic link. Two slower fiber optic links were added for hardware safety signal only. The synchronization of a future 5 MHz buncher to the ALPI 160 MHz fundamental frequency was left out of this scheme.

2. HARDWARE ARCHITECTURE

ECR computer network divides into 4 stations or areas, separated by large distance (greater than 25 m, nominal span of an RS232 cable link [3]) or large potential difference (see figure 1): a) the ALPI-TANDEM console room; b) the platform gate area (at potential $V = 0$), which is just outside the earthed fence that surrounds the platform; b) the isolated platform area, a 4.5 m side square room sustained by 1.7 m high insulating columns, at a potential V_p that may range up to 400 kV; d) the inner ECR equipment area, which is a small floor directly connected to the ECR plasma potential (so that its potential is $V = V_p + V_s$, where typically $V_s = +11$ kV), and allows significant measurement of ECR plasma peripheral pressure, and safe coupling of electrodes into plasma (see elsewhere for experiments with an arc source inside the ECR ion source [4]).

In the ALPI console room, among several equivalent workstations (WS), one can run the ECR control program; here also a dedicated PC will mirror the screen of the LeCroy 9410, sitting on a platform and generally operated as the ion spectrometer acquisition of the ECR ; the redundancy of screen is meant to make operation easier.

At the platform gate, we have the fiber optics/Ethernet interface, a HP34970A acquisition unit for test purposes, (to be substituted with a VME crate, call it 2, for better uniformity with ALPI system), and the 400 kV ps controller; this also is the major node for connection with the safety systems of the laboratory.

Onto the isolated ECR platform, we have

- 1 VME system with a 147-SA CPU card based on a 68030 microprocessor (μp) and with A/D and digital I/O cards (VxWork system)

- two LAT terminal servers with 16 serial ports for local controllers and instrumentation equipped of RS232 interface
- one Digital Unix DEC UDB (Universal Desktop Box), based on 166 MHz alpha μp, for local control.
- one GPIB-Enet National Instruments interface box.

These objects are connected by an ethernet chunk, linked by only one fiber optic pair (L1) to ALPI ethernet (10 Mbit/s). We also have two duplex low speed fiber optics for hardware signals (L2, L3); one line should reflect klystron status; the other should be the request to off the platform voltage.

RS232 controllers and instruments that have been already connected to the ECR control system are:
a) 3 power supplies for 2 solenoids and 1 dipole (Danfysik), respectively named PS1, PS2 and PSD;
b) 1 steerer controller (LNL project), with four bipolar 3 A outputs;
c) 2 Total Pressure Gauge (TPG)controllers, for vacuum monitoring (Balzers);
d)1 Klystron Power Amplifier (KPA) controller (Varian); its output power be P_k. KPA uses two RS232 lines for redundancy;
e) 1 step motor controller for up to 16 motors (LNL project). Motors will act on gas valves and microwave couplers;
f)4 Turbo Pump Controllers (TPC, Balzers);
g) 1 digital teslameter (Danfysik);
h) 4 HV power supplies (Heinzinger); the recent one conforms to SCPI standards.

Current of PS1 and PS2 producing the magnetic field acting on the ECR plasma and P_k are among the major facts determining which ions are emitted from the source, while PSD determines which ion is deflected in the correct beam path to the accelerator (or to the Faraday cup for measure, if the latter is inserted).

2.1 Consideration of interlocks and hardware

Cooperation between hardware implements and computer actions is also important to safe and easy operation of the ion source. First, to facilitate computer tasks, ECR design tries to implement all basic safety interlock with hardware actuated, usually local, systems; as a goal, the computer only reads the status. In a simply solved case, the thermostats measuring cooling water outputs from the 12 pancake coils are series connected and when at least one opens on alarm, this causes the coil ps to stop. A circuitry should detect which individual coil(s) is(are) overheated and make these data available to an I/O board. Flowmeter monitoring is similar. To the other extreme, assume that the platform exit door is opened; then information of that event is easily transmitted to the DEC UDB, that if operative, will transmit (via L1) to the HP34970 (or to VME crate 2) the request to zero the DAC (say DAC1) controlling the 400 kV platform voltage ps. Most compelling than that, line L2 will stop to pulse, causing 400 kV alarm contact to open (and 400 kV ps to stop). After that, a chain released from the door will hit ground. Note that these three actions are necessary: indeed, the chain alone would discharge the platform too rapidly, stressing solid dielectric (columns and power cables) badly.

3. SOFTWARE ARCHITECTURE

The software package for the ECR control systems includes 3 levels of programming:

- a graphic man-machine interface (MMI) based on X11 standard, which may run either locally, on a Digital Unix UDB workstation housed on the HV platform or in remote mode, on a Ultrix RISC workstation (CONSOLE8)

- a TCP/IP and UDP/IP based network communication manager that has in charge of routing commands and data among the MMI, the controllers and the instruments;

- a set of servers on the Ultrix or Digital Unix workstation to manage the communication with the devices connected to terminal servers trough RS232 serial lines and with some VxWorks tasks on a VME based system for analog and digital I/O. Like in the control system for the Tandem-ALPI accelerators the whole ECR software make use of client-server architecture, mainly using TCP/IP channels for network communication and UDP/IP connection-less data stream where necessary.

Where possible ALPI code was re-used, eventually with adjustments to the new needs. Code re-cycling was performed especially in the graphic package, re-using ALPI control windows for the power supply of magnets and steerers. A partial re-use was performed also in the sections of the network manager and servers related to the Danfysik power supplies and steerer controllers.

On workstations, the graphical user interface (GUI) is organized in only two major layers: 1) the ECR global window, a summary panel, with boxes for every major equipment and some help provision (access to text description and specification of GUI itself); each box has with 2 or 3 regulation and 1 alarm with pushbuttons to rise the detail window; 2) the detail windows; usually a window contains one panel associated to an instrument of the platform, and reflects the front panel of that physical instrument as much as possible; a window also includes related additional information, for example the klystron window includes forward and backward microwave reading from diodes near the ECR (reading much more detailed and precise of the klystron meters themselves). So, a detail window controls and surveys a sub-system of the ECR, and the pop-up of dependent windows is needed very seldom, namely to obtain help, or to enter a password authorizing delicate settings.

On the PC, the scope is simulated as a window in the context of Labview Virtual Instruments (controlling two channels for coupling and amplification, the timebase and trigger) and another window deals with the Keithley current amplifier mod.

428. The desired ion spectra acquisition is obtained by not synchronized cooperation of PC, WS and hardware: the faraday cup signal is feed to the 428 and, amplified, to CH1 channel scope; teslameter analog output is feed to CH2, which will act as trigger source; the PC sets the scope and the 428. Then, at one user

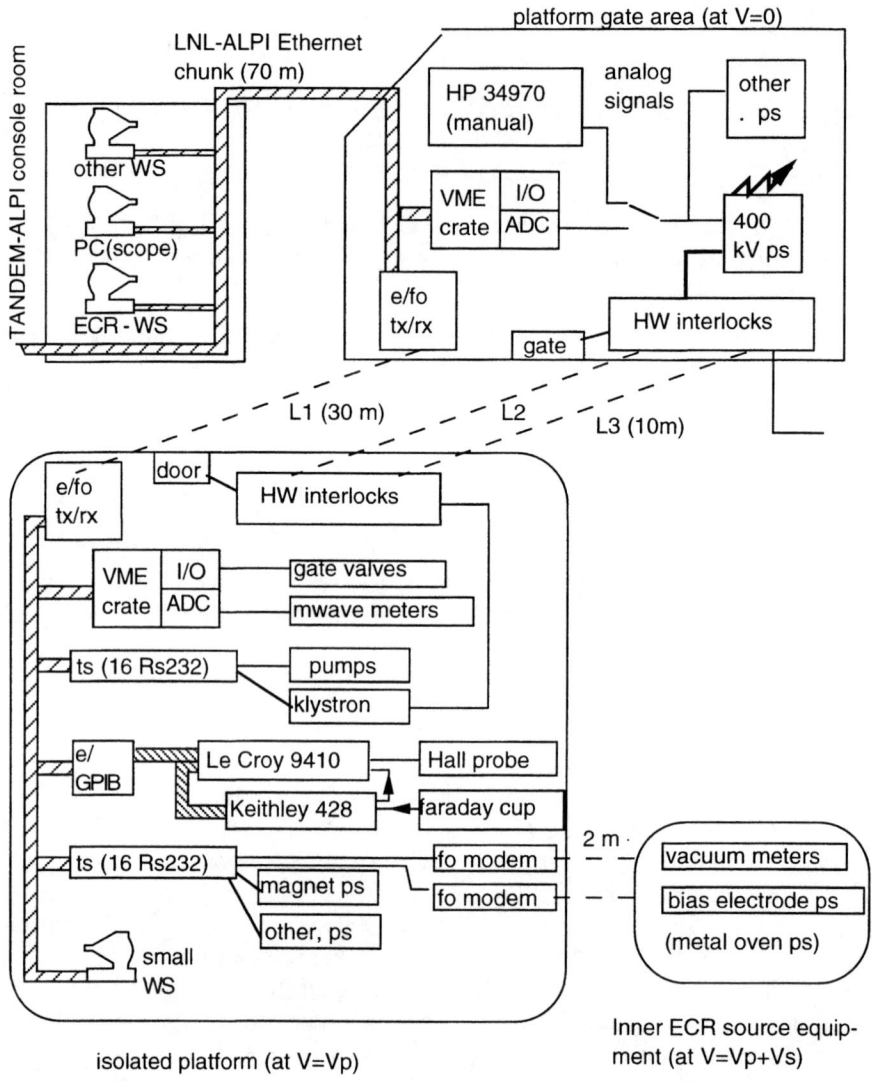

FIGURE 1. Scheme of the control system for ECR and platform (shorthand: ps power supply, WS workstation, HW hardware, e/fo ethernet fiber optics transceiver ; ts terminal server; dashed lines are bidirectional fiber optics links)

keystroke at WS, the WS via the VME I/O board insert the faraday cup in the beam path and via a RS232 port on the VME crate asks the dipole ps to increase the field step by step (each 100 ms) making a predefined ramp (usually 50 s long). Rising magnetic field triggers the scope, which dumps the data on PC after a screen acquisition, for display, storage, and possibly ion species identification (off-line). The workstation can also start spectra sequence each hour, for documenting the ECR status.

4. RS232 EQUIPMENT NOTES

Equipment to be controlled is 2 to 7 years old. Due to the rapid development of digital electronics, equipment interfaces vary largely. We chose to prefer RS232 channels on the basis of their reputed noise immunity; only recently we made a limited use of IEEE488 for speed (scope case) or necessity (amplifier mod. 428). As a comparative advantage of RS232 (or IEEE488) against controlling the 0-10 V input which many instruments have with a DAC, note that we need not to worry that small interferences on driving signal becomes large signals on the high voltage output. To protect the extremely important dipole power supply RS232 interface, we use a fiber optic with two self-powered modems to substitute a cable connection.

It should be noted that RS232 interface reliability and quantity and quality of commands varies largely from one instrument to another, at the manufacturer discretion, ranging from insufficient (and in that case the associated window has inactive fields left for upgrades or supplements information with data from ADC and I/O) to fully redundant (and in that case a choice was done of the more relevant command and simple macros). Most programming work was spent to cope with some malfunctioning.

5. CONCLUSIONS

The ECR control system is expected to be finished by the end of the 1998, at least for the essential functions to be performed on the remote console, this deadline being in time with foreseen commissioning time of the new injector beam line.

REFERENCES

1. Cavenago, M. et al. , these proceeding
2. Pardo R.C., *Nucl. Instr. Meth.*, **B40/41**, pp. 1014-1019, (1989).
3. McNamara J. E. , *Technical aspects of data communication*, Digital Equipment Corp. Press, ISBN 1-55558-007-6 , Bedford MA, 1988.
4. Cavenago, M. , Kulevoy, T. and Vassiliev, A. *Rev. Sci. Instr*, **69**, 795-797 (1998).
5. Bassato, G., Battistella, A. , Bellato, M. and Canella, S. , "Status of the ALPI Control System", Real Time '91 Conference Record pp. 285-291.

THE LEGNARO ECR ION SOURCE PLATFORM

M.Cavenago*, T.Kulevoy*,[†], G.Abrioni*,
S.Canella* and F.Cervellera*

* Laboratori Nazionali di Legnaro-INFN, via Romea n 4, I-35020 Legnaro (PD), Italy
[†] ITEP, 25 Bolshaja Cheremushkinskaja, 111000 Moscow, Russia

Abstract. The Legnaro ECR ion source is installed on high voltage platform, capable of operating up to 350 kV and to be powered by an externally located insulation transformer. The test of the prototype of the high voltage cable termination is described, together with design of the full size cable termination. Tests of other high voltage parts are described. In the extraction beamline from the platform a triplet was added after the acceleration column, so that injection in the following U-bend is simplified. Using empirical fitting formula for prediction of einzel lens cardinal data and perturbative expansion for triplets, understanding of beam lines was considerably improved and optimization for arbitrary ions were possible. A comparison of the performance of the several plasma chambers that were used in this ECR ion source is given, discussing also the effect of a low voltage sputter probe. A comprehensive software program, centralizing remote control of most platform and source regulation, was successfully tested.

1. INTRODUCTION

A small Electron Cyclotron Resonance ion source (for literature on ECR see Ref. [1]) named Alice was designed in our laboratory [2], to extend the mass range of ions accelerated by ALPI, a superconductive linac, well beyond the possibility of the existing 17 MV tandem, so to include heavy elements (lead, bismuth and maybe uranium) and noble gases. To match the fixed input velocity ($v_z = 0.0089\,c$) of the ALPI injector, PIAVE, which is being built, ECR ions need a preacceleration. For beam quality and following some traditional respect for electrostatic acceleration, it was chosen to provide this acceleration by a high voltage platform: let V_p be the platform voltage with respect to ground and V_s be the source voltage with respect to the platform. Installation parameters, especially for v_z value, are similar to ATLAS platform [4], notwithstanding the different injectors: for ATLAS an interdigital QWR linac with superconducting solenoid focusing, instead PIAVE is an RFQ with some QWR booster (both use superconductive cavity technology).

While $V_s \cong 11$ kV is rather constant (depending from current emission of the source and practical consideration), V_p varies largely with the ion to be injected into the RFQ:

$$V_p = -V_s + (A/i)(m_p v_z^2/(2e)) \tag{1}$$

with A the atomic number and i the charge of the ion. Let us define the parameter $R = [1 + (V_p/V_s)]^{1/2}$ that is $R = (A/i)^{1/2}(m_p v_z^2/(2eV_s))^{1/2}$; R plays a major role in beam optics and bunching as seen (easily) in section 4. Note that R is also the ratio of beam speeds, after and before platform acceleration. Since A/i ranges from 2 for helium to 8.5 (U^{28+}) in our application, R ranges from 2.6 to 5.4 .

The ECR ion source Alice was designed to be consistent with practical space available, the choice of a platform and the V_p required (313 kV for uranium) and the power limitation (sew section 3).

Completion of platform received a significant boost in December 1996, when we succeeded to find a suitable insulation transformer, delivered in September 1997. Platform envelope and fence was then put in place; insulation tested (see section 2); power line construction begun (see section 3).

2. PLATFORM INSULATION

Platform insulation was finally tested and does work smoothly up to 400 kV, as measured by 400 kV power supply (ps) setting and readout, see figure 1. Up to 40 kV, 400 kV ps readout were checked against a HP34300A probe. Humidity and temperature were about 64 ± 5 % and $29 \pm 1\,^\circ\mathrm{C}$. Note that current flowing through water is measured (by connecting a current meter to a BNC, permanently shunted by 400 V varistor and with proper probes into water). This is a good measure of water resistivity, found to be 17 Mohm cm (value depends from cleaning system condition and of course deteriorates when water is not flowing). The advantages of water cooling are evident (refrigerator are separated from accelerator building).

Water connection to platform, described elsewhere [2], took some risk in design (creepage distance on tube stands is only 2 m), but evidently works. A reason may be that water tubes weakly help to grade the equipotential along the tube stand. Also, before tests, PVC tubes , stands and the ceramic column were carefully cleaned by common pink alcohol.

400 kV ps is made of two open stacks, one generating voltage, one filtering it within 25 Vpp. To optimize space, 400 kV ps was put under the platform, mounted on a special cart (with very simple wheels), to make movements for possible maintenance simpler. A temporary 6.8 Megaohm resistor connects the filter stack with the platform during tests.

At the platform body envelope the field must be less than air breakdown value E_a (21 kV/cm if dry, 17 kV/cm in normal condition). ECR design features a $R_0 = 40$ cm radius on edges and vertices (each vertex is a true octant of a sphere). This value (obtained by the naive guess $R_0 > V_p/E_a$) is extremely conservative as shown by the rough estimate of maximum field E_3 on a 90 degree bent sheet facing a right corner (see Fig. 3, lower left corner)

$$E_3 = \frac{\sqrt{8}}{3} \frac{V_p}{(b^2 R_0)^{1/3}} \tag{2}$$

where R_0 is the curvature radius of the sheet and b is the distance between the flat parts of the sheet and the corner sides (this estimate follows from approximating the equipotentials with the contours of $\Re[(\sigma - \Sigma_0)^{2/3}]$ with the point $\Sigma_0 = (1 + i)(b + R/3))$ and $\sigma = x' + iy'$, where the axes x', y' are the corner sides).

The platform envelope panels are tied by many M3 round head self-tapping screws, whose heads (diameter 5.5 mm) have a 2.6 mm wide cross socket. The screw protuberance may be inscribed approximately inside an oblate semiellipsoid, major radius $r = 3.5$ mm, height and minor radius $h \cong 2.2$ mm; let us explain why this is not a problem. Field enhances locally by a factor $f_1 = 1/(1 + x_0^2)/(1 - x_0 \arctan(1/x_0))$ with $x_0 = h/(r^2 - h^2)^{1/2}$. This factor (about 2.1 for our screws) multiplies the enhanced field arising from platform edges. Yet we have a local field $E_4 = f_1 E_3 = 7.2$ kV/cm with $b = 1.8$ m, safely withstood by the air . To have an experimental validation, we applied $V = V_g = 38$ kV to a 22 mm gap, with a screw head inside positive plane. Sparks were observed from the 5 mm rounded edges of one plane, not from the screw.

3. POWER TRANSMISSION

Target value of power available to platform is 135 kVA; of this 37 kW are dissipated into the two solenoid coils of ECR source, and 17 kW inside the associated

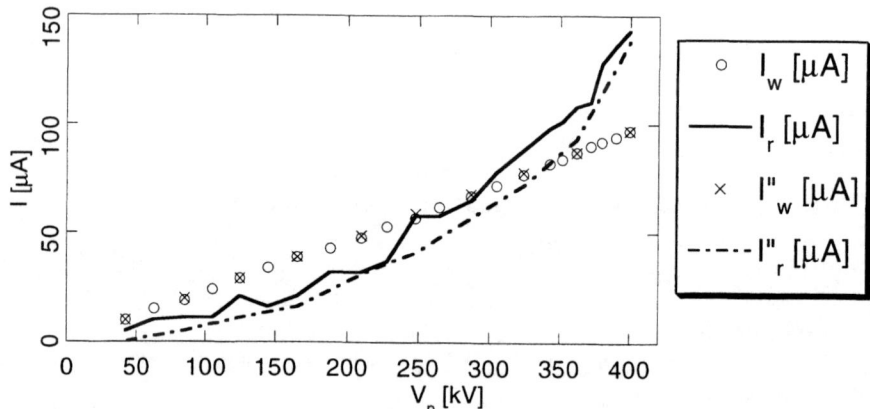

FIGURE 1. Currents drained by the platform versus voltage readout: a) through water, during voltage rise I_w ; b) through water, during voltage fall $I"_w$; c) dispersion through air, columns etc., voltage rising I_r; d) as c, but voltage falling $I"_r$

power supplies (PS1 and PS2, 640 A/29 V each); up to 15 kW between the dipole D1 and its power supply PSD (240A/50V); 10 kW by the klystron, and 5 kW reserved to refrigerate the hot air klystron exhaust; 10 kW by utilities. Even if more power efficient, switching regulators for PS1,PS2 were not used for practical reason (parts exchangeability with ALPI magnet ps, all linearly regulated as PS1,PS2 and PSD to a 1ppm stability); similarly a water cooled klystron was difficult to find (so we use an air cooled klystron, more common and therefore cheaper).

The insulation transformer itself [5] is about a $1.2 \times 2. \times 1.5$ m box from which a 3 m long bushing protrudes (inclination is 28 degrees on horizontal); bushing terminates by a 0.75 m diam sphere, covering secondary winding connections. Transformer insulation is rated $V_1 = 400$ kV DC; input and output voltage are 380 Vrms 50 Hz three phase, with output power rating 135 kVA; winding connection is triangle to star $(\Delta - Y)$, secondary winding neutral point N' is connected to the platform body E' and to the 400 kV DC generator positive pole. Transformer contains a comparatively large (3800 liters) amount of oil for insulation [8].

To transport power from the outside transformer to the platform, 4 unipolar cables were chosen against one channel embedded in plastic; these concepts are described elsewhere [3]. Proper cable termination are needed: construction care was taken by LNL mechanical workshop, which completed the prototype in May 1998 and is producing the full scale version (see scheme "A" in figure 2, where a, b, c, d, g and r are respectively equal to 20.3, 6.7, 925, 1800, 125 and 50 mm for the full scale version and to 3.7, 1.2 ± 0.1, 60, 400, 20 and 10 mm for the prototype).

It is well-known that large fields appears in air near the unshielded jacket (see model "B" in figure 2); in principle, since the same fields exists in the air films between the parts of the termination "A", charge accumulation may occur even there; but, before resorting to oil filling, we tested if air films were fatal. Prototype uses the common RG213 cable; a circular aluminium body and a polyethylene (PE) filling define the geometry and the fields where the shield is separated; four cables are assembled as in the real task. It was tested up to $V_t = 150$ kV for few

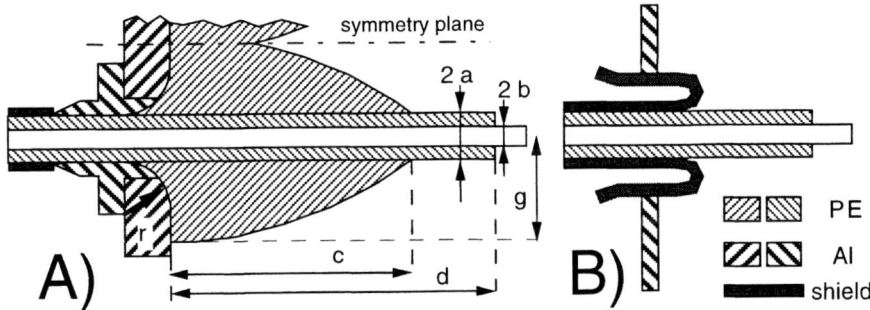

FIGURE 2. Schemes of cable terminations: a) with machined polyethylene (PE) filling; b) simply folding of shield

minutes; this implies a 1.1 MV/cm maximum field inside PE cable jacket (still consistent with the highest fields reported in PE, see [6] and [7]); also V_t was pleasantly larger than the expected threshold for corona effects (estimated about or less $V_t \cong E_a g(\epsilon_r^{-1} \ln(g/b) + 1) \cong 100$ kV). As an inverse proof of termination necessity, we compare with scheme "B" (without any plastic filling), using again RG213 cable; cable punctured at $V_t = 150$ kV, after humming and noticeable surface charge accumulation from $V_t \cong 110$ kV; peak field in PE exceeded 1.2 MV/cm due to different geometry. On these experimental bases, we considered reasonable to build and test the full size power line.

4. BEAM OPTICS

A scheme of beam optics is given in figure 3. The first einzel lens included in the ion source Alice makes the beam parallel inside the analyzing dipole D1, so that a stigmatic image is formed at $z = z_2 = 1.12 - 1.14$ m (be z the magnet exit direction and $z = 0$ the intercept with magnet nominal exit plane). At $z \cong 15$ cm we have a 12 pole magnetic corrector, which is powered by the standard alpi steerer ps (of course, it has more coils, properly connected). At $z = z_1 = 1.13$ m we have the selection slit, while faraday cup inserts at $z = z_3 = 1.18$ m (axis of flange). The accelerating field region extends from $z = z6 = 2.24$ m to $z_7 = z_6 + L$, while einzel lens E1 and E2 midplanes are at $z = z_4 = 1.44$ m and $z = z_5 = 1.80$ m respectively. Note that $L = 1.267$ m, but the actual accelerating tube is 1.467 m long including

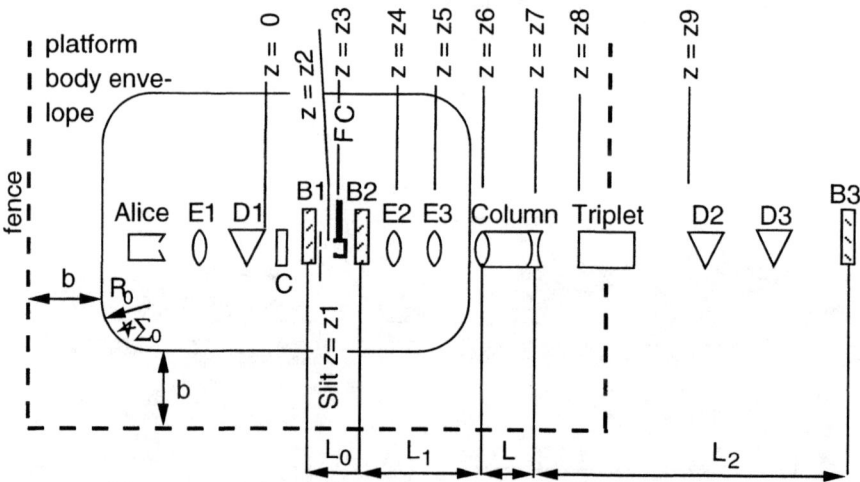

FIGURE 3. Scheme of ECR beamline, with some elements (D2,D3,B3) of the transmission line to RFQ: D are bending dipole, B are bunchers, E are einzel lens, C is a multipole corrector, FC is the faraday cup

flanges. A triplet of electrostatic quadrupoles extends from $z = z_8 = 3.76(8)$m to $z = 4.26(8)$m. This optic line has to match the beam transversally to the entrance of a dipole D2 at $z = z_9 - 5.00(6)$ m and longitudinally to the further away buncher B3 (the distance L_2 from column end $z = z_7$ to B3 midplane is about 14.5 m). Bending radius of D2 and D3 is $R_{D2} = 0.3$ m, of D1 is $R_1 = 0.5$ m.

By fitting the results of some simulations [10], it can be found that focal power q (defined as $q = 1/f$ with f the focal length) and principal plane distance (from midplane) h are of the form

$$q = \frac{1}{f} = \frac{v^2}{q_0 + q_1 v + q_2 v^2} \quad \text{with} \quad v = -V_e/V_s \tag{3}$$

(and similarly for h) with V_e the excitation voltage of the einzel lens, in the interval $-1 < v < 2$. For E2 and E3 (they are equal) we have: $(q_0, q_1, q_2) = (20.2, 18.2, 3.58)$ cm and $(h_0, h_1, h_2) = (-3.31244, -2.13, 0.232)$ cm^{-1}; for first einzel lens E1: $(q_0, q_1, q_2) = (16.9, 15.3, 2.99)$ cm and $(h_0, h_1, h_2) = (-3.99, -2.56, 0.278)$ cm^{-1}.

Equation 3 is of great help in design, since usually first we fix f, neglecting h, then we find what voltage corresponds to f. In other words, we can study the thick lens with no greater difficulty than a thin lens.

About the triplet, let the effective lengths be l_e for lateral quadrupoles, $2l_c$ for the central quad, l_g for the gap. Let $2R_a$ be the electrode aperture and let F_H, (F_V) the distance of focal point in the horizontal plane (vertical plane) from triplet end (note the capital F to distinguish them from focal distances, which are measured from principal planes); let V_c (V_e) the voltage applied to the central (lateral) quadrupole and define the adimensional quantities $v_c = V_c/(V_p + V_s)$ and $v_e = V_e/(V_p + V_s)$. Some indication about voltages needed for stigmatic operation of the triplet comes from considering the simpler case: rays parallel to z axis at midplane. Expressing $1/F_V$ and $1/F_H$ as a function of v_e and v_c, expanding in power series, and solving the series equation $1/F_H - 1/F_V = 0$ for v_e, we find [9]

$$v_e = \frac{l_c}{l_e} v_c + \frac{l_c^3}{15 l_e} \frac{L_F}{R_a^4} v_c^3 + O(v_c^5) \tag{4}$$

$$\frac{1}{F_V} = \frac{(l_c + l_e + 3l_g) l_c^2}{3 R_a^4} v_c^2 + \frac{l_c^4 l_e L_F}{45 R_a^8} v_c^4 + O(v_c^6) \tag{5}$$

where the long factor $L_F = 2l_c^2 + 5l_c l_e + 3l_e^2 + 5l_g(2l_c + 3l_e + 3l_g)$.

Similarly to the einzel lens, first we can determine F_V according to our beam transport strategy and then we insert v_e, v_c (given by eqs. 4 and 5) in any simulation in order to render beam details.

This beamline satisfies the two proposed beam transport strategies: 1) the beam is focused before the dipole D2 entrance at $z = z_9 - 0.6$ m; 2) the beam is focused inside the dipole D2, at about $z = z_9 + R_{D2}$. In both cases, it is desirable that these points stay constant (o move slightly) for any ion mass ratio A/i, that is for any R. On the contrary, focusing property of the accelerating column changes strongly with R (entrance and output lens focal lengths and effective drift length are respectively $4L/(R^2-1)$, $-4LR^2/(R^2-1)$ and $2L/(R+1)$ in the simpler model), so that the requested focal lengths f_2, f_3 of the two einzel lens and the triplet F_V should adapt. Note that we have three parameters to match one condition, so we can also turn off one element and minimize the beam radius in one point (at $z = z_7$ in strategy 2 and the sum of radius at $z = z_6$ and $z = z_7$ in strategy 1, to quote the most common choices).

It was requested that ECR beamline provides space for possible bunchers [11], intending to squeeze a 200 ns beam into an approximately 10-18 ns long pulse at B3, for further compression: choice between a double drift buncher (DDB), made of 5 MHz gap at flange B1 and 10 MHz gap at flange B2, or a different kind (multiharmonic or corrected sawtooth) using B1 only is not yet made. Note that the effective length of the drift between B2 and B3 is

$$L_{II} \cong L_1 + \frac{L_2 - 2R_{D2}}{R^3} + \frac{2L}{R(R+1)} \qquad (6)$$

while the first drift is of course $L_I = L_0$. Note that L_0 and L_1 were the only parameters left open. Moreover, a space constraint is $L_0 + L_1 \leq 1.24$ m. Consider that a 71 % bunching is a well balanced compromise case (that is, more bunching requires more voltage, less bunching is a too large loss of beam) and requires $L_0/(L_0 + L_{II}) = 0.1836$. We choose to attain this condition at about $R = 5$, preferring heavier ions in the R expected range (from 2.6 to 5.4); then $L_0 \leq 0.264$ m. Practically $L_0 = 260$ mm was convenient and it was chosen, as well as $L_1 = 980$ mm.

So we provided a stable position for buncher(s) near the waist and the faraday cup access at the waist, as desired.

5. ECR OPERATIONAL EXPERIENCE; SOFTWARE

To make platform completion possible, ECR experimental program resulted rather limited this year: we delayed metal oven installation and begin to test the New 63 mm internal diameter Removable Plasma Chamber (called NRPC) with noble gases, after completing tests of the old 60 mm internal diameter removable plasma chambers (called RPC).

Test with RPC2 improved Ar^{+12} current up to 660 nA (see figure 4), when total source current was $I_s = 550\mu A$ and PS1,PS2 current was respectively $I_1 = 634$ A and $I_2 = 464$ A (maximum ps current is 640 A); total klystron power was $P_k = 200$ W and peripheral plasma pressure $105\mu Pa$. Gas in use were Ar, O, He, and residuals were present (leaks, wall desorption); plasma composition is estimated to be 40 ± 2 %, 41 ± 5 %, 8 ± 1 % and 11 ± 5 % respectively (from partial pressures, at plasma chamber with plasma off, respectively equal to 3050, 2500, 82 and 735 μPa as measured by closing subsequently gas valves; oxygen larger error accounts for the fact that some residuals may well be oxygen).

NRPC are partially water cooled. One of the new chamber (NRPC2 which has 6 electrodes and therefore will be very adequate for oven) had serious water leakage (which appeared after few days of pumping, due to water dissolving the inclusions in the silver soldering). Preliminary tests, even if supported by extensive helium testing of the water circuit, were inadequate to spot the failure in advance.

NRPC1 (with only two electrodes) did not show significant water leaks, so that ECR experiments were possible. Anyway, we noted a significant reflection of microwave when using quartz window (5.7 mm thick), so we had to go back to kapton window (0.2 mm thick), as used in old RPC, but with different mounting. Vacuum tightness of these mounting is not yet completely demonstrated. We noted the trivial effect of microwave tuners, that, by increasing power arriving to ECR plasma, increased ion yields.

NRPC1 shows several surprising preliminary results: 1) ion production is smaller than RPC2 yield; results show only up to $0.39 - 0.54$ μA of Ar^{9+} at typically $I_1 = 637$ A and $I_2 = 613$ A, with plasma peripheral pressure of $240\,\mu$Pa, with total microwave power of 115 ± 3 W, and second stage absorbtion of 69 W (forward rf 73 W, reflected rf 4 W by the assembly of window, vacuum waveguides and plasma); 2) optimal magnetic configuration seems to be obtained with maximum current or about; 3) plasma is very sensitive to the potential V_B (respect to ECR vessel) of an aluminium disk (15 mm diameter) located on the z axis, oppositely to extraction hole: for example, in preliminary tests, O^{4+} current goes from $0.92\,\mu$A at $V_B = -46$ V to $1.55\,\mu$A at $V_B = +46$ V (at the same time, total current I_s goes from 276 to 309 μA).

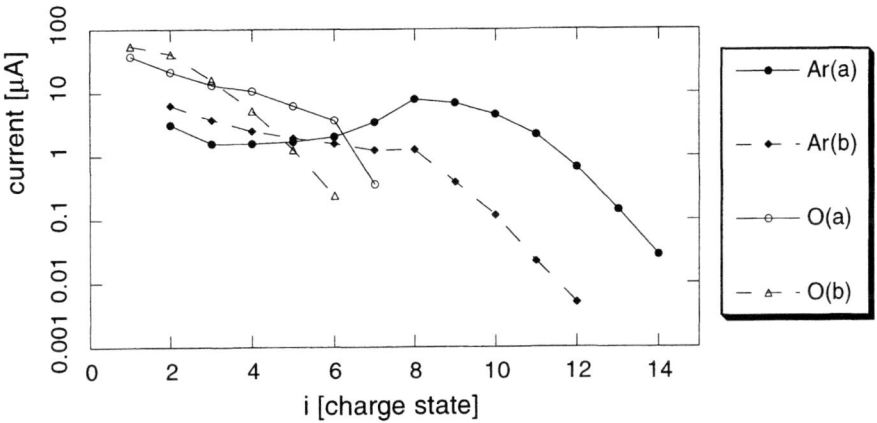

FIGURE 4. Oxigen and Argon output with plasma chambers: a) RPC2 with 60 mm ID) ; b) NRPC1 with 63mm ID . Oxygen charge state 2 and 5 currents are corrected from argon superposed currents, estimated by interpolation

While observation 2) is in perfect agreement with leading theories on ECR (it is well-known that if you increase magnetic field enough, you get better results [1]), 1) and 3) seem to be unexplained or different from usual. Indeed the NRPC attains a far larger magnetic field at wall than RPC2. Further experimental and theoretical work is necessary to clarify NRPC behaviour. Both a total pressure gauge and ion spectrometer, added near the dipole D1 exit, showed interesting large pressure variation during ion scan, due to gas desorption from the walls: typically pressure goes from 3 μPa at ECR plasma and gas off to 30 μPa in steady condition to 300μPa when high current reaches the Faraday Cup.

The final goal of ECR control is to be comprehensive, integrating in one computer system not only magnet and high voltage ps control, but also vacuum and microwave system, as well as acquisition of ion current emitted. Effectiveness and reliability of this system needs extensive tests (and adjustments).

Some operational tests of the ECR control (and acquisition) system begun (see Ref. [12] for a complete system description) in year 1998, with some upgrading of the original hardware. The addition of second terminal server allows to shorten RS232 lines on the platform. Moreover, the addition of a inexpensive workstation on the platform was very useful not only to run the ECR control system, but also as communication tool for bias voltage control and file exchanges. RAM memory had to be increased from 32 Megabyte to 96 before ECR workstation was reasonably fast in logins. Preliminary analysis of the microwave meters showed very informative.

REFERENCES

1. R.Geller, *Electron Cyclotron Resonance Ion Sources and ECR Plasmas*, Institute of Physics Publ.,Bristol,1996.
2. Cavenago M. , *Nucl. Inst. Meth.*,**A382**, 288-291, (1996)
3. Cavenago M. et al. *Rev. Sci. Instr.* **69**, 659 (1998)
4. R.C.Pardo, P.J. Billquist *Rev. Sci. Instr.* **61**, 239 (1990)
5. built by SpecialTrasfo, Cologno Monzese, Italy.
6. Das-Gupta Dilip K. , *IEEE Trans. on Dielectrics and Elec. Ins.*, **4**,149 (1997)
7. Bartnikas R. , *IEEE Trans. on Dielectrics and Elec. Ins.*, **4**,553 (1997) and Ref. 52 herewithin.
8. Chadband W G, *J. Phys. D: Appl. Phys*, **13** , 1299 (1980)
9. Cavenago M., "Optics of ECR line", in preparation
10. Cavenago M., in *Proc. XVIII Int. Linear Accelerator Conf.*, (eds. C.Hill and M.Vretenar, CERN, Geneva, report 96-07, 1996) vol. 1, p. 122
11. Skorka S.J., *3rd Int. Conf. on Electrostatic Acc. Tech.* , (IEEE catalogue n. 81 CH1639-4) , p 130 (1981)
12. Canella S. et al. , elsewhere in this Proceedings

Ion Source Developments for the Relativistic Heavy Ion Collider[1]

D.B. Steski

Alternating Gradient Synchrotron Department, Brookhaven National Laboratory, Upton, New York, 11973

Abstract. When the Relativistic Heavy Ion Collider (RHIC) at Brookhaven National Laboratory is completed in 1999, the Tandem Van de Graaff, using cesium sputter sources, will produce the gold ions. The sputter sources are operated in a unique pulsed beam mode that allows the injection into the tandem of 200 to 300μAmps of negative ions for up to 2 msec. This is several orders of magnitude higher instantaneous current than is possible in the direct current mode. These high current pulses have severe problems with space charge forces. The research to overcome the space charge forces has resulted in a 500μsec pulse containing 9×10^9 ions after acceleration in the tandem. This a three fold increase in intensity in the last three years. There is also growing interest in other ion species specifically iron, silicon, and uranium. The research to produce useable quantities of these ions is also presented.

INTRODUCTION

When the Relativistic Heavy Ion Collider (RHIC) becomes operational in 1999 at Brookhaven National Laboratory the BNL Tandem Van de Graaff will supply the heavy ions. The tandem has been the source of the heavy ions for experiments at the Alternating Gradient Synchrotron (AGS) since 1986. Originally oxygen and silicon ions were injected directly into the AGS(1). However with the completion of the AGS Booster synchrotron in 1992, gold ions were accelerated for the first time. The complete acceleration configuration for RHIC is shown in figure 1. The requirements of the Booster for the injection of short pulses of high charge state ions (2) have resulted in unique modes of operating the tandem.

---

[1] This work was performed under the auspices of the U.S. Department of Energy

The high charge states are obtained by post stripping the gold ions. The ions pass through a thin carbon foil approximately 2μg/cm^2 thickness in the center terminal and are stripped to the +12 charge state. After further acceleration in the high energy end of the tandem, the ions pass through another foil approximately 13μg/cm^2 thick and are stripped to higher charge states. The gold ions are usually injected into the booster in the +31 or +32 charge state.

RHIC ACCELERATION CONFIGURATION

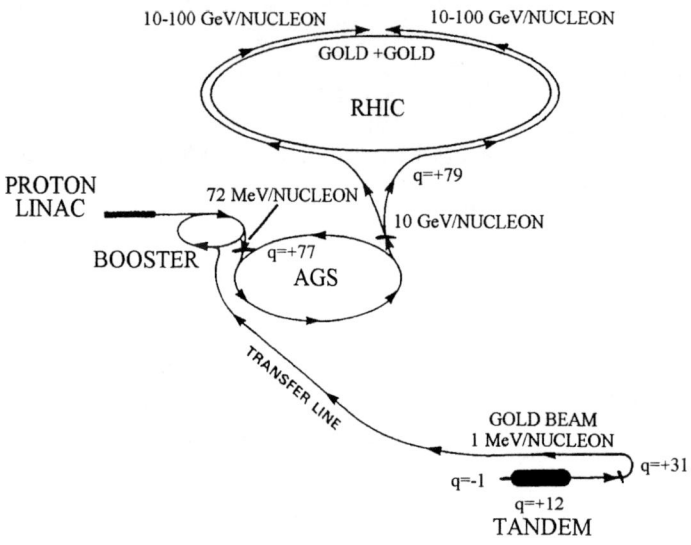

Figure1: The acceleration configuration for RHIC

Operating the cesium sputter sources in a pulsed mode produces the short pulses(3) required by the Booster. The pulsed mode of operation is shown schematically in figure 2. When the cesium acceleration voltage is pulsed from a base DC level, high intensity negative ions are extracted from the ion source. The energy of the negative ions leaving the source remains constant and is defined by the extraction voltage. Both the DC and pulsed cesium acceleration voltages are optimized to inject the tandem.

The pulsed mode of operation allows the injection into the tandem of over 300μA of beam, which is several orders of magnitude higher instantaneous current than is possible in DC operation. The pulse length can be up to 2000μsec although the typical pulse lengths have been between 500μsec and 1000μsec. The combination of high currents and short pulses causes the beam to experience severe space charge forces. The main objective of the ion source development at the tandem has been to understand and overcome the destructive effects of the space charge forces.

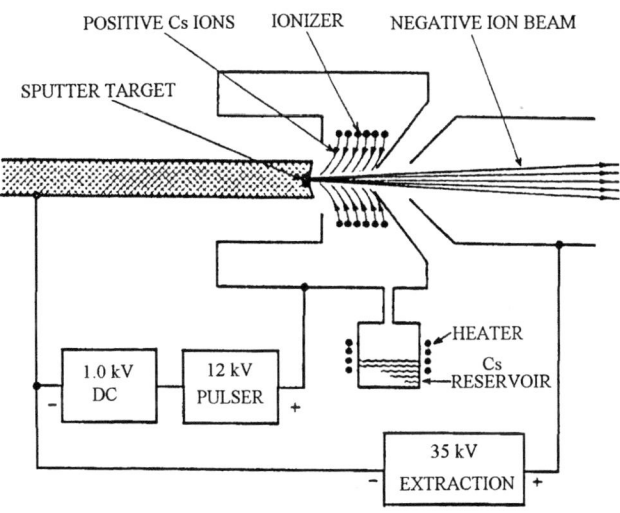

Figure 2: Schematic diagram of the pulse mode of operation for a Cs sputter ion source

OPERATIONAL EXPERIENCE WITH GOLD BEAMS

One of the ways to overcome space charge forces is to increase the extraction voltage of the ion source. The higher beam energy reduces the space charge effects and allows more ions to be transported to the low energy cup in front of the tandem. The ability to run at higher extraction voltages was one of the motivations for replacing the original GI 860 sources(4) with the new PSX-120 ion source(5). The PSX-120 was first used in 1996 and allowed the extraction voltage to be raised to 32kV.

Although there is a definite advantage of using the highest possible extraction voltage, the need for long term stability requires that the source be operated at more modest voltages. Therefore other ways of overcoming the space charge forces have also been investigated. The first attempt was to increase the residual gas pressure by bleeding xenon gas into the ion source region(6). The residual gas was ionized positively by the ion beam and trapped in the potential well of the ion beam, partially neutralizing the space charge forces. The increase in usable beam with the addition of xenon can be seen in figure 3.

However the increased gas pressure also increases the probability of stripping the loosely bound additional electron of the negative ion. By placing a gas restrictor immediately downstream from the ion source the gas load on the vacuum pumps was reduced by a factor of 2 to 3 and the effectiveness of the background gas was increased resulting in more beam being injected into the tandem. This approach was used during the 1996 heavy ion run and resulted in a maximum beam intensity of 52μA of Au^{+32} being delivered to the Booster.

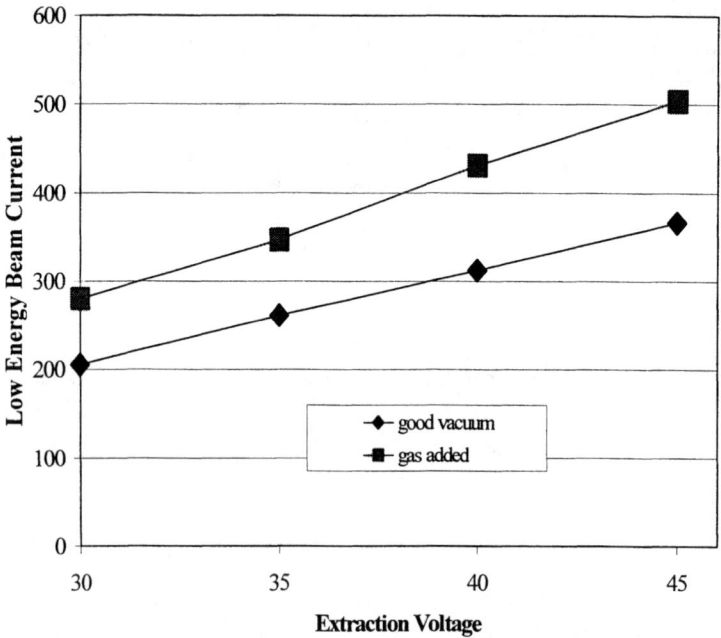

Figure 3: Graph of beam current versus extraction energy with and without adding xenon

For the 1997 run, it was decided to replace the gas bleed system with a more active focusing element that would be easier to tune. A double gridded einzel lens was constructed and is shown in figure 4. The grids were made of tungsten with a transparency of 90%. The entire length of the lens was 51mm long and it fitted inside an existing transition flange. The ID of the center element was 53mm and there was a 32mm diameter aperture in front of the upstream grid to intercept any electrons emanating from the ion source. This lens was much more responsive and allowed easier tuning of the ion beam. The result was an increase in the beam available to the booster. For 231µA of Au- injected into the tandem, 62µA of Au^{+32} was measured at the start of the transfer line.

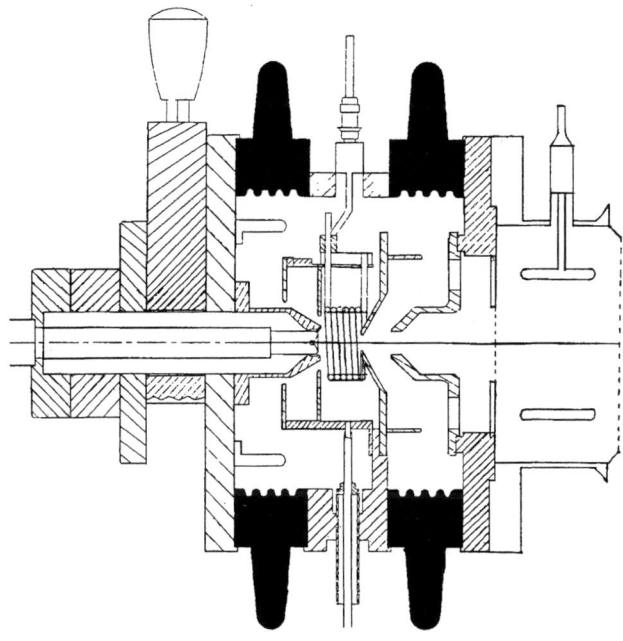

Figure 4: PSX-120 ion source with additional einzel lens

The additional focusing lens was also used successfully during the 1998 heavy ion experiment. The typical intensity delivered to the booster was 32µA of Au^{+31} or 3.2×10^9 particles per pulse for a 500µsec pulse. Immediately following the experiment, several days of dedicated tandem testing were done. During this period a new peak intensity record of 90µA of Au^{+31} was achieved. When the beam was averaged over 600µsec the pulse intensity was 9×10^9 particles per pulse. Figure 5 shows the improvement in peak intensity since the injection of gold ions started.

INTEREST IN OTHER ION SPECIES

In recent years there has been an increased demand for heavy ions other than gold from the AGS. The ions investigated to date are iron and silicon for biological research and uranium for the nuclear physic program at RHIC.

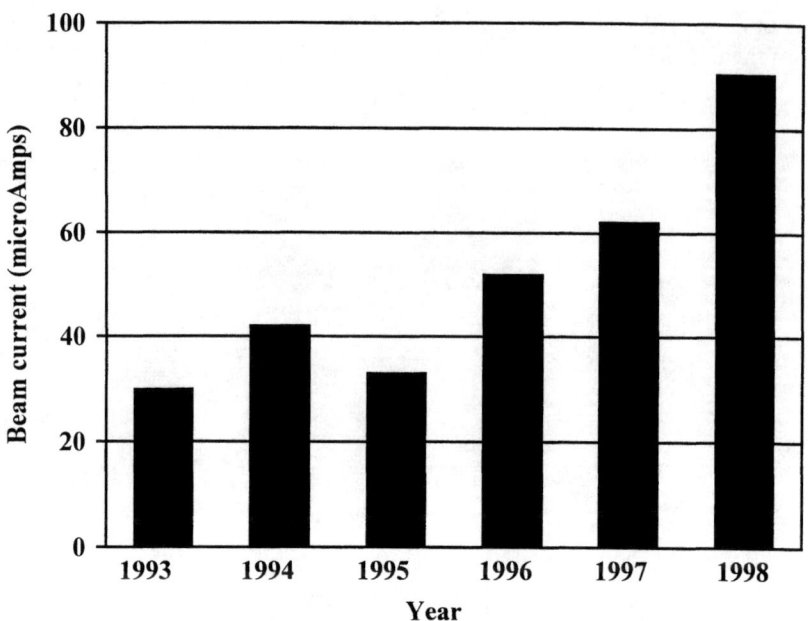

Figure 5: Graph of peak gold intensity versus year

The iron beam has been accelerated since 1996 with one experimental program each in 1996 and 1997, and 2 in 1998. The iron beam was accelerated to the center terminal as the FeO- molecule where it was stripped to the +10 charge state. The iron beam was not post stripped after the tandem, which greatly increased the available beam. Although the intensity of the FeO- beam was lower than the gold intensity, there was a considerable O- beam that contributed to the space charge forces at the exit of the ion source. The same techniques used to compensate for the space charge forces in the gold ion have proved useful in increasing the available iron beam. This can be seen in figure 6, which shows the increase in Fe^{+10} intensity since the iron program began. The highest peak current observed was 142µA which when averaged over 600µsec was 4.5×10^{10} particles per pulse.

Silicon was also demonstrated for possible Booster injection. Because of the relative ease of generating negative silicon ions and because the ion beam was single stripped, large quantities of silicon were available from the tandem. The peak intensity was 230µA Si^{+5} which corresponds to 1.44×10^{11} particles per pulse.

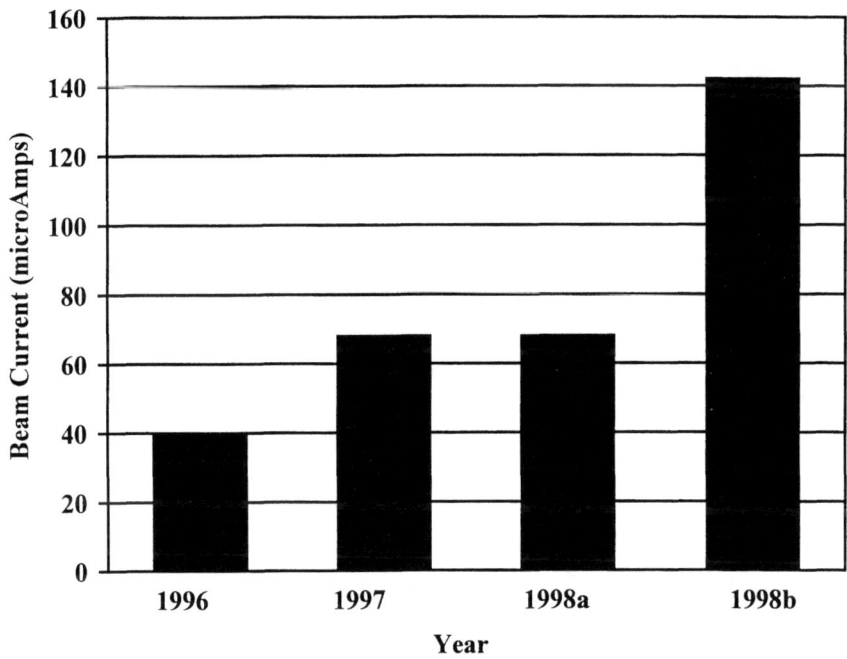

Figure 6: Graph of peak iron intensity versus year

There has been a growing interest in accelerating uranium in RHIC for nuclear physics. Uranium provides a unique challenge because of its low electron affinity and also the need to post strip for injection into the booster. Because the yield of U- ions is small, the focus has been to find a suitable molecule for acceleration. Target materials from several different uranium compounds have been tried and it appears possible to use the UO- molecule. The most successful combination was a solid uranium target with O_2 being sprayed onto the metallic uranium surface. Uranium also requires a different pulsing scheme that increases the DC level to approximately 4 kV one second before the main pulse. The highest UO- pulse generated had a peak amplitude of 34μA. Because this research was performed on the Ion Source Test Bench[7] which has different optics than the tandem, there should be approximately 72μA of UO- available for injection into the tandem. Using the data that has been gathered for the gold, 72μA injected into the tandem should result in approximately 25μA of U^{+31} or 3×10^9 particles per pulse in a 600 μsec pulse available for the Booster. This quantity of uranium ions although less than gold would be usable in the Booster. Because of the preliminary nature of these results further improvements should be possible.

Table 1 summaries the different ion species accelerated in the tandem and the maximum intensities observed.

TABLE 1. Ion species and the maximum intensities observed

Ion	Charge State	Maximum Current	Particles per Pulse
U*	+31	25μA	3×10^9
Au	+31	90μA	9×10^9
Fe	+10	142μA	45×10^9
Si	+5	230μA	144×10^9

* Estimated intensity using data from gold experiments

FUTURE DIRECTIONS

Further increases in the ion beam intensity can be made by improving the extraction insulator. This will allow the extraction energy of the ion source to be reliably raised to 40kV or 45kV. This can be accomplished relatively easily with a minimum investment. Another avenue of research is to combine the gas bleed system with the additional einzel lens. These attempts to further reduce the space charge forces will increase the intensity of all of the ion species accelerated not just gold. Additionally, the research in generating a UO- beam needs to be repeated and confirmed by acceleration through the tandem.

In the longer term, improvements will come from carefully matching the ion beam to the acceptance of the tandem accelerator. The use of a plasma sputter ion source may also lead to an increase in intensity. These studies will require a greater investment in the tandem in terms of time and resources.

ACKNOWLEDGMENTS

The author would like to thank P. Thieberger, C. Carlson, J. Benjamin, M. Manni, and the operations staff for their support and cooperation.

REFERENCES

1. Thieberger P., Barton D.S., Benjamin J., Chasman C., Foelsche H., and Wegner H.E., *Nuclear Instruments and Methods in Physics Research,* **A268,** 513-521, (1988).

2. Prelec K., Rhoades-Brown M.J., Thieberger P., and Wegner H.E., *Nuclear Instruments and Methods in Physics Research,* **A295,** 21-33, (1990).

3. Thieberger P., McKeown M., and Wegner H.E., *IEEE Transactions on Nuclear Science,* **NS-30 No.4,** 2746-2748, (1983).

4. Distributed by High Voltage Engineering Europa, P.O. Box 99, 3800 AB Amerfort, The Netherlands.

5. Distributed by Peabody Scientific, Peabody, MA, 01960.

6. Steski D.B., Zarcone M.J., Smith K.S., and Thieberger P., *Review of Scientific Instruments,* **Vol.67 No.3,** 1221-1223, (1996).

7. Zarcone M.J., "The Ion Source Test Bench for BNL's Tandem Van de Graaff Facility", presented at the Symposium of North Eastern Accelerator Personnel, Santa Fe, New Mexico, (1991)

Design Study of the Extraction System of the 3rd Generation ECR Ion Source

D. Wutte, M. A. Leitner, C. M. Lyneis, C. E. Taylor, Z. Q. Xie

Ernest Orlando Berkeley National Laboratory, University of California at Berkeley, Berkeley, California 94720, USA

Abstract. A design study for the extraction system of the 3rd Generation super conducting ECR ion source at LBNL is presented. The magnetic design of the ion source has a mirror field of 4 T at the injection and 3 T at the extraction side and a radial field of 2.4 T at the plasma chamber wall. Therefore, the ion beam formation takes place in a strong axial magnetic field. Furthermore the axial field drops from 3 T to 0.4 T within the first 30 cm. The influence of the high magnetic field on the ion beam extraction and matching to the beam line is investigated. The extraction system is first simulated with the 2D ion trajectory code IGUN with an estimated mean charge state of the extracted ion beam. These results are then compared with the 2D code AXCEL-INP, which can simulate the extraction of ions with different charge states. Finally, the influence of the strong magnetic hexapole field is studied with the three dimensional ion optics code KOBRA. The introduced tool set can be used to optimize the extraction system of the super conducting ECR ion source.

INTRODUCTION

With the construction of the LBNL 3rd Generation ECR ion source we expect to further enhance the performance of the 88" cyclotron by providing more intense highly charged heavy-ion beams (1). Record high charge states and beam intensities are provided by the AECR-U ion source (2) and usable beams for elements up to mass 200 can be extracted from the cyclotron with sufficient intensities for nuclear structure experiments such as Gammasphere (3). However, for low cross section experiments with the Berkeley Gas-filled Spectrometer (BGS) now coming on line at the 88" cyclotron higher ion beam intensities will be required. With the third Generation ECR ion source we will increase both the maximum charge states and beam intensities for the science programs at the 88" Cyclotron facility.

The magnetic design of the third Generation ECR ion source has a maximum axial field of 4 Tesla at the injection side and 3 Tesla at the extraction side. The maximum hexapole field is 2.4 Tesla at the plasma chamber wall. Figure 1 shows the ion source layout and Figure 2 shows the axial field at full coil excitation. Due to the size of the superconducting coils, the distance from the plasma outlet aperture to the exit of the iron shielding yoke it is about 30 cm. At full coil excitation the axial magnetic field drops from 3 T to 0.4 T within this distance. The axial magnetic field then drops further below 20 G within the next 30 cm. Therefore, the beam formation takes place in a strong magnetic field and has to be included in the ion optics layout.

CP473, *Heavy Ion Accelerator Technology: Eighth International Conference,*
edited by Kenneth W. Shepard
1999 The American Institute of Physics 1-56396-806-1

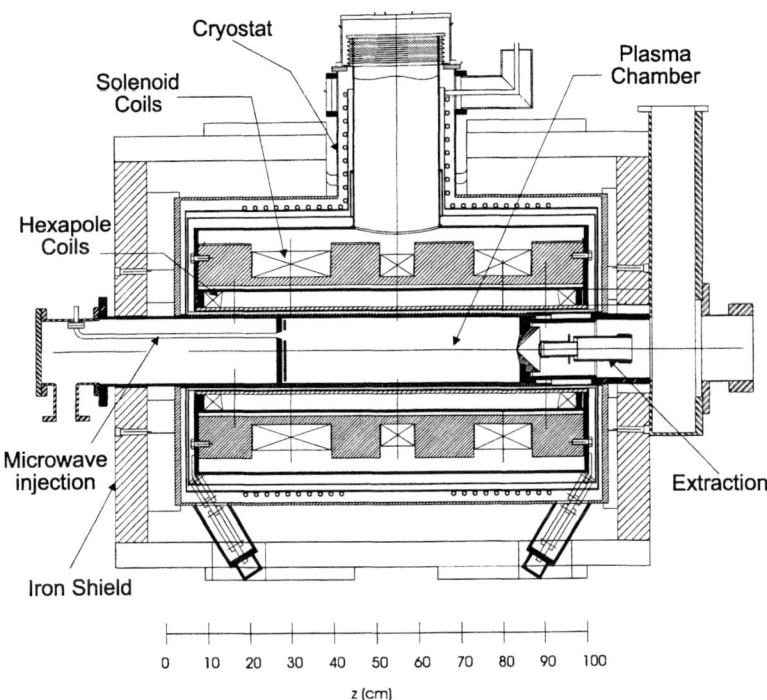

Figure 1. Mechanical layout for the LBNL 3rd Generation ECR ion source, including the iron shield, cryostat, coils, and plasma chamber.

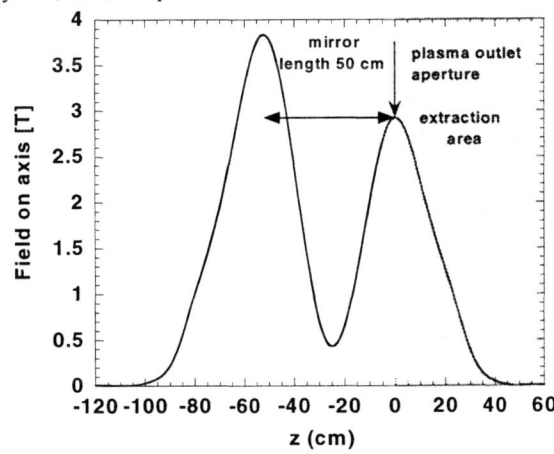

Figure 2. Axial magnet field of the LBNL 3rd Generation ECR ion source.

EXTRACTION SIMULATION OF MULTIPLY CHARGED ION BEAMS IN THE PRESENCE OF STRONG MAGNETIC FIELDS

During the optimization process of an extraction and beam transport system, it is more convenient to simulate only one charge state. It is then possible to concentrate on the transmission of the charge state of interest. Furthermore, computer capacity limits the maximum number of simulated particle trajectories, which can be simulated in reasonable times. By using only one charge state (compared to 20 in the case of argon with oxygen mixing gas) many more particle trajectories can be allocated to the charge state of interest. For these reasons the ion optics code IGUN (4) is used for our first design study. We introduce a physical approximation, which models the extraction of many charge states from an ECR ion source by using only a single charge state. We will compare the IGUN results with AXCEL-INP (5) simulations. AXCEL can simulate the extraction of different charge states simultaneously. Both computer codes use a one-dimensional plasma sheath model and can import axial magnetic field tables.

In the first part of the paper, we describe our model. The second part presents simulation results:

1. IGUN simulations for the extraction gap (2D)
2. Comparison of the IGUN results with AXCEL runs (2D)
3. Influence of the hexapole field on the ion optics (3D)

Finally, we describe the preliminary beam line layout to a Faraday cup after the bending magnet. In that way we are establishing a tool set for further extraction and beam line optimization of the 3rd Generation ECR ion source.

Simulation model

Since IGUN can only simulate one charge state, we will approximate the extraction of many charge states from an ECR ion source by considering only one charge state. It is incorrect to model the beam transport by using the mean charge state (weighted by the current of each charge state). Such a simplification is unphysical, because it does not model the plasma sheath correctly nor the space charge allocation along the beam path. Furthermore, it is not possible to simulate the influence of the magnetic field on the beam envelope for different charge states (see Figure 4). In the strong magnetic field, different charge states have different focal lengths and emittance orientations in phase space (Figure 7). For instance if we calculate the example charge state distribution for Ar^{16+} (Table 1) with a mean mass-to-charge of 12.44, it will result in an emittance pattern within the 20 different emittance pattern for the whole ensemble (see Figure 7). Therefore, no prediction can be made for a particular charge state. The mean charge state approach would give incorrect simulation results and would lead to an non-optimized design.

For our simulations, we have normalized the current of each charge state of each ion to the equivalent current for the charge state of interest by using the Child-Langmuir relation

$$j = 1.72 \cdot \frac{U^{3/2}}{d^2} \sqrt{\frac{q}{M}}, \qquad (1)$$

with j is the current density in (mA/cm^2), q is the charge state, M is the ion mass (amu), U is the extraction voltage (kV), d is the extraction gap width (cm).

The Child-Langmuir relation calculates the maximum extractable current under space charge limited conditions from a plasma for a plane meniscus, which is proportional to $\sqrt{q/m}$. To obtain the same plasma sheath conditions (e.g. a straight plasma sheath at a given extraction voltage) more current must be extracted for the higher charge states q/m than for the lower ones. The procedure of computing the different current contributions of a CSD must consider this behavior. Therefore, we will normalize the contribution of each single charge state q/m of the CSD by multiplying with $\sqrt{m/q} \cdot \sqrt{Q/M}$. Q/M corresponds to the charge state of interest, which will be simulated in the computer model. This approach models the plasma sheath position correctly with only a single charge state and a normalized current value, which includes the current contributions of all the other charge states. In particular, the ion-optical magnetic field influence on the charge state of interest is modeled accurately. Furthermore, the space charge allocation along the beam path is accurate as long as the overall beam envelope does not change considerably. The extraction system can now be optimized for a chosen charge state and plasma condition without including explicitly all the other charge states.

If we want to simulate the transport of an Ar^{16+} ion beam, each charge state of the CSD has to be normalized to Ar^{16+} according to

$$I_{Ar^{n+}_{equivalent}} = I_{Ar^{m+}} \cdot \sqrt{\left(\frac{n}{m}\right)}, \qquad (2)$$

n and m are argon charge states, $I_{Ar^{m+}}$ is the current to be normalized.

For example for the charge state Ar^{10+} with $I_{Ar^{10+}} = 64 e\mu A$ the normalized Ar^{16+} current would be $I_{Ar^{16+}_{equivalent}} = 81 e\mu A$. For the correct simulation of a high charge state distribution, we have included the charge state distribution of the oxygen mixing gas. In that case, the oxygen currents have to be converted to Ar-equivalent currents in the following way

$$I_{Ar^{n+}_{equivalent}} = I_{O^{q+}} \cdot \sqrt{\left(\frac{n}{q}\right)} \sqrt{\left(\frac{M_O}{M_{Ar}}\right)} \quad \text{with} \quad \frac{M_O}{M_{Ar}} = \frac{16}{40}. \qquad (3)$$

INPUT PARAMETERS

As an example, we have modeled two different charge state distributions (CSD) for argon; a high-current medium charge state distribution optimized for Ar^{9+} and a lower-current high-charge state distribution optimized for Ar^{16+} as extracted from the AECR-U. Table 1 summarizes the CSDs as used in the simulations with the ion optics codes ACXEL and IGUN. The axial magnetic fields were calculated with TOSCA3D and imported into the ion trajectory codes.

TABLE 1. A high-current medium charge-state distribution optimized for Ar^{9+} and a lower-current charge-state distribution optimized for Ar^{16+}. For IGUN simulations the total Ar^{9+} or Ar^{16+} equivalent current (indicated by the Σ sign) has been used as input parameter. For AXCEL simulations, each charge state and electrical current have been used as input parameters.

input ion beam distributions						
CSD optimized for Ar^{9+}			CSD optimized for Ar^{16+} with oxygen gas mixing			
q	CSD* [eµA]	Ar^{9+} equiv. curr. [eµA]	CSD* [eµA]	Ar^{16+} equiv. curr. [eµA]	O_2 CSD* [eµA]	O^{n+}, equiv. curr. [eµA]
16	0.48	0.36	17	17		
15	3.2	2.5	45	46		
14	13.3	10.7	69	74		
13	43	36	78	86		
12	107	93	85	98		
11	217	196	75	90		
10	385	365	64	81		
9	483	483	57	76		
8	502	532	50	71		
7	231	262	49	74	48	46
6	183	225	45	73	97	100
5	207	277	40	72	80	91
4	100	150	38	76	72	91
3	100	173	35	81	70	102
2	80	170	30	85	70	125
1	60	180	25	100	70	177
Σ eµA	2715	**3156**	802	**1200**	507	**732**

*typical AECR-U CSD, corrected for transport losses (current estimates were made for the lower charge states below Ar^{5+})

SIMULATION RESULTS

Extraction system layout

Figure 4 shows the layout for the extraction system. The recess in the plasma outlet aperture creates a more uniform electric field equipotential surface at the plasma meniscus (Figure 3). The aperture edge thickness has been chosen as small as practically feasible (e.g. a 0.2mm thickness is a practical value, especially when plasma heat load problems are considered). Generally the thinner the edge of the plasma outlet aperture the smaller are the ion beam losses to the outlet electrode, resulting in a higher extractable ion current at a given plasma density. This behavior is demonstrated in figure 3 with a thick (figure 3a) and a thin (figure 3b) plasma outlet aperture. In the case of a 3 mm thick electrode about 47% of the outward directed current is lost to the electrode. Furthermore the electric equipotential contour lines are less distorted for the thin edge and fewer aberrations are induced to the ion beam (7).

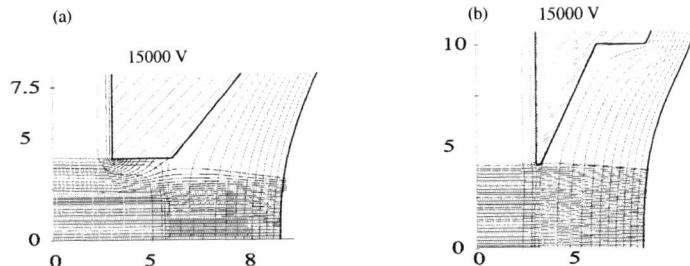

Figure 3. Equipotential contour lines computed at the plasma sheath for a thick (3 mm, figure2a) and a thin (0.2 mm, figure2b) plasma electrode. In both simulations, the distance between the accel electrode and the plasma outlet aperture was 29mm. The density has been adjusted to maintain a flat plasma sheath. In case (a) the total extracted current was 1.5 mA; about 47 % of the available current gets lost in the extraction hole. In case (b) the total extracted current was 3 mA.

Figure 4 shows the influence of the magnetic field on the ion beams of Ar^{1+}, Ar^{5+}, and Ar^{16+}. The equivalent currents for Ar^{1+} (a), Ar^{5+} (b), and Ar^{16+} (c) were computed according to equation 1 and 2 for the lower current CSD (optimized for Ar^{16+}). Each charge state has a different beam envelope in the strong axial magnetic field. For example the waist of Ar^+ is further downstream than for Ar^{16+}. Therefore, the singly charged Ar beam is strongly divergent (Figure 4a) at $z = 15$ cm, whereas Ar^{16+} is convergent (Figure4c).

Figure 5 shows how the location of the rms Ar^{16+} emittance ellipse in phase space can be optimized when the extraction gap spacing is changed. In that way it is possible to match the ion beam extraction to the beam line. Considering the wide range of plasma conditions of this ECR ion source, a movable extraction system will be essential for tuning the beam transport.

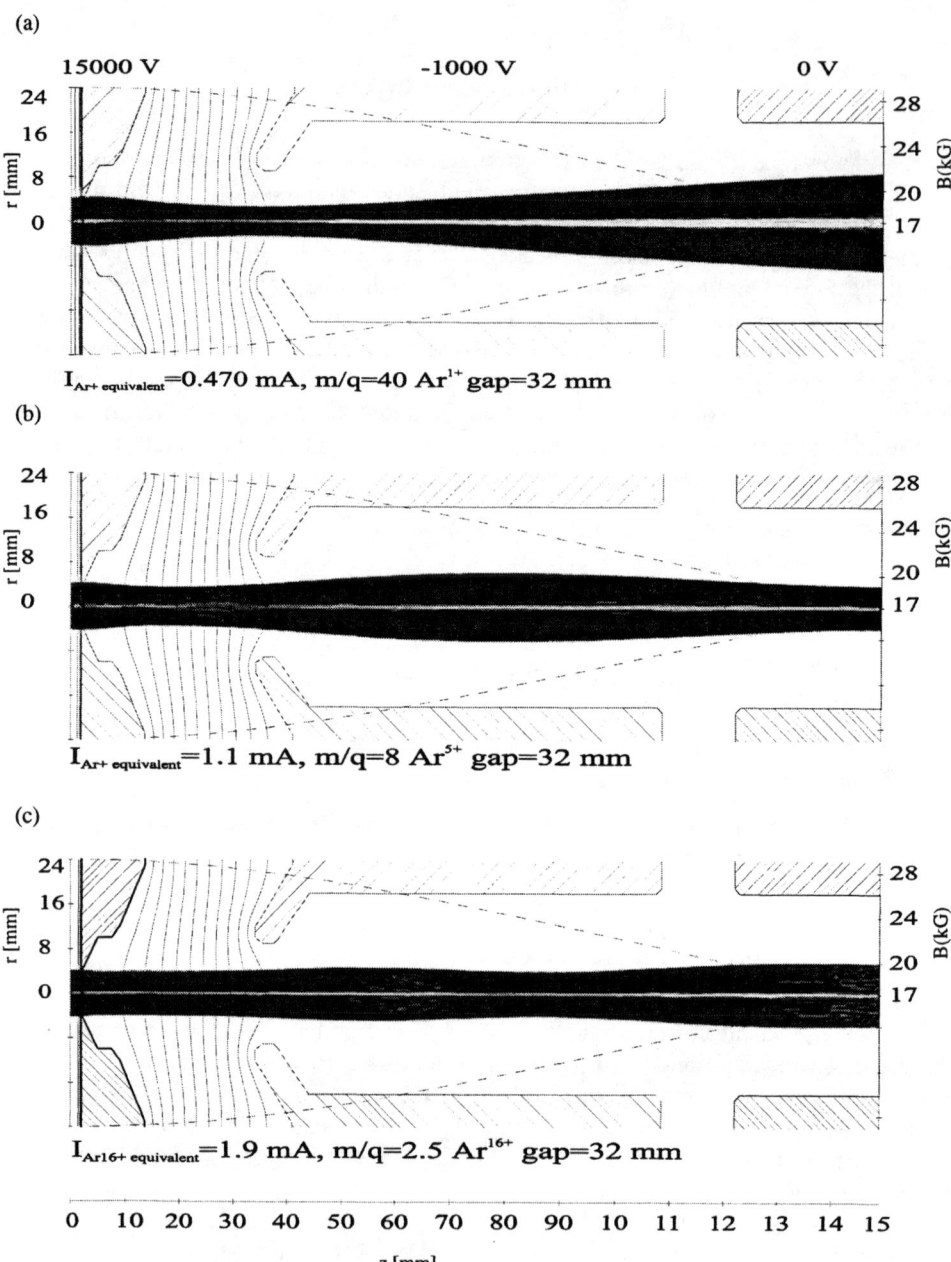

Figure 4. Influence of the magnetic field on the ion beams of Ar^{1+}, Ar^{5+}, and Ar^{16+} as computed with IGUN for the low current CSD (see Table 1).

The lowest emittance for Ar^{16+} has been calculated for an extraction gap of 32 mm. The calculated 100 % π' normalized emittances for this case are 0.108 π mm mrad for an thermal ion energy spread of 0.1 eV and 0.289 π mm mrad for an thermal ion energy spread of 3 eV.

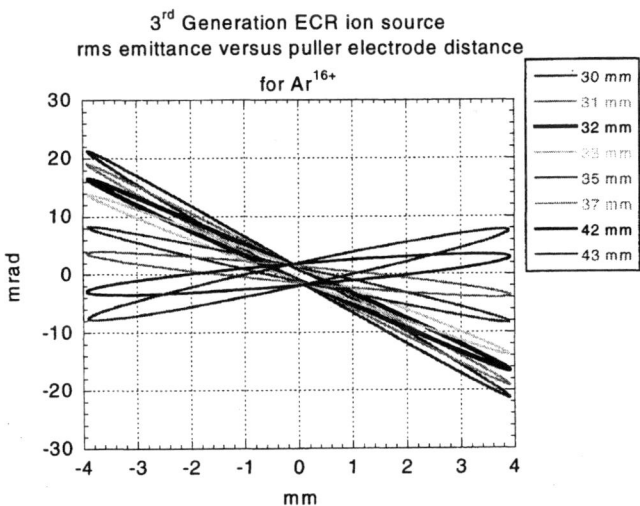

Figure 5. Change of the rms emittance figure versus the distance between the accel-decel system and the plasma outlet aperture for an Ar^{16+} ion beam. The black ellipse indicates the distance with lowest rms emittance for these particular plasma conditions.

Comparison between IGUN and AXCEL simulations

We have compared the simulation results for IGUN and AXCEL at the same input condition. Figure 6 shows the output emittances for both programs for the extraction system as described in Figure 4; the results are in good agreement.

To validate our approach of using a single charge state for the extraction system optimization, we calculated the complete charge state distribution for both CSD as described in Table 1 with AXCEL-INP.

As an example the emittance patterns for all different argon and oxygen charge states for the Ar^{16+} sample CSD is shown in Figure 7. The different focal length for each charge state can be clearly seen.

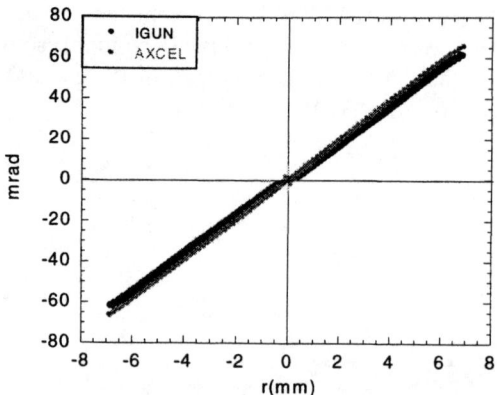

Figure 6. Comparison of an IGUN simulation and with an AXCEL simulation at the same input conditions m/q=4.4444, I_{extr} = 3.1 emA, the 100% normalized rr' emittance is about 0.32 π mm mrad.

As an example the emittance patterns for all different argon and oxygen charge states for the Ar^{16+} sample CSD is shown in Figure 7. The different focal length for each charge state can be clearly seen.

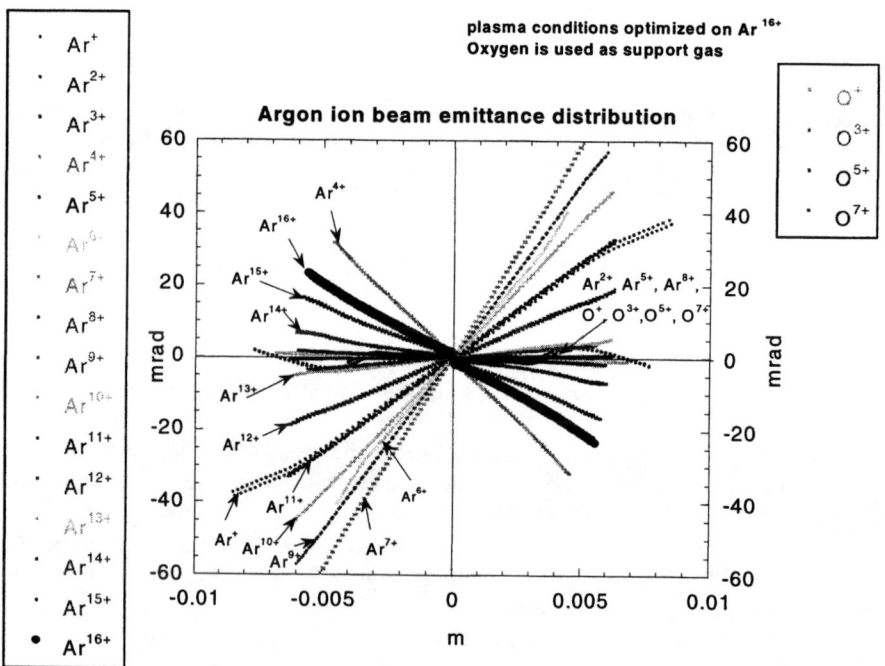

Figure 7. Emittance pattern as calculated with AXCEL. The different charge states are shaded in gray. Ar^{16+} is drawn in black.

We simulated different argon charge states for the low current-medium CSD by using the current equivalent method with IGUN. Figure 8 compares the emittance pattern, the agreement with AXCEL-INP results (considering all charge states, see Figure7) is remarkable. Figure 8b shows this comparison for the high current-medium CSD for Ar^{9+}.

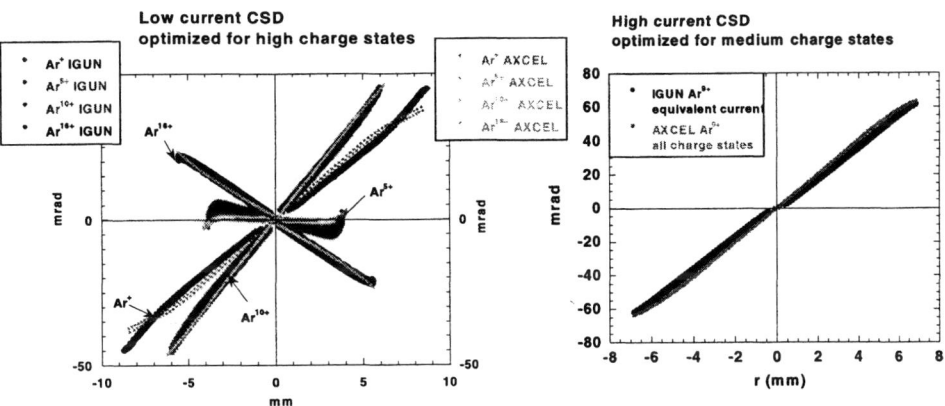

Figure 8. Comparison of the Ar^{16+} emittance pattern from (Figure7) of the AXCEL simulation (considering all charge states) with the IGUN calculation for Ar^{1+}, Ar^{5+}, Ar^{10+}, and Ar^{16+} (equivalent current methode). Figure 8b shows the same comparison for the high current medium charge state distribution. Both simulation result in the same rr' emittance.

Influence of the sextupole field

The influence of the magnetic hexapole field has been studied with the three dimensional ion optics code KOBRA (5). The magnetic input data table has been calculated with TOSCA. The influence of the hexapole field has been found to be negligible. This result is not surprising since the hexapole field strength at 1cm diameter (extraction aperture ∅8mm) does not exceed 150 G. Of course only the influence on the ion optics has been investigated. We have not included any variation of the ion current density across the plasma electrode orifice (caused by plasma density variations due to the magnet field structure).

BEAM LINE

A first layout of the beam line from the 3rd Generation ECR ion source to the Faraday cup after the bending magnet is shown in Figure 9, calculated with TRACE2D (7). The design has a Glazer lens (0.60 m downstream from the plasma outlet aperture) and a double focusing sector magnet (2.4 m downstream from the plasma outlet aperture). The first 30 cm have been calculated with IGUN for Ar^{9+} with an equivalent current for the high current-medium CSD beam (Table 1). The output beam parameter from IGUN at z = 30 cm have been used as input parameter for TRACE 2D. By combining these two computer codes we are able to consistently simulate the beam from the plasma meniscus through the beam line.

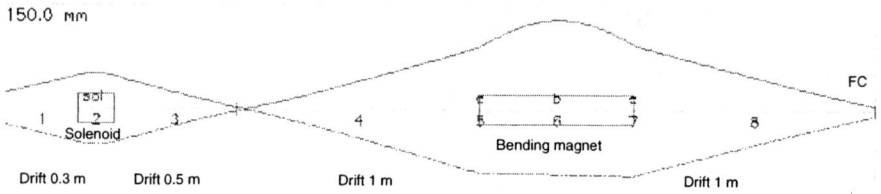

Figure 9. Preliminary layout of the beam line from the 3rd Generation ECR ion source to the Faraday cup after the bending magnet.

CONCLUSION

A tool set for optimizing the extraction system of an ECR ion source in the presence of a strong magnetic field has been introduced. It has been shown that the beam formation of a particular charge state can be modeled by normalizing the charge state current distribution to an equivalent current. A first layout for the extraction system and the ion beam transport line has been presented.

ACKNOWLEDGMENT

We would like to thank S. Lundgren for providing the drawing of the mechanical layout. This work has been supported by the Nuclear Physics Division of the U.S. Department of Energy under contract No. DE-AC03-76SFF00098.

REFERENCES

(1) C. M. Lyneis, Z. Q. Xie, C. E. Taylor, Rev. of Scientific Instruments **69**, 682 (1998)
(2) Z.Q. Xie, Rev. Sci. Instrum. **69**, 625 (1998)
(3) C. M. Lyneis, Z. Q. Xie, D. J. Clark, Proceedings of the 14th International Conference on Cyclotrons and their Applications, Cape Town, S. Africa, Oct. 8-13, 1995
(4) R. Becker, W. B. Hermannsfeldt, RSI **63**, 2756 (1991)
(5) INP, Junkernstr. 99, 65205 Wiesbaden, Germany; http://www.inp-dme.com/
(6) J. R. Coupland, T. S. Green, D. P. Hammond, A. C. Riviere, Rev. Sci. Instrum **44** (1973)
(7) TRACE3D Documentation, K. R. Crandall, D. P. Rusthoi, LA-UR-97-886, Los Alamos National Laboratory Report, May 1997

V
ACCELERATOR MASS SPECTROMETRY

Accelerator Mass Spectrometry – Big and Small

Walter Kutschera

*Vienna Environmental Research Accelerator, Institut für Radiumforschung und Kernphysik
Universität Wien, Währinger Str. 17, A-1090 Wien, Austria*

Abstract. A brief review of the current status of Accelerator Mass Spectrometry is presented, with emphasis on some of the most recent technical developments.

I. INTRODUCTION

Accelerator Mass Spectrometry (AMS) evolved from nuclear physics laboratories some twenty years ago (1-4), when it was realised that long-lived radionuclides, - in particular ^{14}C - can be measured at natural levels by couting atoms directly. It had been noted earlier (5) that during a typical beta decay measurement of ^{14}C lasting two days, only about one out of a million ^{14}C atoms decays (the half-life of ^{14}C is 5730 years). "Waiting around for the decay of these atoms is clearly an inefficient way to count them" (2). With a $^{14}C/^{12}C$ isotopic ratio of 1.2×10^{-12} in modern carbon, one needs a few grams of carbon to obtain enough decays in two days for a statistical uncertainty of 0.5% (corresponding to an age uncertainty of 40 years). In contrast, with a modern AMS facility one can easily obtain counting rates of 50 ^{14}C ions/sec for one hour using only one milligram of carbon in the ion source. This leads in 15 minutes to a counting statistics of 0.5%. In practical terms, the amount of sample material needed is reduced by at least a factor of 1000 and the measuring time by a factor of 100 as compared to beta counting. Such an enormous gain in detection sensitivity ($\sim 10^5$) is similar to the gain in light gathering capability of a very large astronomical telescope as compared to the unequipped eye.

Over the years, AMS has developed into an analytic tool of great versatility, with applications in almost every field of science where the measurement of minute traces of long-lived radioisotope is of interest (6-10). Table 1 gives a summary of fields were AMS measurements are performed. In this table our environment is devided into seven "spheres", each constituting a major domain on Earth and beyond. Measurements of long-lived radionuclides provide important clues for the understanding of chemical and physical processes within each sphere. Even more important, interactions between the spheres in the past and in the presence can also be studied by these AMS measurements. Information gathered in this way will be the basis for extrapolating into our future on Earth, although any of these extrapolations have to be treated with utmost care as to their reliability of firm predictions.

Table 1. The Seven Spheres of the Environment

Sphere	Areas of interest where AMS measurements of long-lived radionuclides can be performed; the respective radionuclides used are given in parenthesis.
Atmosphere	Production and distribution of cosmogenic and anthropogenic radionuclides (^{3}H, ^{7}Be, ^{10}Be, ^{14}C, ^{26}Al, ^{32}Si, ^{36}Cl, ^{39}Ar, ^{81}Kr, ^{85}Kr, ^{129}I) study of trace gases: CO_2, CO, OH, O_3, CH_4 (^{7}Be, ^{10}Be, ^{14}C) transport and origin of aerosols (^{14}C)
Biosphere	dating in archaeology and other fields (^{14}C, ^{41}Ca) ^{14}C calibration studies in tree rings, corals and sediments (^{14}C) in-vivo tracer studies in animals and humans (^{14}C, ^{26}Al, ^{41}Ca, ^{79}Se) studies in forensic medicine through bomb-peak dating (^{14}C)
Hydrosphere	dating of groundwater (^{14}C, ^{36}Cl, ^{39}Ar, ^{81}Kr, ^{129}I) global ocean circulation pattern (^{14}C, ^{39}Ar, ^{129}I) paleoclimatic studies in ocean sediments
Cryosphere	dating of ice cores and paleoclimatic studies in glaciers and polar ice sheets (^{10}Be, ^{14}C, ^{32}Si, ^{36}Cl, ^{39}Ar, ^{81}Kr) variation of cosmic ray intensity with time (^{10}Be, ^{36}Cl) bomb-peak identification (^{36}Cl, ^{41}Ca, ^{129}I)
Lithosphere	exposure dating and erosion studies of surface rocks (^{10}Be, ^{14}C, ^{26}Al, ^{36}Cl) paleoclimatic studies in loess (^{10}Be) tectonic plate subduction studies through volcanos (^{10}Be) platinum group elements in minerals (stable trace isotopes)
Cosmosphere	cosmic ray record in meteorites and lunar materials (^{10}Be, ^{14}C, ^{26}Al, ^{36}Cl, ^{41}Ca, ^{44}Ti, ^{59}Ni, ^{60}Fe, ^{107}Pd, ^{129}I); life on Mars ? (^{14}C) evidence for supernova occurrence through extinct and life radionuclides in meteorites and manganese crusts (^{10}Be, ^{26}Al, ^{36}Cl, ^{41}Ca, ^{60}Fe, ^{107}Pd, ^{146}Sm, ^{244}Pu) geochemical solar neutrino detection (^{98}Tc, ^{205}Pb) search for exotic particles (superheavy elements, fractionally charged particles, strange matter)
Technosphere	releases from nuclear industry (^{14}C, ^{36}Cl, ^{85}Kr, ^{90}Sr, ^{99}Tc, ^{126}Sn, ^{129}I) temperature measurement of fusion plasma (^{26}Al) neutron flux of the Hiroshima bomb (^{36}Cl, ^{41}Ca, ^{63}Ni) characterization of fission material (^{236}U, ^{237}Np, ^{239}Pu, ^{240}Pu, ^{242}Pu, ^{244}Pu) ultra-high purity tests of semiconductor materials (stable trace isotopes)

II. AMS WITH SMALL MACHINES

Measuring cosmogenic radionuclides at natural levels by mass spectrometry means to be capable of measuring radioisotope-to-stable isotope ratios in the range from 10^{-10} to 10^{-16}. For actual applications these extreme isotope ratios have to be measured with a precision of 0.5% for ^{14}C dating purposes, and to a few percent for other radionuclides. This requires to solve three analytical problems: i) separation of the radionuclide from interfering stable atomic isobars (e.g. from ^{14}N for the detection of ^{14}C), ii) separation from interfering stable molecules (e.g. from ^{13}CH and $^{12}CH_2$ for the detection of ^{14}C), and iii) a reliable measurement of extreme isotopes ratios. As it turns out, tandem accelerators offer by far the best conditions for AMS measurements. In particular, the most important long-lived radionuclide in nature, ^{14}C, can be measured with relative ease at tandem accelerators.

In Figure 1 a modern AMS facility is shown, the Vienna Environmental Research Accelerator (VERA)(11,12), which is based on a 3-MV Pelletron tandem accelerator. Since ^{14}N does not form negative ions (1), the otherwise overwhelming background from ^{14}N (2) is completely absent in tandem-based AMS measurements. However, the negative ion spectrum in Figure 2a measured before the entrance into the tandem accelerator shows a very large background of molecular ions of mass 14, which completely masks the ^{14}C signal. An important step in the consecutive acceleration in the tandem is therefore the stripping process in the terminal, which dissociates the $^{13}CH^-$ and $^{12}CH_2^-$ molecules very effectively when sufficient electrons are stripped off. For twenty years it was believed that $^{14}C^{3+}$ ions (or a higher charge states) must be selected to break up the molecules for sure. With the high energy analysing magnet set to select $^{14}C^{3+}$ ions, one observes the energy spectrum shown in Figure 2b. Although the residual ^{12}C and ^{13}C peaks are greatly reduced in intensity, there is still a large number of background peaks which happen to have the same magnetic rigidity as $^{14}C^{3+}$. However, Figure 2c shows that the $^{14}C^{3+}$ ions can be cleanly selected by sending the mix of ions in figure 2b through a Wien filter (see fig. 1) set to the velocity of $^{14}C^{3+}$. For details of $^{14}C/^{12}C$ ratio measurements at VERA the reader is referred to references (12-14).

It is interesting to note that the "dogma" of stripping to at least the 3+ charge state for obtaining a clean ^{14}C signal was only recently revised, although indications for a deviation were reported much earlier (15). Using a sufficiently thick stripper it is possible to destroy the molecules in the 2+ charge state at about 1 MeV (16-18). Even 1+ stripping looked feasible for obtaining a reasonable ^{14}C separation (16). The latter assumption was recently proven to work very well using a 0.5 MV Pelletron tandem in a collaborative effort of NEC and the AMS laboratory of the ETH/PSI Zurich (19). The most surprising result was the measurement of $^{14}C/^{14}C$ ratios down to a level corresponding to a radiocarbon age of 48,000 years. It therefore looks feasible to perform ^{14}C dating measurements with much smaller tandem accelerators than presently in use, approaching essentially the size of table top machines.

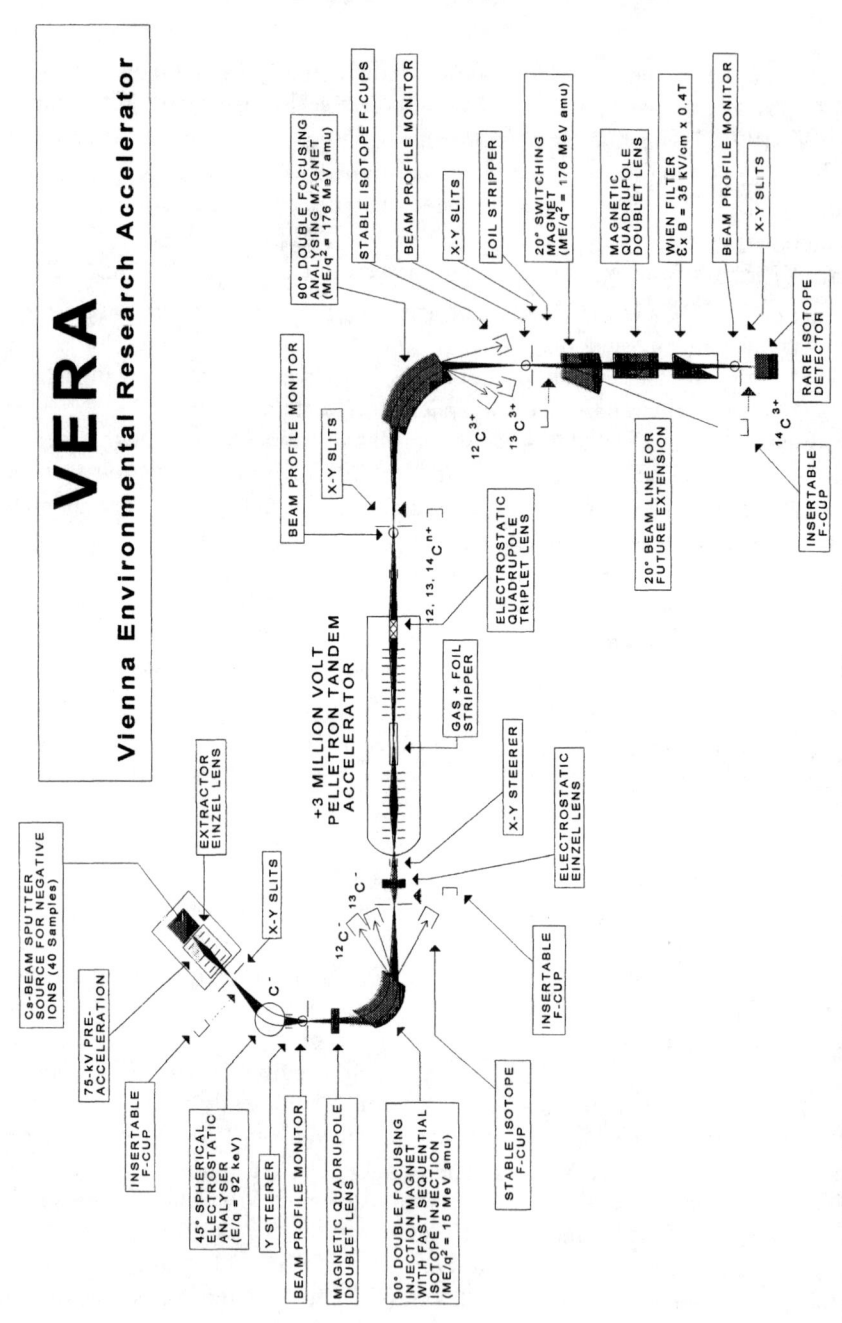

Figure 1. Schematic layout of VERA showing the essential features of the AMS system. ^{14}C measurements are typically performed at a terminal voltage of 2.7 MV.

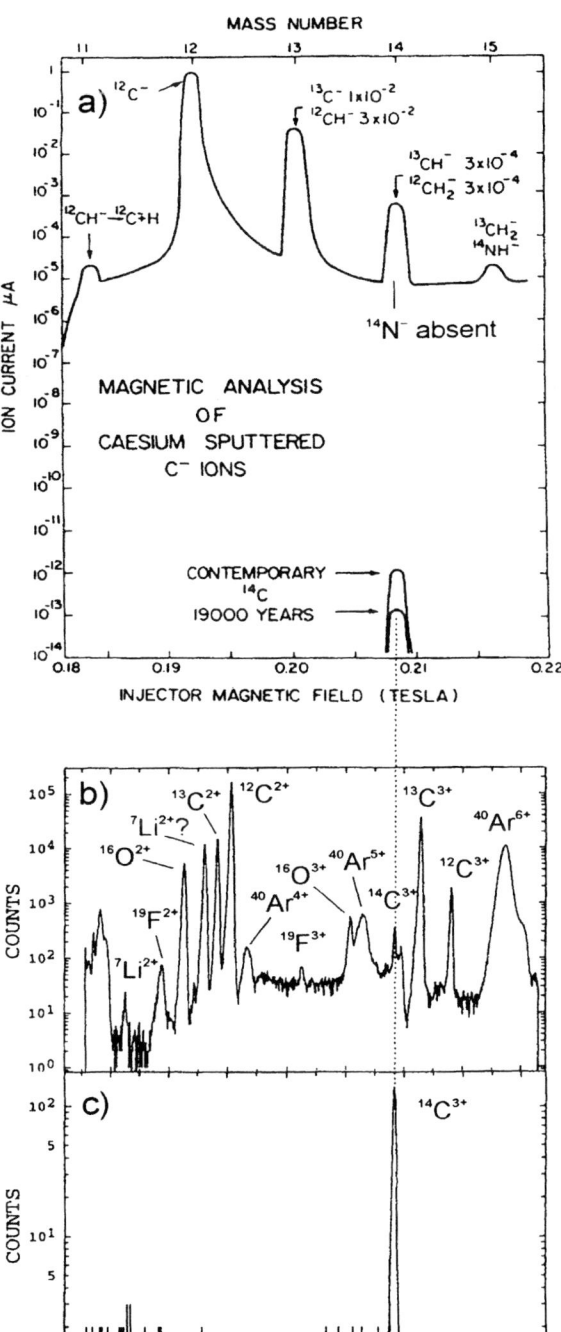

Figure 2. The three steps in the detection of ^{14}C with AMS.

a) In the negative ion mass spectrum after the ion source the ^{14}C signal is buried under an enormous background of mass-14 molecules (15). Most important however, is the absence of $^{14}N^-$ since nitrogen does not form negative ions

b) After acceleration through the tandem and analysis in the high-energy magnet the energy spectrum measured in a Si surface barrier detector still shows many background peaks.

c) After a final analysis through a velocity filter (Wien filter) a clean ^{14}C signal emerges

III. AMS WITH BIG MACHINES

In contrast to the previous section, sometimes very big accelerators are necessary to measure particular radionuclides. This was the case in developing an AMS method for measuring cosmogenic ^{81}Kr ($t_{1/2}$ = 230,000 yr) in the atmosphere (21, 22), and in ground water. Tandem accelerators cannot be used, because Kr does not form negative ions. Therefore, the experiments were performed at a positive ion machine, the K1200 superconducting cyclotron at Michigan State University (22). In order to get rid of the stable isobar ^{81}Br which strongly interfers with ^{81}Kr ($\Delta M/M$ = 3.7x10^{-6}), 17+ ions from the superconducting ECR source were accelerated to an energy of 45 MeV/nucleon (3.65 GeV). At this high energy, 80% of the ^{81}Kr ions can be fully stripped to the 36+ charge state and separated in a magnetic spectrometer from fully stripped ^{81}Br, which can only acquire a maximum charge of 35+. In order to measure small Kr gas samples (~0.4 cm^3 STP), a special gas handling system was developed (21, 22) and a comparison with pre- and post nuclear krypton was performed (Collon 1998). Since no difference between the two Kr sources was found, a first ^{81}Kr dating of groundwater from the Great Artesian Basin in Australia, the largest groundwater system in the world, was attempted. Four samples of 16,000 l of groundwater each were degassed in the field and the extracted gas (320,000 cm^3/sample) were subjected to a rigorous separation procedure at the University of

Table 2. Preliminary results of ^{81}Kr-dating of groundwater from the Great Artesian Basin in Australia

Sample	^{81}Kr/Kr [10^{-13}]	Age [yr]
Atmospheric Krypton	5.20 ± 0.70 [a]	0
Raspberry Creek	2.63 ± 0.32	225,000 ± 42,000
Duck Hole	2.19 ± 0.28	287,000 ± 38,000
Oodnadatta	1.78 ± 0.26	354,000 ± 50,000
Watson Creek	1.54 ± 0.22	402,000 ± 51,000

[a] Reference value for natural atmospheric krypton

Bern. This resulted in 0.4 cm^3 Kr /sample containing approximately 3 million ^{81}Kr atoms. Typically, 60 to 100 ^{81}Kr^{36+} ions could be counted in the final detection system, resulting in an overall efficiency of ~2x10^{-5} (atoms detected/atoms in the sample). In Table 2, preliminary results for the measured groundwater ages are listed. Although the overall efficiency is a factor of 1000 lower than the typical one achieved for ^{14}C measurements, it was possible to obtain a definite result for very old groundwater samples. Clearly, a substantial improvement in efficiency would be desirable to start "routine" measurements for groundwater samples.

IV. AMS FACILITIES WORLD-WIDE

As mentioned above, AMS originally developed at accelerators in nuclear physics laboratories. A few years after the initiation of AMS, the first generation of small dedicated AMS facilities (Tandetrons) appeared on the market (23). Eventually, a second and third generation of small machines (3 MV terminal voltage) were developed, which became the workhorse for ^{14}C measurements. Recently, several new AMS facilities based on 5-MV Pelletron tandems were established. Parallel to this development a number of larger nuclear physics tandem accelerators were upgraded for AMS measurements. Sometimes, these tandem accelerators were shipped around the world to be assembled as dedicated AMS facilities in a new location: the EN tandem from Oxford went to Peking University, the FN tandem from Rutgers University went to ANSTO in Sydney; the EN tandem from Canberra went to Lower Hutt in New Zealand, and the FN tandem from Washington University went to Livermore. In Table 3, a summary of AMS facilities around the world is given. These 47 facilities measure an estimated total of well over 100,000 samples per year, approximately 90% of it for ^{14}C. Although ^{14}C is by far the most used radionuclide with AMS, many others are gradually increasing in importance (see Table 1). It is foreseeable that eventually all long-lived radionuclides with half-lives longer than approximately 100 years will be subject to AMS measurements.

Comparing big and small AMS facility, there are a few points to be mentioned:

- ^{14}C can be well measured with tandem accelerators at TV = 2 -3 MV. The newest development (19) indicates that it is possible to use also much lower terminal voltages. True radiocarbon dating seems feasible at TV = 0.5 MV.. It may even be possible to measure ^{26}Al ($t_{1/2}$ = 7.1x10^5 yr) and ^{129}I (1.7x10^7 yr) with these mini-tandems because the respective stable isobars, ^{26}Mg and ^{129}Xe, do not form negative ions. In addition, actinides seem to be another group of radionuclides suited for small tandem accelerators because there are no stable isobars in this mass region (24).

- Isobar separation is the dominant analytic problem in AMS measurements, whenever the stable isobars do form negative ions. Here, higher energy helps greatly, and larger tandem accelerators can more easily perform measurements for interesting radionuclides such as ^{10}Be ($t_{1/2} = 1.5 \times 10^6$ yr), ^{32}Si (135 yr), ^{36}Cl (3.0×10^5 yr), ^{41}Ca (1.0×10^5 yr), ^{44}Ti (59 yr), ^{53}Mn (3.7×10^6 yr), ^{59}Ni (9.2×10^4 yr), ^{63}Ni (100 yr), ^{60}Fe (1.5×10^6 yr), ^{90}Sr (29 yr), ^{98}Tc (4.2×10^6 yr), ^{126}Sn (2.3×10^5 yr), ^{205}Pb (1.5×10^7 yr), and others.

- Noble gases can only be measured with positive-ion accelerators. As discussed above, a very big machine was necessary to remove the stable isobar ^{81}Br for the ^{81}Kr measurements.

- Finally, a possible solution to the *isobar* separation problem in connection with small accelerators may come from combining the power of elemental separation through laser ion sources (25) with a small accelerator (e.g. a cyclotron) supplying the necessary *isotope* separation.

Table 3. Facilities for Accelerator Mass Spectrometry (1998)

Country	No.	Accelerator	Location
North America			
Canada	1	2.5 MV Tandetron	University of Toronto, Toronto
USA	8	ATLAS Linac	Argonne National Laboratory, Chicago [a]
		2.5 MV Tandetron	University of Arizona, Tucson
		9.5 MV FN Tandem	Lawrence Livermore Nat. Lab, Livermore
		K1200 Cyclotron	Michigan State Univ., East Lansing [a]
		3 MV Pelletron	Naval Research Lab, Washington D. C. [a]
		3 MV Pelletron	University of North Texas, Denton
		9 MV FN Tandem	Purdue University, West Lafayette
		2.5 MV Tandetron	Woodshole Oceanographic Institution
Europe			
Austria	1	3 MV Pelletron	University of Vienna, Vienna
Denmark	1	6 MV EN Tandem	University of Aarhus, Aarhus
England	2	2.5 MV Tandetron	University of Oxford, Oxford
		5 MV Pelletron	University of York, Sand Hutton [b],
France	1	2.5 MV Tandetron	Nat. Sci. Research Center, Gif-sur-Yvette,
Germany	4	6 MV EN Tandem	Univ. of Erlangen-Nuernberg, Erlangen
		3 MV Tandetron	University of Kiel, Kiel
		14 MV MP Tandem	Tech. Univ. & Univ. of Munich, Garching
		3 MV Tandem	Forschungszentrum Rossendorf [a],
Israel	1	14 MV Pelletron	Weizmann Institute of Science, Rehovot
Italy	1	3 MV Tandem	University of Napels [a]
Netherlands	2	3 MV Tandetron	University of Groningen, Groningen
		6 MV EN Tandem	University of Utrecht, Utrecht

Table 3. (continued)

Country	No.	Accelerator	Location
Sweden	2	6 MV EN Tandem	University of Uppsala, Uppsala
		3 MV Pelletron	University of Lund, Lund
Switzerland	2	6 MV EN Tandem	Swiss Federal Inst. of Technology Zürich
		0.5 MV Pelletron	Swiss Federal Inst. of Technology Zürich [b],
Asia			
China	4	14 MV MP Tandem	Chinese Inst. of Atomic Energy, Bejing
		6 MV EN Tandem	Peking University, Beijing
		6 MV Tandem	Shanghai Inst. of Nucl. Res., Shanghai
		Mini Cyclotron	Shanghai Inst. of Nucl. Res., Shanghai [b]
Japan	9	3 MV Tandetron	Japan Atomic Energy Res. Inst., Mutsu [b]
		5 MV Pelletron	Japan Nucl. Cycle Develop. Inst., Toki [b]
		8 MV Pelletron	Kyoto University, Kyoto [a]
		10 MV Tandem	Kyushu University, Fukuoka [a]
		2.5 MV Tandetron	Nagoya University, Nagoya
		3 MV Tandetron	Nagoya University, Nagoya [b]
		5 MV Pelletron	University of Tokyo, Tokyo
		5 MV Pelletron	Nat. Inst. for Environ. Studies, Tsukuba
		12 MV Pelletron	University of Tsukuba [a]
India	1	3 MV Pelletron	Institute of Physics, Bhubaneswar [a],
Korea	1	3 MV Tandetron	Seoul National University, Seoul [b],
Australia & New Zealand			
Australia	3	8 MV FN Tandem	Nucl. Sci. and Technol. Organ., Sydney
		2.5 MV Tandetron	Comm. Sci. and Industr. Res. Organ., Sydney
		14 MV Pelletron	Australian National University, Canberra
New Zealand	1	6 MV EN Tandem	Inst. of Geolog.& Nucl. Sci., Lower Hutt
South America			
Argentina	1	20 MV Pelletron	Nat. Atomic. Energy Comm., Buenos Aires [a]
Brazil	1	9 MV Pelletron	University of Sao Paulo, Sao Paulo [a]

[a] AMS development at existing accelerators
[b] New dedicated AMS facilities in test operation

V. CONCLUSIONS

Table 3 clearly shows that tandem accelerators dominate the field of AMS. The size of these machines vary from very small (TV = 0.5 MV) to very large (TV = 20 MV). New machines are generally on the small side of the spectrum (TV = 3 to 5 MV). Often, AMS facilities developed around accelerators which simply were available. It is probably fair to say that with enough technical upgrading and modification, almost any accelerator can be used for AMS. This makes AMS a universally available technique.

The breadth of information which can be gathered with AMS in the seven spheres of our environment is enormous. Combined with high-precision stable isotope measurements, this constitutes the "isotope language", which may allow us one day to disentangle even the most complex processes in the environment. Since we can reasonably expect that the power of both AMS and stable isotope MS will increase with time, a bright future for this field lies ahead of us.

REFERENCES

1. Purser, K. H., Liebert, R. B., Litherland, A. E., Beukens, R. P., Gove, H. E., Bennet, C. L., Clover, H. R., and Sondheim, W. E., "An attempt to detect stable N⁻ ions from a sputter ion source and some implications of the results for the design of tandems for ultrasensitive carbon analysis", *Revue de Physique Appliquée* **12**, 1487-1492 (1977).

2. Muller, R. A., "Radioisotope dating with a cyclotron", *Science* **196**, 489-494 (1977).

3. Nelson, D. E., Korteling, R. G. and Stott, W. R., "Carbon-14: Direct detection at natural concentrations", *Science* **198**, 507-508 (1977).

4. Bennet, C. L., Beukens, R. P., Clover, M. R., Gove, H. E., Liebert, R. B., Litherland, A. E., Purser, K. H.,and Sondheim, W., "Radiocarbon dating using electrostatic accelerators: negative ions provide the key", *Science*, **198**, 508-510.

5. Oeschger, H., Houtermans, J., Loosli, H., and Wahlen, M., "The constancy of cosmic radiation from isotope studies in meteorites and on the Earth", in *12th Nobel Symposium on Radiocarbon Variations and Absolute Chronology*, ed. Olssen, I. U., New York: John Wiley & Sons, 1970, pp. 471-498.

6. Litherland, A. E., "Ultrasensitive mass spectrometry with accelerators", *Ann. Rev. Nucl. Part. Sci.* **30**, 437-473 (1980).

7. Elmore, D. and Philiips, F. M., "Accelerator mass spectrometry for measurement of long-lived isotopes", *Science*, 236, 543-550 (1987).

8. Kutschera, W. and Paul, M., "Accelerator mass spectrometry in nuclear physics and astrophysics", *Ann. Rev. Nucl. Part. Sci.*, **40**, 411-438 (1990).

9. Finkel, R. C. and Suter M., "AMS in the earth sciences: techniques and applications", *Advances in Anal. Geochem.* **1**, 1-114 (1993).

10. Tuniz, C., Bird, J. R., Fink, D., and Herzog, G. F., *Accelerator Mass Spectrometry: Ultrasensitive Analysis for Global Science*, Boca Raton: CRC Press, 1998, pp. 1-371.

11. Kutschera, W., Collon, P., Friedmann, H., Golser, R., Hille., P., Priller, A., Rom, W., Steier, P., Tagesen, S., Wallner, A., Wild, E. and Winkler, G., "VERA: A new AMS facility in Vienna", *Nucl. Instr. Meth.* B **123**, 47-50 (1997).

12. Priller, A., Golser, R., Hille, P., Kutschera, W., Rom, W., Steier, P., Wallner, A. and Wild, E., "First performance tests of VERA", *Nucl. Instr. Meth.* B **123**, 193-198 (1997).

13. Rom, W., Golser, R., Kutschera, W., Priller, A., Steier, P. and Wild, E., "Systematic investigations of ^{14}C measurements at the Vienna Environmental Research Accelerator", *Radiocarbon*, **40/1**, 255-263 (1998).

14. Wild, E., Golser, R., Hille, P., Kutschera, W., Priller A., Puchegger, S., Rom, W. and Steier, P., "First ^{14}C results from archaeological and forensic studies at the Vienna Environmental Research Accelerator", *Radiocarbon*, **40/1**, 273-281 (1998).

15. Litherland, A. E., "Accelerator mass spectrometry", *Nucl. Instr. Meth.* B5, 100-108 (1984).

16. Suter, M., Jacob, St. and Synal, H. A., "AMS of ^{14}C at low energies", *Nucl. Instr. Meth.* B **123**, 148-152 (1997).

17. Mous, D. J. W., Purser, K. H., Fokker, W., van den Broek, R. and Koopmans, R. B., "A compact ^{14}C isotope ratio mass spectrometer for biomedical applications", *Nucl. Instr. Meth.* B **123**, 153-158 (1997).

18. Mous, D. J. W., Fokker, W., van den Broek, R. and Koopmans, R. B., "An ion source for the HVEE ^{14}C isotope ratio mass spectrometer for biomedical applications", *Radiocarbon* **40/1**, 283-288 (1998).

19. Suter, M., Huber, R.., Jacob, S. A. W., Synal, H. A. and Schroeder, J. B., "A new small accelerator for radiocarbon dating ", to be published in the Proceedings of the Denton Conference 1998.

20. Kutschera, W., Paul, M., Ahmad, I., Antaya, T. A., Billquist, P. J., Glagola, B. G., Harkewicz, R., Hellstrom, M., Morrissey, D. J., Pardo, R. C., Rehm, K. E., Sherrill, B. M., and Steiner, M., "Long-lived noble gas radionuclides", *Nucl. Instr. Meth.* B **92**, 241-248 (1994)

21. Collon, P., Cole, D., Davids, B., Fauerbach, M., Harkewicz, R., Kutschera, W., Morrissey, D. J., Pardo, R., Paul, M., Sherrill, B. M. and Steiner, M., "Measurement of the long-lived radionuclide 81Kr in pre-nulcear and present-day atmospheric krypton", to be published in Radiochimica Acta (1999).

22. Collon, P., Antaya, T. A., Davids, B., Fauerbach, M., Harkewicz, R., Hellstrom, M., Kutschera, W., Morrissey, D. J., Pardo, R. C., Paul, M., Sherrill, B. M., and Steiner, M., "Measurement of ^{81}Kr in the atmosphere ", *Nucl. Instr. Meth.* B **123**, 122-127 (1997).

23. Purser, K. H., Schneider, R. J., Dobbs, J. McG. and Post, R., "A preliminary description of a dedicated commercial ultra-sensitive mass spectrometer for direct atom counting of ^{14}C", *Proc. Sypmosium on* National Laboratory Report ANL/PHY-81-1, 1981, pp. 431-462.

24. Zhao, X.-L., Nadeau, M.-J., Kilius, L. R. and A. E. Litherland, "The first detection of naturally-occurring ^{236}U with accelerator mass spectrometry", *Nucl. Instr. Meth.* B **92**, 249-253 (1994).

25. Van Duppen, P., "Laser ion sources for on-line isotope separators", *Nucl. Instr. Meth.* B **126**, 66-72 (1997).

High-precision Measurements of ^{14}C as a Circulation Tracer in the Pacific, Indian, and Southern Oceans with Accelerator Mass Spectrometry (AMS)

Karl F. von Reden, John C. Peden, Robert J. Schneider, Mary Bellino, Joanne Donoghue, Kathryn L. Elder, Alan R. Gagnon, Patricia Long, Ann P. McNichol, Tracey Morin, Dana Stuart, and John M. Hayes

NOSAMS Facility, Woods Hole Oceanographic Institution, Woods Hole, MA 02543

Robert M. Key

Princeton University, Princeton, NJ 08544

Abstract. The National Ocean Sciences Accelerator Mass Spectrometry Facility (NOSAMS) has completed the carbon isotope analysis of a major fraction of the 13,500 sea water samples collected in the framework of the World Ocean Circulation Experiment (WOCE) from three of the major world oceans between 1991 and 1996. We will describe the AMS technique employed at NOSAMS and, using 3-D data visualization techniques we will demonstrate the present status of the data set and offer some preliminary conclusions about the distribution of natural and anthropogenic ^{14}C in the oceans. In particular, we will be able to compare some of the data with results from the Geochemical Ocean Sections Study (GEOSECS, 1972-1978) to obtain information about the time dependence of oceanic circulation processes, tracing the ^{14}C signal introduced into the oceans during the atmospheric nuclear bomb tests in the 1950's and 1960's.

Introduction

In the late 1980's, a project solicitation by the National Science Foundation for establishing an accelerator mass spectrometry facility for the ocean sciences resulted in the award of this project to the Woods Hole Oceanographic Institution on Cape Cod, Massachusetts (1). The charge to the facility was "*to analyze radiocarbon and other tracer samples, collected during upcoming field programs of major Global Geosciences projects and individual research grants*". Led by Glenn Jones, the Woods Hole AMS group installed a state-of-the-art accelerator mass spectrometer system (2) which became fully operational in 1991. At that time one of the major recent geosciences projects, the World Ocean Circulation Experiment (WOCE) had already started collecting seawater samples for radiocarbon analysis from the Pacific Ocean. Since then, about 13,000 samples have been collected, 9000 of which have been analyzed so far. From the beginning, it was clear that the first decade of operation of this facility would be dominated by the task of analyzing the WOCE samples for radiocarbon. Plans to measure other isotopes (e.g., ^{10}Be, ^{26}Al) were put on hold and R&D concentrated on the refinement and further development of ^{14}C AMS. Interest in radiocarbon AMS analyses has increased over the last ten years, especially in light of the development of techniques to reduce AMS sample sizes from the current mg C level to less than 0.1 mg (3). This development allows the analysis of carbon specimens that are extremely difficult to assemble in large quantities such as microfossils in sediment cores. Another example for very small sample sizes is the product of the chemical isolation of individual organic compounds from natural matrices using preparative capillary gas chromatography (PCGC), currently under development at WHOI (4). In this paper we will describe the AMS method as applied at the NOSAMS facility and discuss our major program in more detail: the ocean circulation tracer analyses.

METHODS

Accelerator mass spectrometry has been established over the last two decades as a competitive tool for high-precision ratio measurements on elements with extremely rare isotopes. General descriptions and reviews of the method can be found elsewhere (5). Here we will concentrate on the specific implementation of a ^{14}C-dedicated AMS facility at the Woods Hole Oceanographic Institution. The requirements for this system were given by the need to resolve changes as small as 4 ‰ in the ^{14}C concentration down the water column and across the basins of major oceans. Logistic considerations limited the sample size to 500 mL to enable reasonable procedures for collection and handling on and off shipboard, including shipping and storage. At prevailing seawater CO_2 concentrations of about 2 mmol/kg, the weight of carbon in these samples is about 12 mg, more than enough for AMS analysis. In normal operation, less than half of the sample is analyzed while the remainder is archived as CO_2 in a flame-sealed glass container. For a description of the shipboard and sample preparation procedures refer to (6). The final step of the chemical sample preparation consists of the catalytic reduction of CO_2 to elemental carbon, using fine mesh dendritic reduced iron powder as catalyst at a gravimetric ratio C:Fe of about 0.7. A C:Fe mix of less than 1 appears to be favorable for a stable and reproducible output from our AMS ion sources. The C/Fe powder is compacted into aluminum cartridges for loading into the ion source in batches of 58 samples at a time.

AMS Apparatus

The NOSAMS facility's AMS system is the first of its generation, based on the Tandetron design (2), a parallel-fed 3 MV Cockcroft-Walton type tandem accelerator with solid state RF driver. Figure 1 shows a schematic layout of our system. A more detailed description can be found in (7). Unique to NOSAMS are the two cesium sputter ion sources and injectors which allow us to operate the system virtually without interruption, alternatively running or maintaining each ion source. This setup has proven to be helpful when one of the ion sources has needed extensive maintenance or

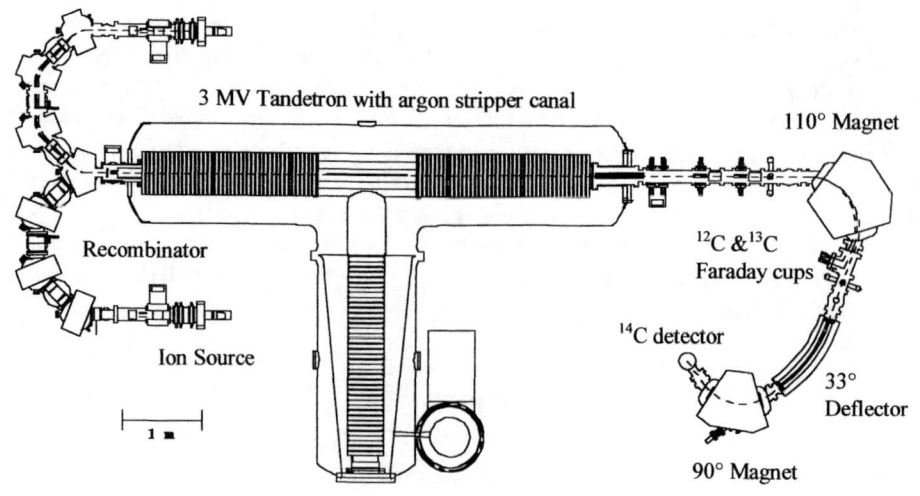

FIGURE 1: Layout of the NOSAMS ^{14}C system

repair. Rather than having to shut down the whole system we are able to continue with a one-source operation until the other source is back on-line. Another valuable feature of this setup will be the ability to develop and test new ion source designs on one injector while maintaining a near-normal operation on the other.

The negative ion sources are of the spherical-ionizer inverted-cathode design described by Middleton (8). Similar types are used in many heavy-ion accelerator labs worldwide. A discussion of our sources can be found in (9). The injectors are of the "recombinator" type (10) and incorporate a low-resolution mass filter, allowing simultaneous injection of ions in a mass range of A = 12 - 14 amu (in our case). They effectively cut down on the injected beam current by removing the C_n, the Cs_nC_n-clusters (11) and other compounds outside of the mass range. The NOSAMS system also is set up to chop the ^{12}C beam by a factor of about 90 bringing it down to the level of the ^{13}C beam. These measures taken together lead to a beam reduction by a factor of about 200 from the ion source to the entrance of the accelerator. Injected beams are of the order of 1 – 2 µA.

The accelerator tubes have inclined-field electrodes and are fitted with permanent magnets for electron suppression. At a terminal voltage of 2.5 MV, the efficiency of the argon gas stripper canal is close to 50% for stripping to the 3+ charge state and 100% for dissociating molecular ions. A major effort in the design of this system has gone into removing particles from the analyzed ion beams that have undergone more than one charge exchange (12). Beginning in the second stage of the accelerator (after the stripper canal) the tube electrodes are arranged in a specific inclined-field pattern to create a filter, first described by VandeGraaff (13): The main ion beam is slightly deflected off the axis and then brought back on axis before leaving the accelerator. Ions that do not have the correct kinematics remain off-axis and are stopped before entering the high-energy mass spectrometer. The ions of biggest concern would be those that would have the right kinematics for entering the ^{14}C channel of the analyzed beam (ME/q^2 = 15.56 MeV amu, where M is the atomic mass number, E is the kinetic energy of the particles, and q is the charge state). An example would be $^{13}C^{4+}$, charge-exchanging within the high-energy tube to q = 3+ to gain a total energy of 10.81 MeV. Given that even for modern samples the $^{14}C/^{12}C$ ratio is about 1×10^{-12}, scattering tails from ^{12}C- or ^{13}C-related beam fractions can easily overwhelm the ^{14}C channel. As it is, a substantial number of unwanted higher energy ions still enter the 33° electrostatic deflector after the main analyzing magnet. A pulse height spectrum (figure 2) taken at normal device settings for ^{14}C analysis with a surface barrier detector in place of the gas ionization chamber revealed a second strong peak at 13.1 MeV, next to the strongest 10 MeV ^{14}C peak. Other significant peaks were at lower energy and had no

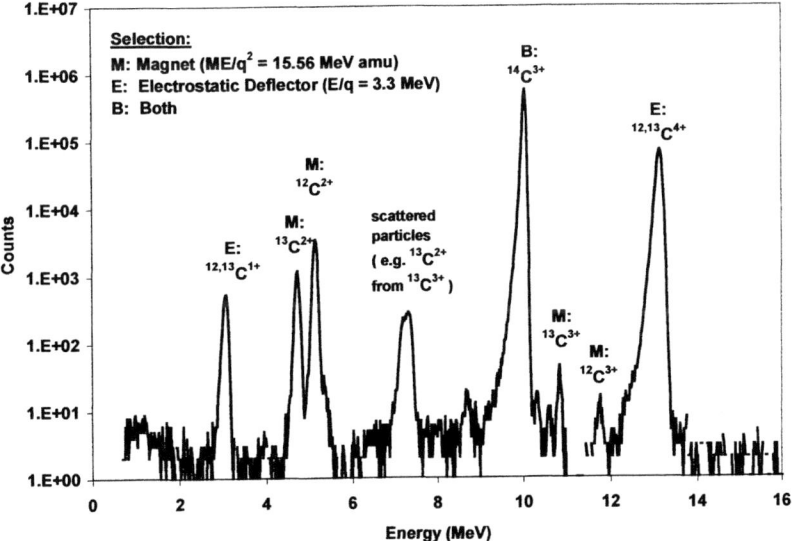

FIGURE 2: Surface barrier detector spectrum. The accelerator terminal voltage is 2.5 MV. The most probable particle properties are indicated in the figure.

possible effect on the ^{14}C channel. The 13.1 MeV signal has been identified as caused by ^{12}C or ^{13}C ions (5+ to 4+, 13.3 MeV), fulfilling the E/q condition of the deflector and inelastically scattered within the deflector. Under normal circumstances this contaminant beam (momentum ~580 MeV/c) will not be transmitted through the following 90° magnet, selecting particles of 511 MeV/c momentum at 3+ charge state. However, a final charge exchange to 3+ of the contaminant particles can bring them back into the main beam path.

As it was installed in our system, the 33° deflector had a modification in the exit area, presumably designed to stop this type of contaminant ion beam (see figure 3). After about two years of operation, the deflector began discharging heavily and did not hold full high voltage any longer. We found a beam spot on the modified deflector plate with a discharge mirror image on the opposite plate. We decided to remove the modification from the deflector plate in order to improve the long term stability of the

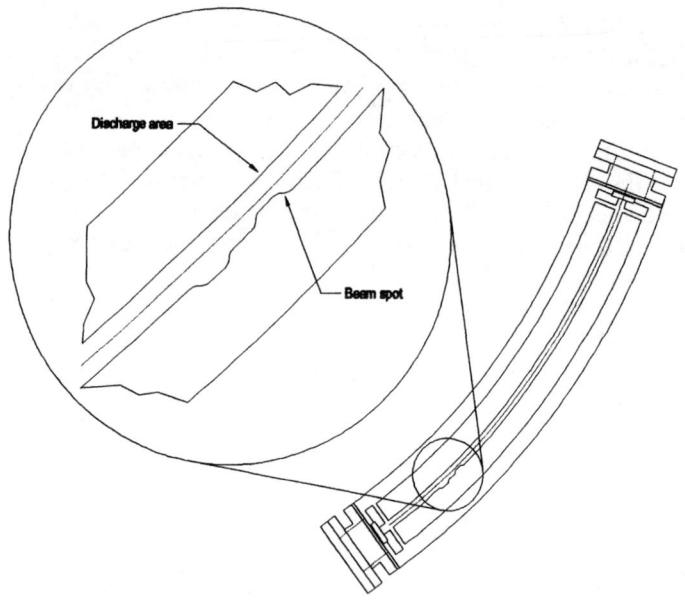

FIGURE 3: 33° electrostatic deflector in its original configuration with discharge locations marked.

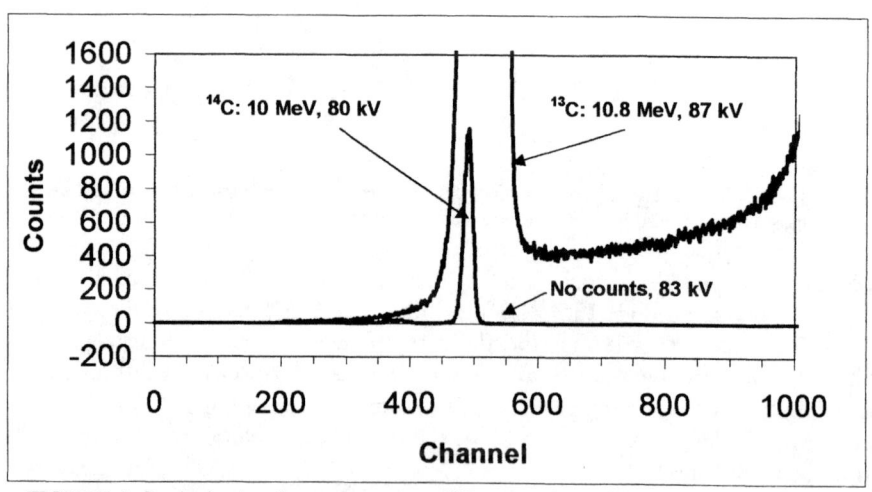

FIGURE 4: Gas ionization detector spectra at different settings of the electrostatic deflector

electric field in the deflector. Figure 4 shows three gas ionization detector spectra, acquired in 200 s with different settings of the deflector voltage in the present configuration. Clearly, the deflector still efficiently removes the higher energy particles from the beam when operated at the normal ^{14}C setting (80 kV). Only a deliberate increase of the voltage by 8 % will allow a large number of mostly ^{13}C ions to enter the final detector. Notice that at the intermediate setting (4% above normal) no particles are observed in the detector. Using a commercial carbon powder (14) as AMS sample we routinely achieve radiocarbon ages of 65,000 years or older, suggesting a background suppression in the AMS of 5×10^{15} or better. In the following section we will describe the main application of our system, analysis of dissolved inorganic CO_2 in seawater.

^{14}C AS OCEAN CIRCULATION TRACER

During the GEOSECS program a total of about 1200 samples were collected for ^{14}C analysis in the Pacific, Southern, and Indian Oceans (15). Given the sparse distribution of data points the analysis and interpretation had to rely heavily on interpolation and extrapolation across vast areas of these ocean basins. Nevertheless, the project generated a remarkable advance in our understanding of the global carbon cycle (16). In particular, the "unintentional" spiking of the oceans with ^{14}C and other radioisotopes during the atmospheric nuclear bomb tests in the 1950's and 60's resulted in a long term tracer "experiment" that is still ongoing. The advent of AMS allowed a dramatic reduction in sample size and enabled WOCE scientists to collect over 10 times as many samples in the early 1990's, which are expected to result in a data set with more specific information about large parts of the ocean basins. One of the main questions to be asked is how the bomb signal has proceeded to penetrate the water column to greater depths. Figure 5 shows a comparison of two profiles taken approximately at the same location in the South Pacific, 17 years apart. The conventional way of displaying ^{14}C concentration is used in this figure, namely as deviation from a theoretical 1950 wood standard, defined as the "modern" reference (17). It can be seen that over the 17 years the surface water down to 100 m depth has declined in ^{14}C by over 30 ‰, following the trend seen in other measurements over this time period (18). In depths between 100 and about 1200 m the signal strongly increased by as much as 80 ‰. Below 1200 m down to 4700 m no significant change has occurred, in line with the assumption that mid-depth Pacific waters (~2500 m depth) may have an apparent "age" or ventilation time of more than 2000 years. Interestingly, this figure also suggests a decadal variation in the bottom water of 5 – 15 ‰. Whether the observations in this figure can be generalized to other areas of the ocean will be discussed next, compiling the data from the local profiles into contours to obtain regional and global information about the dynamics of $^{14}CO_2$ in the ocean.

Figure 6 shows a perspective "satellite" view of ^{14}C in the upper 1200 m of the Pacific, Southern, and Indian ocean basins, looking from above the North-American continent southwest in the direction of Australia (the gap in the data in the upper part

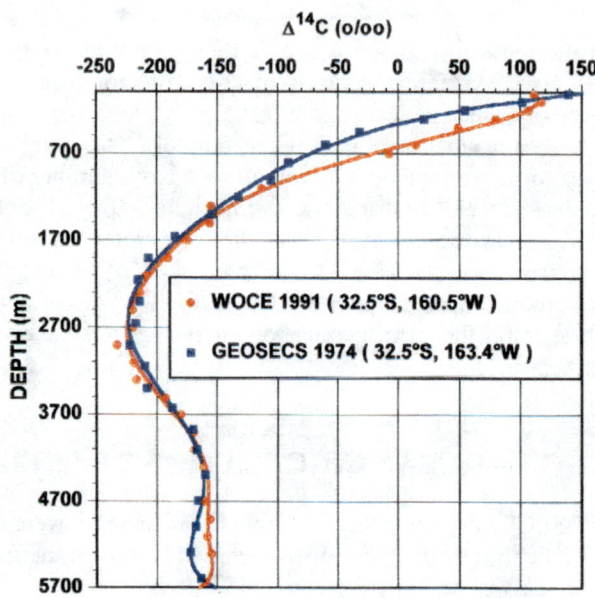

FIGURE 5: Comparison of $\Delta^{14}C$ profiles, taken at about the same location in 1974 and 1991. In this figure, GEOSECS data are from large-volume samples (200 L), WOCE data from small-volume samples (500 mL). The dot size reflects the average uncertainty of the data. The lines are drawn to guide the eye.

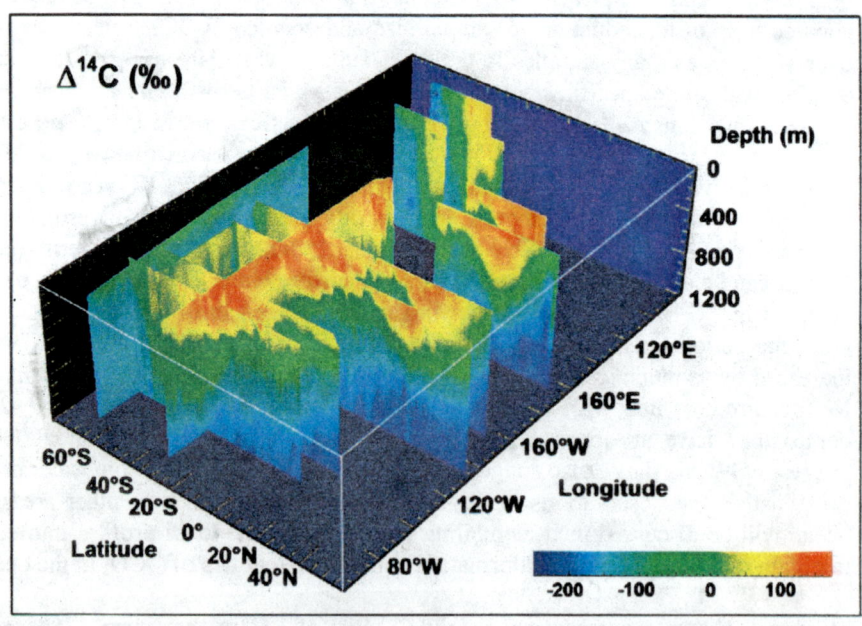

FIGURE 6: ^{14}C distribution in the upper 1200 m of the Pacific, Southern, and Indian Oceans.

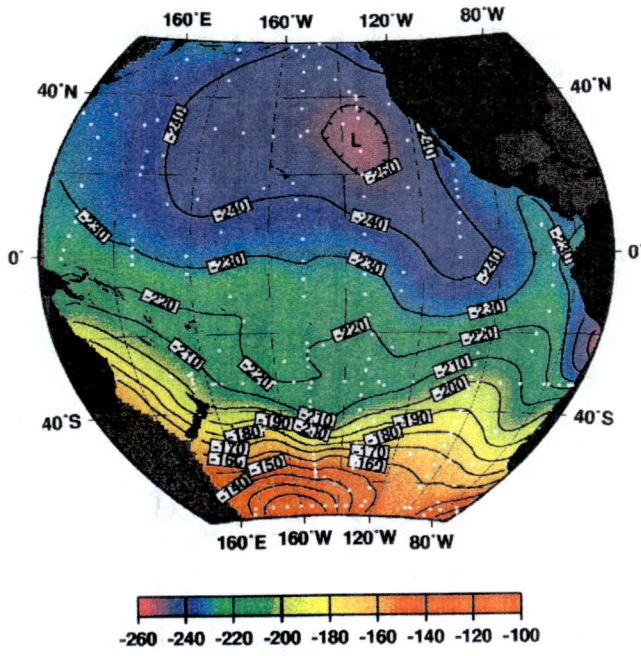

FIGURE 7: $\Delta^{14}C$ concentration at relative density $\sigma_2 = 1036.9$ kg/m^3 (~2400 m depth).

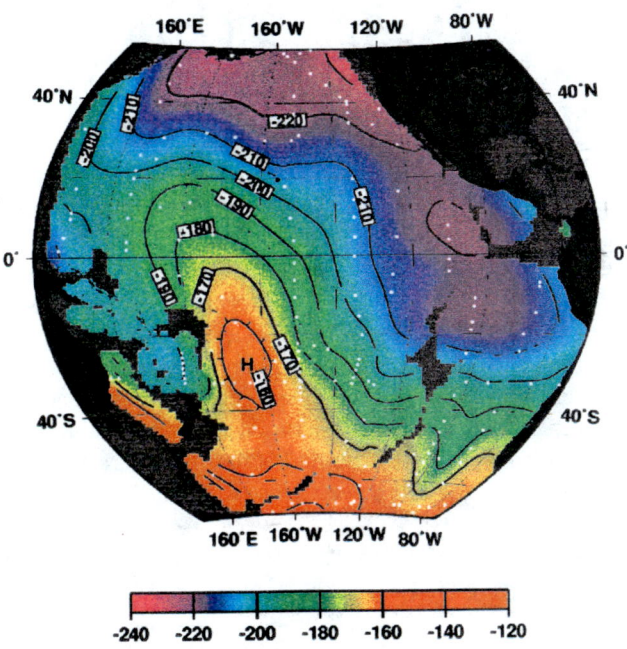

FIGURE 8: Near-bottom $\Delta^{14}C$ concentration in the Pacific and Southern Ocean.

of the figure reflects the Australian shelf). This figure is an extension of the data discussed in (19). Notice the high concentration values at the low latitudes of the northern and southern hemisphere with emphasis on the northwest and the southeast Pacific. The transition between the green and the blue contours approximately marks the penetration depth of bomb introduced ^{14}C at the time of sampling (1991-96).

In addition to the small volume AMS samples, several hundred large volume samples were taken mostly from the deep water column and analyzed by beta-counting at other laboratories (20), both to supplement and to verify the AMS analyses for a number of stations (21). The following discussion includes both types of samples from the Pacific and Southern Oceans. For compilations and discussions of some of the data see (22, 23, 24). Contour plots of the ^{14}C concentration at two levels in the water column will help illuminate the probable large-scale circulation patterns in these oceans. Because of its relatively long half-life, even naturally produced radiocarbon is a useful tracer of deep and bottom water circulation patterns and rates. Figure 7 shows Δ^{14}C on a surface of constant density (referenced to a pressure of 2000dB) which occurs at a depth of approximately 2400 meters in the North Pacific. This is near the level of minimum concentration and the level at which deep water is thought to return southward toward the Antarctic. The oldest waters, that is, the water that has been removed from contact with the atmosphere for the longest time are found in the northeastern Pacific. The return flow pathway appears to be in two regions: one just west of South America and adjacent to the continental slope and the other centered around 160°W. The very small east to west gradients in this figure demonstrate the need for high precision results.

Figure 8 shows Δ^{14}C near the ocean bottom for regions where the water depth is at least 3500 meters. While this sort of map has no quantitative use, it very clearly emphasizes the flow pathway of Southern Ocean water as it moves from the circumpolar circulation regime northward. The main flow is centered around 170°W until it crosses the equator. At that point the bottom flow is clockwise around the North Pacific basin. This circulation when coupled with upwelling and mixing is the reason for the distribution seen in Figure 7. With the exception of very near Antarctica, where new bottom waters are formed, the evidence of bottom water contamination by bomb produced radiocarbon is weak. So far the implication of bottom water contamination in figure 5 has not been verified on a larger scale in the present AMS data set. This preliminary result is in contrast to other tracer measurements (chlorofluorocarbons) which do show an anthropogenic signal in the main bottom water flow path in the South Pacific. This may change when the data are fully analyzed, but the signal will be very close to the measurement precision.

The new radiocarbon data are providing very stringent constraints on numerical ocean models. For an example for the current state of general ocean circulation models see (25). The distribution of naturally produced radiocarbon is used to gauge model performance in deep and bottom waters and on long time scales while the bomb produced component serves a similar function for decadal time scale processes in the upper water column. Preliminary investigations have indicated significant problems with the model simulations particularly in the Southern Ocean. Most of the numerical models also appear to have a problem in the air-sea flux parameterization. These models incorporated the bomb signal into the surface ocean much too rapidly and in the 30 years since the peak in the atmospheric signal that excess is carried into the upper thermocline. This leads to model concentrations that have the same general distribution as measured, but the levels are significantly in error. Graphical and more sophisticated comparisons between the new data and model calculations are expected to result in model improvement. This improvement is critical to model based predictions of global climate change. Much of expected improvement is directly attributable to the high precision and relatively high density of the new AMS data set.

CONCLUSIONS

After more than 2/3 of the U.S. WOCE samples have been analyzed, the ability of small sample AMS to provide high-precision data, comparable to large-volume beta-counting results, is not questioned any longer. The previously existing database on ^{14}C in the largest ocean basins on earth has been expanded considerably and new focus has been applied to the Southern Ocean, considered to be of great importance for the understanding of the global carbon cycle and the related climate change implications. The evolution of the bomb ^{14}C signal since its generation 40 years ago continues to be an important source of knowledge on the decadal time scale about the exchange processes in the upper water column and will eventually provide information about the deep water formation processes as well. The next few years will bring the data analysis to a conclusion and allow the results to be incorporated into refined ocean circulation models. Perhaps, as an outcome of this decade-long endeavor, a need for repeated more limited, more localized sampling will arise in areas of specific importance for exchange processes between ocean basins or atmosphere and oceans.

ACKNOWLEDGMENTS

This work has been supported by the National Science Foundation (NSF) Cooperative Agreements OC-8801015 and OC-9807266 and other grants by NSF and the National Oceanic and Atmospheric Administration (NOAA). Some of the AMS samples were collected under supervision of WOCE ^{14}C Principal Investigators Peter Schlosser (Lamont-Doherty Earth Observatory, Palisades, NY) and Paul Quay (University of Washington, Seattle, WA).

REFERENCES

1. Jones, G. A., McNichol, A. P., von Reden, K. F., Schneider, R. J., *Nuclear Instruments and Methods in Physics Research* **B52**, 278-284 (1990).
2. Purser, K. H., Smick, T. H., Purser, R. K., *Nuclear Instruments and Methods in Physics Research* **B52**, 263-268 (1990); Purser, K. H., *Radiocarbon* **34**(3), 458-467 (1992).
3. Brown, T. A., Southon, J. R., *Nuclear Instruments and Methods in Physics Research* **B123**, 208-213 (1997); Pearson, A., McNichol, A. P., Schneider, R. J., von Reden, K. F., Zheng, Y., *Radiocarbon* **40**(1), 61-75 (1998); von Reden, K. F., McNichol, A. P., Pearson, A., Schneider, R. J., *Radiocarbon* **40**(1), 247-253 (1998).
4. Eglinton, T. I., Aluwihare, L. I., Bauer, J. E., Druffel, E. R. M., McNichol, A. P., *Analytical Chemistry* **68**(5), 904-912 (1996)
5. Suter, M., *Nuclear Instruments and Methods in Physics Research* **B52**, 211-223 (1990); Davis, J. C., *Nuclear Instruments and Methods in Physics Research* **B92**, 1-6 (1994); Kutschera, W., *Nuclear Instruments and Methods in Physics Research* **B123**, 594-598 (1997).
6. McNichol, A. P., Jones, G. A., in *World Ocean Circulation Experiment – Operations Manual* **3.1.3**, (1991) *WOCE Report* **68/91**, Woods Hole Oceanographic Institution, Woods Hole, MA 02543; McNichol, A. P., Jones, G. A., Hutton, D. L., Gagnon, A. R., *Radiocarbon* **36**(2), 237-246 (1994).
7. Von Reden, K. F., Schneider, R. J., Cohen, G. J., Jones, G. A, *Nuclear Instruments and Methods in Physics Research* **B92**, 7-11 (1994); von Reden, K. F., Jones, G. A., Schneider, R. J., McNichol, A. P., Cohen, G. J., Purser, K. H., *Radiocarbon* **34**(3) 478-482 (1992).
8. Middleton, R., *IEEE Transactions of Nuclear Science* NS-**23**, 1098-1103 (1976).
9. Von Reden, K. F., " The Hemispherical Ionizer Sputter Sources at the Woods Hole AMS Facility – A Performance Review", presented at the 29th Symposium of Northeastern Accelerator Personnel, Durham, NC, Oct 12-14, 1995; URL: http://nosams.whoi.edu/papers/finsneap.pdf.
10. Litherland, A. E., Kilius, L. R., *Nuclear Instruments and Methods in Physics Research* **B52**, 375-377 (1990); Southon, J. R., Nelson, D. E., Vogel, J. S., *ibid.* 370-374 (1990).
11. Middleton, R., Klein, J., *Nuclear Instruments and Methods in Physics Research* **B123**, 532-538 (1997).
12. Purser, K. H., Litherland, A. E., *Nuclear Instruments and Methods in Physics Research* **B52**, 424-427 (1990).
13. Van de Graaff, R. J., Rose, P. H., Wittkower, A. B., *Nature* **195**, 1293 (1965); Purser, K. H., Galejs, A., Rose, P. H., Wittkower, A. B., Van de Graaff, R. J., *Review of Scientific Instruments* **36**, 453-461 (1965).
14. Alfa Aesar Johnson Matthey Co., Ward Hill, MA 01835, Graphite Powder, ultra superior 200 mesh, 99.9999%.
15. GEOSECS Atlantic, Pacific, and Indian Ocean Expeditions, Volume 7, Shorebased Data and Graphics, ed. Östlund, H. G., Craig, H., Broecker, W. S., Spencer, D., US Government Printing Office, Washington, D.C. 20402 (1987).
16. Broecker, W. S., Peng, T., *Tracers in the Sea*, 690 pp., Lamont-Doherty Geological Observatory, Palisades, NY (1982); Broecker, W. S., Sutherland, S., Smethie, W., Peng, T., Östlund, G., *Global Biogeochemical Cycles* **9**(2), 263-288 (1995).
17. Stuiver, M., Polach, H. A., *Radiocarbon* **19**(3) 355-363 (1977).
18. *Transient Tracers in the Ocean* (TTO 1980-82), Physical and Chemical Oceanographic Data Facility, SIO Ref No. 86-15/16 and *South Atlantic Ventilation Experiment* (SAVE 1987-89), Ocean Data Facility, SIO Ref Nos. 92-9/10, Scripps Institute of Oceanography, La Jolla, CA (1992).
19. Von Reden, K. F., McNichol, A. P., Peden, J. C., Elder, K. L., Gagnon, A. R., Schneider, R. J., *Nuclear Instruments and Methods in Physics Research* **B123**, 438-442 (1997).
20. University of Miami Tritium Laboratory (H. G. Östlund); University of Washington Quaternary Isotope Laboratory (M. Stuiver).
21. Key, R.M., *Radiocarbon* **38**(3), 415-423 (1996).
22. Key, R.M., Quay, P. D., Jones, G. A., McNichol, A. P., von Reden, K. F., Schneider, R. J., *Radiocarbon* **38**(3) 425-518 (1996).

23. Stuiver M., Östlund, H. G., Key, R. M., Reimer, P. J., *Radiocarbon* **38**(3) 519-561 (1996).
24. WOCE Hydrographic Data, Electronic Release (URL) http://whpo.ucsd.edu/whp_data.html.
25. Toggweiler, J.R., K. Dixon and K. Bryan, *J. Geophys. Res.* **94** 8217-8242 (1989); Toggweiler, J.R., K. Dixon and K. Bryan, *J. Geophys. Res.* **94** 8243-8264 (1989).

Application of a Compact Microwave Ion Source to Radiocarbon Analysis

R. J. Schneider, K. F. von Reden, and J. M. Hayes

NOSAMS Facility, Woods Hole Oceanographic Institution, Woods Hole, MA 02543

J. S. C. Wills

AECL Chalk River Laboratories, Chalk River, Ontario, Canada K0J 1J0

Abstract. The compact, high current, 2.45 GHz microwave-driven plasma ion source which was built for the Chalk River TASCC facility is presently being adapted for testing as a gas ion source for accelerator mass spectrometry, at the Woods Hole Oceanographic Institution accelerator mass spectrometer. The special requirements for producing carbon-ion beams from micromole quantities of carbon dioxide produced from environmental samples will be discussed. These samples will be introduced into the ion source by means of argon carrier gas and a silicon capillary injection system. Following the extraction of positive ions from the source, negative ion formation in a charge exchange vapor will effectively remove the argon from the carbon beam. Simultaneous injection of the three carbon isotopes into the accelerator is planned.

INTRODUCTION

The techniques utilized in accelerator mass spectrometry have been developed over the last twenty years, combining advances in the stabilization of high voltage electrostatic accelerators, high intensity sputter ion sources, particle detection and identification from nuclear physics, and chemical extraction of smaller and smaller amounts of carbon containing compounds for radiocarbon analysis. Samples are routinely prepared by converting an environmental, archeological or other scientific sample to CO_2 and then converting the gas to graphite in a small reactor tube, at high temperature. A catalyst is used, either powdered iron or cobalt. The resulting sample-catalyst mixture is then pressed to form a sputter target, for use in one of several types of cesium sputter ion sources. The usual amount of graphite in these targets is about one milligram. All the seawater samples described in an accompanying paper (1) contain about this amount. In addition, there are many environmental analyses being developed with much smaller sample requirements. For instance, atmospheric aerosol samples often contain less than a micromole of carbon, or carbon-containing compounds. Sediment samples are being analyzed for mixtures of compounds having both marine and terrestrial origins. (2) By measuring the radiocarbon ages of the different members of an alkane series, for example, it was determined that organic carbon was contributed from both fossil and contemporary sources. These compounds are typically separated into isolates by gas chromatography, where each molecular peak in the spectrum may contain only nanomoles of sample. Through heroic efforts, multiple cycles through these instruments, running up to a hundred passes, have been successively trapped to produce up to 20 µg of carbon. With special handling techniques, the dedicated graduate student is able to produce graphite targets from these and standard reference materials. The technology of making silica capillary columns with diameters up to 0.5 mm (called megabore columns) and in bundles of up to 900 separate 0.040 mm capillaries in a single column (Multi-Cap^{TM} column, Alltech Associates, Inc.) is rapidly advancing. These will provide higher sample loadings (several micrograms per pass) and reduce the necessity for so many trapping cycles. Equally dedicated AMS system operators can then run such targets and produce viable radiocarbon data from the small currents produced! Since a radiocarbon age depends on the ratio of ^{14}C counts from the sample to a modern reference standard, it is necessary to match the sizes of targets made from each, in order to cancel out any beam intensity-dependent effects. The accelerated ion currents are both dominated by the ^{12}C currents, of course. As in conventional mass spectrometry, ratios of ion currents are measured for ^{13}C and ^{12}C. The statistical precision for radiocarbon, however, is dominated by the number of detected ^{14}C ions. Since the abundance of the latter is a maximum of 10^{-12} for a modern sample, it is necessary to detect them with as high an efficiency as possible. Typically 1% of the atoms in a sputter target can be detected by a well designed AMS system, if the target is totally consumed.

Faced with the above considerations, we began looking into a microwave ion source starting in 1996. Our first experiments with the AECL Chalk River permanent magnet ion source have been described recently (3,4). We designed a special laminar flow device out of silica capillaries, to inject small pulses of sample gas (CH_4 or CO_2) into a carrier gas at constant pressure. It was found that a stable plasma discharge and stable ion beams of singly charged carbon could be produced this way. This work came to an abrupt halt, when the TASCC laboratory was shut down in March of 1997, by the Canadian government. Through the courtesy and efforts of the University of Toronto, we have been able to resume our experiments at Woods Hole, with the loan of the compact permanent magnet microwave plasma ion source.

METHODS

A detailed description of the ion source, as it was designed for high currents (60 mA of hydrogen) has recently been published by Wills, et al.(5). Basically it consists of a simple plasma generator with an inexpensive 2.45 GHz magnetron-based microwave drive system, a solenoidal magnetic confinement field, and a high-perveance accel-decel extraction column. We have substituted a small 250 W magnetron power supply (Astex Model S-250) for the original 1kW power supply. The solenoidal field is supplied by a permanent magnet array, consisting of 132 neodymium iron boron magnets. The field on axis is 930 G, and it is uniform to 1% over 5 cm diameter. A correction coil is also used to locally trim the field. A tapered ridged waveguide leads into the water-cooled copper plasma chamber through an aluminum nitride window. A choice of plasma chamber diameters is available, up to 5 cm. The length is also variable, up to 7 cm. We are presently using a configuration having a gas volume of 137 cc in the copper chamber. As a result of the earlier experiments, wherein we injected rectangular pulses of gas and found a certain amount of tailing in the ion pulses, we have decided to minimize the volume of the plasma chamber, and may use a quartz insert to confine the gas volume.

We plan to make use of a window chamber to visualize the plasma discharge under various conditions. For this, it will temporarily be necessary to substitute field coils for the permanent magnet array, in order to leave space for the viewing ports. We have recently set up the ion source in our AMS laboratory at Woods Hole, and are presently seeking an einzel lens and a suitable magnet to analyze the extracted beams. Up until now, we have been limited to detecting the plasma discharge by using a residual gas analyzer. This instrument (Ametek Dycor Model 150) consists of a small quadrupole mass analyser, located in the pumping system for the ion source. It is possible to scan the mass range of the gas molecules emerging from the plasma chamber, as power is switched on and off. With CO_2 flowing into the chamber, at a rate of 1 ml/m, a clear indication of the breakup of the mass 44 peak, and growth of the mass 12, 28 and 32 peaks appeared. This showed the presence of carbon, carbon monoxide and (recombined) oxygen, products of the dissociation of the carbon dioxide molecule.

Once we have set up the system with a lens and an analyzing magnet, and activated the extraction column, the gas-sample injector (rebuilt from the experiments at Chalk River) will be coupled to the ion source and the response to square waves of CO_2 injected into the carrier gas will again be determined. The magnetic analysis system will be utilized to separate the carbon ions produced from the argon ions of the carrier gas. We hope to reproduce or exceed the 14% conversion efficiency for carbon ions produced per atom, which we obtained in 1997. Following production of beams of C^+ an alkali vapor charge exchange cell will be installed to measure the efficiency of producing negative ions from the analyzed positive ion beam. (In future operation for AMS, the charge-exchange canal may instead be located immediately following the ion source.) Testing to minimize holdup times in the ion source, with memory effects being examined by use of ^{13}C-labeled inputs, will then commence. Some of the parameters which are expected to influence these effects include the surface conditions and materials of the plasma chamber, with and without a quartz confinement tube. The operating temperature and gas volume of the source will also be optimized.

SUMMARY

When these developments are completed, one of the existing Cs-sputter ion sources will be removed from the AMS and the compact gas ion source will be installed in its place. The recombinator injector for the AMS system operates at 40 keV energy for the carbon ions to be suitably injected into the accelerator. We will use this extraction voltage for the ion source. If an overall efficiency of 1% (C ions collected per molecules of CO_2 introduced) can be obtained, it should be possible at the 95% confidence level to distinguish between two 50-nmol samples of C (600 ng C each), one with fraction modern = 1.000 and the other with fraction modern = 0.843. Standard deviations obtained in actual measurements of 10-µmol (120µg C) samples of graphite using the Cs-sputter source are 5-10‰. It is hard to beat the performance of the sputter source for samples of this size or larger. However, a successful gas source would be applicable to 100-nmol samples of C, a range in which even dilution techniques fail. For samples smaller than 3µmol C, it is likely that a gas source would be advantageous in terms of speed and numbers of analyses that could be contemplated. Also, for samples large enough that ion source efficiency was not an issue, the direct analysis of CO_2 would reduce costs by avoiding graphitization.

ACKNOWLEDGMENTS

We thank the AECL, Chalk River Laboratories and Isotrace Lab, of the University of Toronto, for making the ion source available to us. This work is partially supported by a Senior Technical Staff Award to one of us (RJS) by Woods Hole Oceanographic Institution. The NOSAMS (National Ocean Sciences AMS) facility is supported by the National Science Foundation, under Cooperative Agreement OCE-9807266.

REFERENCES

1. von Reden, K. F., Peden, J. C., Schneider, R. J., et al. These proceedings.
2. Eglinton, T. I., Benitez-Nelson, B. C., Pearson, A., McNichol, A. P., Bauer, J. E., and Druffel, E. R. M., *Science* 277, 796-799 (1997).
3. Schneider, R. J., von Reden, K. F., Wills, J. S. C., Diamond, W. T., Lewis, R., Savard, G., and Schmeing, H., *Nuclear Instruments and Methods in Physics Research* B123, 554-557 (1997).
4. Schneider, R. J., Hayes, J. M., von Reden, K. F., McNichol, A. P., Eglinton, T. I. and Wills, J. S. C., *Radiocarbon* 40, 95-102 (1998).
5. Wills, J. S. C., Lewis, R. A., Diserens, J., Schmeing, H., and Taylor, T. *Review of Scientific Instruments* 69, 65-68 (1998).

Neutral Injection for Radioactive Ion Beams and Accelerator Mass Spectrometry

A.E. Litherland, K.H. Purser and H.E. Gove

University of Toronto, Toronto, Ontario, M5S 1A7, CANADA

ABSTRACT

Neutral Injection was first used forty years ago for the acceleration of He ions by tandem accelerators, before the advent of the negative He ion source. Later, almost universal sputter-negative-ion sources were developed and the use of neutral injection fell into disuse for nuclear physics with tandem accelerators. With the advent of Accelerator Mass Spectrometry (AMS) in 1977 the idea of neutral injection was revived briefly. Now that AMS has matured and Radioactive Ion Beam (RIB) acceleration is being undertaken for nuclear physics, we propose the revival of a modern form of Neutral Injection. The use of resonant electron transfer for neutralization is discussed, as is the scattering degradation of the ion beam during neutralization and re-ionization. We show that the process is suitable for some types of AMS and the acceleration for RIB of positive ions, that form negative ions with difficulty, to a few MeV.

INTRODUCTION

Many similar problems are encountered in the use of Radioactive Ion Beams (RIB) (1) for nuclear physics and in the measurement of the abundance of long-lived radioisotopes by Accelerator Mass Spectrometry (AMS) (2). In both cases it is necessary to maximize the accelerated ion currents used, so that in RIB, low cross section nuclear reactions can be studied and, in AMS, to allow the very low abundance of the radioactive species to be measured. For nuclear reactions, a flux of at least 10^{12}/s would be highly desirable to get an adequate counting rate per second of reaction particles. Such a flux in AMS would give about 1/s for natural ^{14}C in contemporary materials, although this is barely adequate by today's AMS standards.

A second point of similarity is the need to separate atomic isobars. Radioactive species usually have at least one stable isobar and in the study of nuclear reactions, these must be attenuated in the RIB by some factor, which depends upon the relative cross sections of the isobars with the target nucleus. A factor of 10^2 would be acceptable in many cases. In

AMS the separation of isobars must be much higher and often must be well over a factor of 10^6. In AMS (3) such separation methods are one of the main problems encountered.

Finally, in both cases the separation of molecular isobars is essential because of their interference with the measurements. In the case of RIB the use of higher charge states for acceleration ensures that this problem is minimized, but not necessarily eliminated; in the case of AMS this problem is one that must be solved as well as possible. This implies that an accelerator should be incorporated into the mass spectrometry; small tandems have been found to be sufficient in many cases. Some of the problems encountered in RIB and AMS are rather different and it is instructive to review them, as such a discussion can suggest new ways of finding solutions. Small tandems are very useful for the initial stages of acceleration because they can accelerate all masses to several MeV, unlike Radio Frequency Quadrupoles (RFQ). However present tandems need the injection of negative ions and many atomic species are difficult to convert efficiently to negative ions. In addition, much higher energies are often needed for RIB, so that in general, tandems will be useful only for the first stage of acceleration.

In this paper we will explore the recent proposal for the re-introduction of the use of Neutral Injection as an option for AMS (4) and its relevance to the production of ion beams for RIB. Neutral Injection was introduced for nuclear physics uses many years ago (5,6 and 7) but was dropped following the successful development of powerful negative helium ion sources using charge changing in Li, Na, K and Cs, together with 'Universal' sputter negative ion sources (8,9,10 and 11).

The central advantage of neutral injection over negative ion injection is that it becomes possible to produce efficiently continuous beams of positive ions of all species with energies ranging from 2.5 MeV for light ions to 5 MeV for heavy ions.

NEUTRAL INJECTION FOR NUCLEAR PHYSICS AND AMS

The first tandem accelerators could not accelerate helium ions for nuclear physics until the development of high-current He^- ion sources, and so, initially He was accelerated using the neutral injection mode. This method was used for only for a few years because of the development of suitable negative ion sources. Also, because for neutral/positive operation, the injection had to be carried out at energies to maximize the high charge states of the final ion currents. This made the study of nuclear reactions possible. 400 keV helium and 900 keV neon neutrals were required to do this. Using these energies, the neutral atoms reaching the stripping canal of the tandem accelerator could become efficiently doubly charged helium ions and in the case of neon, triply charged ions. These became 10 MeV He^{++} and 15 MeV Ne^{+++} after further acceleration by a 5 MV terminal voltage. The neutral injector used was a 1 MV single ended Van de Graaff (12) and the positive ions were neutralized in hydrogen or some other low-pressure gas. The neutralization of He^+ and Ne^+ at such high energies has a much lower cross section, than at 50 to 100 keV, and their stripping in the terminal stripper gas also required enough gas to reach charge state equilibrium. Such a procedure, together with the long distance to the tandem terminal, combines to make this particle transport process somewhat inefficient, but it was satisfactory until the development of suitable negative ion sources of He⁻.

The long distance to the tandem terminal implies that the ion optics must have very low aberrations, as a large magnification has to be used. In addition, the multiple scattering during the re-ionization is large because of the need to generate high positive ion charge states. High charge states generally exploit lower cross sections and consequently more stripper gas pressure must be used. It is therefore not surprising that neutral injection for nuclear physics was dropped when powerful negative ion sources became available.

In contrast to this early nuclear physics experience, AMS does not always require high energy (13), but only those energies necessary to destroy adequately any molecular interference. In a number of cases, a small 3 MV terminal voltage tandem is sufficient for this purpose, provided the isobar separation problem can be solved by the use of negative ions, such as is the case for ^{14}C, ^{26}Al and ^{129}I. In the case of AMS measurements for stable atoms, 3 MV is suitable for nearly all of the periodic table. In addition, the neutralization can then be at the optimum energy for the production of neutrals, which is anywhere between 50 and 250 keV. The re-ionization can be done efficiently in the tandem terminal to produce singly-charged or even the doubly-charged ions, which then generate MeV energy ions suitable for the efficient destruction of the molecules outside the accelerator. Finally, in such a machine there is a requirement for high voltage gradients because the shorter the distance to the stripping canal, the smaller the effect of ion optical aberrations will have on neutral beam transmission.

Such a specialized Neutral Injection Tandem for AMS (4) can be optimized far better than one for nuclear physics use. As a result it could also, in principle, be used as a pre-accelerator for all radioactive ion beams for further acceleration to energies suitable for nuclear reaction studies. These ideas will be amplified in the following sections.

SCATTERING DURING THE NEUTRALIZATION OF POSITIVE AND NEGATIVE IONS

The cross sections for the formation of neutral atoms from singly charged ions are very large, when ions collide with the stable atoms of the same species. This is the process known as resonant neutralization (14 and 15). However, provided the binding energies of the electrons to the projectile ion are close to those of the ground or an excited state of the bombarding or bombarded species the cross sections are also very large. An example of the latter is given in (16), where Ar^+ becomes Ar^0 in Rb vapor. The cross section leading to the meta-stable state in Ar at ~ 11.548 eV is very large. As the lifetime of this meta-stable state is long, the re-ionization in this case to produce positive ions will be particularly easy.

The cross sections for these resonant processes have been discussed theoretically and compared with experiment in recent publications (15 and 17). They are often in the vicinity of $5 \times 10^{-15} cm^2$/atom, which implies that a large fraction, $\sim 63\%$ of the ions, will be neutralized in a vapor with an areal density of only 2×10^{14} atoms/cm^2. At the scattering angles of interest to the accelerator designer (see next section), neutralization at such low pressures, will be accompanied by mostly single scattering, and not multiple scattering. Unfortunately few, if any, detailed studies have been made of the angular distributions accompanying the neutralization. However, H^+ on H (18 and 19) and Ar^+ on Ar (16) are the exceptions. The study of Ar^+ on Ar at 75 keV is particularly instructive. The angular distribution of the neutralized particle shows no measurable angular distribution compared to that in the channels leading to Ar^+, Ar^{++}, Ar^{+++} etc. In fact the sum of the differential cross sections of all charged channels is equal to that calculated for the angular distribution of screened Coulomb scattering of Ar^+ by Ar. These charged channels are relevant only when more than one scattering is important and they affect mainly the contribution at the smaller angles. The scattering of the ions during resonant neutralization (15 and 17) is probably very small.

Because of the fact that the neutrals cannot be focused and continue in straight lines from the point of neutralization to the stripping canal, focusing is essential before neutralization to direct them through the stripper. This inevitably involves the introduction of some degree of magnification between source and stripper. For example if the focusing lens has to focus on a stripping canal 3 m away, then a magnification of about three would be needed if the object of the lens were 1 m from the lens. Small errors in focusing then become important as an angular error of 1 mr in the lens becomes 3 mm at the stripping canal. Likewise any small angular spread introduced by the neutralization becomes magnified so that a 1 mr half-angle for the total intensity becomes, at 3 m, a half width of 3 mm. It is consequently very important to minimize the drift distance, scattering and ion optical errors; a good design will do this. Estimates of these effects have been given earlier (4). With a 3 m long "acceleration" tube a voltage of 3 MV should be readily attainable. Using re-ionization to the 1^+ charge state, 3 MeV ions will be readily available.

No experimental studies adequate for the detailed design of such systems have been made, but the use of near resonant electron transfer is expected to be a less significant scattering problem than that for the later re-ionization. The re-ionization cross sections are lower and consequently more gas is needed, leading to greater scattering. Fortunately, the further acceleration of the ions from say 100 keV to 3 MeV multiplies the energy by about a factor of 30 which reduces the final angular spreads by about a factor of 5.5.

The optimal design of a neutral injection tandem, to maximize the useful voltage gradient, will probably preclude the possibility of accelerating charged particles through first half of the tandem. Because of this, when it is advantageous to produce negative ions initially, e.g. when interfering isobars have electron affinities less than or equal to zero, it is worth noting that the cross sections for neutralization of negative ions are even larger than are those for positive ions (20). Consequently, neutral injection is still a very viable option for isotopes such as ^{14}C.

RE-IONIZATION OF THE FAST NEUTRAL ATOMS

In this section the theory of multiple, plural and single scattering will be discussed, as it is particularly relevant to the problem of re-ionization.

The fast neutrals must be re-ionized in the stripping canal of the tandem accelerator if the ions are to be further accelerated above the many keV, needed for neutral production, to the MeV energies needed for RIB and AMS measurements. The multiple, plural and single scattering of such neutral atoms, and their resultant ions during the re-ionization process in the stripping canal gas now becomes important. While all basic scattering studies usually start with ions, and in some cases these have been very detailed (21), there are no direct studies of such scattering starting with neutrals.

The amount of scattering during re-ionization of the fast neutral atoms generated by a neutral injection system is an important quantity to estimate because the cross sections are smaller, by about a factor of ten, than are those for resonant neutralization. As a result, more gas must be used in the stripping canal than is needed for neutralization, and consequently there will be more scattering. The angular distributions for the individual collisions for both neutralization and re-ionization collisions are, in general, unknown, so they can be assumed to be approximately given by those of screened Coulomb scattering. This effect has been estimated in order to provide some guidance to the design of a neutral injection system for Accelerator Mass Spectrometry (AMS). The basic theory of Sigmund and Winterbon (22) has been used together with guidance from the valuable papers from the groups at Aarhus (23, 24 and 25).

The single scattering of ions in the energy regime used by a neutral injector, which is from about 50 keV to maybe 150 keV, has been shown to be described approximately by a power law somewhere in between that of Thomas-Fermi and Lenz-Jensen (26). In fact, a potential proportional to the inverse fourth power of the distance is a good approximation to the data and gives an angular distribution, which varies as the inverse 5/2 power of the angle, which is singular at the origin. This implies that multiple scattering dominates at the smaller angles. The contributions to the scattering at larger angles come from both multiple small angle scattering and plural medium angle scattering. At larger angles the contribution will be mainly from single scattering. All these contributions from the scattering must be taken into account if an estimate is to be made of the losses due to the use of charge changing collisions at all angles. Fortunately, the energies and angles of importance to neutral injection can be treated another way. From theory (22) it can be shown that if α_R, the reduced angle of scattering, is larger than the reduced thickness τ, the scattering is dominated by single screened Coulomb scattering.

From a neutral injector designer's point of view, the angular distribution is less useful than the integrated angular distribution over the azimuthal angles. This integration has been done for both the Thomas-Fermi and the Lenz-Jensen scattering and a graph, shown in Figure 1, has been prepared to help in the estimation of the losses due to multiple scattering. The axes are the reduced scattering angle and the reduced thickness of the scattering gas. The quantity F is the fraction of the ions lost in scattering through all angles greater than α_R.

The reduced scattering angle, α_R, is given by:

$$\alpha_R = \left[\frac{Ea}{(2Z_1 Z_2 \varepsilon^2)}\right]\alpha$$

The reduced thickness is given by:

$$\tau = \pi a^2 Nt$$

In this equation, the screening radius, a, is given by:

$$a = \frac{0.0468}{\left(Z_1^{2/3} + Z_2^{2/3}\right)^{1/2}}$$

Here a is in nanometers; E, the ion energy is in eV; α is the laboratory angle of scattering in radians, beyond which the fraction of the ions, F, is lost after traversal through of gas stripper of reduced thickness τ. Z_1 and Z_2 are the nuclear charge of the projectile and the target and Nt is the areal density in units of 10^{14} atoms/cm^2. ε^2 is proportional to the square of the electronic charge and is given by:

$$\varepsilon^2 = 1.44 \text{ eV nm.}$$

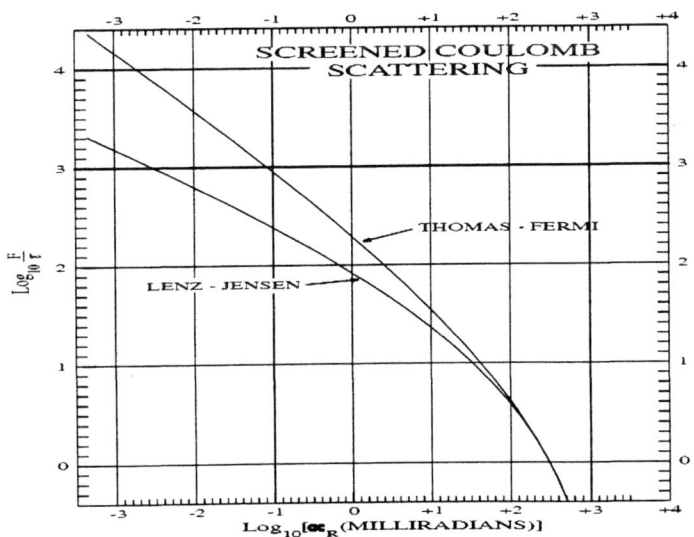

FIGURE 1.

The use of the graph can be illustrated by an example. If the cross section for the re-ionization of U neutrals to U^+ is about 2×10^{-15} cm^2/ molecule, in oxygen at 100 keV (27) then, for an areal density of 5×10^{14} /cm^2, about 9% of the ions would be lost by single scattering beyond 6.7 mr for the laboratory angle. For this calculation, the reduced angle is assumed to be 0.003, and the reduced thickness 0.0014. The scattering will then be dominated at 6.7 mr by single scattering. The losses depend sensitively upon the form of the potential and range from 6.8% for Lenz-Jensen to 12% for Thomas-Fermi. Observation indicates that the actual potential shows shell effects and is nearer that of Lenz-Jensen. Note that the further acceleration from say 100 keV to 250 keV drops the angle from 6.7 to 4.2 mr and from 100 keV to 3MeV from 6.7 to 1.2 mr.

Re-ionization after neutral injection certainly introduces an angular spread and the estimates that can be made with the help of Figure 1 are useful for exploring the various possibilities. However the detailed distribution, of the re-ionized U atoms in the above example, within the 6.7 mr cone cannot be deduced from Figure 1 and the more detailed distributions given by Sigmund and Winterbon (22) must be used.

An approximate expression for F_{out}/τ, using a Thomas-Fermi screened Coulomb potential, is given by

$$\frac{F_{out}}{\tau} = \frac{2.733}{\left(a^{0.622} + 7.176 a^{1.244} + 3.749 a^{1.866}\right)}$$

This expression is valid for $a<0.1$.

Using a Lenz-Jensen potential,

$$\frac{F_{out}}{\tau} = \frac{7.644}{\left(a^{0.382} + 3.0956 a^{0.764} + 10.925 a^{1.146}\right)}$$

This is valid for $a<0.01$

In both cases, most regions of interest are well within these limits. The leading term is related to the exponent of the power law of the potential. In this case V(r) is proportional to $1/r^{2m}$. The parameter m is 0.311 for Thomas-Fermi and 0.191 for Lenz-Jensen.

NEUTRAL INJECTION FOR AMS AND RIB

In previous sections it has been shown that the scattering during the two charge changes necessary for accelerating the ions to higher energy using Neutral Injection (NI) can be controlled so that losses are acceptable. This is done, by ensuring that large cross section processes are used; thereby minimizing the number of gas atoms encountered (27). Such losses also occur in AMS using negative ions where the transition from negative to positive is accompanied by both charge division and losses due to multiple scattering, especially for the heavy elements. However, these losses, which can be as high as 50% for heavy ions are not disastrous and can be reduced by improvements in stripping canal design.

Care has to be taken also with the subsequent mass spectrometry design (29) because of the need to remove the molecular ion fragments. In some cases further isobar separation is possible after acceleration using range or energy loss measurements (3,13). In others, the use of negative ions can separate isobars near the ion source due the instability of the negative ion of the stable isobar; ^{14}C, ^{26}Al and ^{129}I are three well-known examples (28). ^{202}Pb is another, which has not yet been exploited. If these negative ions are derived from positive ions, by charge changing in vapors, another scattering occurs. However, this is not necessarily a disadvantage as it implies only another optimization with some significant advantages conferred by the use of positive ions initially. One of these advantages is that the negative ion sputtering of insulators gives stable beams (30), unlike the use of Cs$^+$ sputtering. Charge changing to negative ions in Li vapor has been used for ^{26}Al detection (31).

The use of neutral injection can avoid the need to make negative ions. Of course, for AMS, this re-introduces the problem of isobar separation for the isotopes listed above. Neutral injection also lowers the final energy, making the use of range or rate of energy loss methods less effective. However, all AMS measurements do not require isobar separation. Two well-known examples are the study of rare elements such as Platinum Group Elements (PGE) in minerals, and the study of the actinides. In the former case, the level of the PGE is rarely below ppt, and the main problem encountered is the interfering molecular ions. In the latter, the actinides themselves are the longest lived radioactive elements, and their isobars are the shorter lived elements. There are exceptions, of course, such as ^{236}U and ^{236}Np, but these are few and are usually not important. While the example quoted could be a formidable one to solve, fortunately the isotope ^{236}Np was created only during a thermonuclear weapons explosion whereas ^{236}U is created mainly in reactors, although it does occur naturally at low concentration (32).

An examination of the equilibrium fraction compilation by Wittkower and Betz (33), shows that while the most efficient use of NI AMS results in the generation of singly charged ions, a substantial flux (>25%) of doubly charged ions is available for the heavier elements. An optimal strategy is to first carry out an energy and mass analysis of the singly charged ions after the accelerator, and then charge change to triply charged ions in some suitable gas such as He. This can be followed by further energy and mass analysis prior to ion detection, to eliminate the molecular interference, provided the usual precautions are taken.

Atomic isobars will, of course, still be a problem, but the use of resonant electron transfer already attenuates the isobar because of the different electron binding energy. Strategies for enhancing this attenuation have already been discussed (34). It is, of course, possible to neutralize negative ions easily and so take advantage of the isobar separation provided by their use.

Neutral Injection for RIB studies, with ions having charge of 1^+, would lead to only up to 3 MeV ions. This implies that only the lightest ions could be used for nuclear reaction studies. However, as an injector for further acceleration, by, for example, a Radio Frequency Quadrupole prior to a Super Conducting Linac, NI could be very useful especially for the heavier ions. This could make the acceleration of the heavier neutron rich nuclei for nuclear reaction studies somewhat easier. Detailed studies of this option are needed. However the NI AMS system can also be used for the analysis of the rare nuclei generated by nuclear reactions by detecting the reaction products. This option also requires study but all the techniques of AMS can be applied to the study of such beams and the frontiers of research extended. For example the nuclei near the neutron drip line, which are generated by fission, have lower Q-values and so become rare very quickly. These could be prepared for further study by exploiting the techniques of NI AMS, which does not require the frequently difficult stage of converting a positive ion into a negative ion. The fission by photons as a source of neutron rich nuclei has been proposed recently (35). In this process the extra energy provided by the photon should enhance the nuclei near the neutron drip line, which then in turn could be prepared for further analysis by NI AMS.

CONCLUSIONS

There are strong indications that a tandem especially designed for neutral injection could be an efficient way to accelerate ions to energies of about 3 MeV, irrespective of mass. Such a procedure makes it possible to accelerate ions from an ion source near ground potential, a scheme that has all the advantages of the negative ion source of the conventional tandem. The efficiency per positive ion created in the ion source for generating 3 MeV heavy ions could be near 40%, a result of the losses in two charge changes. However, these charge changes also will also attenuate the molecular ions that accompany the atomic ions, thanks to the structural changes that take place as the number of electrons is changed, and thanks to the Franck-Condon principle. If further charge changing to triply charged ions is undertaken, a further loss of efficiency to about 16% per positive ion will occur, but now there are no molecular ions accompanying the triply positive atomic ions. Molecular dissociation will be complete so that high-mass resolution is again unnecessary. The use of doubly charged atomic ions will ensure the disintegration of more molecules than singly charged and may be an option for the intermediate abundance elements, say down to 1ppt or so.

The efficient acceleration of positive ions in a mass independent way may make the Neutral Injection Tandem a competitor for the early stages of heavy ion acceleration for RIB.

REFERENCES

1. Gelletly, W., "Entering a new era: beams of radioactive nuclei", *Phil. Trans. R. Soc. London*, **A356**, 1951-1954 (1998)
2. Elmore, D. and Phillips, F. M., "Accelerator Mass Spectrometry for measurement of long-lived isotopes", *Science,* **236**, 543-550 (1987)
3. Litherland, A. E., "Fundamentals of accelerator mass spectrometry", *Phil. Trans. R. Soc. London,* **A323**, 5-21 (1987)
4. Litherland, A. E. and Kilius, L.R., "Neutral Injection for AMS", *Nucl. Instr. and Meth.* **B123**, 18-21 (1997)
5. Van de Graaff, R.J., "Tandem Electrostatic Accelerators", *Nucl. Instr. and Meth.*, **8**, 195-202 (1960)
6. Van de Graaff, R.J., *Bull. Am. Phys. Soc.*, **4**, 54 (1959)
7. Rose, P.H., "The Production of Intense Neutral and Negative Beams", *Nucl. Instr. and Meth.*, **28**, 146-153 (1964)
8. Carter, E.B. and Davis, R.H., "He$^-$ and other Negative Ions from a Duoplasmatron", *Rev. Sci. Instr.* **34**, 93-96 (1963)
9. Rose, F.A., Tollesfrud, P.B. and Richards, H.T., *Bull. Am. Phys. Soc.,* **12**, 29 (1967)
10. Donnally, B.L. and Thoeming, G., "Helium Negative Ions from Metastable Helium Atoms", *Phys. Rev.* **159**, 87-90 (1967)
11. Middleton, R. and Adams, C.T., "A close to universal negative ion source", *Nucl. Instr. and Meth.,* **118**, 329-336 (1977)
12. Rose, P.H., Bastide, R.P., Wittkower, A.B., Webb, D.L., Goldie, C.H., and Shaw, J., "A Neutral Helium Injector for Tandems", *Rev. Sci. Inst.*, **31**, 1052 (1960)
13. Litherland, A. E., "Ultra sensitive mass spectrometry with accelerators", *Ann. Rev. Nucl. Part. Sci.,* **30**, 437-473 (1980)
14. Rapp, D. and Francis, W. E., "Charge Exchange between Gaseous Ions and Atoms", Journ. Chem. Phys., **37**, 2631-2645 (1962)
15. Sakabe, S. and Izawa, Y., "Cross Sections for Resonant Charge Transfer between Atoms and their positive Ions: Collision Velocity < 1 a.u.", *Atom. Data and Nucl. Data Tables,* **49**, 257 – 314 (1991)
16. Massey, H. S. W., Burhop, E. H. S. and Gilbody, H. B., "Electronic and Ionic Impact Phenomena, IV", Oxford, 1974, see Fig.24.97 for Ar^+ + Rb on p. 2796 and Fig. 24.53 for Ar^+ + Ar on p. 2734
17. Copeland, F. B. M. and Crothers, D. S. F., "Cross Sections for Resonant Charge Transfer between Atoms and their Positive Ions", *Atom. Data and Nucl. Data Tables,* **65**, 273-288 (1997)
18. Gealy, M. and Van Zyl, B., "Cross sections for electron capture and loss. I. H^+ and H^- impact on H and H_2", *Phys. Rev.* **36** #7, 3091-3099 (1987)
19. Wittkower, A.B., Rose, P.H., Bastide, R.P., and Brooks, N.B., "Small Angle Scattering Observed in the formation of Neutral Atoms from 10-55 keV Positive Ion Beams", *Phys. Rev.*, **136**, A1254-A1259 (1964)

20. Bydin, Yu. F., "Resonance charge exchange of negative alkali metal ions", *Sov. Phys. JETP,* **19 #5**, 1091-1094 (1964)
21. Sidenius, G. and Andersen, N., "Multiple scattering of keV ions: lateral distributions in argon and nitrogen", *Nucl. Instr. and Meth.,* **131**, 387-389 (1975)
22. Sigmund, P. and Winterbon, K. B., "Small-angle multiple scattering of ions in the screened coulomb region: I Angular distributions", *Nucl. Inst. and Meth.,* **119**, 541-557 (1974) and 'Erratum' *Nucl. Instr. and Meth.,* **125**, 491 (1975)
23. Heinemeier, J. and Hvelplund, P., "Production of 10-80 keV negative heavy ions by charge exchange in Na vapour", *Nucl. Instr. and Meth.,* **148**, 425- 429 (1978)
24. Heinemeier, J. and Hvelplund, P., "Production of 15-90 keV negative heavy ions by charge exchange with Mg vapour", *Nucl. Instr. and Meth.,* **148**, 65-75 (1978)
25. Sigmund, P., Heinemeier, J., Besenbacher, F., Hvelplund, P. and Knudsen, H., "Small-angle multiple scattering of ions in the screened-Coulomb region. III. Combined angular and lateral spread", *Nucl. Instr. and Meth.,* **150**, 221-231 (1978)
26. Knudsen, H., Besenbacher, F., Heinemeier, J. and Hvelplund, P., "Lateral multiple scattering of energetic ions in very thin targets: Z_1 oscillations", *Phys. Rev.,* **A13 # 6**, 2095-2105 (1976)
27. Dehmel, R. C., Chau, H. K. and Fleischmann, H. H. "Experimental stripping Cross Sections for Atoms and Ions in Gases, 1950 – 1970", *Atomic Data,* **5**, 231-289 (1973)
28. Nadeau, M-J., Garwan, M. A., Zhao, X-L. and Litherland, A. E., "A negative ion survey; towards the completion of the periodic table of the negative ions", *Nucl. Instr. and Meth.,* **B123**, 521-526 (1997)
29. Kilius, L. R., Zhao, X-L., Litherland, A. E. and Purser, K. H., "Molecular fragment problems in heavy element AMS", *Nucl. Instr. and Meth.,* **B123**, 10-17 (1997)
30. Ishikawa,J., "Negative-ion sources for modification of materials", *Rev. Sci. Instrum.,* **67(3)**, 1410-1415 (1996)
31. Aardsma, G. E., "Accelerator Mass Spectrometry of ^{26}Al", Ph.D. Thesis, University of Toronto (1983)
32. Purser, K. H., Kilius, L. R., Litherland, A. E., and Zhao, X-L., "Detection of ^{236}U: a possible 100-million year flux integrator", *Nucl. Instr. and Meth.,* **B113**, 445-452 (1996)
33. Wittkower, A. B. and Betz, H.D., "Equilibrium charge state distributions of energetic ions in gaseous and solid media", *Atomic data,* **5**, 113-166 (1973)
34. Litherland, A. E., "New frontiers in accelerator mass spectrometry: isobar separation methods at low energy", *Nucl. Instr. and Meth.,* **B92**, 207-212 (1994)
35. Diamond, W. T. "A Radioactive Ion Beam Facility using Photo-fission", AECL-11949 (1998) (unpublished).

VI
EXOTIC ION BEAMS

An Overview of ISAC

P. W. Schmor

TRIUMF,
4004 Wesbrook Mall, University of British Columbia, Vancouver, British Columbia, Canada, V6T 2A3

Abstract. Construction has begun on ISAC, a radioactive ion beam (RIB) and accelerator facility that utilizes the ISOL (on-line isotope separation) production method. A five-year budget for this new radioactive beam facility at TRIUMF was approved in June 1995. ISAC includes: a new building with 5000 m^2 of floor space, a beam line with adequate shielding to transport up to 100 µA of 500 MeV protons from the TRIUMF cyclotron to two target/ion-source stations, remote handling facilities for the targets, a high-resolution mass-separator, linear accelerators and experimental facilities. The ISAC target/ion source station permits the production of nuclei far from stability over a large isotopic range with high luminosity. Ions from the target/ion-source will be transported at energies up to 60 keV through a low-resolution pre-separator magnet followed by a high-acceptance, high-resolution mass-separator magnet to a variety of low energy experimental stations. Alternatively, ions with q/A ≥ 1/30 and an energy of 2 keV/amu can be bunched in the low energy beam transport line prior to a RFQ accelerator. The final energy will be variable from 0.15 to 1.5 MeV/amu. The accelerated beams will be used primarily for nuclear astrophysics studies. The buildings are now complete and the commissioning of the target/ion source, mass separator and RFQ systems has started. The TRIUMF neutral atom trap (TRINAT) is scheduled to begin using the low energy ISAC beam in November 1998. The full energy beam will become available for the DRAGON recoil spectrometer at the end of 2000. A new five-year plan will be presented to the Canadian Government before the end of 1998. The plan includes an upgrade of ISAC to permit acceleration of the RIB up to 6.5 MeV/amu for masses up to 150.

INTRODUCTION

There have been more than ten years of experience with radioactive ion beams at TRIUMF. Two Canadian workshops and a subsequent 'Proposal for an Intense Radioactive Beams Facility' in 1985 led to the construction of the TRIUMF ISOL (TISOL) facility in 1987 (1). TISOL is a first generation ISOL type facility that has not only provided useful beams for scientific research but has also provided valuable information on targets, ion sources and remote handling requirements for a second generation facility. Although the proton beam line is capable of 10 µA, inadequate shielding above the target and primitive remote handling procedures for servicing irradiated components around the target have limited the proton beam intensity at TISOL to about 2 µA. TISOL will continue to be useful in the future as a facility to test new targets and ion sources, at lower currents, prior to production running in ISAC. In 1995, it was decided to exploit the full potential of the TRIUMF site by constructing ISAC-I (first stage of the Isotope Separator and Accelerator) in a new building to the north of the cyclotron. This facility required a new proton beam line (BL2A) and adequate shielding to be capable of dealing with targets exposed to 100 µA of 500 MeV protons. Protons are extracted from the cyclotron by stripping the circulating H⁻ beam at the appropriate cyclotron radius and azimuth. By adjusting the

CP473, *Heavy Ion Accelerator Technology: Eighth International Conference,*
edited by Kenneth W. Shepard
© 1999 The American Institute of Physics 1-56396-806-1/99/$15.00

width of the stripper foils, the locations of the strippers in the cyclotron and the depth that strippers are inserted into the circulating beam, it is possible to extract a number of independent proton beams simultaneously at different energies and different intensities from the TRIUMF cyclotron. The ISAC-I accelerators will accelerate singly charged isotopes with masses less than or equal to 30 up to a final energy of 1.5 MeV/amu. The TRIUMF site would accommodate a future upgrade of ISAC (ISAC-II) in both final energy and maximum mass to include the full scope of scientific interest. The funds for the construction of ISAC-I came from the Canadian federal government as one component of a five-year budget for the period 1995 to 2000. An additional CDN\$9.7M for the construction of the buildings came from the Province of British Columbia. The design specifications for the ISAC I RIB (radioactive ion beam) delivery system are listed in table 1.

Table 1. Design Parameters for ISAC I

ISOTOPE PRODUCTION SYSTEM		
Driver	Projectile	Protons
	Energy	470 - 510 MeV
	Intensity	$\leq 100\ \mu A$
	Beam Size	$\leq 1\ cm^2$
	Time Structure	Cw
Target	Length	$\leq 20\ cm$
	Power Deposited	$\leq 20\ kW$
RADIOACTIVE ION BEAM SYSTEM		
Ion Source	Surface, ECR, Laser, FEBIAD, etc.	
	Max. Energy	$\leq 60\ keV$
Mass Separator	Mass/δ(Mass)	≤ 10000
	Mass	≤ 240
	Vacuum	$\leq 2 \times 10^{-7}$ torr
ACCELERATOR SYSTEM		
Pre-Buncher	Fundamental Frequency	11.67 MHz
	Harmonics	3
RFQ	Input Energy	2 keV/nucleon
	Input charge/mass	$\geq 1/30$
	Output Energy	150 keV/nucleon
	Frequency	35 MHz
DTL	Input charge/mass	$\geq 1/6$
	Output Energy	0.15 to 1.5 MeV/nucleon
	Frequency	105 MHz
	$\Delta E/E$	< 0.1%

ISAC CIVIL CONSTRUCTION

ISAC-I required that a new building be erected to the north of the cyclotron. The first major ISAC-I construction contract (for the relocation of site services) was awarded in May, 1996 followed by the primary structure contract commencing in September 1996 immediately following receipt of a building permit. In July 1997,

TRIUMF personnel began occupying the experimental hall for installation of experimental apparatus. By the end of 1997, most trades were at the stage of addressing deficiencies, slightly behind the initial milestone for construction completion by November 1997. The project was completed within budget. Figure 1 is a sketch of the building layout showing through cutouts the main features of ISAC-I.

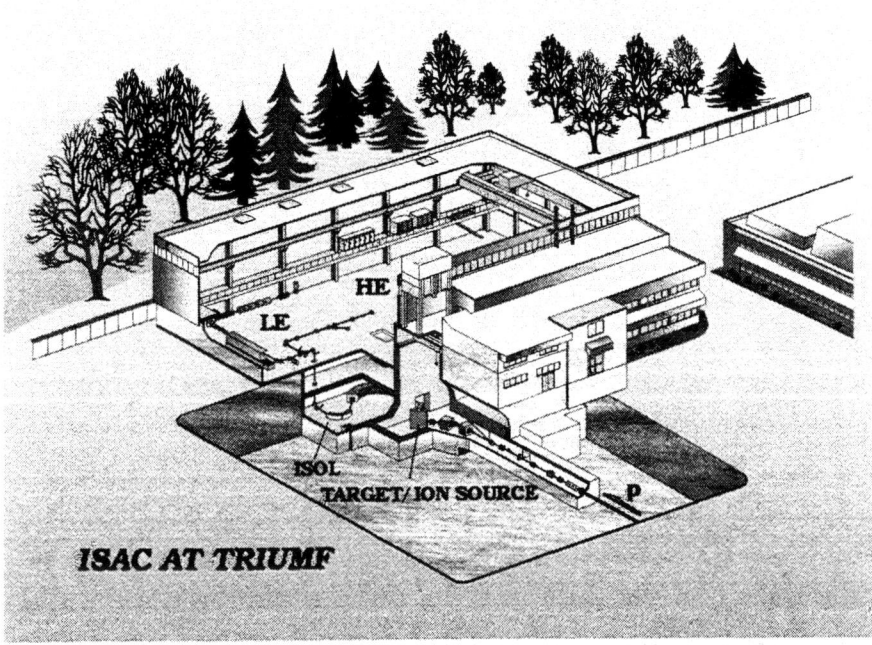

Figure 1. A sketch of the ISAC-I facility showing the proton beam line, the target vault, the target service building, the mass separator vault, the TRINAT mezzanine (above the mass separator vault) and the accelerator/experimental hall.

The facility contains about 53,000 ft^2 of useable floor space. Electrical services to the facility could eventually accommodate up to 6 MW of power, although only about half this power is required for ISAC-I. A hole had to be bored through the cyclotron vault wall and a tunnel built to transport the high intensity proton beam from the cyclotron to the ISAC target stations. A four-story target-service wing was built which contains a chemistry laboratory, an assembly laboratory, a meeting room, a service elevator, active sumps and two open areas for offices. A heavily-shielded target-vault includes two target stations, shielding for two 50 kW beam dumps, two hot cells and a target storage pit. The entire target vault can be accessed with a remotely-controlled 20 ton rolling-crane. A penthouse was constructed above the target vault houses the HEPA-filtered exhaust-system, the building mechanical services and the heat exchanger. A shielded room is located at beam level next to the target vault for a high-resolution mass-separator system. A large grade-level hall to the north of the target vault contains the accelerators and experimental stations. The entire hall can be accessed with an overhead 35 ton rolling-crane. The experimental hall also contains two experimental counting rooms and an accelerator control room.

ISAC TECHNICAL FACILITY

Proton Beam Line

A new beam line was built to take the proton beam from the cyclotron to the ISAC-I target. In May 1997, beams were extracted over the energy range 472 to 510 MeV confirming the validity of the theoretical beam dynamics calculations. The entire beam line was completed to the ISAC-I beam dump by the spring 1998 and protons were transported to the ISAC beam dump in May 1998. Figure 2 indicates that the beam shape is well matched to the target dimensions.

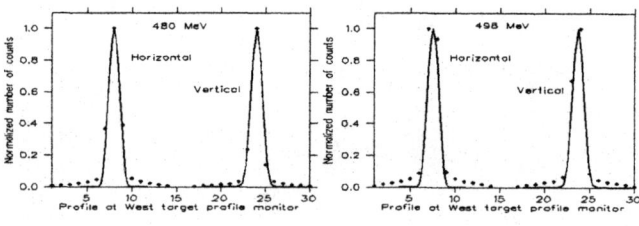

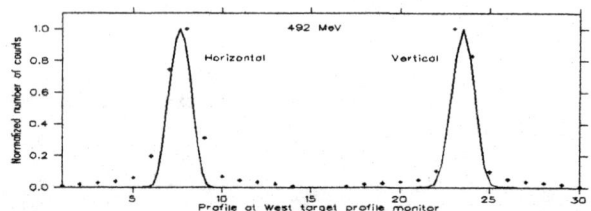

Figure 2. The calculated (solid) and experimental (diamonds) profiles from profile monitor located about 1 m upstream of the target dump & 60 cm upstream of target centre. (Wire Spacing = 2 mm)

Target

The target hall vault has three main areas; a target area, a storage area and a hot cell area. The entire vault is serviced by a 20 ton overhead crane that can be operated remotely from the target service building. In order to achieve efficient ISAC operation, it was necessary to shield the target area such that personnel could work in the target vault during beam operation. The required shielding was estimated assuming a 100 µA proton beam impinging on a high z target such as uranium. Ninety-six ten-ton blocks of recycled reactor steel were placed around the target vacuum tanks. Voids in the corners surrounding the target area were filled with railroad rails and concrete. As a result there is at least 6 m of high density shielding between the target and the occupied regions. The shielding surrounds two, large, T-shaped, steel vacuum-tanks which each contain five removable modules. Two of the modules (an entrance module that contains proton beam diagnostics and an exit module that houses the proton beam dump) are required for the proton beam. The remaining three modules (a target module which holds the target, ion source and ion beam extraction electrodes and two exit modules which contain the ion beam optical elements and the ion beam diagnostics) are required for the radioactive ion beam. All of the modules are designed with manual quick-disconnects for attaching and removing the required external services. Elastomer O-ring vacuum seals have been located at the top of the modules with approximately 2 m of steel shielding separating the seals from the target. The three modules related to the radioactive ion beam have all of the ion optical elements placed inside of a primary vacuum vessel. The primary vacuum system is contained within the secondary vacuum space of the T-shaped tank. The dual vacuum system is designed to prevent radioactive contamination from migrating to either the outside walls of the modules or to the entrance and dump modules. The primary vacuum system is maintained between the three modules by air-actuated 'pillow'-seals that are retracted when a module is removed for servicing. The entrance and dump modules are located in the secondary vacuum of the large tank. Water-cooled windows are used where the proton beam enters and exits the target module. An additional water-cooled window separates the vacuum of the proton beam line from the secondary vacuum of the target tank. The volatile compounds that are pumped from the target tank are stored in tanks, monitored for activity and allowed to decay to acceptable levels prior to release to the atmosphere through HEPA filters. Although both target tanks were manufactured, initially only the west target tank and five modules are operational. The second tank will be used to condition, store and leak check spare modules. The servicing philosophy of the components in the tanks is based on years of successful experience with the meson production targets. The services are manually disconnected from a module requiring servicing and then the overhead crane is used to pick up that module and transport it to either a storage vessel or to the hot cell where manipulators are used to remove and replace components. The storage area contains spare modules and the 'pigs' that are used to store the used targets and ion sources. The storage area serves as a module staging/assembly area and also contains the alignment fixture for confirming alignment of modules prior to insertion into the vacuum tanks. A future target/ion source conditioning facility is planned for this area as well.

Initially the beam current on target will be only a few microamperes. During this stage it is necessary to provide ohmic heating in order to achieve the required target temperature. These targets are similar to those presently used in TISOL. As the proton beam current (beam power) is increased it will eventually be necessary to cool the target. Numerical simulations have been performed for various design concepts of a high power target. The designs attempt to achieve a temperature of about 2000° ± 100° C over the entire target. The optimal design will have to be determined empirically in ISAC.

Ion Source

The operational target and ion-source system requires frequent servicing to meet the experimental requests for isotopes. It is expensive and not cost effective to develop new ion source systems in a radioactive environment. A test stand was built to test and validate the ISAC-I approach which has the entire ion source system immersed in the target vacuum. With this approach it is unnecessary to use O-ring seals in the high radiation area surrounding the target. To simplify operation the ion source extraction-system was designed with multiple electrodes that avoid the need for movable electrodes. The mechanism that allows the movement has consistently been a source of problems at TISOL. The first ion source to be approved for ISAC-I experiments is a surface ion source for alkali elements. Although similar ion sources had been used for years at ISOLDE and at TISOL, no experimental information could be found on the beam emittance that could be expected. Moreover both laboratories had been using a simple single-gap ion-extraction electrode. This approach resulted in a very divergent ion beam that was difficult to fully recapture in a subsequent ion optical system. An alternate double-gap extraction system was designed for ISAC and tested. The tests were successful and the design was approved for use in the ISAC target module. The beam could be extracted over a wide range of masses and energies without intensity or emittance degradation. As predicted by the simulations, it was unnecessary to adjust the electrode position each time a new mass or new ion energy was selected. The ion source and target have been designed to operate up to a maximum bias of 60 kV.

The TRINAT (TRIUMF neutral atom trap) will be the first experimental station to receive a radioactive ion-beam from ISAC-I. The first experiment approved by the TRIUMF Experimental Evaluation Committee (EEC) requires ^{37,38m}K beams. CaO powder targets were being used in TISOL to produce K beams. ISAC proposed to use compressed CaO pellets. The ISAC pellet target was successfully tested in TISOL in the spring of 1998. The beam intensity was as high as the previous best target and the lifetime exceeded previous targets. Other isotopes requiring different ion sources are required quickly. An electron-cyclotron-resonance (ECR) ion source has been very useful in ionizing gaseous elements in TISOL. This source is too large and complex to be operated reliably in ISAC. Development has started on a compact microwave ECR ion source. A prototype version is now installed in the ion source test stand. ECR ion sources produce ions with energy spreads of tens of eV. It is not possible to achieve a mass resolution (isobaric rejection) greater than 10,000, unless the energy spread is kept below about 2 eV. The initial experiments, requiring an ECR ion source, have not requested a high mass-resolution. Other ion sources or techniques will be required when a high resolving power is required. Provision has been made so that the ISAC system can accommodate laser ion sources, FEBIAD ion sources and Cusp ion sources. These sources will have to be developed for ISAC as required by the scientific programs.

Mass Separator

The ISAC mass separator system includes a pre-separator magnet, three matching sections using electrostatic optics, an acceleration column, a mass analyzing magnet and a deceleration column. The pre-separator stage uses a 'Y' shaped magnet that is designed to accept the RIB from either one of the target stations. It is used to select out most of the unwanted radioactive ions and to deposit this activity on slits that can be removed remotely to the hot cells by the crane in the target hall. The main mass separator magnet was acquired from Chalk River in August 1997, following the closure of the TASCC laboratory. The mass separator system is located on a high voltage platform that can be raised in potential up to 60 kV. By varying the potential of the mass separator it will be possible to scan and select a particular isotope without readjusting the magnetic field of the magnet. The system has been designed to provide a mass resolving power of 10,000 for an emittance of 30 π mm mrad. An outline drawing showing the configuration of the mass separator system is given in figure 3.

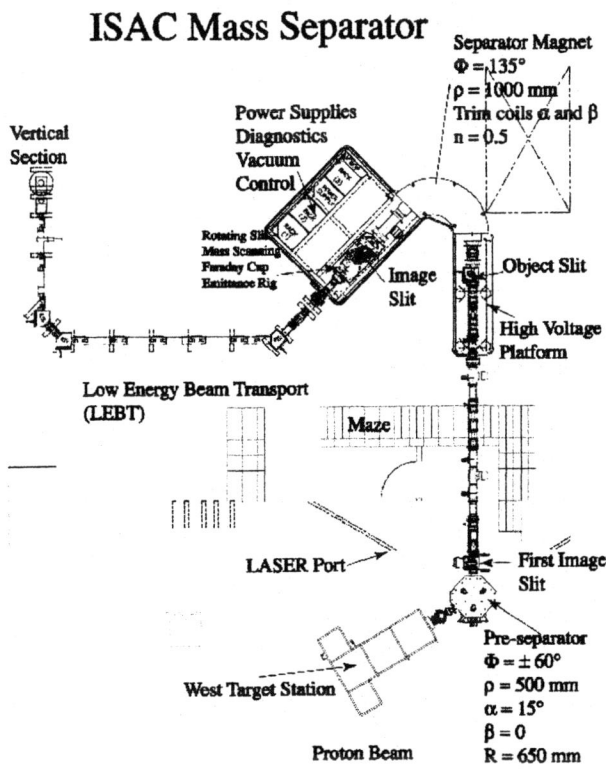

Figure 3. Layout of the beam transport system at the target elevation from the target station through the mass separator system to LEBT

Low Energy Beam Transport

The ion beam from the mass separator can be transported either to the low energy experimental area or to the RFQ accelerator. At the same time, it is desirable to have an off-line ion source (OLIS) to provide a beam of stable isotopes to the same two areas, although its primary purpose is for commissioning the accelerators. A switchyard has been designed and built to meet these goals. All of the optics in the low-energy-beam-transport (LEBT) is electrostatic. The line is made up of a number of similar modules. Spherical bends are used to achieve focusing in both transverse planes. The RFQ, having no bunching section, requires an external buncher in the LEBT. The section of the LEBT that connects the mass separator to the rest of the low energy beam line is now being assembled for the October 1998 beam commissioning to TRINAT. The LEBT from the OLIS to the RFQ was commissioned early in 1998. The installation of the OLIS began in July 1997 and the first ion beam was extracted in November of the same year. The LEBT and buncher were commissioned from OLIS to the entrance of the RFQ by March 1998.

Accelerators

Beams having a mass to charge ratio less than or equal to 30 are to be accelerated from the injection energy of 2 keV/amu up to a final energy of 1.5 MeV/amu. The accelerating system consists of a pre-buncher, a cw RFQ, a medium energy beam transport (MEBT) section, an electron stripper, a re-buncher, a cw drift tube linac (2). The pre-buncher provides a pseudo saw tooth velocity profile at a fundamental frequency of 11.67 MHz, thereby providing approximately 86 nS between beam bursts. Bunched beam from the pre-buncher fills every third bucket of the 35 MHz, cw, 8 m long RFQ. The beam out of the RFQ, at energy 0.15 Mev/amu, is first focused and stripped to a higher charge state. The beam is then magnetically bent to select only those isotopes having a mass to charge ratio less than or equal to 6 and then re-bunched prior to injection into the first tank of the DTL. The DTL must provide a bunched beam that can be continuously varied in energy from 0.15 to 1.5 MeV/amu. To achieve this a separated-function structure with five DTL tanks and three split-ring bunchers has been designed. As the DTL system operates cw at 105 MHz, only 1 in 9 rf buckets are used to accelerate beam.

The RFQ design is unique in many ways. In order to demonstrate that the concept would operate cw at 35 MHz, a 1 m long (3 rings) prototype was successfully tested without beam but at full power in November of 1996. Following this result, drawings for the first part of the RFQ were released for manufacture. The RFQ accelerator requires precise alignment to operate with the predicted acceptance. In order to demonstrate that the alignment had indeed been achieved and that the numerical simulations were correct, a RFQ tank with only the first 7 of the eventual 19 rings was commissioned with beam in 1998. The tests were successful and manufacture of the remaining 12 rings has started. The quadrupoles required for the MEBT were acquired in 1998 from the TASCC facility. The MEBT dipoles have been designed and are ready to be sent out for manufacture. A prototype spiral re-buncher was built in 1998 and is being tested at signal level. The first tank of the DTL with electrodes has been manufactured and will be tested at power before the end of 1998. As with the RFQ, the engineering challenge is to remove the heat when the DTL is operating at full voltage. If the tests are successful the remaining four tanks will be sent out for manufacture, early in 1999. The first DTL buncher was built in collaboration with the Institute for Nuclear Research by the rf group at INR, Troitsk, Russia. This buncher was delivered to TRIUMF in August 1998. The detail design has started on the magnetic triplet required to achieve transverse focusing between each DTL tank.

ISAC SCIENCE FACILITIES

The initial experimental program at ISAC includes several experimental stations. In particular, the experimental stations include; the TRIUMF Neutral Atom Trap (TRINAT), a Yield Station, the Low Temperature Nuclear Orientation system (LTNO), a General Purpose & Lifetime measurement station, the recoil mass spectrometer (DRAGON) facility and a ^{8}Li polarizer for the β-NMR station. The hall can accommodate additional stations. TRINAT is located below grade in a well-shielded mezzanine above the mass separator. The layout of the stations is given in figure 4.

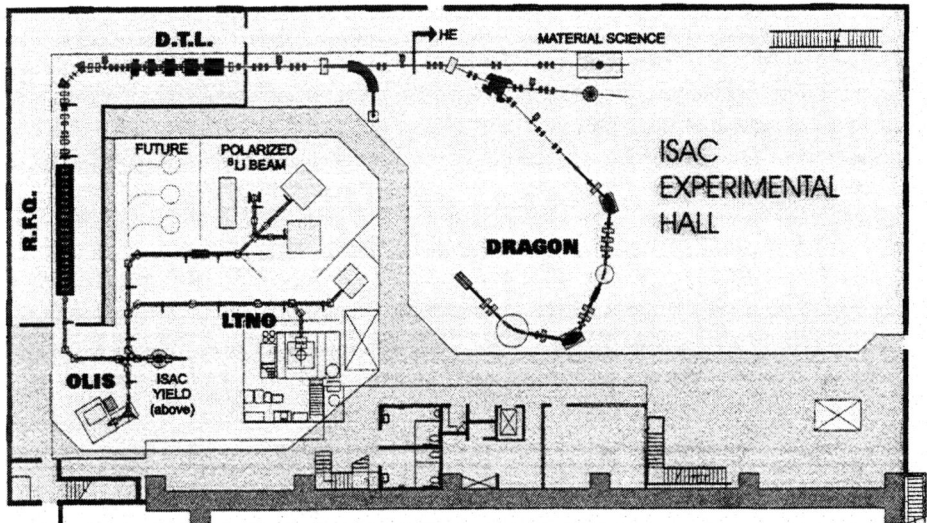

Figure 4. The layout of the experimental/accelerator hall in ISAC-I shows the location of the various experimental stations. The lifetime station will initially be located to the right of the yield station. This figure also indicates the accelerator locations.

TRINAT has been trapping radioactive atoms at the TISOL facility in a Zeeman optical trap. Detection of the decay products from these trapped atoms permits sensitive tests of the Standard Model. The trap has recently moved from TISOL to ISAC where the higher beam flux should yield increased sensitivity. In nuclear astrophysics, many of the reactions of interest involve proton or alpha captures. The ISAC nuclear astrophysical experiments will be carried out with reverse kinematics by bombarding a high-pressure windowless gas-target with the accelerated radioactive ion beams. The reaction products are detected using a recoil mass fragment detector, DRAGON, designed to have a high detection-selectivity at the ISAC energies. The nuclear orientation facility (LTNO) has been shipped from ORNL to TRIUMF and is being reassembled in the low energy experimental area of ISAC. The initial scientific program for the LTNO will focus on nuclear structure in the mass range 80 to 100. In addition the facility will use nuclear magnetic resonance on oriented nuclei (NMRON) as a sensitive probe of condensed matter. The μSR group at TRIUMF is building a β-NMR facility that initially will be using a polarized ^{8}Li beam to enhance their condensed matter program at TRIUMF. The longer lifetime of the radioactive isotope, compared to that of the muon, makes it the preferred probe for relaxation studies in solids.

COMMISSIONING/OPERATION

A proton beam has been transported from the cyclotron to the ISAC beam dump. A stable alkali beam was produced from alkali salts placed in a heated target without using the proton beam. The stable alkali ion beam has been transported from the surface ion source to the Faraday cup at the exit of the target vacuum tank (exit module 2). Commissioning of the mass separator system with stable beams will commence in mid October. The schedule predicts that the stable beam will be delivered to TRINAT before the end of October 1998. Early in November the proton beam will be used to produce a radioactive potassium beam from a CaO target for TRINAT and the ISAC yield station. Operation of the low energy beam will expand to the other low-energy experimental-stations as they become available in 1999. The off-line ion source allows the RFQ, MEBT, DTL and HEBT commissioning to take place at the same time as the radioactive beam is being delivered to the low energy experiments. The schedule anticipates that by the end of 2000 the accelerators will be operating and that the high-energy experiments will be ready for an accelerated radioactive ion beam.

FUTURE PLANS

The five-year funding cycle, which was used to build ISAC-I, began in June 1995 and ends in April 2000. A new proposal for the next five years is being prepared for submission to the Canadian Government in November 1998. In response to

TRIUMF's request for input into the new five-year plan, the Canadian user community made a strong submission to have ISAC-I augmented. These submissions were used to produce some generalized facility specifications (3). The proposed ISAC-II facility, shown in figure 5, will provide beams of short-lived exotic isotopes at energies up to 6.5 MeV/amu for isotopes having masses less than 150. Additionally, an additional request to provide slightly higher energies for the lighter masses is possible with this design. A modest grade-level building expansion will be required to accommodate the additional accelerators and experimental stations. As requested, construction of ISAC-II could be staged in order that the nuclear physics program can begin prior to facility completion. To achieve this early start to an experimental program, it is proposed to carry out the building construction concurrently with the assembly of small superconducting rf cavities. These cavities would then be initially installed, as they become available, downstream of the DTL1 in the ISAC-I experimental hall.

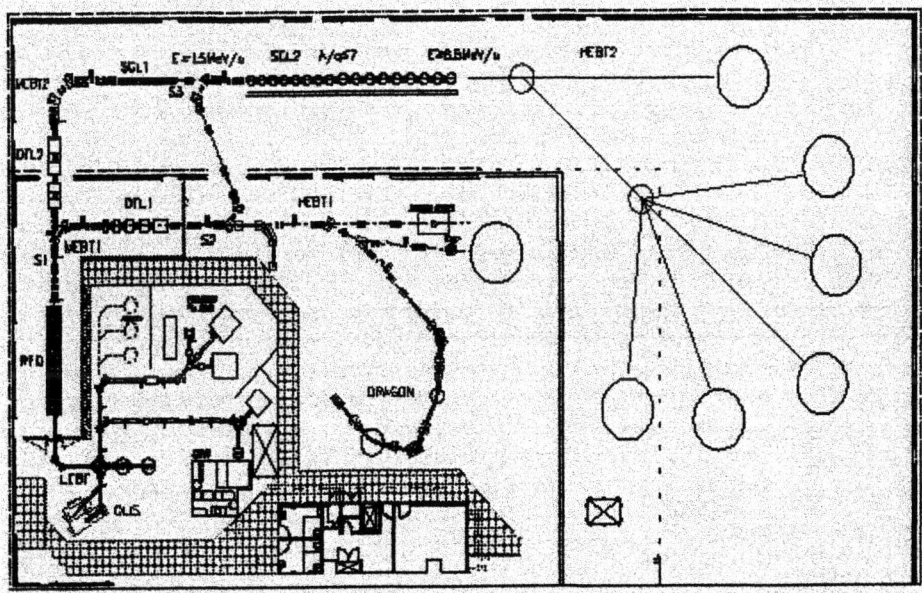

Figure 5. The proposed layout for ISAC-II indicates how the accelerators and experimental stations would be added to ISAC-I. The dotted line indicates the wall of the existing ISAC-I building.

The ISAC-I facility has the capability of accelerating radioactive ions (with $q = \pm 1$ and $A \leq 30$) up to 1.5 MeV/amu. In order for experimenters to reach the Coulomb barrier (roughly 6.5 MeV/amu) with masses up to 150, it is necessary to increase both the length of the ISAC-I accelerating system and the mass range which can be accelerated. In fact, ions of any mass can already be accelerated through the ISAC-I RFQ provided that their charge/mass ratio is greater than 1/30. The first logical step for increasing the available mass range is to increase the charge-state of ions entering the RFQ. Electron-cyclotron-resonance (ECRIS) ion sources easily reach the required $q/A \geq 1/30$ for many elements (4). The installation of a charge-state-booster in ISAC would allow, in principle, acceleration of all masses to 1.5 MeV/amu. However, the beam intensities for masses beyond A ≈ 70 would be too low to be useful if the ions require further stripping before injection into the DTL1. The stripper in ISAC-I is located at 0.15 MeV/amu and reasonable beam intensities with a q/A of 1/6 cannot be achieved for masses beyond about 70 at this low velocity. The optimum stripping energy for 30 ≤ A ≤ 150 with q/A ≥ 1/7 is about 0.4 MeV/amu. To accelerate the ion beam with q/A ≥ 1/30 from 0.15 to 0.4 MeV/amu, requires a new linac (DTL2). A DTL very similar to the DTL1 would be about 5 m long assuming an average gradient of 1.5 MV/m. After the DTL2, the beam would go through a beam transport system consisting of a short matching section, a stripping foil, a 90° bend for charge selection, a rebuncher and finally a matching section to a post-stripper linac (SCL1). To reach 6.5 MeV/amu from 0.4 MeV/amu with A/q = 7 requires a total voltage gain of 42.7 MV. The preferred linac would use many short (2 to 4 cells) superconducting cavities each with an accelerating gradient of 3 MV/m. Short modules have the advantage that the ions do not have to rigorously follow a fixed velocity profile to stay in phase with the rf. Consequently the maximum energy of a particle depends on its charge to mass ratio. For example, an energy greater than 15 MeV/amu can be attained for particles with q/A = 1/3. (Stripping at 400 keV/amu can efficiently produce ions with q/A = 1/3 for A < 30.) Even for the highest masses, higher energies would be possible at a cost in intensity with the addition of an intermediate stripping station. The flexibility of short superconducting modules has other advantages as well. Since they are short and have wide velocity acceptance, they can be built and used to accelerate radioactive ion beams even before the building addition for ISAC-II is completed. In order to satisfy the request of the Canadian Nuclear Physics Community for accelerated beam prior to 2005, it would be possible to install modules downstream of the DTL1 as they become available. High-energy (E ≥ 4.5 MeV/amu) experiments could start towards the end of 2002. Masses up to 30 and energies up to 5 MeV/amu would be available for these initial experiments. With the addition of the charge-state-booster, masses up to A = 62 would become available. In order to reach the design energy (6.5 MeV/amu) before 2005, the super-conducting modules could be moved to the new building extension near the end of 2003 and a beam transfer line installed to take the beam from ISAC-I to the completed final section of the superconducting accelerator (SCL2). This would allow additional experiments to start in the extension to the experimental hall. The higher energy beam would be available in mid 2004. Finally, SCL1, DTL2 and MEBT2 would be installed. Full energy and mass would optimistically be available about five years after the start of the five-year plan in this staged approach.

REFERENCES

1. J.M. D'Auria et al., *Nuclear Instruments and Methods in Physics Research* **B 126**, 7-11 (1997)
2. R.E. Laxdal, "RFQ-IH Radioactive Beam Linac for ISAC", these proceedings
3. R. Baartman et al., "Long Range Plan proposal for an Extension to ISAC" Proceedings of the 1998 Linac Conference, Chicago, to be published
4. T. Lamy et al., *Review of Scientific Instruments* **69**, 741 (1998)

The RB Facility at KEK-Tanashi

M. Tomizawa, S. Arai, Y. Arakaki, A. Imanishi,
M. Okada, K. Niki, Y. Takeda and E. Tojyo

KEK-Tanashi, 3-2-1 Midori-cho, Tanashi-shi, Tokyo 188-8501, Japan

Abstract. An ISOL-based radioactive nuclear beam facility at KEK-Tanashi has been constructed. The linac complex to accelerate radioactive beams comprises a 25.5 MHz split coaxial RFQ (SCRFQ) and a 51 MHz interdigital-H (IH) linac, and accelerates heavy ions up to 1 MeV/u. Beam tests using stable nuclear beam have been done in order to examine the performance of the linacs. Acceleration tests of a radioactive nuclear beam was also performed. This facility is a prototype for the exotic nuclei arena (E-arena) of the proposed Japanese Hadron Facility (JHF), in which 3 GeV, 10 μA protons is used as a primary beam, and a radioactive nuclear beam is accelerated up 6.5 MeV/u by an extension of the IH linac. In this paper, outline, present status and future plan of the facility are reported with emphasis on the heavy ion linacs.

INTRODUCTION

An ISOL-based radioactive nuclear beam facility at KEK-Tanashi has been constructed. Radioactive nuclei, produced by bombarding a thick target with protons or light ions from an SF cyclotron, are ionized in an ion source [1–3], mass-analyzed by an isotope separator on line (ISOL) [4], and transported to a heavy ion linac complex through a 60 m long beam line [4]. The linac complex comprises a 25.5

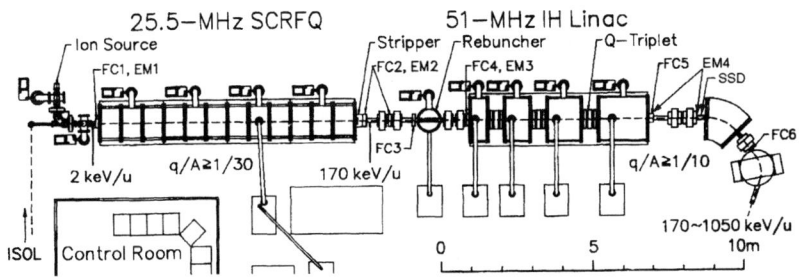

FIGURE 1. Layout of the linac complex.

MHz split coaxial RFQ (SCRFQ) and a 51 MHz interdigital-H (IH) linac, and accelerates heavy ions up to 1 MeV/u (see Fig. 1).

This facility is a prototype for the exotic nuclei arena (E-arena) of the Japanese Hadron Facility (JHF), in which 3 GeV,10 µA protons is used as a primary beam, and a radioactive nuclear beam is accelerated up 6.5 MeV/u by an extension of the IH linac [5]. The main purpose of the prototype facility is to study various technical problems for the E-arena in the JHF and to perform pioneering works with respect to nuclear astrophysics. In this paper, outline, present status and future plan of the facility are reported with emphasis on the heavy ion linacs.

OUTLINE OF THE LINAC COMPLEX

Split Coaxial RFQ

A split coaxial RFQ (SCRFQ) has been developed to accelerate heavy ions with a small charge to mass ratio (q/A) [6]. Design parameters of the SCRFQ are summarized in Table 1. The SCRFQ accelerates ions with q/A greater than 1/30 from 2 to 172 keV/u. The resonant frequency is chosen to be 25.5 MHz to accelerate the ions with $q/A \geq 1/60$ considering future extension. The duty factor can be 30% for ions with q/A=1/30 and 100% for $q/A \geq 1/16$. The cavity, 0.9 m in inner diameter and 8.6 m in length, comprises four unit cavities, and each of which is composed of three modules (see Fig 2). Obtained flatness of the longitudinal field distribution is within ±1%. Unloaded Q-value is 5800. The resonance resistance ($=V^2/2P$) is 24.55 ± 0.44 kΩ, which is obtained from the endpoint energy of X-rays generated from the cavity.

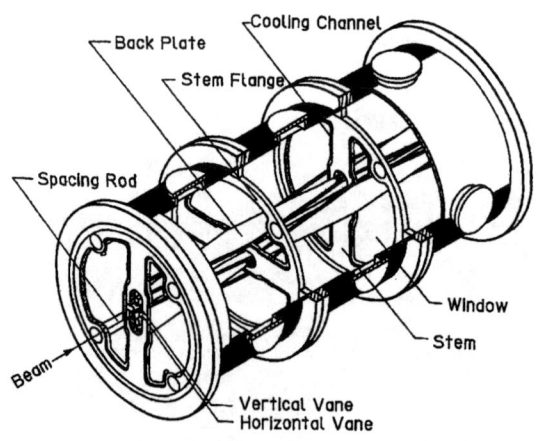

FIGURE 2. SCRFQ view.

TABLE 1. Design parameters of the SCRFQ

Frequency	25.5 MHz
Charge-to-mass ratio	$\geq 1/30$
Energy	$2 \to 172$ KeV/u
Input emittance	291 π mm·mrad
Normalized emittance	0.6 π mm·mrad
Vane length	8.585 m
Number of cells (radial matcher)	172(20)
Max. Intervane voltage	108.6 kV
Max. surface field	178.2 kV/cm
	(2.49 Kilpatrick)
Mean aperture radius (r_0)	0.9846 cm
Minimum aperture radius (a_{min})	0.5388 cm
Max. modulation index (m_{max})	2.53
Margin of bore radius (a_{min}/a_{beam})	1.2
Final synchronous phase	-30°
Focusing strength (B)	5.5
Max. defocusing strength (Δ_b)	-0.17
Transmission (0 mA input)	91.4 %

Transport Line between RFQ and IH linac

A transport system between the RFQ and the IH linac comprises a charge stripper (C-foil), a rebuncher and two sets of quadrupole doublets [7]. The charge stripper is used to increase the charge state of the ions with a small q/A. The rebuncher is a 25.5-MHz double coaxial quarter wave resonator with six gaps. The power consumption in the cavity is less than 1.5 kW even in a maximum operation. Main parameters of the rebuncher is listed in Table 2.

TABLE 2. Parameters of the rebuncher

frequency	25.5 MHz
incident energy	172 KeV/u
gap number	6
bore radius of drift tubes	3 cm
gap voltage	34 kV
unloaded-Q	6000
eff. shunt impedance	29 MΩ/m
power	1.46 KW

Drift Tube Linac

The ions with $q/A \geq 1/10$ are accelerated from 172 keV/u to 1 MeV/u by a 51 MHz interdigital-H (IH) linac. The view of the IH linac is shown Fig. 3. To obtain a high shunt impedance, π-π drift tubes without transverse focusing element were adopted. The linac has four separated tanks. The output energy can be continuously varied in the whole energy range from 172 keV/u to 1 MeV/u by adjusting rf power levels and rf phases. Three sets of quadrupole triplets are placed between tanks. The design parameters of the IH linac are listed in Table 3 together with unloaded-Q and effective shunt impedance (Z_{eff}) obtained from low power measurements. The power consumptions were estimated from the effective shunt impedance measured by a bead-pull method [8].

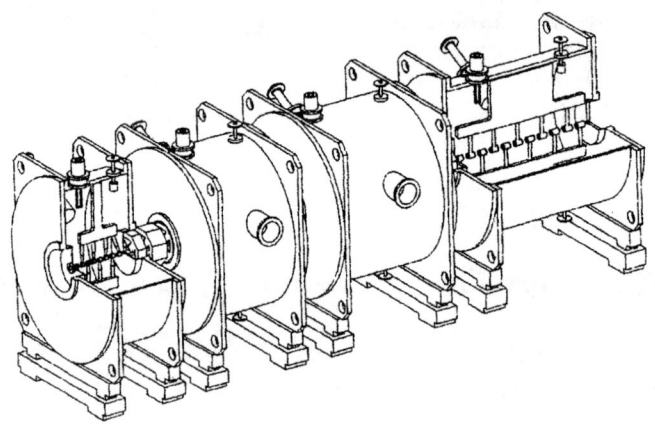

FIGURE 3. IH linac view.

SUMMARY OF ACCELERATION TESTS

Acceleration Tests of Stable Nuclear Beams

Tests using stable nuclear beams have been done in order to know the performance of the linac complex. The stable nuclear beams were produced by a 2.45-GHz ECR ion source with permanent magnets, mass-analyzed by a bending magnet and two quadrupole magnets, and injected into the SCRFQ.

TABLE 3. Main parameters of the IH linac

	tank1	tank2	tank3	tank4
f(MHz)	51	51	51	51
max. q/A	1/10	1/10	1/10	1/10
T_{out}(MeV/u)	0.294	0.475	0.725	1.053
L_{tank}(m)	0.68	0.90	1.16	1.53
D_{tank}(m)	1.49	1.49	1.49	1.34
D_{bore}(cm)	2.0	2.4	2.8	3.2
D_{tube}(cm)	3.8	4.4	4.6	5.2
L_{gap}(cm)	2.9	3.7	4.5	5.3
Cell No.	9	10	11	12
V_{gap}(kV)	200	250	313	370
unloaded-Q	10681	15387	16230	18490
Z_{eff}(MΩ/m)	264	289	268	218
P(kW)	10.5	15	25	39

RFQ beam tests

The stable nuclear beams, H_2^+, He^+, N^{2+}, Ne^{2+}, N^+, Ne^+, Ar^{2+} and N_2^+ were successfully accelerated by the RFQ so far. The intervane voltages corresponding to these accelerations range from 7.2 to 101 kV. The output energy measured by a time of flight method is 170.9±1.3 keV/u, which agrees with the design value within the measuring error. Figure 4 shows the beam transmission of the RFQ as a function of intervane voltages. At the design intervane voltages, the measured transmission efficiencies for the accelerated ions (solid circle) are about 90%, which agree with the prediction by the PARMTEQ-H simulation (solid line).

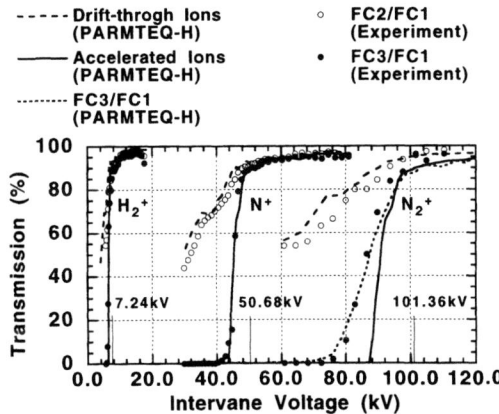

FIGURE 4. Transmission efficiency as a function of the intervane voltage.

Beam transmission of the linac complex

Figure 5 shows the transmission efficiency of $^{14}N^{2+}$ beam accelerated at 1 MeV/u. The effect of Particles not accelerated by the RFQ are included at RFQ OUT. The actual transmission of the transport line between the RFQ and the IH linac is higher than the ratio of RFQ OUT and DTL IN.

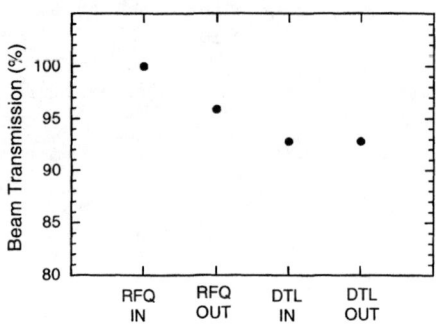

FIGURE 5. Transmission efficiency of the linac complex.

Energy Variable Mode of the IH linac

A variability of the output energy was confirmed in $^{14}N^{2+}$ beam test. Figure 6 shows energy spectra measured by using a momentum-analyzing magnet. The peak indicated at the arrow is the nominal energy of the each tank. For other peaks, the rf power and its phase were set to the values determined by the simulation. The measured output energies well agree with the simulated ones. The measured energy spreads for several output energies were less than the simulated ones. The transmission efficiencies for all energies were over 80%, though the tuning of the quadrupole magnets were not enough.

Deceleration by the IH linac

It is possible to decelerate the beam from the RFQ by the first IH tank. The beam bunch from the RFQ is injected into a deceleration phase of the first IH-tank. In the deceleration test of He^+, the rf phase of the first IH-tank was set to +135 degree. Obtained output energy was 134 KeV/u for the design gap voltage ($V_n=1$) of the first IH tank, and 112 KeV/u for $V_n=1.8$.

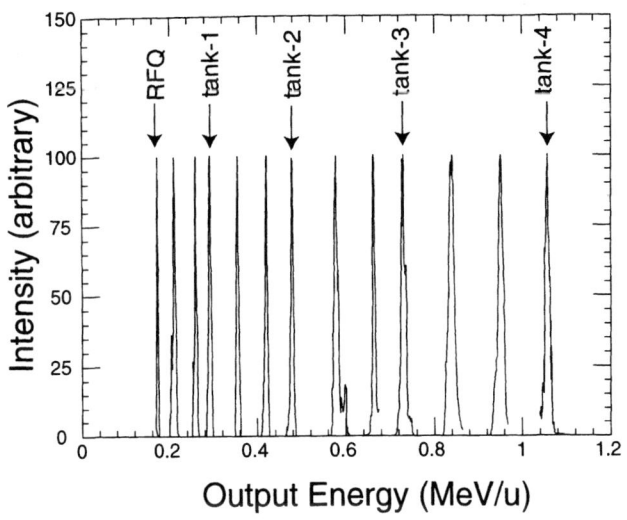

FIGURE 6. Energy spectra measured by using a bending magnet.

Beam test of a stripping foil

The test of a carbon stripping foil was done by using $^{14}N^{2+}$ beam. A result of the beam test are summarized in Table 4. The thickness of the carbon foils is 10 and 15 $\mu g/cm^2$. The energy loss and straggling well agree with the estimated values. The beam transmission efficiency was ~60% of that expected for a most probable charge state ($q = 3+$). A fine tuning of the transverse focusing elements is necessary to increase it. At present, carbon foils with a thickness of 5 $\mu g/cm^2$ are available [9]. In the next test, a growth rate of the transverse emittance will be also measured.

TABLE 4. Measured energy loss and its straggling

Thickness ($\mu g/cm^2$)	0	10	15
energy loss (KeV/u)	0.0	7.7	11.9
energy spread (KeV/u)	2.9	3.9	4.2
energy straggling (KeV/u)	0.0	1.0	1.3

Acceleration Test of RI beam

Acceleration tests of $^{19}Ne^{2+}$ ($T_{1/2}$=17.3 s) have been conducted [8]. The ^{19}Ne beam was produced using $^{19}F(p,n)$ reaction with 30 MeV-protons from the SF cyclotron. The production rate in the target is estimated to be 2×10^9 pps with 1 μA, 30 MeV protons. The ISOL-ECR ion source was in a pulse operation, 2.0 ms in width and 100 Hz in repetition rate, which were determined by the operation of the linac complex. The IH tank-1 through tank-3 were operated to accelerate ^{19}Ne up to 0.52~0.72 MeV/u. The ^{19}Ne-transmission efficiency of the linac complex is estimated to be over 80%. The intensity delivered to a secondary target is not yet enough (1.4×10^6 pps at 1μA protons). Further improvements will be done to perform experiments using accelerated radioactive nuclei.

RECENT PROBLEMS AND THEIR IMPROVEMENTS OF THE LINAC COMPLEX

Asymmetric Excitation of the Quadrupole Magnet

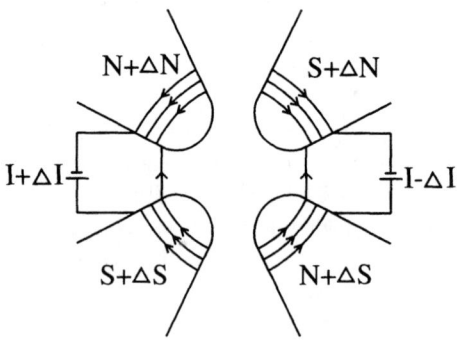

FIGURE 7. Method to generate steering field in a quadrupole magnet.

The large horizontal shift of the beam position was observed in a profile measurement at the exit of the IH linac. We checked alignment of the quadrupole magnets of the IH linac. As a result, we found that the first magnet of the triplet between the tank-1 and tank-2 was horizontally shifted by about 1 mm. This was also confirmed by the following beam test: beam narrowed by a double slits was injected to the IH linac, and one quadrupole magnet of the triplets was excited and the profile was measured at the exit of the IH linac. This process was repeated for other magnets. The relative shift for the magnets was obtained from the measured beam positions.

We estimate that the alignment error has been caused by the deformation of the fittings to fix the magnet on the tank. We tried to compensate beam deflection due to this alignment error by the following method. In a normal quadrupole magnet,

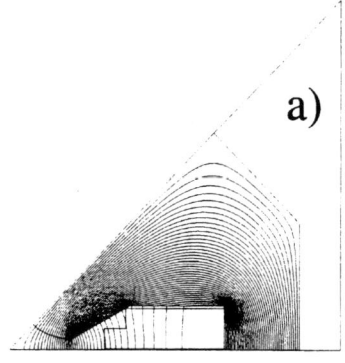 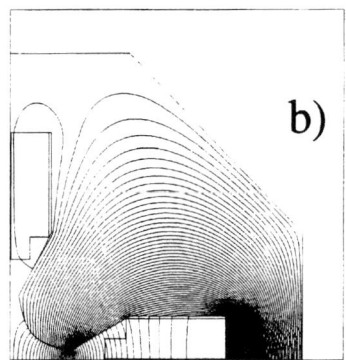

FIGURE 8. Field plots for the quadrupole a) and the steering b) excitations of the quadrupole magnet.

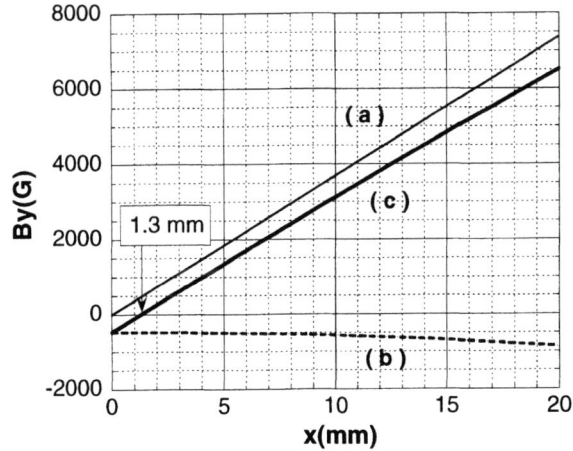

FIGURE 9. Field distribution along x-axis.

four poles are excited at the same current. We can add a steering field (dipole like) to this quadrupole field, when currents different by ΔI for I are asymmetrically excited on the poles by two sets of power supply as shown in Figure 7. Figure 8 shows field plots for the quadrupole and steering excitations of the quadrupole magnet calculated by the POISSON. Bore radius of the quadrupole magnet is 20 mm. Figure 9 shows the magnetic field distribution along horizontal direction. In this figure, (a) shows the field for the quadrupole excitation ($I = 180$ A, $\Delta I = 0$ A), (b) for the steering excitation ($I = 0$ A, $\Delta I = 20.24$ A), and (c) is the added field ($I = 180$ A, $\Delta I = 20.24$ A). The (c) field is horizontally shifted from (a) field by 1.3 mm. Obtained steering field is not ideal dipole field but includes multipole fields. The multipole components for added field are listed in Table 5. We have examined the effect of such multipole fields on the beam by a tracking code. In this code, the multipole fields in Table 5 are taken into account as a thin lense. When the horizontal shift of 1.3 mm for the magnet is assumed in the simulation, the beam deflection due to this horizontal shift is well compensated by the steering components in Table 5. The simulation predicts that emittance growth due to the steering field is almost zero for both horizontal and vertical directions.

Next we performed a beam test. ^{20}Ne^{2+} beam was accelerated up to 0.47 MeV/u (tank-2 mode). Figure 10 shows the emittance measured at the exit of the IH linac. Beam is well adjusted on center in the horizontal(x)-phase space by the steering field of the quadrupole magnet(see a)). On the other hand, the beam is shifted without the steering field (see b)). In case of a), the beam transmission through the IH linac was near 100%. It was difficult to get such a high transmission keeping the beam on center without this steering field, even if the steering magnets placed upstream of the IH linac were used.

TABLE 5. Multipole components for I=180 A, ΔI=20.24 A

Multipole Field	
B_y [kG]	-4.7785×10^{-1}
dB_y/dx [kG/cm]	$+3.6905$
$d^2 B_y/dx^2/2!$ [kG/cm^2]	-9.3093×10^{-2}
$d^4 B_y/dx^4/4!$ [kG/cm^4]	-3.0313×10^{-3}
$d^5 B_y/dx^5/5!$ [kG/cm^5]	$+4.6831 \times 10^{-4}$
$d^6 B_y/dx^6/6!$ [kG/cm^6]	$+5.9441 \times 10^{-4}$
$d^8 B_y/dx^8/8!$ [kG/cm^8]	$+1.6565 \times 10^{-5}$
$d^9 B_y/dx^9/9!$ [kG/cm^9]	$+4.9045 \times 10^{-5}$
$d^{10} B_y/dx^{10}/10!$ [kG/cm^{10}]	-7.5786×10^{-6}
$d^{12} B_y/dx^{12}/12!$ [kG/cm^{12}]	-2.3360×10^{-7}
$d^{13} B_y/dx^{13}/13!$ [kG/cm^{13}]	$+4.3640 \times 10^{-6}$
$d^{14} B_y/dx^{14}/14!$ [kG/cm^{14}]	$+7.0306 \times 10^{-8}$
$d^{16} B_y/dx^{16}/16!$ [kG/cm^{16}]	-5.1835×10^{-9}
$d^{17} B_y/dx^{17}/17!$ [kG/cm^{17}]	$+2.9843 \times 10^{-7}$
$d^{18} B_y/dx^{18}/18!$ [kG/cm^{18}]	-3.2048×10^{-9}

Emittance Monitor

In the linac complex, transverse emittance monitors have been equipped at the entrance and the exit of the SCRFQ and the IH linac. These monitors are conventional double slits ones. A shorter measuring time is desired for the efficient beam

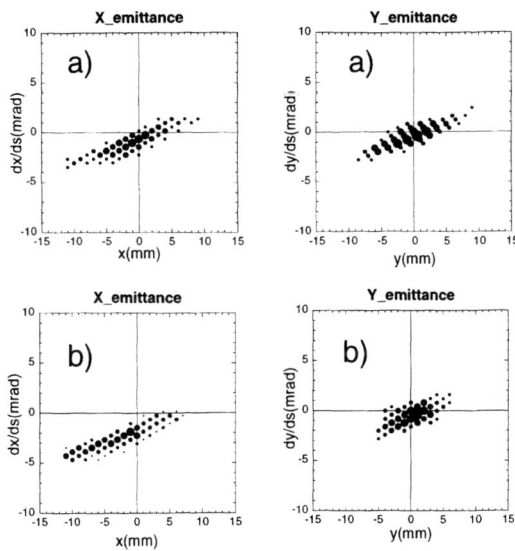

FIGURE 10. Emittances measured at the exit of the IH linac.

tuning. To shorten it, the current meter and the GPIB board were replaced to new ones, and the software for the emittance measurement was changed from the MSDOS based one to the LabVIEW. As a result, the measuring time was improved to half of the previous one. Still, It takes, for instance, about 35 min. for the measurement of 40×80 points. Most of the measuring time is spent to move the 1st and 2nd slits by stepping motors. We have tested the following monitor; the 1st and 2nd slit are moved at the same time keeping the same position, and in order to measure the angular distribution, the beam is scanned by changing the field of a magnet placed between the 1st and the 2nd slits (see Fig. 11). The time for the beam scan can be much shorter. A beam test was performed using the double slits and the steering magnet placed at the exit of the SCRFQ. The measuring time was improved to 4 min. for the measurement of 40×80 points. The beam center in the phase space measured by the beam scan was slightly different from that by the double slits. This is caused by changing the field fast. This problem could be solved by replacing the block magnet to a laminated one.

Phase Stability of the Cavities

It is very important for our linacs to keep rf phase of each cavity constant, since the output energy is sensitive to it. The phase stability is mainly determined by

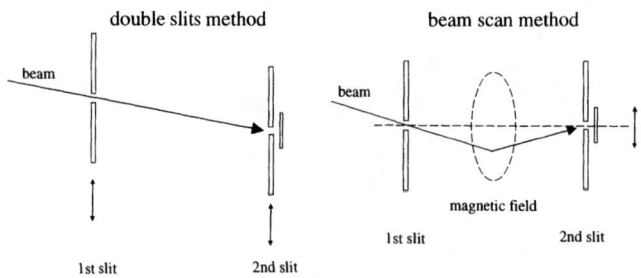

FIGURE 11. Schematic diagram of the emittance monitors

temperature change of the cooling water for the cavities. Inductive piston tuners to compensate frequency shift due to the temperature change have been equipped in the cavities. These tuners are automatically moved so as to maximize the signal level from the cavities by a personal computer (see tuner control in Fig. 12). But they are not enough to keep the phase within an acceptable range.

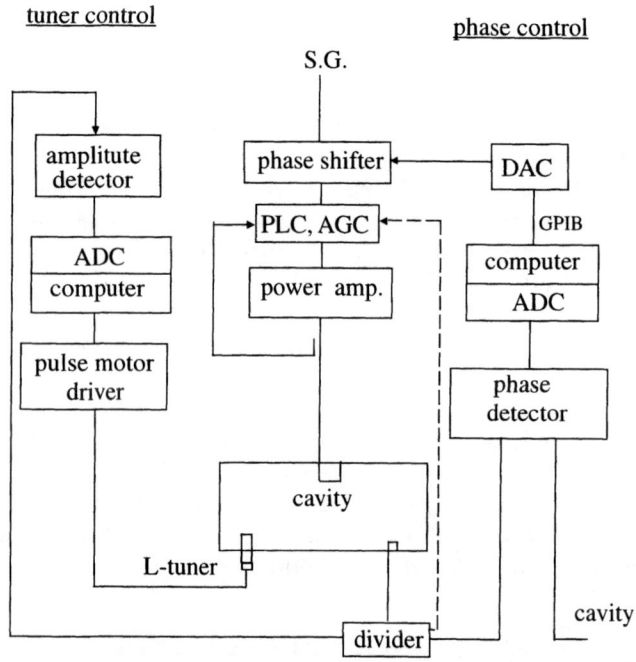

FIGURE 12. Block diagram of the tuner and phase control system

A phase control system to improve the phase stability is now under test. In this system, the phase between adjacent cavities is measured by a phase detector, and its signal is taken by a personal computer through an AD converter. The DC signal from the personal computer is fed into a phase shifter placed in front of a phase lock circuit (PLC) in the rf source (see phase control in Fig. 12). We expect the phase stability is improved within ±0.2 deg. by this system.

In the present rf sources, a forward signal from the final amplifier are used as a feedback signal to the PLC. We are also testing to feed the pick-up signal from the cavity into a new low-level PLC for the RFQ (see dotted line in Fig. 12). In this case, a fast feedback time are also expected.

FUTURE PLAN

The E-arena in proposed Japanese Hadron Facility (JHF) is an upgrade of the RB facility at Tanashi, and provides radioactive beams with broader mass range and higher energies above the Coulomb barrier [10]. In the E-arena, the radioactive beam is produced by 3 GeV, 10 μA protons [11]. The existing 25.5-MHz SCRFQ and 51-MHz IH linac (IH-1) are used to accelerate the radioactive beam up to 1 MeV/u. The output energy is boosted up to 6.5 MeV/u by a newly constructed IH linac (IH-2). A prebuncher to improve a longitudinal emittance is installed upstream of the SCRFQ. Main Parameters of the E-arena is shown in Table 6.

TABLE 6. Main Parameters of the E-arena

primary accelerators		linac and synchrotron
	beam	protons
	Intensity	10 μA
	energy	3 GeV
secondary accelerators		
SCRFQ	frequency	25.5 MHz
	A/q	30
	duty factor	30%
	energy	0.002~0.17 MeV/u
IH-1	frequency	51 MHz
	A/q	10
	duty factor	30~100%
	energy	0.17~1 MeV/u
IH-2	frequency	102 MHz
	A/q	7
	duty factor	30~100%
	energy	1~6.5 MeV/u

Carbon-stripping foils are used for the acceleration of the beam with a heavy mass. Figure 13 shows the relation between a mass to charge ratio (A/Q) and an atomic number (Z) of the beam at 0.17 and 1 MeV/u, where Q is the most probable charge under an equilibrium state, A is a mass number of the beam with the proton number Z. We here assume that the radioactive beam are near the stability line

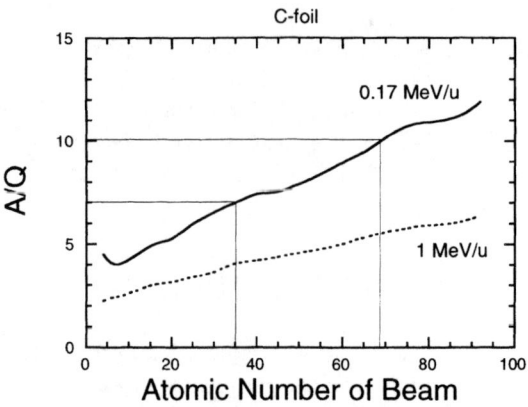

FIGURE 13. Relation between mass to mass to charge ratio A/q and atomic number Z

in a nuclear chart. Reference [12] was used to obtain the most probable charge state of ions stripped on a carbon foil. Stripping schemes for the acceleration by the IH-2 are as followed; (1) No stripping foil is used for light ions with $A/q \leq 7$. (2) A stripping foil at 1 MeV/u is used for ions with $7 \leq A/q \leq 10$. (3) Stripping at 0.17 MeV/u. As seen in Fig. 13, ions up to $Z \sim 35$ corresponding to $A/Q=7$ are accelerated. (4) Two stage stripping at 0.17 and 1 MeV/u. Heavy ions up to $Z \sim 65$ are accelerated. In this case, ions with a high charge must be produced by the ion source. Beam intensity is reduced to % level by the two stage stripping.

Rf frequency of the IH-2 linac is chosen to be 102 MHz, which is twice of that of the IH-1. The IH-2 has the same acceleration scheme as that of the IH-1. In the present design, the IH-2 consists of 11 tanks and 10 quadrupole triplets between tanks. Each tank has 16 gaps at which 300 kV is supplied. Total length of the IH-2 is about 30 m and effective acceleration voltage is 40 MV.

According to the beam simulation, transverse and longitudinal acceptances of the IH-2 are 1.4π mm·mrad(normalized) and 550π KeV/u·deg at 102 MHz, respectively. They would be enough to accept the beam from the IH-1. Intertank length for the quadrupole triplet is chosen to 60 cm. Field strength required for the quadrupole magnet is about 55 T/m for bore radius of 15 mm.

Preliminary cavity analysis was performed by MAFIA code. The tank radius is 0.35 through 0.4 m. The effective shunt impedance is 100 to 200 MΩ/m, where we assume that real surface conductivity of the cavities is 70% of the ideal one. Total power loss in the cavities is about 700 kW. In the present design, power deposit per unit length for several tanks in low energy side is slightly too high for 100%-duty operation. The design gap voltage of their tanks will be reduced to 200 to 250 kV. Optimization study in order to obtain a flat field distribution and a higher shunt impedance will be done.

ACKNOWLEDGMENTS

We express our thanks to T. Nomura for his continuous encouragement. The target, ISOL-ion source, ISOL, 60 m LEBT,HEBT and RMS have been constructed and serviced by the E-arena group in Tanashi.

REFERENCES

1. Y. Shirakabe et al., 'Bunching of Radioactive Ion Beam in the Millisecond Region -a Pulsed Gating-Potential Method in an Ion Source for On-Line Isotope Separation', Nucl. Instr. and Meth., **A337**(1993)11.
2. S.C. Jeong et al., 'Beam-Bunching in an ECR Ion Source by the Pulsed Gating-Potential Method', Nucl. Instr. and Meth., **B114**(1996)154.
3. T. Nomura et al., 'Simple On-Line Method of Measuring the Absolute Ionization Efficiency of an Ion Source', Nucl. Instr. and Meth., **B93**(1994)492.
4. M. Wada et al., 'High-Resolution Mass Separator and 60 m Beam Transport Line for the Radioactive Nuclear Facility at INS', Nucl. Instr. and Meth. B, to be published.
5. T. Nomura, 'Exotic Nuclei Arena in Japanese Hadron Project', INS-Report-780(1989).
6. S. Arai et al., 'Construction and Beam Tests of a 25.5-MHz Split Coaxial RFQ for Radioactive Nuclei', Nucl. Instr. and Meth. **A390**(1997)9-24.
7. K. Niki et al., 'Beam Transport Design for the Linac System in the INS Radioactive Beam Facility', Proc. of 1994 Int. Linac Conf., 1994, p.725.
8. M. Tomizawa et al., 'Progress Report of the Interdigital-H Linac for Radioactive Nuclei at INS', Proc. of the 5th European Particle Accelerator Conf., 1996, p.780.
9. I. Sugai,Private communication.
10. JHF Project Office, 'Proposal for Japan Hadron Facility', KEK Report 97-3.
11. JHF Project Office, 'JHF Accelerator Design Study Report', KEK Report 97-16.
12. K. Shima et al., 'Equilibrium Charge Fraction of Ions of Z=4-92(0.02-6 MeV/u) and Z=4-20 (up to 40 MeV/u) Emerging from a Carbon Foil', NIFS-DATA-10 (1991).

Plans for Constructing a Next-generation ISOL Facility at ORNL

J. D. Garrett,* G. D. Alton,* R. L. Auble,* C. Baktash,* J. R. Beene,*
F. E. Bertrand,* J. D. Fox,* R. A. Gough,[†] M. L. Halbert,*
J. G. Kalnins,[†] Y. Liu,* M. W. Ogan,[#] F. Plasil,* D. Shapira,*
P. T. Spampinato,[¶] J. W. Staples,[†] H. Wollnik,[‡] and M. S. Zisman[†]

* Physics Division, Oak Ridge National Laboratory, Oak Ridge, Tennessee, 37831
[†] Accelerator and Fusion Research Division, Lawrence Berkeley National Laboratory,
Berkeley, California 94720
[#] Engineering Division, Oak Ridge National Laboratory, Oak Ridge, Tennessee, 37831
[¶] Robotics and Process Systems Division, Oak Ridge National Laboratory,
Oak Ridge, Tennessee, 37831
[‡] University of Giessen, Giessen D-6300, Germany

Abstract. The U.S. Nuclear Science Community in its 1996 Long Range Plan identified an advanced radioactive ion beam (RIB) facility based on the ISOL technique as the next major facility to be constructed for U.S. nuclear physics. The proposed Spallation Neutron Source (SNS) for Oak Ridge National Laboratory, whose construction design funds have recently been appropriated, offers a unique opportunity for the construction of this new facility. Plans for extracting a proton beam from the SNS, transporting it to the RIB facility, and constructing the new RIB facility at the SNS site are discussed, as are the ISOL targets, radiation handling, isobaric separation, acceleration of beams of radioactive ions, and the layout of the experimental areas.

BACKGROUND

The Nuclear Science Advisory Committee (NSAC), an official DOE/NSF advisory body, has identified an Isotope-Separator-On-Line (ISOL) Radioactive Ion Beam (RIB) Facility in its 1996 Long Range Plan for U.S. Nuclear Science as the next major facility to be constructed under the auspices of DOE's Nuclear Physics Program Office. The Long Range Plan states: "The scientific opportunities made available by world-class radioactive beams are extremely compelling and merit very high priority. The U.S. is well-positioned for a leadership role in this important area. ... We strongly recommend development of a cost-effective plan for a next generation ISOL-type facility and its construction when current major construction activities are substantially complete."

The on-line isotope-separator technology (1), developed at CERN and at other facilities, has progressed to the extent that it is feasible to produce and accelerate beams of short-lived isotopes with sufficient intensity to address a large variety of new nuclear structure, nuclear astrophysics, and materials science problems (2-6). Several first-generation ISOL facilities, e.g., the Holifield Radioactive Ion Beam Facility (7) (HRIBF) at ORNL, are being constructed or are in operation in North America, Europe, and Japan, based on existing accelerators and reactors.

A Workshop on the Science for an Advanced ISOL Facility, organized jointly by ORNL and ANL, was held at Ohio State University in Columbus, Ohio, July 30 -- August 1, 1997. The scientific case developed at this workshop was summarized in a White Paper (6), which was submitted to the Nuclear Physics Program Office at DOE in November 1997. This White Paper has been used as the basis of a "Mission Need" document which has been prepared by the Nuclear Physics Program Office at DOE. Current DOE plans calls for starting the construction of this Next-generation ISOL Facility (NISOL) after the peak of the Spallation Neutron Source (SNS) construction budget which is projected to be reached in FY 2002.

The proposed NISOL Facility must provide the large variety of very intense radioactive ion beams necessary to take full advantage of the scientific opportunities afforded by this new interdisciplinary research tool. Indeed, in 1990 a North American Steering Committee for Radioactive Ion Beam Physics was established. After canvassing the nuclear physics and nuclear astrophysics communities, this committee proposed (2) an advanced ISOL facility in 1991, in which the radioactive atoms were produced using 1-GeV protons. Due to the cost of constructing a 1-GeV proton accelerator (estimated to be in excess of $400 million for the SNS more recent considerations (8) of the NISOL Facility have been based on 200-250 MeV accelerated protons and deuterons. However, the recent funding of the SNS (9) with its beam of 1-GeV protons provides the unique opportunity to utilize the advantages of higher-energy protons for producing radioactive ions. The higher-proton energy spreads the energy loss through a thicker target. Thus, for the same power dissipated in a given target, the 1-GeV proton beam will produce a smaller density of energy deposited per unit volume than a less energetic proton or heavier-ion beam. The higher-energy proton beam will also provide a greatly increased production yield of both neutron-rich fission products and predominately proton-rich spallation products. Siting the NISOL Facility at the SNS will provide a "green field" for the remainder of the facility in which the RIB accelerator and experimental equipment can be optimized with the money saved by using proton beams from the SNS linac to produce the radionuclides. Likewise, other cost efficiencies, e.g., in facility operation and ESH, are being investigated. Construction at the SNS site will also minimize the impact on the HRIBF operation during the construction. During this period the HRIBF will be the only ISOL facility in the U.S., and its technical and scientific program will be crucial in defining the early program for the NISOL Facility.

SCIENTIFIC MOTIVATION

A description of the multidisciplinary scientific topics that a NISOL Facility can address is documented in a number of reports (2-8). The most recent such document, the OSU Symposium White Paper (6) provided to the U.S. DOE in November 1997, is being used to justify the construction of a NISOL Facility. Such a facility will afford research opportunities to a variety of scientific disciplines as diverse as nuclear structure physics, astrophysics, standard model tests, materials science, tribology, and biological and medical sciences. Siting the NISOL Facility together with the SNS would provide radioactive ions for materials science and biological studies at the same location as the premier neutron scattering facility.

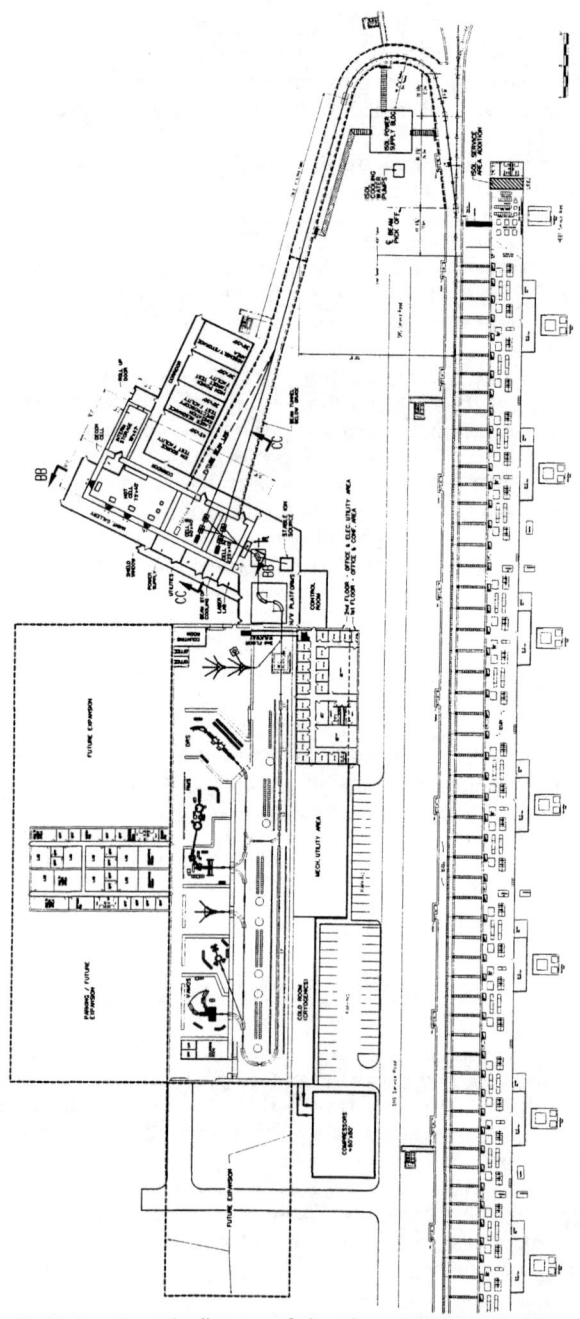

FIGURE 1. Preliminary schematic diagram of the planned NISOL Facility at the SNS site at ORNL. Enlarged diagrams of the target, ion source, and radiation areas and the accelerator and experimental areas of this planned facility are shown in Figures 4 and 9, respectively.

The NISOL facility proposed for the SNS site at ORNL will be capable of providing a broad range of intense proton- and neutron-rich beams of radioactive ions to a large scientific user community. The new facility will produce intense beams of many neutron- and proton-rich isotopes whose halflives are greater than about one second and that are sufficiently volatile to diffuse out of a heated target. Both proton- and neutron-rich RIBs will be accelerated from tens of keV, for materials- and biological-science studies, nuclear mass measurements, and radioactive target preparation, to energies near and above the Coulomb barrier for nuclear structure, nuclear reaction, and nuclear astrophysics studies.

THE NISOL FACILITY AT THE SNS

The general layout of the NISOL Facility which could be constructed at the SNS Site is shown in Figure 1 and an artistic conception of what this facility might look like is shown in Figure 2. A time-averaged beam intensity of about 100 µA, i.e., 10% of the SNS's initial designed beam intensity, would be added to the SNS H⁻ beam. This added intensity would be extracted in the High Energy Beam Transport (HEBT) line downstream from the end of the SNS H⁻ linac and provided to the NISOL facility. The extraction will be accomplished by deflecting the H⁻ beam by about 2.5° as shown in Figure 3. After deflection a thin foil (or perhaps a wire or a grid) would be inserted in the edge of the H⁻ beam to strip the desired portion of the H⁻ beam to H⁺, i.e., to protons. The proton beam will be separated from the H⁻ beam in a separator magnet. Then it will be transported to the NISOL facility through a high-energy proton transport line, see Figure 1. The unstripped H⁻ beam (>90% of the total beam) will be deflected in the opposite direction from the proton beam in the separator magnet and returned to the HEBT by another small magnet. Preliminary calculations indicate that these three magnets should have fields below about 0.3 Tesla. The intensity of the stripped proton beam can be varied by moving the foil; fine tuning can be accomplished by "tweaking" the deflection magnet upstream of the stripper. Stripping only a limited area of the H⁻ beam ensures that the emittance of the extracted proton beam will be comparable to or less than that provided to the SNS. Using a relatively thick stripper foil (≈ 1 mg/cm^2) minimizes the production of H° atoms. Nevertheless, it will be necessary to provide a low-intensity (<10 Watts of 1 GeV H°) beam stop to safely dispose of this small fraction of the H⁻ beam that remains as H° after the stripping.

The proton beam will enter the NISOL facility on the lower level where all the target stations and the other high radiation areas will be located. Two sets of two closely spaced target stations are indicated in Figures 4, 5 and 6. Options for operating the two stations on a specific beam line by allowing the proton beam to simultaneously transverse both targets or by time sharing the beam between these two targets are being investigated. An additional target area to the right of the two shown also is being reserved for future developments.

The ion sources would be located above, but perhaps offset, from the target. This slight offset would reduce the line of sight radiation from the target up the beam line. The ionized beams would be focused upward through a series of separated vacuum envelopes to a low–resolution ($\Delta m/m \approx 1/500$) preseparator which would bend the beam into the horizontal plane (see Figure 7). The low–resolution separator would remove the radioactive ions with masses other than that of the isotope of interest thereby reducing the radioactivity deposited in the remaining separators and beam–handling devices. It might be possible to recover a portion of these radionuclides for use as targets or other purposes. All of the services to the target and ion source and the associated vacuum enclosures (electrical, cooling water, control, He, laser channel, etc.) would be provided through the vertical tube. The vacuum to be provided by

FIGURE 2. Artist's conception of the planned NISOL Facility at the SNS site at ORNL.

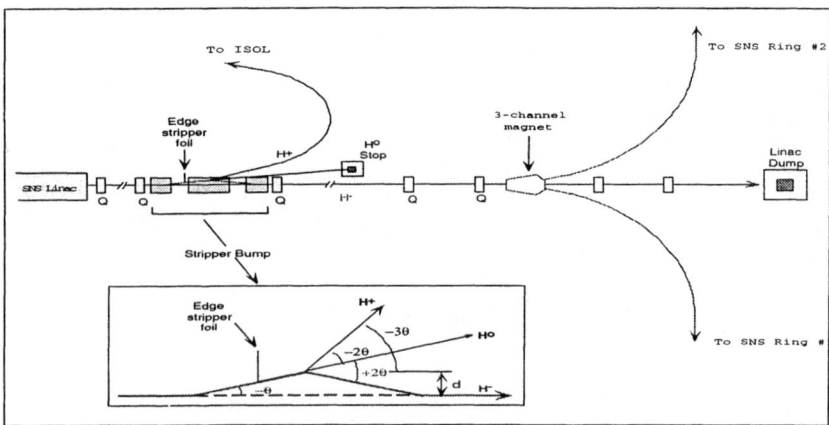

FIGURE 3. Schematic diagram illustrating the technique for extracting up to ≈ 10% of the H⁻ beam after the SNS linac, converting it to protons by stripping, and providing it to the NISOL site while returning the remainder of the beam to the HEBT of the SNS.

cryopanels should be sufficient for ion source operation ($\approx 10^{-6}$ Torr). The three separated vacuum envelopes of the vertical tube and preseparator are designed to maximize the localization of the radioactivity in the lower portions of this assembly. The whole target ion source assembly together with the three-stage vacuum envelopes (and perhaps the low–resolution separator magnet) would be removed as a unit for service. This assembly would be removed vertically by a crane which would insert it into shielded hot cells for servicing the target and ion source assemblies--See Figure 6. Target and ion source assembly, servicing, testing, and used target storage areas would be located near the RIB production target areas as indicated in Figure 4.

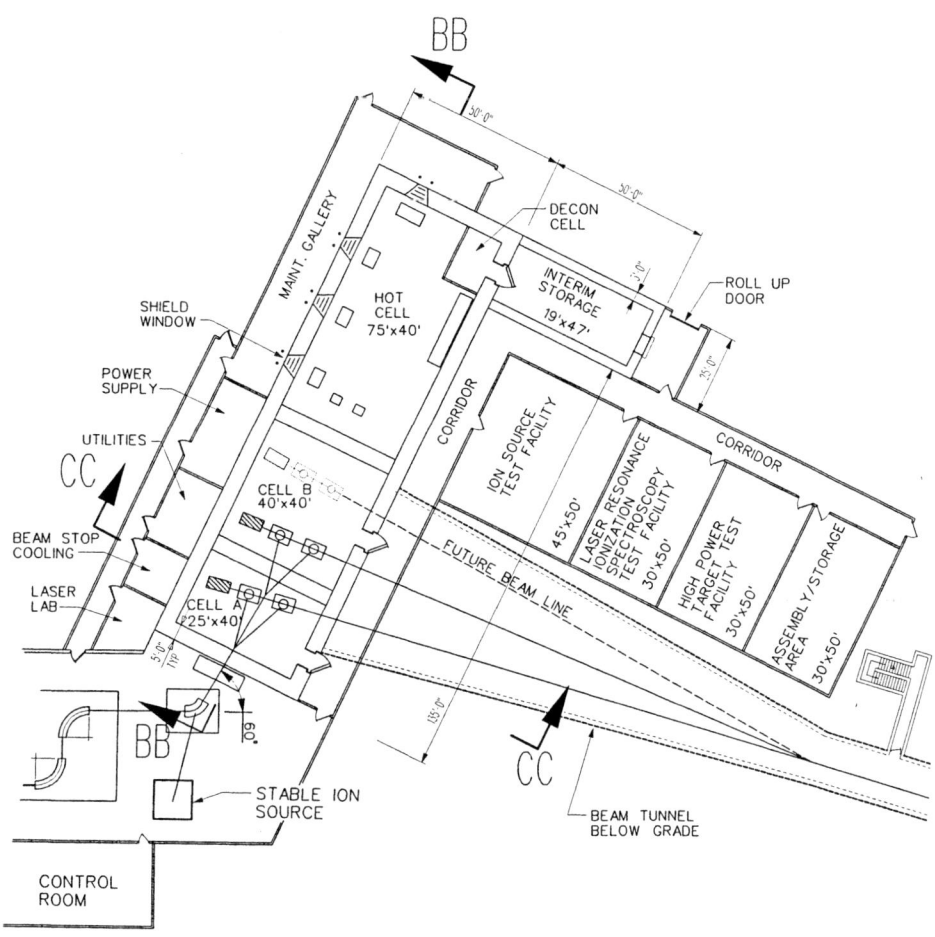

FIGURE 4. Expanded portion of Figure 1 showing the target, ion source, radiation handling, and hot cell areas for the planned NISOL facility at the SNS site.

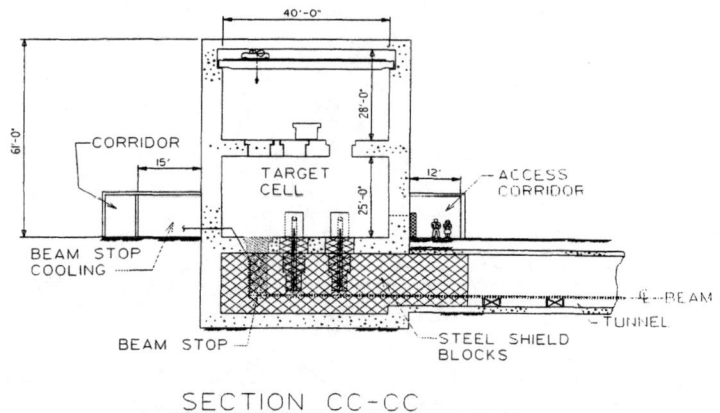

FIGURE 5. Cross-sectional side view along the beam direction of the proton beam line showing the target and beam stop areas, see Figure 4.

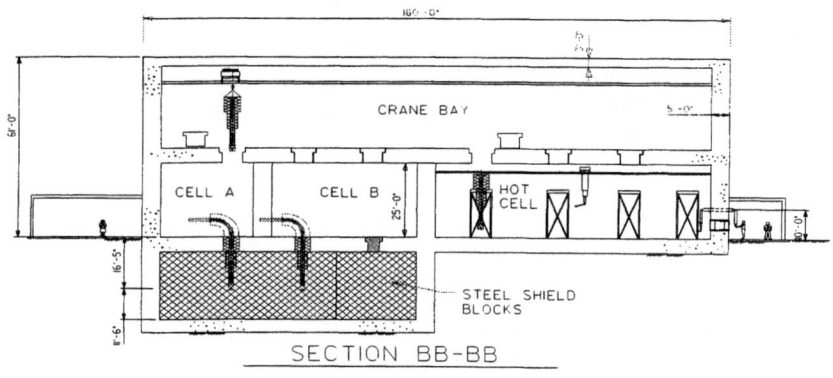

FIGURE 6. Cross-sectional side view perpendicular to the beam line showing a target area, the radiation handling, and the hot cell areas, see Figure 4. Note a target ion source assembly in transit (in the crane bay) between target cell A. Likewise, similar assemblies are shown in the target positions and being serviced in the hot cell. An expanded view of the target ion source assembly is shown in Figure 7.

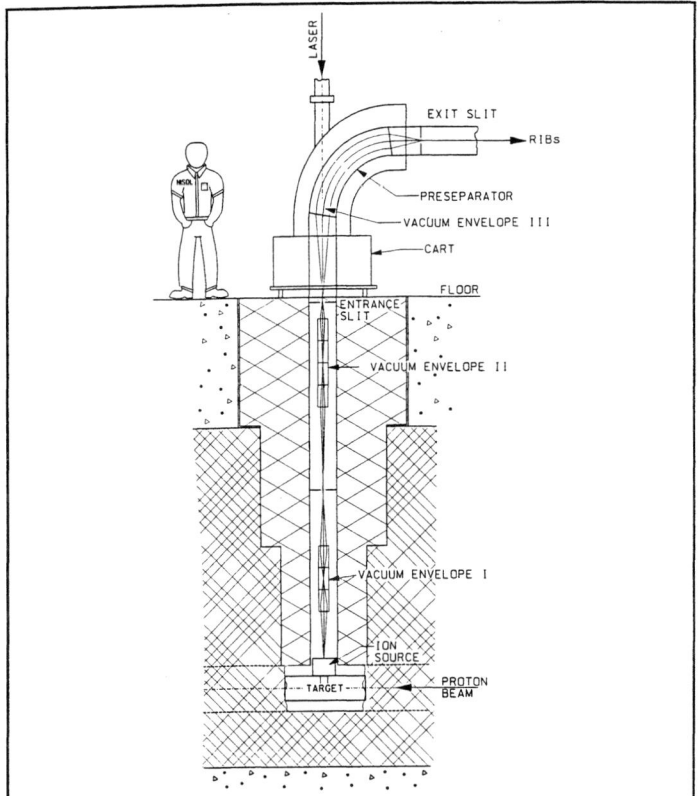

FIGURE 7. Expanded view of the target, ion source, vertical beam transport, and preseparator showing the three-stage vacuum envelopes maximizing radiation containment in the shielded areas. This assembly would be remotely removed vertically and serviced in the hot cell shown in Figures 4 and 6.

Ion sources under consideration for the NISOL facility include electron cyclotron resonance (ECR), electron beam plasma (EBP) (i.e., modifications of the sources used for ISOLDE (1) and the HRIBF (7), and laser sources. A stable ion source for providing test beams of stable ions also is shown in Figures 1 and 4.

A three–stage isobar separation system with a resolution of $\Delta m/m \approx 1/20,000$ is shown in Figures 1 and 4. The low–resolution vertical-to-horizontal bending preseparator (described above) is the first stage of this system. The second and third stages are an improved version of the present HRIBF isobaric analysis system (7) with a single–magnet second stage ($\Delta m/m \approx 1/2500$) and a double–bend third stage ($\Delta m/m \approx 1/20,000$). To provide maximum flexibility for a wide variety of masses and charge states, all these magnets will be at high voltage as will the RFQs. Space will be reserved before the second stage isobaric analyzer for the installation of a beam cooler. Likewise space also will be reserved between the isobaric analyzer and the RFQ for the future installation of a charge state enforcer, e.g., an electron beam ion trap (10) (EBIT), if this technology proves to be useful.

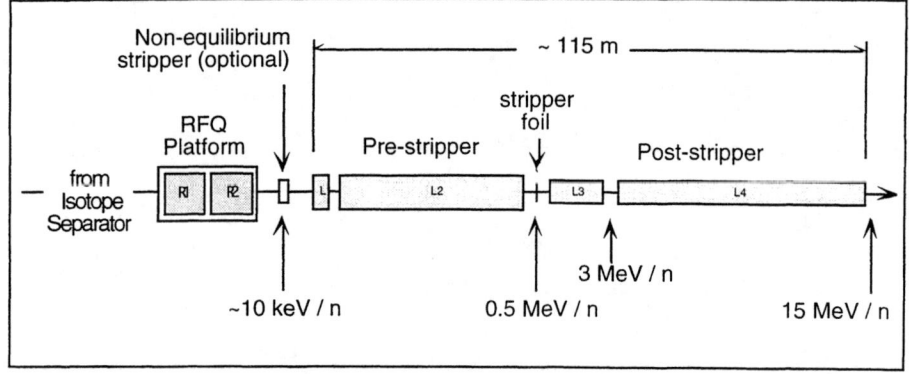

FIGURE 8. Schematic diagram of a possible layout of the RFQs and the superconducting linac for the NISOL facility.

A layout for a two–stage RFQ and a four–stage superconducting heavy ion linac, capable of accelerating single–charged ions of A≈140 to 15 MeV/u is shown in Figures 1, 8 and 9. In this scheme the heavy ions would be stripped to an equilibrium charge state after the first two stages of the heavy-ion linac. The heaviest ions (i.e., masses greater than about 140) would be extracted as q = 2 from the ion source or stripped to q = 2 after the first RFQ stage. The RFQ will probably be of the split coaxial type (SCRFQ) now operating (11) at the KEK/Tanashi Facility (previously the Institute of Nuclear Studies of the University of Tokyo) in Tokyo. To achieve the very low q/m values (≈1/140) the SCRFQ would operate at about 12.5 MHz.

A large, versatile experimental area, allowing a nearly continuous range of beam energies (from ≈100 keV to 15 MeV/u) to be directed to a variety of experimental apparatus, is shown in Figures 1 and 9. From left to right the experimental apparatus shown for illustrating the size and the versatility of this area are:
- A highly efficient mass separator (e.g., VAMOS proposed for GANIL);
- a versatile time-of-flight setup for nuclear reactions studies;
- a series of general purpose beam lines for customized small experimental setups;
- the present HRIBF Recoil Mass Separator (RMS) with a large germanium array (e.g., GRETA) at the target position and a battery of focal–plane detectors;
- an upgraded version of the Daresbury Recoil Separator (DRS) optimized for astrophysics experiments; and
- an "unaccelerated" beam area including mass traps and other apparatus for materials science and biomedical studies with unaccelerated beams.

For more information on the experimental equipment applicable to the NISOL Facility the reader is referred to the proceedings of a recent workshop on this topic (12).

To take advantage of the "green field" approach available at the SNS site the area at the high–energy end of the heavy–ion linac is reserved for future upgrades as are areas adjacent to the experimental hall. Likewise additional space is provided for an upgraded RIB target station, since targets and ion sources are considered to be the area that future technical breakthroughs are most likely to occur.

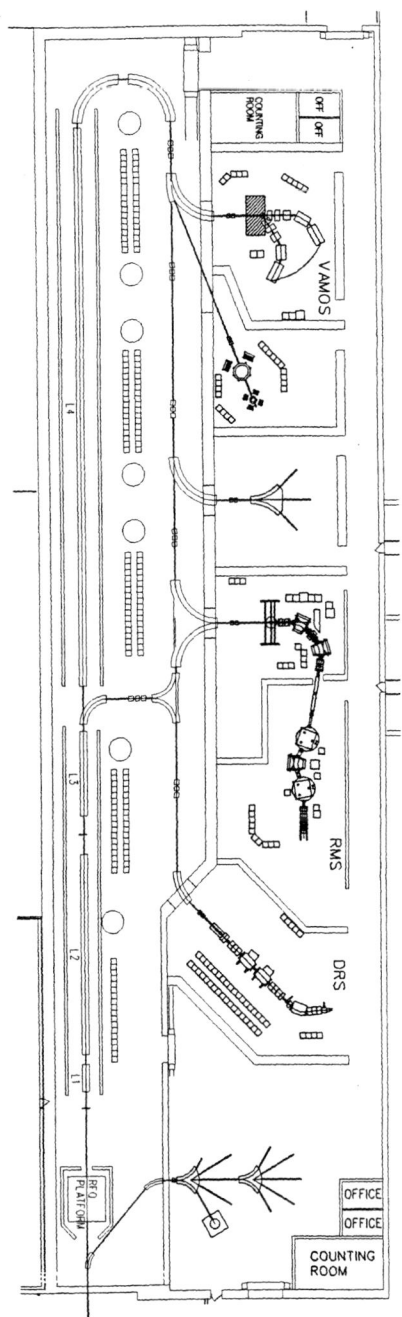

FIGURE 9. Expanded portion of Figure 1 showing RFQs and a superconducting linac capable of accelerating q=1 ions of A=140 to 15 MeV/u. A layout of a variety of experimental equipment described in the text also is shown to illustrate the dimensions of the experimental area.

ACKNOWLEDGMENTS

The authors acknowledge many informative discussions with Jose Alonso and other members of the SNS technical staff. This research is sponsored by the Laboratory Directed Research and Development Program of Oak Ridge National Laboratory, managed by Lockheed Martin Energy Research Corp. for the U.S. Department of Energy under contract DE-AC05-96-OR22464. The LBNL portions of this work also were partially supported by the U.S. Department of Energy under contract No. DE-AC03-76SF 00098.

REFERENCES

1. Ravn, H. L. and Allardyce, B. W, "On-Line Mass Separators," *Treatise on Heavy Ion Science*, ed. Bromley, D.A. (Plenum Press, 1989, New York, Vol. **8**, pg. 363).

2. Casten, R. F., D'Auria, J. M., Davids, C. N., Garrett, J. D., Nitschke, J. M., Sherrill, B. M., Vieira, D. J., Wiescher, M. and Zganjar, E. F., "The Iso-Spin Laboratory: Research Opportunities with Radioactive Nuclear Beams," Los Alamos National Laboratory Report LALP91-51.

3. Casten, R. F., D'Auria, J. M., Davids, C. N., Garrett, J. D., Nitschke, J. M., Sherrill, B. M., Vieira, D. J., Wiescher, M. and Zganjar, E. F., "Overview of Research Opportunities with Radioactive Nuclear Beams," February 1995, available from R. F. Casten, Yale University.

4. "European Radioactive Beam Facilities," NuPECC Report, May 1993.

5. Tanihata, I., "RI Beam Project," *Nuclear Physics* **A616**, pp. 56c-68c (1997) and "RI Beam Factory: Basic Science," RIKEN Accelerator Research Facility Report, August 1994.

6. Casten, R.F. et al., "Scientific Opportunities with an Advanced ISOL Facility," OSU Symposium White Paper, November 1997, available from J.D. Garrett, ORNL, or from the NISOL website (http://www.phy.ornl.gov/nisol.html).

7. Garrett, J. D., *Nuclear Physics* **A616**, pp. 3c-10c (1997) and references therein.

8. Rehm, K. E. et al., "Concept for an Advanced Exotic Beam Facility based on ATLAS," Physics Division Annual Report, Argonne National Laboratory, April 1, 1994-March 31, 1995, ANL-95/14, p. 56.

9. "The Spallation Neutron Source Conceptual Design Report," http://www.ornl.gov/~nsns/CDRDocuments/CDR.html.

10. EBIT, Marrs, R. E. et al., *Phys. Rev. Lett.* **60**, 1715 (1988); and Levine, M. A. et al., *Phys. Scr.* **T22**, 157 (1988).

11. Arai, S. et al., *Nuclear Instruments and Methods* **A390** (1997) pp. 9-24.

12. *Proceedings of the Workshop on the Experimental Equipment for an Advanced ISOL Facility*, edited by I. Y. Lee, Nuclear Science Division, Lawrence Berkeley National Laboratory, July 22-25, 1998.

An Advanced ISOL Facility Based on ATLAS

J. A. Nolen, K. W. Shepard, R. C. Pardo, G. Savard, K. E. Rehm,
J. P. Schiffer, W. F. Henning, C.-L. Jiang, I. Ahmad, B. B. Back,
R. A. Kaye, M. Petra, M. Portillo, J. P. Greene, B. E. Clifft,
J. R. Specht, R. V. F. Janssens, R. H. Siemssen,
I. Gomes,* C. B. Reed,* A. M. Hassanein[†]

Physics Division, *Technology Development Division, [†]Energy Technology Division
Argonne National Laboratory, Argonne, IL 60439 USA

Abstract. The Argonne concept for an accelerator complex for efficiently producing high-quality radioactive beams from ion source energy up to 6-15 MeV/u is described. The Isotope-Separator-On-Line (ISOL) method is used. A high-power driver accelerator produces radionuclides in a target that is closely coupled to an ion source and mass separator. By using a driver accelerator which can deliver a variety of beams and energies the radionuclide production mechanisms can be chosen to optimize yields for the species of interest. To effectively utilize the high beam power of the driver two-step target/ion source geometries are proposed: (1) Neutron production with intermediate energy deuterons on a primary target to produce neutron-rich fission products in a secondary ^{238}U target, and (2) Fragmentation of neutron-rich heavy ion beams such as ^{18}O in a target/catcher geometry. Heavy ion beams with total energies in the 1-10 GeV range are also available for radionuclide production via high-energy spallation reactions. At the present time R&D is in progress to develop superconducting resonator structures for a driver linac to cover the energy range up to 100 MeV per nucleon for heavy ions and 200 MeV for protons. The post accelerator scheme is based on using existing ISOL-type 1+ ion source technology followed by CW Radio Frequency Quadrupole (RFQ) accelerators and superconducting linacs including the present ATLAS accelerator. A full-scale prototype of the first-stage RFQ has been successfully tested with RF at full design voltage and tests with ion beams are in progress. A benchmark beam, ^{132}Sn @ 7 MeV/u, requires two stripping stages, one a gas stripper at very low velocity after the first RFQ section, and one a foil stripper at higher velocity after a superconducting-linac injector.

INTRODUCTION

There is much enthusiasm in the nuclear physics community for the research opportunities that would be enabled by an advanced, high intensity accelerated radioactive beam facility based on the isotope-separator on-line (ISOL) method. There are recent reports from both North American and European study groups (1,2). A group including many ANL Physics Division staff and ATLAS outside users has discussed the research possibilities and prepared a working paper entitled "Concept for an Advanced Exotic Beam Facility Based on ATLAS." The working paper is available on the World Wide Web at the ANL Physics Division home page

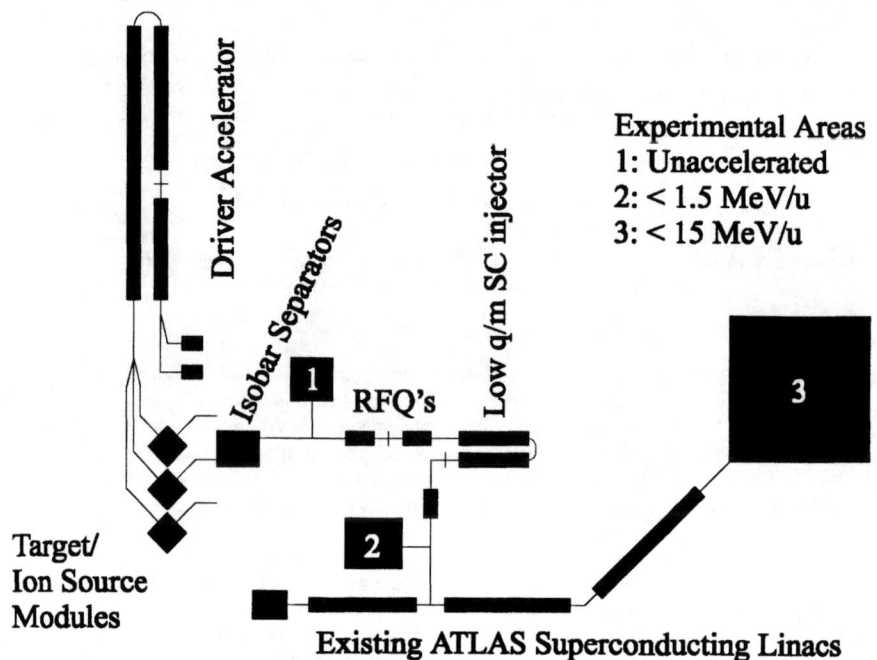

FIGURE 1. A block diagram of the proposed Advanced ISOL Facility based on ATLAS at Argonne. The major components of the complex, including the three different experimental areas, are indicated. The details of the low energy beam switchyard that connects the three target modules to the various experimental areas and/or post accelerators are not shown.

(http://www.phy.anl.gov). The U.S. Nuclear Science Advisory Committee (NSAC) included a strong recommendation for the construction of an Advanced ISOL Facility in its 1996 Long Range Plan. The scientific case for such a facility was summarized recently in a White Paper (3). The present paper is a status report of the Argonne concept originally presented in the working paper in 1995. An alternative concept for an Advanced ISOL Facility to be located at the Spallation Neutron Source is being developed at ORNL and is described by J. D. Garrett at this conference (4).

A schematic block diagram of the facility conceived by ANL is shown in Fig. 1 and a more detailed layout of the proposed complex is shown in Fig. 2. The present ATLAS complex produces state-of-the-art heavy ion beams covering the entire mass range from protons to uranium for nuclear physics research at energies above the Coulomb barrier. The new facility would build on the existing expertise in accelerator and nuclear physics at ATLAS. In Fig. 2 the existing ATLAS accelerators and experimental facilities are in the lower part of the figure, and the proposed driver

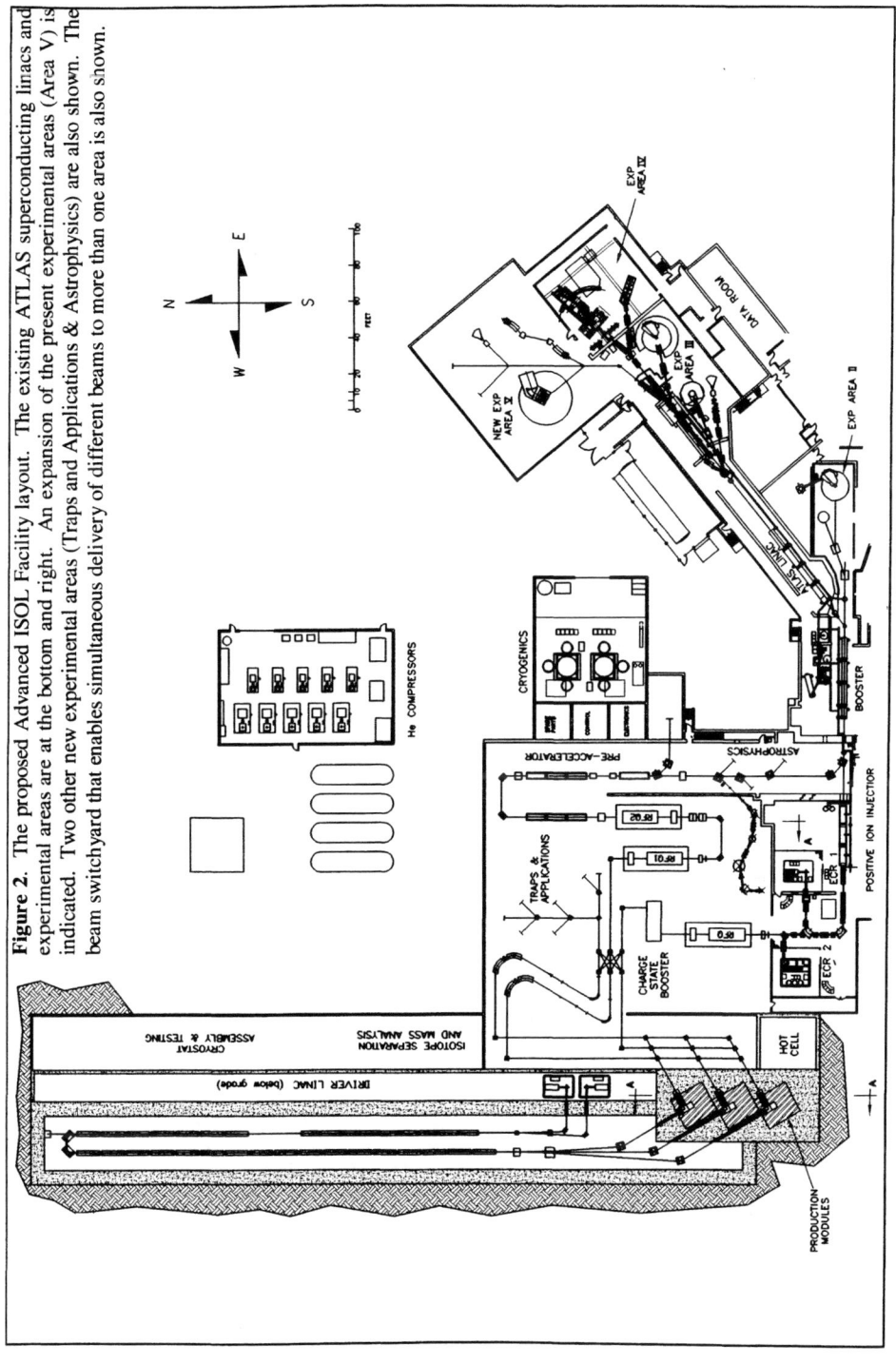

Figure 2. The proposed Advanced ISOL Facility layout. The existing ATLAS superconducting linacs and experimental areas are at the bottom and right. An expansion of the present experimental areas (Area V) is indicated. Two other new experimental areas (Traps and Applications & Astrophysics) are also shown. The beam switchyard that enables simultaneous delivery of different beams to more than one area is also shown.

linac, production target area, mass separators, new low energy experimental areas, and post accelerator injector are in the upper part. The new components for the radioactive beam laboratory will be constructed in the area just north of the present ATLAS facility. The capability of ATLAS to accelerate stable beams will remain independent of the added radioactive beam capability, both during construction and afterwards. The major components of the proposed facility are described in the following sections.

PRODUCTION MECHANISMS

The Argonne concept for the Advanced ISOL Facility utilizes a high power driver accelerator that can deliver a variety of beams and energies such that the radionuclide production mechanisms can be chosen to optimize yields for the species of interest. The production mechanisms include: (1) Neutron-induced fission in a secondary target following neutron production by deuterons on a primary target; (2) Fragmentation of neutron-rich heavy ion beams such as ^{18}O in a two-step target/catcher geometry; (3) Spallation reactions induced by heavy ions with total energies of 1 to 10 GeV; and (4) Compound nucleus reactions at lower beam energies when appropriate. This approach permits the choice of the optimal production method for specific isotopes while minimizing the production of unwanted byproducts. An Advanced ISOL Facility based on the multi-beam driver concept will address all thirteen of the physics areas spelled out in the Columbus White Paper (3).

DRIVER LINAC

The high-power, multi-beam driver being proposed is a superconducting linac capable of delivering over 100 kW of beam power for a variety of light and heavy ions. The preliminary design of this linac is based on using two-cell, 350 MHz spoke-type niobium resonators which are currently being developed (5). These new resonators are required at the high-velocity end of the linac ($\beta \sim 0.4$). More conventional, lower frequency inter-digital and quarter-wave resonators can be used at

TABLE 1. Typical Beams Available From the Driver Linac

Ion	Mass	Intensity	Energy	Energy	Power
		part./sec.	MeV/u	GeV	kW
Proton	1	8.0E+15	212	0.2	300*
Deuteron	2	7.0E+15	123	0.25	300*
Oxygen	16	1.0E+15	118	2	300*
Argon	40	1.7E+14	97	4	110
Krypton	82	1.2E+13	82	7	16
Xenon	132	2.4E+12	75	10	4

*The actual beam power limit will be set by the linac RF design specification.

the low velocity end of the linac. The ion source/injector for this driver accelerator can be a high-intensity version of the ECR ion source/high-voltage platform system currently used at ATLAS (6). Present-day ECR ion sources such as the AECR at LBNL (7) and SERSE at Catania (8) can provide DC beams of heavy ions such as those listed in Table 1 at intensities necessary to achieve the indicated beam powers with a single foil stripping at 10 MeV per nucleon. For the lighter ions such as protons and deuterons DC currents of well over a milliampere are available from conventional ion sources and no stripping is required. For the lighter ions the maximum beam power available will be set by the RF power design choice and will not be limited by the ion sources. In Fig. 2 the driver is shown with two ion source/injector platforms at the low energy end and three beam lines feeding target modules at the high-energy end. At either or both ends, RF switching can be used to simultaneously accelerate beams from alternate sources and irradiate multiple target modules.

TARGET COMPLEX

High-power beams from the driver linac are used to produce radionuclides in well-shielded target/ion source modules that are located below grade level to assist with shielding prompt neutron radiation. A preliminary investigation of the radiological issues to be encountered in the target complex of an Advanced ISOL Facility was carried out at LBNL (9) assuming 100 microamperes of 600 MeV protons. General radiation protection issues for such facilities were also reviewed recently by L. Moritz (10) of TRIUMF. The conclusions are that the level of radiological issues of the Advanced ISOL Facility is well within the realm of experience previously and currently encountered at meson factories such as TRIUMF and LAMPF, as well as at many non-reactor nuclear facilities at several DOE national laboratories such as Argonne.

As shown in Figs. 1 and 2, three target modules are included in this proposal. By using RF switching from the driver linac beams can irradiate targets in all three modules simultaneously. This enables, for example, target and ion source development to be carried out in one module while two are used as sources of radioactive beams for the nuclear physics research program.

Shielding and Remote Handling

The target complex shielding and remote handling designs worked out by the TRIUMF group for the ISAC facility can be applied to the present proposal. Paul Schmor described the ISAC facility, currently being commissioned at TRIUMF, at this conference (11).

High Power Targets

The production targets for the Advanced ISOL Facility must work at high beam power and, at the same time, be coupled efficiently to ion sources for the production of the secondary beams of short-lived isotopes. To effectively utilize beam powers of 100 kW or more involves extrapolation beyond current experience at any ISOL-type facility. For the Argonne multi-beam driver approach there are currently three basic concepts being developed in order to utilize the variety of radionuclide production mechanisms mentioned above. Based on preliminary engineering designs all three approaches can be used at beam powers up to 100 kW or more. The design problem can be viewed as one of minimizing target size and geometry while still handling the beam power. Minimizing target size and geometry is essential to optimize diffusion and effusion efficiencies for very short-lived radionuclides. Generally, the short-lived isotopes tend to be the most interesting and to have lower intrinsic production cross sections due to being further from stability.

Two of the three concepts being developed involve the so-called two-step geometry. In conventional ISOL production schemes the target and ion source are integrally coupled so that the primary target can not be massively cooled without interfering with the ion source performance. With the two-step schemes these functions are physically separated as described below.

The fusion power community has developed liquid lithium cooling loops for a variety of applications. Engineering designs have been carried out for stopping 10-MW deuteron beams in windowless flowing liquid lithium targets (12). Pumps for recirculating liquid lithium are commercially available and are quite small compared to those required for similar mass flow rates of helium gas (13). The two-step target concepts described below can both utilize this technology.

Two-step, Neutron-generator Targets

Figure 3 illustrates schematically a simple, two-step geometry for the production of large yields of neutron-rich fission products. In this example a primary beam of protons or deuterons impinges upon a well-cooled tungsten target to produce an intense flux of secondary neutrons. The secondary neutrons are relatively low energy and approximately isotropic. To optimize the solid angle of the secondary uranium target a cylindrical geometry is indicated. The length and diameter of the secondary target are kept to a minimum for optimal extraction of short-lived products. Using thicker and longer secondary targets can increase the yields of longer-lived isotopes. With the Los Alamos LAHET Code System (LAHET and MCNP), the fission fragment production rates can be calculated for realistic geometries (14). The calculated yield of the doubly magic nucleus ^{132}Sn (40 sec half-life) in a compact secondary target (2 cm radial thickness by 10 cm axial length) is 2×10^{11} per second with a 500-microampere, 200-MeV deuteron beam.

The geometry of Fig. 3 is well suited to the production of radionuclides that have high yields in low-energy-neutron-induced fission. This conclusion is based on

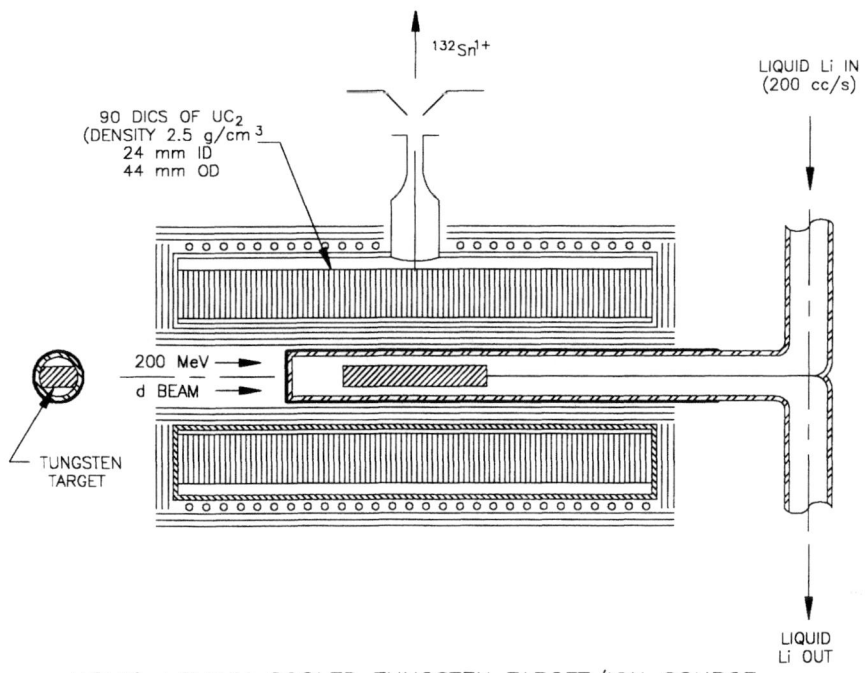

Figure 3. Schematic of a two-step, neutron-generator-type, high-power target for the production of fission fragments. A deuteron beam on a tungsten primary target produces secondary neutrons. Circulating liquid lithium cools the primary target. The secondary uranium carbide target is heated to 2000 C by a combination of the fission power and supplemental electrical power.

calculations of thin target yields of neutron-rich fission products for a range of neutron and proton energies from a few MeV up to 100 MeV (15). The best yields of neutron-rich isotopes of elements such as Kr, Rb, Xe, and Cs are from 2-20 MeV neutron-induced fission of ^{238}U. Harder neutron spectra, such as produced via the originally proposed d+Be reaction (16), are likely to produce more radionuclides outside the standard low-energy, asymmetric fission mass range.

Fragmentation/catcher Targets

A schematic diagram of the second type of two-step target is shown in Fig. 4. Here a 100 kW heavy ion beam (e.g. ^{18}O) is stopped in flowing liquid lithium while the neutron-rich fragments (e.g. ^{11}Li), which have a longer range and are kinematically forward peaked, go on and stop in the catcher/ion source that is physically separated downstream of the primary target. The purpose of the graphite in the primary target of Fig. 4 is to reduce the stopping thickness that would be required with pure lithium due to its very low density of 0.5 g/cm^3. The target is designed to have the high-power-

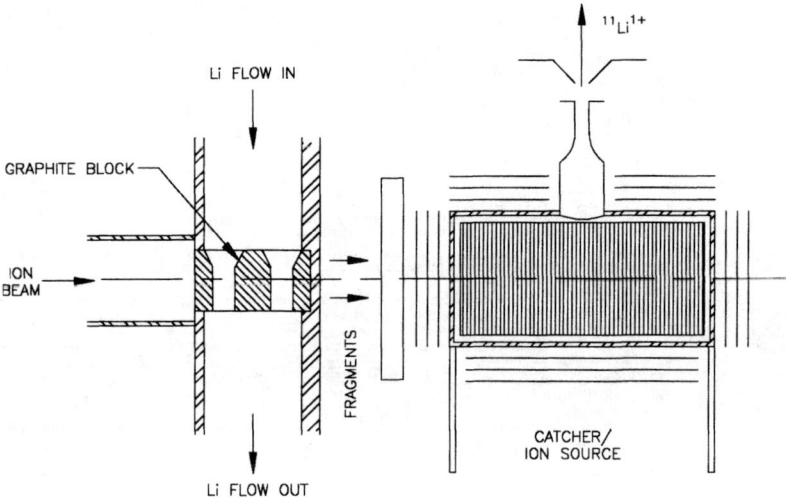

Figure 4. Schematic of a two-step, fragmentation/catcher target combination. A high-power, neutron-rich heavy ion beam, such as ^{18}O, stops in the primary target that consists of a combination of graphite and liquid lithium coolant. The neutron-rich, lower-Z fragments, such as ^{11}Li, have a much longer range than the primary beam and are stopped in the downstream catcher. The catcher is close-coupled to the ion source and is optimized for rapid release of the short-lived fragments.

density Bragg peak occur in the flowing liquid lithium. For heavier ions such as Kr and Xe the ranges are short enough to use pure liquid lithium primary targets.

With the target geometry of Fig. 4, known fragmentation cross sections for ^{11}Li (17), and graphite catchers optimized for rapid release of alkalies (18), mass separated beams of 5×10^7 ^{11}Li per second are predicted. This is a very high intensity for such a short-lived isotope (9 ms half-life).

A new development that is currently being tested in conjunction with ion accumulation for the Canadian Penning Trap (19) may lead to significant improvement over the concept illustrated in Fig. 4. By stopping the fragmentation products in a high-pressure gas cell of pure helium or argon, and using a combination of rf and static electric fields as in (19), the radionuclides come to rest as 1+ ions. They are quickly extracted from the gas volume with high efficiency independent of the chemical properties.

Direct Irradiation Spallation Targets

A third target design concept is being developed to utilize an important radionuclide production mechanism, heavy-ion-induced spallation of heavy targets such as uranium. Large yields of very neutron-rich sodium isotopes, for example, were observed in the pioneering work with 86 MeV per nucleon (0.9 GeV total energy) carbon beams at the ISOLDE facility at the CERN Synchrocyclotron (20, 21).

Based on these ISOLDE yield measurements, scaled to 100 kW of 1.4 GeV carbon beam on a UC_x target, mass separated beams of ^{31}Na (17 ms half-life) could be obtained with very high intensities, over 5×10^6 per second. However, this requires a direct irradiation of the UC_x target with the intense carbon beam; the two-step geometries are not applicable. Direct irradiation of the standard low-density porous UC_x used at ISOLDE (22) can not be scaled to high beam power due to the very low thermal conductivity of this form of the material (23). Hence, a new geometry for irradiating a large area sheet of thin (~100 μm to 1 mm thick) higher density UC_x is being considered. A large area sheet tilted at a large angle to the driver beam can dissipate the large beam power by black body radiation while the higher thermal conductivity combined with the thin sample dimensions prevent high internal temperatures within the sheet. This geometry leads to a very open geometry that should yield very short effusion times. Analysis of this scheme taking into account the overall diffusion from the UC_x and the effusion form the open geometry target chamber will be carried out using methods similar to those discussed by Roger Bennett (24) and Will Talbert (25) at this conference. New measurements of the thermal conductivity and radionuclide release properties of UC_x prepared at various densities will have to be made. With beams of 1.4 GeV carbon, 2 GeV neon, and 4 GeV argon from the heavy ion driver, this reaction mechanism and target concept could prove to be extremely productive.

Ion Sources

The present proposal is based on the ISOL-type ion source technology that has been continuously developed and refined at various laboratories over the past 30 years. See (26) for a review and many references on these ion sources. These types of ion sources produce a wide variety of atomic species with high efficiency and varying degrees of chemical selectivity, usually in the 1+ charge state. Chemically selective, laser-driven ion sources have developed rapidly in recent years and have been shown to be very useful at ISOL facilities (27). Major advantages of standard 1+ ISOL-type ion sources are their high ionization efficiencies and their relatively radiation-resistant, simple construction. Furthermore, designing the overall post-accelerator scheme to work with 1+ ions enables the implementation of new concepts such as the gas fragment catcher discussed above.

ISOBAR SEPARATOR

The goal of the Advanced ISOL Facility is to produce energetic beams of short-lived isotopes far from the valley of nuclear stability. The beams are to be of high quality (excellent emittances) and as free of contaminants as possible. For some elements there are chemically selective ionization processes, such as surface ionization for the alkalies and the laser resonance ionization process for others. And for other elements physical separation processes are possible, such as cold transfer lines for the noble gases. In these cases it is adequate to produce mass separated beams with resolution m/Δm ~ 500. However, in non-selective cases a specific radionuclide of interest may be associated with a much more intense neighboring isobar. Separation of isobars requires mass separation with resolution of 20,000 or more. Important parameters of the isobar separator are the emittance acceptance and energy spread compensation. The present proposal is to include one or two large isobar separators (see Fig. 2) with emittance acceptances of 10π mm-mr at 100 keV ion beam energy and the ability to compensate for ion source energy spread or voltage ripple of at least 10-20 eV. The ion-optical scheme of using two magnetic separators with ion energy deceleration in between, as originally proposed by H. Wollnik; will be used. See (28) for another example of this type of isobar separator. In the future it may be possible to enhance the performance of the isobar separator by incorporating new technology which is on the horizon. The new concept is to effect cooling of the 1+ ion beams via a buffer-gas/ion guide scheme (29).

POST ACCELERATOR

An essential feature of the post accelerator is to preserve the excellent beam quality currently available at ATLAS for stable beams of any mass up to uranium in the energy range from 6-15 MeV per nucleon. The post accelerator starts with mass separated, 1+ radioactive ion beams at initial energies of 50-100 keV. A high overall efficiency of the acceleration process is of the utmost importance.

Stripping Scheme

The science program of the Advanced ISOL Facility requires delivering mass separated ion beams at three qualitatively different energies: unaccelerated beams at ion source energy, beams in the 1 MeV per nucleon energy range, and beams accelerated to 6-15 MeV per nucleon. For ions of the lower energies and atomic masses it is possible to deliver the beams with very high efficiency, directly in the 1+ charge state. However, economic considerations necessitate the use of one or two stages of ion stripping for the higher mass ions to be accelerated to the higher energies. The present proposal involves stripping in a thin helium gas cell, to 2+ or 3+, after the initial acceleration in the RFQ, for masses above 70. The efficiency for this process has been shown to be 40-50% for a broad range of heavy ions (30). For acceleration to energies above 600 keV per nucleon a foil stripper is used to increase the charge-to-mass ratio to > 0.15, with a typical efficiency of about 20%. Hence, the overall efficiency of the process varies with both mass and energy depending on the need for: no stripping (100%), gas stripping only (~40-50%), foil stripping only (~20%), or both strippers (~8-10%).

Various groups around the world are developing alternates to gas or foil strippers, the so-called charge breeders (31, 32, 33). As indicated in Fig. 2, one of these charge breeder concepts could be incorporated into the present proposal. In the scheme shown in Fig. 2 the charge breeder is used for the beams with mass A > 70, for acceleration to energies of 6-15 MeV per nucleon. This is the category of beams for which the breeders are likely to be most competitive with the stripping alternative (8-10%), at least initially. Incorporation of the breeder scheme into this proposal in this way takes advantage of the present injector stage of ATLAS, and permits the simultaneous delivery of radioactive beams to the high energy experimental areas and one or both of the lower energy areas.

First Accelerator Section: CW RFQ

The capability of the Advanced ISOL Facility to deliver high quality beams with high efficiency over a broad range of masses and energies from 1+ ion sources depends critically on the first stage of acceleration. The present proposal is based on the use of a low-frequency, normally conducting, CW radio-frequency quadrupole (RFQ) acceleration section that is currently under development at Argonne (34), with a status report at this conference. The acceptance of this type of RFQ, which would be operated on a negative high voltage platform, is well matched to typical ISOL-type ion source emittances and is expected to be useful for initial acceleration of 1+ ions with masses up to ~200. This type of RFQ is also adaptable for matching a charge breeder to the present ATLAS superconducting injector, as indicated schematically in Fig. 2.

Superconducting Linacs

Following the normally conducting RFQ accelerator sections described above, all of the radioactive beam acceleration is done with superconducting linear accelerators. [see (35) for a description of this injector] The present proposal is to add a high charge-to-mass ratio injector section (m/q < 70) for the initial acceleration up to 600 keV per nucleon. The superconducting resonators of this section are of the same type currently in use for the ATLAS low-velocity injector ($v \geq 0.008c$). However, due to the higher mass-to-charge ratio of the new injector, the transverse focussing requirements are more demanding. The present concept is to use high-gradient superconducting quadrupole triplets for focussing in the new injector as opposed to the superconducting solenoid focussing elements used in the present ATLAS injector. Beams from the high m/q injector can be delivered, without further stripping, at energies up to 600 keV/u to either the intermediate energy experimental areas or transported without further acceleration to the apparatus in the high-energy experimental areas. For acceleration to energies above 600 keV/u the high m/q injector is followed by the foil stripper and a short superconducting matching section. The matching section provides beams at up to 1.2 MeV/u in the intermediate energy experimental areas or delivers the beam for further acceleration by the present ATLAS linacs. Energies up to the 6-15 MeV per nucleon range, depending on ion mass, are available in the high-energy experimental areas.

EXPERIMENTAL AREAS AND INSTRUMENTATION

The research enabled by the Advanced ISOL Facility is discussed in the Columbus White Paper (3). As pointed out in the discussion of the stripping scheme above, there are three qualitatively different energy regimes required for this research program. The associated experimental areas are indicated and labeled in the block diagram of Fig. 1 and shown in more detail in Fig. 2. In addition to the areas indicated there is space for further expansion as future needs dictate. The experimental apparatus deemed necessary to carry out the proposed research program was the subject of a workshop sponsored jointly by LBNL, ORNL, and ANL in the summer of 1998 (36). A report detailing the recommendations of the working groups from this workshop is being prepared.

ACKNOWLEDGMENTS

The help of T. Barlow, C. Batson, P. Billquist, P. Decrock, A. Geraci, M. Kedzie, A. Ruthenberg, Dale L. Smith, and H. Zaim with several aspects of this work is appreciated.

This work is supported by the U. S. DOE Nuclear Physics Division under contract W-31-109-ENG-38.

REFERENCES

1. "The IsoSpin Laboratory, Research Opportunities with Radioactive Beams," LALP report 91-51 (1991).
2. "European Radioactive Beam Facilities," Report of the NuPECC Study Group (1993).
3. "Scientific Opportunities with an Advanced ISOL Facility," R.F. Casten, et al., eds., November, 1997. (Available at http://www.er.doe.gov/production/henp/isolpaper.pdf).
4. Garrett, J. D., this conference.
5. Shepard, K. W., et al., "Development of Niobium Spoke Cavities for a Superconducting Light-Ion Linac," Proceedings of the 1998 Linear Accelerator Conference, August 23-29, 1998, Chicago, IL.
6. Bollinger, L. M., et al., Nucl. Instr. Methods **B79** (1993) 753.
7. Xie, Z. Q., Rev. Sci. Instr. **69** (1998) 625.
8. Gammino, S., this conference.
9. Donahue, R. J., et al., "Radiation Problems in the Design of a Radioactive Nuclear Beam Facility," Proc. Specialists Meeting on Shielding Aspects of Accelerators, Targets, and Irradiation Facilities, Arlington, TX, April 28-29, 1994 (LBL-35459).
10. Moritz, L. E., Cyclotron Conf., Caen, France, June, 1998.
11. Schmor, P., this conference.
12. Hassanein, A., Journal of Nuclear Materials, **233-237** (1996) 1547.
13. Reed, C. B., Technology Development Division, ANL, private communication.
14. The LCS calculations and post-processing were carried out by I. C. Gomes, Technology Development Division, Argonne National Laboratory.
15. Gomes, I. C., Nolen, J. A., "Influence of the Incident Particle Energy on the Fission Product Mass Distribution," Proc. AccApp'98 (2nd International Topical Meeting on Nuclear Applications of Accelerator Technology), Gatlinburg, TN USA, September 20-23, 1998.
16. Nolen, J., "A Target Concept for Intense Radioactive Beams in the ^{132}Sn Region," Proc. Third Inter. Conf. On Radioactive Nuclear Beams, Ed. D. J. Morrissey, East Lansing, MI, May 24-27, 1993.
17. Kubo, T., et al., Nucl. Instr. Meth., **B70** (1992) 309.
18. Dax, A., et al., GSI Scientific Report 1997, GSI 98-1, p. 95.
19. Savard, G., et al., Nucl. Phys. **A626** (1997) 353c.
20. Bjornstad, T., et al., Z. Phys. A 303 (1981) 227.
21. Saint Simon, M. de, et al., Phys. Rev. C26 (1982) 2447.
22. Evensen, A. H. M., et al., Nucl. Instr. Meth. In Phys. Res. **B126** (1997) 160.
23. Taylor, R.E., et al., Thermophysical Properties Research Laboratory, Report 1825, March, 1997.
24. Bennett, J. R. J., this conf.
25. Talbert, W., this conf.
26. Ravn, H. L., "Sources for Production of Radioactive Ion Beams," Proceedings of the 1995 Particle Accelerator Conference, p. 858.
27. Lettry, J., et al., Rev. Sci. Instr. **69** (1998) 761.
28. Ciavola, G., et al., Nucl. Instr. Meth. In Phys. Res. B126 (1997) 17.
29. Lunney, M. D., et al., Report CSNSM 97-02, Orsay, 1997 (unpublished)
30. Decrock, P., et al., Rev. Sci. Instr. **68** (1997) 2322.
31. Habs, D., et al., Nucl. Instr. Meth. In Phys. Res. **B126** (1997) 218.
32. Marrs, R. E., Slaughter, D. R., "Charge State Boosting with a High Intensity Electron Beam Ion Trap," 15th Inter. Conf. on the Appl. of Accel. Res. and Industry, November 4-7, 1998, Denton, TX.
33. Geller, R., this conference.
34. Kaye, R. A., et al., this conference.
35. Shepard, K. W., Kim, J.-W., "A Low-charge-state Injector Linac for ATLAS," Proceedings of the 1995 Particle Accelerator Conference, p. 1128.
36. "Experimental Equipment for an Advanced ISOL Facility," Workshop held at LBNL, Berkeley, CA, July 22-25, 1998.

A Radioactive Ion Beam Facility, SIRIUS, at ISIS

J R J Bennett
For members of the SIRIUS collaboration in the UK Universities and CLRC.

CLRC Rutherford Appleton Laboratory, Chilton, Didcot, Oxon, OX11 0QX, U.K.

Abstract. A description is given of the radioactive ion beam facility, SIRIUS, proposed to be built at ISIS, the world leading pulsed neutron source. Up to 100 µA of 800 MeV protons from the ISIS synchrotron will be taken down a new beam line into a target station complex where radioactive nuclei will be formed as a result of the interaction with the target. The nuclei are subsequently ionised and accelerated to 200 kV. The beam passes through a broad range spectrometer to provide several beams of different ion species simultaneously. One beam passes through a high resolution spectrometer and is accelerated in an RFQ and a superconducting linear accelerator to 10 MeV/amu, for nuclear physics research. The other beams are taken to a low energy experimental area.

INTRODUCTION

Radioactive nuclear beams (RNB) have increasing uses in a wide range of sciences: nuclear physics and astrophysics, solid state and atomic physics, biology and medicine. The need for more intense sources of RNB and in particular of the rare and, as yet undiscovered, shorter lived isotopes, is apparent. A team from the UK universities and CLRC have carried out a design study for a RNB Facility, SIRIUS, to be based on the 800 MeV proton beam from the synchrotron of the world leading pulsed neutron source, ISIS [1].

The radioactive ions are produced as a result of nuclear reactions by the energetic proton beam hitting a suitable target. This method is preferred over a two step process, as proposed by Argonne [2], where protons or deuterons hit a primary target to produce neutrons, which in turn are used to produce the unstable isotopes in a second target. It is considered that a wider range of radioactive particles can be produced by direct proton impact and the system is simpler.

The specification is for a proton current on the target of up to 100 µA, about 50 times the intensity of the present ISOLDE [3] facility at CERN and comparable with that of ISAC [4] being built at TRIUMF, Vancouver. The proton beam is brought out of the synchrotron into a target station where singly charged particles are accelerated to 200 kV, and passed through a wide range spectrometer to produce several beams of different masses. One of these beams enters a high resolution magnetic spectrometer

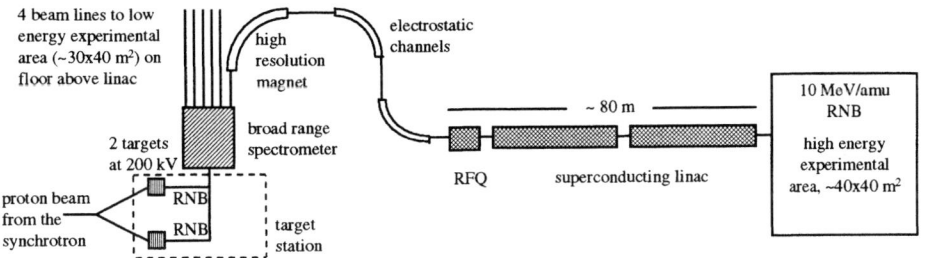

FIGURE 1. Schematic plan view of the SIRIUS Facility, showing the main components.

(1/30000) and then is accelerated in a linear accelerator to an energy of up to 10 MeV/nucleon for nuclear physics studies. The other beams are taken into an experimental hall for low energy experiments. Figure 1 shows the overall layout of the facility adjacent to the ISIS synchrotron ring.

The ISIS proton beam is pulsed at 50 Hz and will be shared between the RNB and the neutron target stations on a pulse to pulse basis with up to 1 in 2 pulses for the RNB targets.

TARGET

A tantalum target (5) has been developed for the full 100 μA proton current. The target tube is 20 cm long, 4 cm in diameter and is partly filled with 25 μm thick tantalum foil discs spaced apart by 25 μm. The target is radiation cooled, small fins increasing the effective thermal emissivity of the surface, allowing the power (20 kW) to be dissipated at temperatures of ~2300 K. This target will be used initially, but subsequent targets will be developed for other materials.

The first source will be a simple hot tungsten tube thermal ioniser, but again, other sources will be developed to allow a full range of ions to be produced. The present plan is to produce singly charged ions. To improve the acceleration efficiency in the high energy linac, suitable stripping stages are incorporated to increase the charge state.

Existing RNB targets need to be changed every few weeks. At the power levels to be operated at SIRIUS, the life could be even shorter. Therefore, target changing is a very important consideration. To allow almost continuous operation of the facility, there will be 2 targets and the proton beam can be switched to either station. For even greater flexibility to provide beams of different species from different targets, it will be possible to switch from one target to another on alternate pulses.

Target development will be an important factor in the successful operation of the facility and it is envisaged that there will be an active on-going programme to improve the understanding of the mechanisms going on in the targets, their yields, reliability, life and range of materials, including molten materials.

TARGET STATION

The target station is a complex and sophisticated system providing radiation shielding and remote handling of the active targets and beam lines up to, and including, the wide range analyser magnet; also, it provides a high voltage enclosure. There are 2 installed target positions spaced 1.3 m apart. Each target requires a water cooled copper beam stop, since the RNB targets are not thick enough to stop the protons. Each target and beam stop is contained within its own separate vacuum vessel, with water cooled proton beam windows. The vacuum vessels of the targets and beam stops are contained within a common vacuum void vessel connected to the beam line via a vacuum valve. Each target is removed with its own vacuum vessel to provide a safe enclosure containing any possible contamination. Target changes are expected to take up to 8 hours. Figure 2 shows the layout of the front end containing the target and ion source, attached to the stepped shield plug and Figure 3 shows the layout of the target station.

The beam from the ioniser is accelerated and focused to 200 kV by electrodes close to the target. The beam then passes through a 5 cm diameter tube through the steel shield plug, 3 m long. Focusing for the beam is not required within the plug. The front end is supported from the shield plug and all services (water pipes, high voltage cables and electrical leads) pass through the plug.

For many low energy experiments an energy of 200-300 keV is beneficial. A design has been made with 200-300 kV on the target, but the subsequent electrostatic bending and focusing elements are relatively large and require correspondingly high voltages. An alternative using acceleration to only 30 kV, followed by an RFQ accelerating to 200-300 keV in one of the beam lines after the broad range analyser magnet, may be more cost effective.

The target station, shown in Figure 3, is mainly above ground level; the two targets and their beam stops are vertically above one another. The target station incorporates a remote handling cell with access from two sides and storage within the shield walls for 6 target modules. One of the stored target modules can be used off-line to provide stable or long lived beams to the separator system. The targets and shield plugs are moved horizontally and vertically on a special trolley.

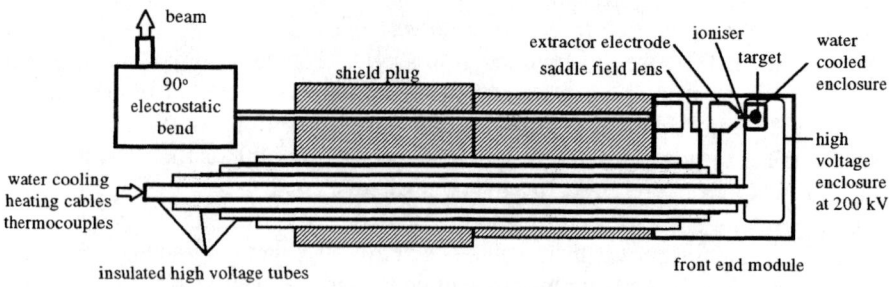

FIGURE 2. Schematic section of the front end target module mounted on the steel shield plug

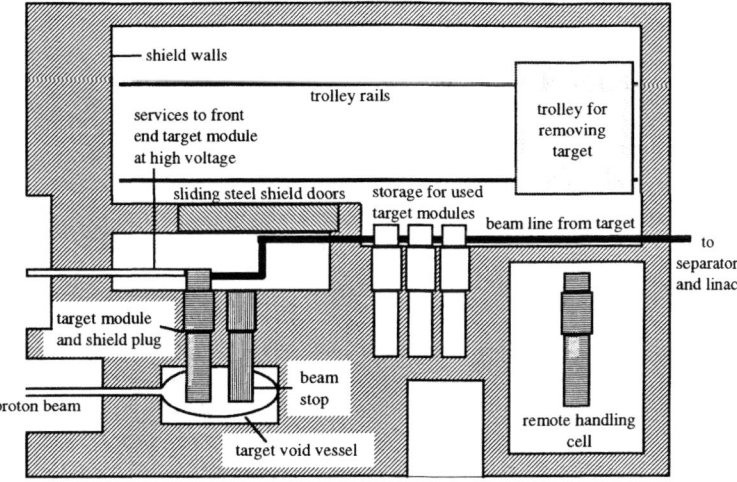

FIGURE 3. Schematic plan section through the target station.

MASS SEPARATORS

The broad range separator is designed to deliver four independent beams of different masses simultaneously. The device has a long focal plane simultaneously providing separated beams of mass 5-240. Ray-tracing indicates a minimum dispersion of ~ 6 mm per amu at A = 200. This is adequate, provided that slits are incorporated in the device in the focal plane where the transmitted beam is to be defined, and that unwanted masses are prevented from painting the contents and walls of the vacuum chamber and its devices.

A high resolution separator is required to operate at resolutions of $\Delta m/m < 1/10000$, and as near to 1/30000 as possible to provide mass separation of isobars for the reduction of contamination and for injection into the linac. Extensive calculations have been carried out, indicating that resolutions of 1 in 30000 are possible by careful shaping of the pole entrance and exit faces, with realistic beam emittance and energy spread (± 3 parts in 10^5). The magnet is followed by two electrostatic channels to improve the resolution in the event that the energy spread is significantly worse.

THE RFQ AND SUPERCONDUCTING LINAC

The SIRIUS post-accelerator is designed to accelerate radioactive ions to an energy of 10 MeV per nucleon over a mass range of 10-240 amu. Figure 4 shows the system schematically. Both the RFQ and the cavities are based on designs in use or planned at the Argonne National Laboratory (2). The cavities themselves have been described and costed in reference (6) (a design study for a machine at GANIL).

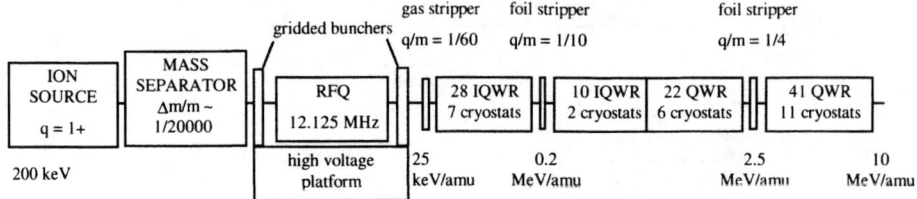

FIGURE 4. Schematic of the linac accelerator.

The singly charged ions emerge from the ion-source at an energy of 200 keV and pass through the high-resolution spectrometer before injection into the RFQ section. They are matched into the RFQ by setting the RFQ on a high voltage platform to give an injection energy into the RFQ of 1 keV/amu for the entire mass range. The RFQ is preceded by a gridded buncher, operating at the RFQ frequency of 12.125 MHz, which compresses the beam and significantly shortens the RFQ. The RFQ is ~ 5 m long and accelerates the full range of particles by varying the voltage on the vanes. The device is of the split-coaxial type because of the low frequency. Following the RFQ there is a second gridded buncher. The RFQ accelerates the ions to 25 keV/amu from where they are injected into the linac after passing through a gas stripper to raise the charge state.

The beam is then injected into a number of super-conducting accelerating cavities. The devices are based on the Argonne (2) design at a frequency of 12.125 MHz and its multiples. The IQWR are described as inter-digital quarter wave resonators in reference (6). They are essentially combinations of pairs of the two gap quarter wave resonators (QWR) driven by a common rf source. As the particle velocity increases, each cavity (QWR) is independently fed with rf and its phase adjusted to optimise the acceleration. Further stripping in thin foils is used at optimal points (at energies of 0.2 MeV/amu and 2.5 MeV/amu) along the linac to improve the accelerating efficiency.

REFERENCES

1. Finney, J.L., *Europhysics News* **20**,119 (1989).
2. *Concept for an advanced Exotic beams facility Based on Atlas*, Physics Division, Argonne National Laboratory, February 1995.
3. Kluge, H.-J., editor, *ISOLDE User's Guide*, CERN 86-05 (1986).
4. Bricault, P. G., Dombsky, M., Schmor, P. W., and Stanford, G., *Nucl. Instr. and Meth. B* **126**, 231-235 (1997).
5. Bennett, J. R. J., Densham, C.,V., Drumm, P. V., Evans, W. R., Holding, M., Murdoch, G.R. and Panteleev, V., *Nucl. Instr.and Meth. B* **126**, 117-120 (1997).
6. *Proposition d'un ensemble de production et d'acceleration d'ions secondaires a GANIL*, GANIL R 92 04.

High Power Targets for Production of Intense Radioactive Ion Beams

W. L. Talbert*, D. M. Drake*, M. T. Wilson*, J. J. Walker* and J. W. Lenz[†]

Amparo Corporation, Santa Fe, New Mexico 87504
[†] *John W. Lenz and Associates, Waxahachie, Texas 75165*

Abstract: Issues are discussed in producing intense Radioactive Ion Beams (RIB) using the Isotope Separator On-Line (ISOL) approach, based on the use of thick targets employed at existing facilities. Some new physics studies may possibly be addressed by improving the performance of these existing targets through improvements in release and effusion properties to optimize the RIB yields. It is, however, acknowledged that many desired physics objectives using RIB can be met only by employing production beams of energetic light ions or protons with currents up to 100 µA. Development of targets that use such intense production beams needs to address the requirement to control operational temperatures derived from internal production beam interactions with the target materials. In addition, issues arise for target materials in terms of their thermal characteristics, such as thermal conductivity and thermo-mechanical properties. A target concept is described for an in-beam test of a prototype target for actual thermal behavior under RIB production conditions. For such a test, a high-power test facility is needed; fortunately, the prototypical production beam currents required exist at the TRIUMF accelerator facility. An experimental proposal has been approved for such a test.

INTRODUCTION

An emerging application of heavy ion accelerators is for the acceleration of RIB produced using the ISOL approach, where a thick target is irradiated by an energetic light particle beam (for example, 500-MeV to 1-GeV protons). The radioisotopes produced are transported to an ion source and subsequently mass separated (isotopically purified) before introduction into a heavy ion accelerator for use in physics experiments. An important component of this chain of processes is the RIB production target, which must produce and release the radioactive species of interest. RIB production targets are usually operated at elevated temperatures to enhance the release of produced radioactive species. The temperatures must be controlled to avoid thermal conditions where chemical and/or mechanical stability is compromised.

As physics issues using RIB are more fully identified, the attainment of high intensity RIB is ever more desired. Many of the targets in use today provide rather intense RIB for some species, even though the production beam current is typically of the order of 1 µA. In practice, these targets are typically heated in an oven to the desired operating temperatures, because the internal beam heating of the target is insufficient to provide adequate operational temperatures.

CP473, *Heavy Ion Accelerator Technology: Eighth International Conference,*
edited by Kenneth W. Shepard
© 1999 The American Institute of Physics 1-56396-806-1/99/$15.00

For RIB species that are less prolifically available, the production rates can only be increased by increasing the production beam current. It is generally assumed that advanced RIB/ISOL facilities will employ production beams of energetic light ions with intensities up to 100 µA. At these production beam intensities, the internal heating caused by the beam interactions with the target material is more than that required to raise the target material to the appropriate operating temperature; therefore the target needs to be cooled so that it is thermally and mechanically stable. Although there exist many different target types in current use, the requirement to remove the heat produced by the production beam limits what target materials can be used (they must be able to conduct heat to a cooling system) and poses a variety of issues to be addressed in the development of cooling approaches.

Related Studies

There have been some studies in the past on possible cooling approaches for so-called high power targets that can tolerate intense incident production beams (1,2). Experimental studies have not been carried out to develop RIB targets exposed to intense high-energy light ion beams. However, studies have been conducted at Louvain-la-Neuve (3) or GANIL (4) for incident low-energy light ion or energetic heavy ion beams. These studies have focused on targets in which the incident beams are stopped, in contrast to the targets for high-energy light ion production beams, where the beam passes through the target which is typically of one interaction length (5).

Activities in target development also exist in laboratories where new RIB facilities are under construction, such as at GANIL (6), ISOLDE (7), TRIUMF (8), Catania (9), and the Japanese Hadron Project (10). In addition, development of a high power RIST (Radioactive Ion Source Test) target employing radiative cooling has been carried out at the Rutherford Appleton Laboratory (11). Only one of these new facilities, TRIUMF, will employ energetic proton beams up to 100 µA.

There is also much interest in planning construction of one or more "next-generation" RIB facilities which would employ energetic, intense light ion beams to produce the desired RIB. This work addresses an approach to develop targets for use at these new intense RIB facilities, using the TRIUMF facility as an example.

HIGH POWER TARGET ISSUES

There are a number of technical issues that must be addressed during the development of high power RIB targets. In some cases these issues carry over from development of existing targets. For example, material compatibilties at high temperatures have been addressed successfully at existing facilities for many target systems. New issues specific to high power targets include the thermal properties, particularly thermal conductivity of the target and container materials and assessment of target robustness for useful length of service.

This work is devoted to the conceptual design of a prototype target to be tested for thermal behavior at the ISAC facility (Isotope Separator ACcellerator) at the TRIUMF accelerator (8). It is not intended that this test also be used to study the release properties of the target design. Similar studies using the RIST target design have shown that the approach has improved release times of the metallic target used at the ISOLDE facility (12).

Target Material

The choice of suitable target materials for high power RIB targets is rather limited compared to existing targets at low-intensity facilities. In order to obtain good release of the produced radioactive species, operation at elevated temperatures is imperative, with temperatures chosen to respect the limits of target material volatility (usually to avoid vapor pressures larger than 10^{-4} Torr). This dictates that a chosen target material should have good refractory properties such as low vapor pressure, high melting point, and benign chemical effects at high operating temperatures. In order to remove the heat generated internally, the material must exhibit good thermal conductivity at operating temperatures. The operating temperatures for the classes of target materials vary from a few hundred °C to over 2000 °C, depending on the material properties and the desired radioactive species to be released.

The combination of required properties for target materials limits the choice of these materials to metallic foils, molten metals, or thermally conductive refractory powders or powder-like materials. The last category of materials has not been studied in much detail, and may not prove to be a viable form in practice. A possible exception could be for uncooled graphite fibers, which have not yet been tested at high powers. Accordingly, the initial studies of high power targets have concentrated on metallic foils, but possibilities for molten metals seem to be a viable alternative (13).

Problems of existing targets are much more tractable because, being heated externally, they are operated at uniform (and controlled) temperatures. A target heated internally by an intense beam will exhibit non-uniform temperature distribution, reflecting the beam heating profile as it traverses the target. Any control on the temperature distribution is much more limited with external cooling because of the limitations posed by thermal conduction from the regions of heat deposition. Furthermore, design of a cooling approach may be valid for only one incident production beam current (and internal heating rate).

Cooling Requirements

In general, beam heating with a 100-µA proton beam results in a heat load on the target of the order of one kW/cm throughout the length of the target. This heat load can vary by as much as a factor of two depending on the density of the target material. For a nominal 20-cm long target, therefore, a cooling system will be required to remove tens of kW from the target.

Another issue is that the heat load per unit length should be as uniform as possible. With equal total heat generated per unit length along the target, it may be possible to control the hottest temperature within the target material as a function of target axial position. Such uniform heat load can be accomplished with a simple density gradation along the target length, following the example of the RIST target developed at Rutherford Appleton Laboratory (11). An example of the axial heat load for a graded density vanadium target is shown in Fig. 1, where the vanadium target material density varies from 37.5% to 50% of solid density from the front to the back of the target. The discrete changes of density within the target can be seen in the figure. This energy deposition calculation was made using the LAHET Code System (14), for a cylindrical target of 0.9-cm radius and 20-cm length with graded target material density.

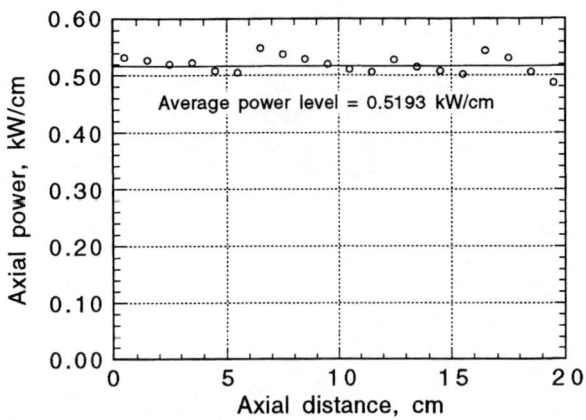

FIGURE 1. Axial power distribution for a vanadium target of graded density with an incident 500-MeV proton beam intensity of 100 μA.

The average axial heat load of 0.519 kW/cm is consistent with the relatively low density of the target, ranging from 2.24 g/cm^3 at the beginning to 2.99 g/cm^3 at the end of the target for a total target density of 49.8 g/cm^2.

APPROACHES FOR TARGET COOLING

Most approaches to target cooling incorporate cooling from the cylindrical surface of the thick target. The early studies (1,2) set the stage for more realistic approaches, ranging from heat conduction through a thermal barrier surrounding the target surface to augmenting radiative cooling with conductive cooling pathways. An issue associated with temperature control is the need to provide supplementary heating for production beam currents lower than the design goal. The internal heating will then be insufficient to provide the necessary operational temperature in the target material and additional, externally applied heating will be required for desired target performance.

Of the possible cooling schemes, three have been pursued in simulations -- radiative cooling, conductive cooling through fins attached to the target, and helium cooling. Only that of radiative cooling has resulted to date in actual experiments. Conductive cooling has been the subject of extensive work over the past five years. The helium cooling scheme is attractive because it also offers the possibility of supplementary heating, but this approach appears to be quite expensive and hasn't been studied extensively.

Many target materials have desired operating temperatures that are much lower than those required for radiative cooling. For this reason, each target type must be treated separately, with appropriate conductive cooling tailored to the particular temperature limit and incident current intensity.

High Temperature Targets -- Radiative Cooling

Radiative cooling is an attractive, passive approach for very refractory metal target materials, such as tantalum (11). In the RIST design, the outer surface area of the target is increased by frequent "fins" resulting in increased radiation efficiency. In this design, supplementary heating is achieved (and has been tested) by means of electron bombardment from separated filaments running the length of the target. The RIST target is elegant in its design, but difficult to manufacture because the cooling surfaces represent an integral extension of the target material and the complete assembly (made of thousands of stacked foils and spacers) is diffusion bonded together to form a solid, vacuum tight assembly.

Radiative cooling has application to a very limited set of target materials that can operate at temperatures where cooling efficiency is high (≥ 2000 °C). While representing an attractive approach for such targets, radiative cooling is not appropriate for most target materials that operate at lower temperatures. For such materials, conductive or convective (in the case of molten target materials) cooling are the only possibilities.

Studies of Thermal Barrier Approaches

The first mention of a thermal barrier for providing cooling and thermal isolation of the target material was made with a gas-filled annulus surrounding the target (2). Appropriate thermal barriers are also possible from granulated spacers, porous metals, refractory insulators, or even contact thermal resistance between surfaces. A comprehensive set of studies has been made for thermal barriers that could be used that resulted in possibilities for thermal barriers for targets operating above about 800 °C (15).

Thermal barriers can be tailored to a wide range of heat loads and target operating temperatures and wide experience is available on their application. However, the stability of the conductive nature of thermal barriers is not fully known under conditions of continued use and changing dimensions of targets as they rise in temperature during irradiation. More experimentation is required before thermal barriers can be applied to the high power RIB target concept with confidence.

Other Conductive Cooling Approaches

Recent work at Lawrence Berkeley National Laboratory (16) has investigated the use of helium cooling and the use of longitudinal fins connecting the target to a surrounding cooling manifold. Helium cooling was studied for a manifold of pathways for helium distributed azimuthally along the length of the target. The results of the study indicated that this approach to cooling would be versatile and also allow for heating of the target by introducing heated helium gas into the manifold.

Also at Berkeley, a suggestion was advanced (17) for cooling by means of 24 longitudinal segmented fins that were spaced azimuthally on the cylindrical surface of the target. The fins were segmented to reflect the lesser cooling requirements toward the end of the target compared to the entrance of the target. By judicious selection of the segment widths (which decreased toward the end of the target), good temperature uniformity was achieved throughout most of the target material. However, the target analyzed was of tantalum, and radiative cooling from the cylinder surface contributed significantly to the total cooling of the fin system.

The use of radial fins as a cooling approach (akin to "weak coupling" cooling) has recently been a subject of intense study. The location of the fins and determination of their thicknesses to match the cooling rate requirements along the length of the target has been shown in simulations to be an attractive approach, as shown below. The attachment of the fins to the target container must be thermally robust, and could be accomplished either by diffusion bonding or by making the target container and fins from a single piece of metal.

Molten Targets

At existing facilities, molten targets have proven to be quite useful and robust. Such targets are quite dense, however, and internal heat generation is considerable. Because of the natural processes of convection and evaporation, transport of heat from the location of the beam and removal to a cooling surface should be very efficient. The use of a refluxing chimney in the line for transport of radioactive species from the target to the ion source has proven to be successful in suppressing the effects of large heat transients resulting from intense pulsed beams (13).

The application of this technique to high power molten metal targets appears attractive in using the molten metal to provide transport of the internally generated heat to the cooled reflux chimney. Despite large rates of evaporation, if the chimney is properly designed the molten target "working fluid" should be confined to the target system, while the produced radioactive species can be transported to the ion source (it is assumed that the released radioactive species are more volatile than the molten target material).

ENERGY DEPOSITION

For most studies carried out to date, energy deposition rates have been predicted for energetic protons using the LAHET Code System (14), which is a generalized particle transport code embodying charged-particle transport and generation and transport of secondary particles arising from intranuclear cascade processes.

To verify the simulations that have been carried out, a prototype target has been conceived for testing of thermal behavior at the ISAC facility of the TRIUMF accelerator, using up to 100 µA of 500-MeV protons. The target is schematically shown in Fig. 2, where the target cooling approach using radial fins arrayed along the length of the target has been chosen. The figure shows only a short length (~3 cm) of the target, and also illustrates thermocouple connections for the purpose of temperature measurements in a test experiment. The vanadium target material is a cylinder of dimension 0.95 cm radius and length 20 cm, and has graded density as mentioned above. Figure 1 shows the axial power load for this target. The target is contained in a niobium cylinder of thickness 0.05 cm, and the niobium fins extend out to a radius of 2.5 cm. A possible future connection of a transfer line to an ion source is indicated.

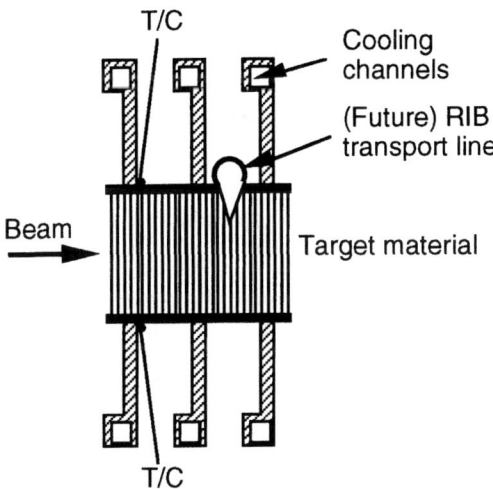

FIGURE 2. Schematic view of prototype ISAC target.

From use of the LAHET Code System (14), radial energy deposition rate profiles for intervals of 0.1 cm have been generated for the ISAC target irradiated with a 100-µA proton beam, shown in Fig. 3.

Note that at the entrance to the target, the energy deposition rate follows the incident beam profile, which was parabolic with a radial width of 0.28 cm. As the beam progresses along the target, scattering processes cause it to spread out until at the end of the target, the heating is only about an order of magnitude less at the edges of the target than in the middle. This figure illustrates the proton heating in the target material that also results in a nearly uniform axial heat load shown in Fig. 1.

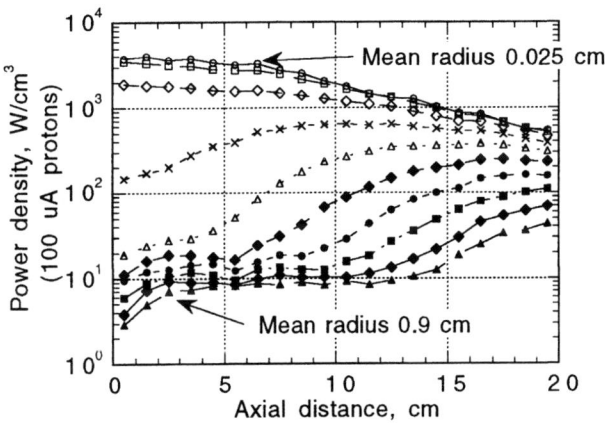

FIGURE 3. Energy density deposition rate profiles in ISAC target.

THERMAL ANALYSES

The beam heating profile generated by the LAHET Code System is used as heat source terms for finite element analysis of the thermal behavior of the target system, where the outer edges of the radial fins are assumed to be at 100 °C. The finite element analysis is accomplished with codes such as ANSYS (18) or COSMOS (19). For the ISAC target, the temperature profiles are shown in Fig. 4 for various radii in the target, where fin thicknesses have been varied to establish a constant axial temperature of ~1700 °C (the operating temperature limit for the vanadium target material). The presence of the fins is apparent; at their locations (at 0.125, 0.875, 1.875,...,19.875 cm) there are a local decreases in the target outer radius temperatures. The fin thicknesses were 0.0875, 0.25, 0.25, 0.2, 0.2, 0.1875, 0.1875, 0.175, 0.175, 0.15, 0.15, 0.1375, 0.1375, 0.125, 0.125, 0.1125, 0.1125, 0.11, 0.11, 0.11, and 0.06 cm, from front to rear of the target. Note that the temperatures at the outer surface of the ISAC target are insufficient for radiative cooling to contribute effectively to the heat transfer out of the target.

The large temperature gradient within the target material at the entrance compared to the exit of the target is a serious problem in any target heated by an axial beam and reflects the dispersion of the beam as it progresses along the target. Recall that existing targets, heated externally, are uniform in temperature. The gradients are a result of the finite thermal conductivity of the target material, even for metallic foils. Future development programs will have to address this situation; for purposes of a thermal test, the target container axial thermal gradient is a fortuitous occurrence, providing an observable that can be used to validate the simulation approaches.

Another cooling approach recently studied is that of using longitudinal fins instead of radial fins. This approach has the possible advantage that thermal stresses resulting from temperature gradients in the fins are not of concern. While the approach looks promising, it is too early in the studies to make judgment of the suitability for the use of

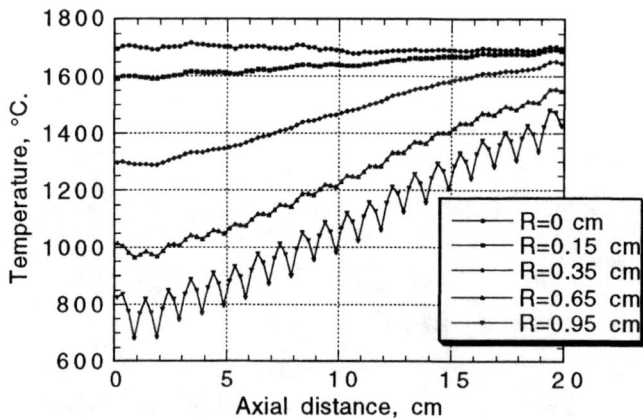

FIGURE 4. Temperature profiles in ISAC target material (the curve for 0.95 cm radius is for the inside surface of niobium target container).

longitudinal fins.

Among the issues to be decided are: How many fins -- two, four, or eight? Recall that the Berkeley group had suggested the use of 24 fins, which may be difficult to fabricate. Also, how are the longitudinal temperature variations to be accommodated, through fin thickness adjustment, through fin height adjustment, or through fin segmenting (as suggested in Ref. 17)? Preliminary analyses indicate that a linear fin height adjustment (shorter at the front of the target, longer at the end) is effective.

With the use of longitudinal fins, the preliminary temperature profiles in the target material are similar to those for radial fins, shown in Fig. 4, with azimuthal variations evident at the outer surface of the target between fins rather than the axial variations shown between radial fins.

IN-BEAM THERMAL TEST

For testing in the ISAC facility at TRIUMF, the above analysis has been applied to generate temperature profiles along the target container as a function of incident current, with no other changes to the target configuration. The resulting temperature profiles are shown in Fig. 5, where it is seen that the observable temperatures depend greatly on the incident beam current. Accordingly, the axial temperature of the target changes from ~1700 °C for 100 µA to ~1350 °C at 75 µA, to ~980 °C at 50 µA, to ~570 °C at 25 µA, and to ~295 °C at 10 µA.

The thermal test has been proposed at TRIUMF and has been accepted. However, because of the ongoing construction of the ISAC facility, it appears that the test is not possible until November of 1999. Even for the short irradiation times needed to establish thermal equilibrium, the target activation will be such as to require remote handling of the target assembly after the test, and the remote handling will not be available for some time (ISAC will operate at reduced beam currents in the meantime).

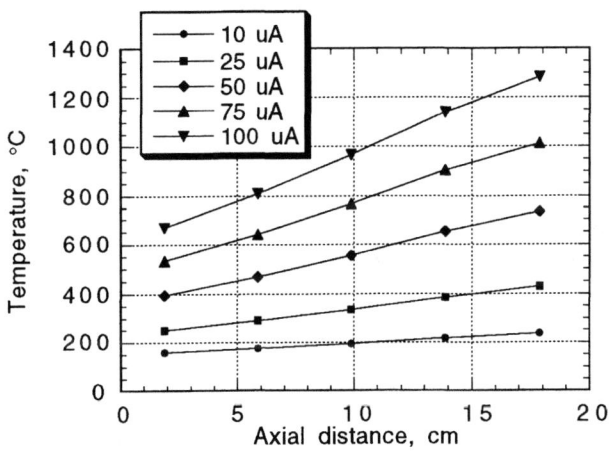

FIGURE 5. Temperature profiles along target container length for various beam currents for the ISAC experiment.

FUTURE STUDIES

Clearly, the analyses of the ISAC target show a deficiency in target material temperature uniformity. This could impact the performance of the target in terms of providing the proper high-temperature environment for rapid release and transport of the radioactive species produced. Analyses have shown that instead of an axially-incident beam, use of an annular beam can help reduce the large thermal gradients (15). In the case of an annular beam, the heat deposition occurs near the edge of the target throughout its length, reducing the radial thermal gradients. Furthermore, the absence of beam in the center of the target allows provision of a central channel through which radioactive species could be transported more easily to an ion source. Further study is needed for this alternative irradiation approach.

As mentioned above, the use of molten target materials can possibly provide a robust target with built-in temperature uniformity. Such targets are limited by the availability of low melting point metals, which produce radioactive species that are more volatile than the target material. While representing a valuable class of targets, molten metal targets are not widely available and may have limited utility.

The target concept developed for the thermal test includes no provision for external heating of the target to maintain a desired operation temperature at lower beam currents. In an operational RIB target, this provision would need to be included. With the use of separated radial fins, it may be possible to accomplish external heating by passing a current through the target container material, but this possibility needs to be studied further.

Finally, studies are needed to investigate the possibility of enhancing the thermal conductivity of powder-based targets. Studies exist that illustrate the effect of adding graphite fibers or boron nitride platelets to a powder to enhance the thermal conductivity of the composite material. The possibility that such enhancement can be utilized for high power target materials will require an ambitious program of study.

REFERENCES

1. Eaton, T. W., Ravn, H. L., and the ISOLDE Collaboration, *Nucl. Instr. Meth.*, **B26**, 190 (1987).
2. Talbert, W. L., Hsu, H.-H., and Prenger, F. C., *Nucl. Instr. and Meth.*, **B70**, 175 (1992).
3. Gaelens, M. et al., *Nucl. Inst. Meth.*, **B126**, 125 (1997).
4. Putaux, J. C. et al., *Nucl. Instr. Meth.*, **B126**, 113 (1997).
5. Talbert, W. L., and Hsu, H.-H., *Nucl. Instr. Meth.*, **A362**, 229 (1995).
6. Villari, A.C.C. et al., *Nucl. Phys.*, **A616**, 21c (1997).
7. Habs, D. et al., *Nucl. Phys.*, **A616**, 29c (1997).
8. Bricault, P. et al., *Nucl. Instr. Meth.*, **B126**, 231 (1997).
9. Ciavola, G. et al., *Nucl. Instr. Meth.*, **B126**, 258 (1997).
10. Kubono, S. et al., *Nucl. Phys.*, **A616**, 11c (1997).
11. Bennett, J. R. J. et al., *Nucl. Instr. Meth.*, **B126**, 117 (1997).
12. Drumm, P. V. et al., *Nucl.. Instr. Meth.*, **B126**, 121 (1997).
13. Lettry, J. et al., *Nucl Instr. Meth.*, **B126**, 170 (1997).
14. Prael, R. E. and Lichtenstein, H., *User Guide to LCS - the LAHET Code System*, Los Alamos National Laboratory report LALP 91-51 (1991).
15. Talbert, W. L., Hodges, T. A., Hsu, H.-H. and Fikani, M. M., *Rev. Sci. Instrum.*, **69**, 3019 (1997).
16. Maddi, J. A. et al., *Nucl. Instr. Meth.*, **A397**, 209 (1997).
17. J. M. Nitschke, private communication (1995)
18. *ANSYS User's Guide for Revision 5.0* (Swanson Analysis Systems Inc., Houston, PA. 1993).
19. *The COSMOS/M Code* (Structural Research and Analysis Corporation, Los Angeles, CA. 1988).

Isobar Separators for Radioactive Ion Beam Facilities

H. Wollnik[*] and J. Garrett[†]

[*] II. Physikalisches Institut, Universität Giessen, 35392 Giessen, Germany
[†] Physics Division, Oak Ridge National Laboratory, Oak Ridge, Tennessee, 37831, USA

Abstract. A radioactive ion beam facility - in short a RIB facility - produces ions of short-lived nuclei and accelerates them to energies of 0.1…10 MeV per nucleon or even higher. In this process it is important that the resulting RIB beams are free from nuclei of neighboring isobars or of neighboring elements. This task requires the production and ionization of the nuclei of interest as well as separating them from all others with a high-mass resolving power and small-mass cross contaminations. When constructing such a facility it also is very important to find ways that allow the accelerated ions to be provided to different experiments at least quasi simultaneously.

Introduction

To produce radioactive ion beams one usually fragments heavy nuclei by reacting energetic protons or neutrons with a thick target, diffuses the fragments out of the target at elevated temperatures, and then ionizes and accelerates them to the required energies. In the initial fragmentation process usually not only the desired nuclei of Z_0 protons and N_0 neutrons are formed. Therefore, it is necessary to remove the nuclei with neighboring Z- and N-values. Since the unwanted neighboring nuclei are often produced abundantly, this task can be difficult. In some cases this separation can be achieved by the accelerator itself, which in case of a linear accelerator acts as a multiple velocity analyzer or in case of a cyclotron as a multiple time-of-flight mass analyzer. However, this method is problematic for the acceleration of ions of short-lived nuclei because:

1. the precise tuning of an accelerator to the ions of interest becomes difficult if the intensity of these ions is small as compared to the intensity of undesired ions of neighboring nuclei. In this case it is not simple to record the intensity of the ions of interest. Thus no signal or at least no good signal is available for optimization.
2. In a RIB facility the undesired neighboring nuclei are radioactive and will contaminate the accelerator structure considerably.

For these reasons it is advantageous to precede the accelerator by a high performance isotope separator which, not only effectively removes all undesired neighboring nuclei and delivers a pure ion beam to the accelerator, but also does this with only small mass cross contaminations.

The isotope separation is usually made using singly-charged ions because on-line ion sources provide only singly-charged ions efficiently. However, it could be useful to feed the purified beam of singly-charged ions into a special ion source [1] a "charge booster" which increases the charge states of these ions so that they can be accelerated more efficiently. For such multi-charged ions the acceleration to ground already can yield high ion energies.

The structure of a RIB isobar separator

In principle the mass resolving power $m/\Delta m$ of a magnetic sector field separator is [2] the ratio between the mass dispersion D and the widths of the entrance and exit slits $2x_0$ and $2x_1$. Thus for a given sector magnet the mass resolving power can be increased if the slit widths $2x_0$ and $2x_1$ are reduced as long as the image aberrations remain small. Since for a radioactive ion beam facility the intensity of the ions of interest is usually low, one is reluctant to reducing the slit widths to less than width of the beam delivered by the ion source, since this would further reduce the beam intensity.

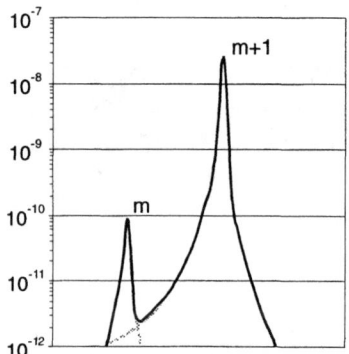

Fig. 1. Experimentally recorded intensities [8] of ions of isobars of masses m and $m+1$ with m=84 amu. Note the tail of low energy ions of mass $m+1$ that extends into the position of the regular ions of mass m.

The properties of an electromagnetic isotope separator are best described by its Q-value [2,3], i.e. the product of the momentum-to-charge resolving power $p/\Delta p = 2m/\Delta m$ and the radial phase space area $4\varepsilon_x = 2x_0 2\alpha_0$ of the ion beam. Here α_0 is the maximal inclination of ion trajectories at the entrance slit. This Q-value is directly proportional to the size of the instrument, thus large low-field magnets, which unfortunately are difficult to build for high precision systems, are desirable. There are side effects, however, that independent of the ion optics may cause mass cross contaminations and which may reduce the finally purity achieved for the desired ions of mass m_0 and energy K_0. Thus the design of the overall system should be characterized by:

1. Small lateral image aberrations. Here it usually is better to choose a system design that has low aberrations from the beginning [4] than to choose an otherwise advantageous design whose aberrations must be corrected by multipole elements. Such multipole elements will usually cause other higher-order image aberrations. To separate neighboring elements of the same isobar from each other, one must postulate resolving powers at and above $m/\Delta m$ =20,000 (see Fig.4) if one assumes that the isobar under consideration contains elements of mass ≈100 amu and if the mass difference between the two elements is Q_β≈5MeV. Though 5MeV is a very high Q_β-value for nuclei near β-stability, even higher Q_β-values are quite common for short-lived nuclei.

2. The reduction of the background of ions with energy K_0 and undesired mass values $m_1=m_0\pm\Delta m$. Such ions can cause mass cross contaminations after they have been scattered on residual gas atoms or on slits giving rise to long tails of intense mass lines. The magnitude of such tails is often [5] about 0.01% of the peak intensity even at the position of a neighboring isobar. This can be very detrimental if – as is often the case – the undesired ions are 10^6 or more times abundant [6] than the ions of interest. However, undesired ions can be removed efficiently by the use of multiple separator stages [7] whose purification factors multiply.

3. The ability to reduce the background of ions formed in charge exchange-processes in the acceleration region after the ion source. In this region ions of higher mass $m_1=m_0+\Delta m$ can be formed at a potential that is slightly lower than that of the ion source. Thus these ions attain only the energy $K_1=K_0-\Delta K$ which can give them the same energy mass product K_0m_0. Thus they are undistinguishable from the ions of interest, with mass m_0 and energy K_0, for any magnetic field. Also tails of these ions are observed with intensities of about 0.01% of the peak intensity (see Fig.1). These ions can be removed, however, by placing the different separator stages at different electrostatic potentials [7,9,10].

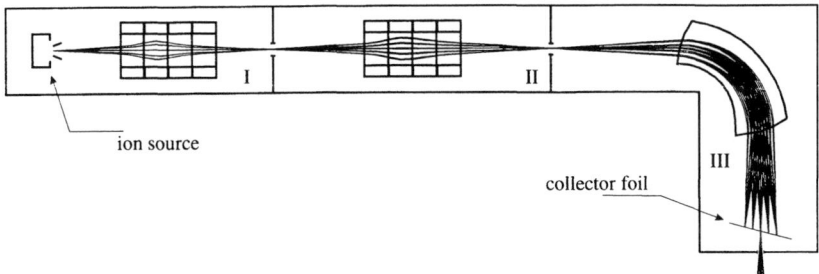

Fig.2. A feasible preseparator. Note the intermediate images with small intermediate slits and orifices before the actual separator. This structure should allow a greatly reduced yield of radioactive neutrals in stages II and III. Note also the collector foils located on both sides of the exit slit. Note further that this unit could be housed in a separate vacuum tank that only is opened in a "hot cell."

A feasible RIB isobar separator design

One way to build an efficient isobar separator installation is to construct it from three stages at different electrostatic potentials [7,9,10]. The first three stages should each contain a sector field separator (see also ref [11]) while the fourth stage should contain the first section of the RIB accelerator [9,10]. In detail such a system (see Fig.3) would contain:

1. A "preseparator" at potential V_0 that achieves a mass resolving power $m/\Delta m \approx 400$. In order that the ions form a reasonably good beam in this preseparator, one should place the ion source at some potential V_{00} so that $V_{00} - V_0 \approx 20-60$ kV. The main task of this preseparator is to extract the ion beam from the source, separate the ions of mass m_0 and energy K_0 roughly from the ions of neighboring isobars and at the same time ensure that the radioactive but neutral atoms are retained in the system efficiently. One way to do this is shown in Fig.2. Note here that
   1.1 there are intermediate images formed by electrostatic quadrupoles that allow all ions to pass through narrow orifices and slits which efficiently retain the neutral atoms, the bulk of the produced radioactivity
   1.2 the ions of neighboring isobars are implanted into foils on both sides of the preseparator exit slit.

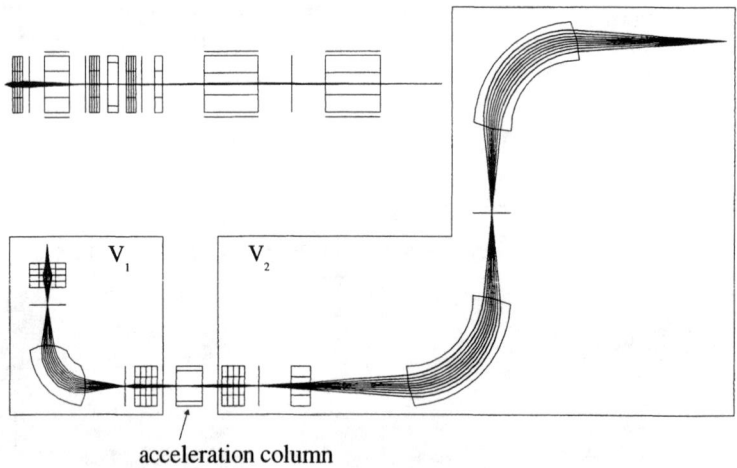

Fig. 3 An energy-achromatic arrangement of a double-stage separator with the two separator stages being at different electrostatic potentials V_1 and V_2. The ion trajectories are shown projected both onto the plane of symmetry and onto a perpendicular surface. Note that ions of equal masses but different energies are refocused to the same position, as illustrated in Fig. 4. Note further that the arrangement of two separator stages shown also can be reversed.

2. A "beam cooler" at a potential ≈50 V below V_{00}. Such "beam coolers" are not common features in on-line isotope separators yet. However, for special applications gas-filled RF-quadrupoles have been used as beam coolers [12] in which the ions are dragged through a gas of ≈0.1mbar and thus equilibrate with the gas atoms. This not only reduces the energy distributions but also the full phase-space area of the ion beam. After a new acceleration thus quite narrow ion beams can be formed.

3. The main isotope separator, consisting of two stages (see Fig.3)
   3.1 the 1st-stage separator at potential V_1 that achieves a mass resolving power $m/\Delta m$ ≈2000. The task of this separator is to purify the beam of ions of mass m_0 and energy K_0 reasonably well from the undesired ions of neighboring isobars.
   3.2 the 2nd-stage separator at potential V_2 that achieves a mass resolving power $m/\Delta m$ ≈20,000. The task of this separator is to further purify the beam of ions of mass m_0 and energy K_0 from the undesired ions of neighboring elements within the same isobar.

By adjusting the potential difference V_2-V_1 properly, it is possible to achieve an energy-achromatic mass separation as outlined in ref. [13] (see also refs. [11,14]) like in a classical Aston mass spectrometer [15]. Fig. 4 indicates that this really can be expected. Here mass lines of ions are shown whose mass-to-charge values differ by 0.005% and whose kinetic energies vary by ±0.005% in Fig. 4a and by ±0.05% in Fig.4b. Both cases have been calculated with an initial phase-space area of ε_x =±$x_0\alpha_0$=±0.2mm*20mrad.

Note here also that the order of the two separator stages can also be reversed.

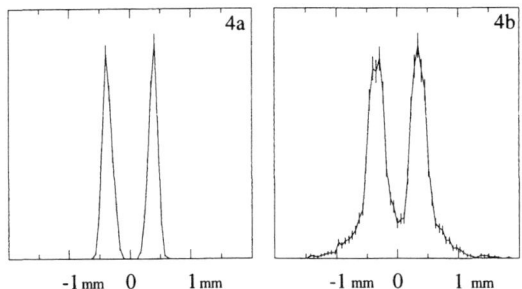

Fig. 4 The calculated mass spectrum of ions whose masses differ by 0.005% and whose phase space area is ε_x=$x_0\alpha_0$=±0.2mm*20mrad. For Figs. 4a and 4b it was assumed that the ions have energy spreads of 0.005% and 0.05%, respectively. This would correspond to energy spreads of ±3eV and ±30eV for V_{00} - V_0=60,000 V. Note that even for these relatively large energy spreads a mass resolving power of $m/\Delta m$ =20,000 is calculated.

4. A "charge booster" [1] at a potential of ≈ 10 V below V_{00} which will increase the charge states of the ions of interest considerably and thus make any subsequent acceleration more efficient.

5. The first accelerator section at potential V_3. This potential must be chosen such that the ion velocity is always the same whatever ion mass value m_0 has been chosen, since only ions of a fixed velocity can be accelerated in the RFQ-accelerator probably used here.

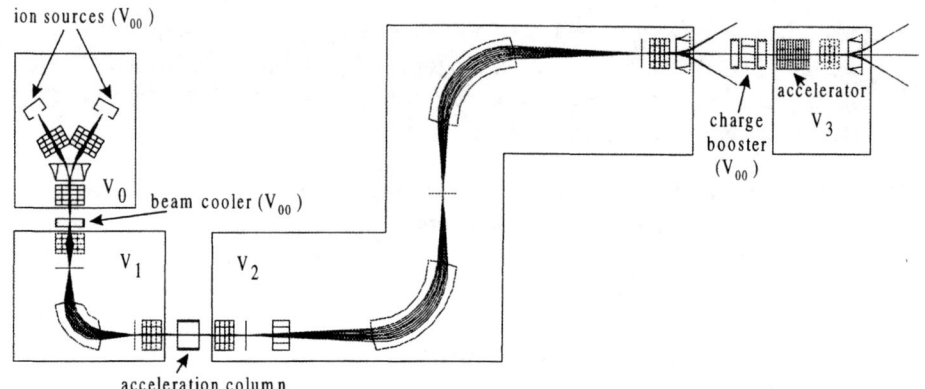

Fig. 5. Overview of a multi-user RIB-isobar separator. Note that two ion sources and two preseparators are assumed feeding the double-stage main separator at potentials V_1 and V_2. Note also that the first-stage of the RIB-accelerator is at a potential V_3. Also note the "beam cooler [12]" located before the entrance to the double-stage separator and the "charge booster [1]" after the exit slit of this separator. Note finally that the beam is assumed to be switchable to some experiments immediately after the 2nd-stage separator and after the first accelerator section but that this beam also can be provided to other accelerator sections.

A multi-user RIB facility

In order to use the beams of a RIB facility most efficiently, it may be advisable to distribute the ion beam to different experiments in a time-shared fashion. This may require that at least for the first part of the accelerator structure be based on non-superconducting technology to allow for fast switching. Furthermore it may be advisable to feed ions into the system from different "pulsed" ion sources and possibly different preseparators. Perhaps this can be accomplished with no (or minimal) ion losses, since it has been shown [16] that between extraction pulses the ions can be stored in the ion source plasma for ≈ 10msec before they are extracted during ≈ 2msec. One can do this in an interweaved fashion for two on-line sources. Perhaps a third or even fourth properly pulsed ion source providing, for instance a stable ion beam, could be integrated into such a multi-user facility.

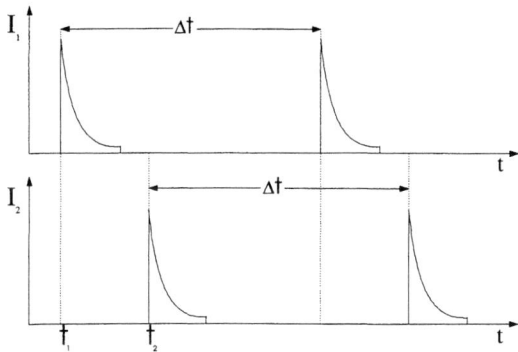

Fig. 6 An illustration of the interweaved operation of two ion sources. At the beginning of each extraction pulse of possibly 2msec the intensity is considerably higher than at the end of this pulse at which time the ion intensity has been reduced to approximately the dc-extraction value. In total the ion intensity extracted in 2msec can even be higher than that extracted in 10msec [15].

Acknowledgements

Discussions with G. D. Alton, J.R.Beene, F.E.Bertrand, J.D.Fox as well as with W.Talbert and M.Winkler are acknowledged. This research was sponsored by the German Minister für Forschung und Technologie under contract GI 849 I, the Laboratory Directed Research, the Development Program of the Oak Ridge National Laboratory, managed by Lockheed Martin Research Corporation for the U.S. Department of Energy under contract DE-AC05-96-OR22464, and the Joint Institute for Heavy Ion Research in Oak Ridge.

References

1. R. Geller, in these proceedings
2. H.Wollnik, "Optics of Charged Particles", 1987 Acad. Press, Orlando
3. H.Wollnik, Nucl. Instr. and Meth., **95**(1971)453
4. H.Wollnik, Nucl. Instr. and Meth., **56/57**(1991)1096
5. H.J.Freeman, Nucl. Instr. and Meth., **28**(1965)49
6. H.L.Ravn et al., Nucl. Instr. and Meth., **B126**(1997)176
7. H.Wollnik, in "Radioactive Beams", ed. J.D. Garrett, 1992, p. 213
8. E.P. Chamberlin, private communication
9. H.Wollnik, Particle Accelerators, **47**(1994)241
10. H.Wollnik, Nucl. Instr. and Meth., **A363**(1995)393
11. G.Ciavola et al., Nucl. Instr. and Meth. **B126**(1996)17
12. V.Koslowsky et al., Int. J. Mass Spectr. And Ion Proc, in print
13. H.Wollnik, Nucl.Instr. and Meth., in print
14. M.Yavor, Nucl. Instr. and Meth. **B126**(1996)266
15. F.W.Aston, Philosoph. Mag. **38**(1919)709
16. Y.Shirakabe et al., Nucl. Instr. and Meth. **A 337**(1993)11

The Physical Basis of the Release Curve for RIB Foil Targets

J R J Bennett

CLRC, Rutherford Appleton Laboratory, Chilton, Didcot, Oxon, OX11 0QX, UK.

Abstract. Both diffusion and effusion in radioactive ion beam foil targets can be described mathematically by Fick's Law of diffusion. An approximate solution to the equation with the appropriate boundary conditions is presented for the RIST geometry target. This solution is expressed in terms of the physical and geometrical properties of the target and ioniser: the diffusion constant of the foil material, the conductance and the dimensions of the target and ioniser. The results of fitting the equation to measured data are shown. The equation can be used to optimise the design of targets for different radioactive ion beams.

INTRODUCTION

The time dependant release of radioactive ions from a hot target, bombarded with high energy protons, and followed by a suitable ioniser, generally fit a characteristic release curve (1). A short (relative to the radioactive decay time of the ion species and their diffusion and effusion time constants in the target and ioniser) pulse of protons on the target gives a current of radioactive ions which initially rises from zero to a peak and then falls, as shown in Figure 1. This release curve can be used to give an indication of the properties of the target and its suitability for the production of rare short lived isotopes. Also, it is hoped that the decay curve can be used to help in the design of better targets through an understanding of the fundamental process occurring in the target such as diffusion and effusion.

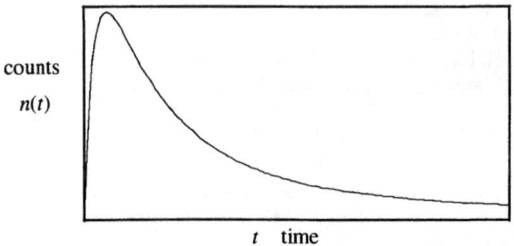

FIGURE 1. A typical release curve.

CP473, *Heavy Ion Accelerator Technology: Eighth International Conference,*
edited by Kenneth W. Shepard
© 1999 The American Institute of Physics 1-56396-806-1/99/$15.00

Measurements (1) of the release curves from radioactive beam targets have been taken at ISOLDE (2) by putting a single pulse of protons onto the target, transporting a time gated portion of the beam to a tape station and counting the beta decay particles. The number of beta counts, $n(t)$, is plotted as a function of the release time, t, of the gated radioactive beam, measured from the arrival of the proton pulse on target ($t = 0$).

2. RNB PRODUCTION

When an energetic proton hits the target it will produce, amongst other reactions, radioactive particles. Once the radioactive particles have been formed, they start to decay at the appropriate fractional rate, λ. The particles also start to diffuse out of the target material. On reaching the surface they are, in principle, free to leave the surface and will in general hit another surface where they have the choice of sticking for some time before leaving or being reabsorbed and diffusing into the target material. The diffusion rates depend on the geometry of the target material and the diffusion constant, which is temperature dependant, hence the requirement for hot targets. Particles reach the surface more quickly if the material is thin, prompting the use of powders and thin foils.

Assume that the target is devoid of particles initially when the short proton pulse hits the target. Immediately following the pulse the particles at the surface of the foils start to fill the target void volume. Some of the particles eventually reach the opening into the ioniser, which may be connected via a tube. This tube causes an additional delay to the particles before they emerge from the ioniser. Hence the release curve rises from zero at time $t = 0$ to reach a peak and then falls as the diffusion decreases. The effusion of particles within the target and the ioniser is dependant on the temperature and geometry.

For a given element in a given target and ion source at a particular temperature, it is possible to define a probability, the release function (1), $p(t)$, for an atom, generated at time $t = 0$, to be released from the ion source. Some of these atoms will be released as ions, due to the action of the ioniser. If the ionisation efficiency for the particular species is ε, the ion current as a function of time will be of the form,

$$i(t) = \varepsilon \cdot \sigma \cdot n_p \cdot g \cdot p(t) \cdot e^{-\lambda \cdot t} \tag{1}$$

where σ is the effective cross section for production of the isotope by the proton beam as it passes through the target (assuming only a direct formation of the isotope by the beam), n_p is the number of protons in a pulse, assumed to be much shorter than the decay time of the isotope, and g is a geometry factor for the target and proton beam to take account of scattering and resultant beam interaction with the target material. Note that all the diffusion and effusion, including temperature, geometry effects and any reabsorption are included in the term $p(t)$. By definition, the integral of $p(t)$, from 0 to ∞, is 1.

Because $p(t)$ is closely related to the diffusion and effusion in the target and both have exponential forms for their release functions (3), it is reasonable to fit the observed curve with three exponential terms in the form (1),

$$p(t) = A\left(1 - e^{-\lambda_r t}\right)\left[Be^{-\lambda_f t} + (1-E)e^{-\lambda_s t}\right] \qquad (2)$$

where $1/\lambda_r$, $1/\lambda_f$ and $1/\lambda_s$ are the rise, fast fall and slow fall time constants respectively and A and B are constants. This formulation fits the data very well in most cases.

DIFFUSION AND EFFUSION IN THE TARGET AND IONISER

Both the diffusion and effusion of particles in the target can be treated mathematically as a diffusion process and can be described by Fick's Law (4), which states that the rate of transfer per unit cross sectional area, F, is proportional to the concentration gradient, $\dfrac{\partial C}{\partial x}$, measured normal to the section,

$$F = -D\frac{\partial C}{\partial x} \qquad (3)$$

where D is the diffusion constant. The second law of diffusion follows from the first, above:

$$\frac{\partial C}{\partial t} = D\frac{\partial^2 C}{\partial x^2} \qquad (4)$$

Consider the case of the RIST target (5), shown in Figure 2. The target tube is filled with thin tantalum foil discs, $2d$ thick, and regularly spaced apart by $2d$. A central hole in the discs allows the passage of the radioactive ions to the ioniser tube which is placed centrally along the length of the target.

Assume the target contains no (radioactive) particles of a particular species initially, at time $t < 0$. A very short pulse of protons then irradiates the target at $t = 0$, producing a uniform density in the foils of particles of the species under consideration. The solution can be represented by the diffusion of particles within the foils at a concentration C_0 at time $t = 0$ with diffusion constant D, followed by diffusion (effusion), with diffusion constant E, through the target and ioniser (see Fig. 3), obeying the appropriate boundary conditions. It is reasonable to assume that the

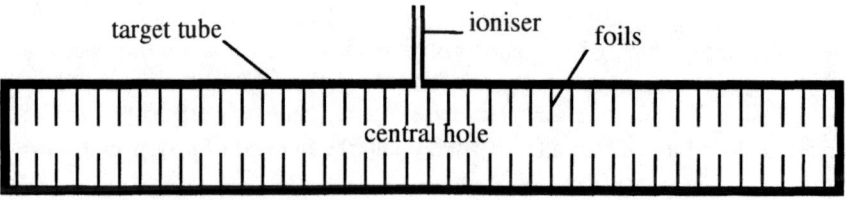

FIGURE 2. Schematic section through the RIST target and ioniser.

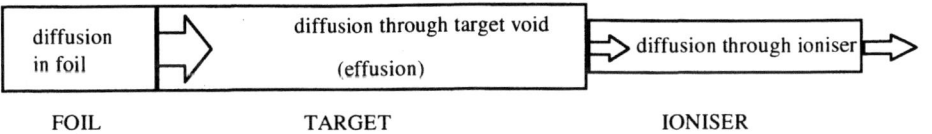

FIGURE 3. The schematic diagram for the flow from one representative foil through the target and the ioniser.

concentration at the exit of the ioniser is very low, the vacuum pressure is typically below 10^{-6} mbar, so that the concentration can be put at zero. However, the concentration at the surface of the foils is unknown; sticky particles are likely to result in high surface concentrations while gases and other volatile substances are likely to give rise to relatively low surface concentrations. Furthermore, the concentration on the surface may well be effected by the concentration in the target void, i.e. the partial pressure of the particles in the target. The material of the foils and the temperature will also affect the surface concentrations. The effects of any re-absorption are ignored.

The solution of diffusion in a composite medium, with two or more different diffusion constants, is complicated. This, coupled to the uncertainty in the relationship of the concentration on and adjacent to the surface of the foils, results in difficulties in solving the diffusion equation analytically. Monte Carlo or numerical methods can be used (although this does not help with the boundary conditions), but a better understanding of the physics can often be obtained by an analytic solution, even if it is only approximate.

The effusion in the target is generally rapid compared to the diffusion in the foil, so the concentration of particles in the target void is small compared to that in the foils. Assume that the concentration on the surface of the foils is zero, or that they follow the same relationship but with a smaller effective diffusion constant to mimic the reluctance of particles to leave the surface. The same effect will be experienced at every surface with which the particles come into contact as they bounce around the target void on the way to the ioniser. This reduces the effective effusion rate.

The concentration of a particular species within a foil of thickness $2d$, with zero concentration at the surfaces, is given by (6),

$$C(x,t) = C_{d0}\left[1 - \sum_{n=-\infty}^{\infty}(-1)^n\left(1 - \mathrm{erf}\left(\frac{2nd - x}{\sqrt{4Dt}}\right)\right)\right] \quad (5)$$

where C_{d0} is the initial concentration at $t = 0$, assumed uniform throughout the target, x is the distance through the foil taking the origin, $x = 0$, at one surface and $x = -2d$ at the other surface, and D is the diffusion constant at the operating temperature, assumed uniform throughout the target. The short time form of the solution (6) to the diffusion equation is given in all cases since a relative small number of sum terms (n) are required in the evaluation for the time period of interest, i.e. where $t \leq D/d^2$. Use of the long time solution gives serious errors as t approaches zero even for very large number of sum terms. The production of the particles will not be uniform within the target due to changes in the energy of the protons as they pass through the foils and any transverse variation in the proton current density across the target. However, the model should be reasonably accurate as the conductance of the gaps between foils is high and hence the effect of the variation of particle formation with radial position will be small.

The flow of particles out of the surface is,

$$F_d(t) = C_{d0} D \sum_{n=-\infty}^{\infty} (-1)^n \frac{e^{-\frac{(nd)^2}{Dt}}}{\sqrt{\pi Dt}} \qquad (6)$$

Every surface of the foils represents a sheet source of particles entering the target void. If the number of particles per unit area is M at time T in time interval dT then,

$$M(T) = F_d(T) dT \qquad (7)$$

The concentration of particles diffusing from a thin sheet, having M particles per unit area, through a medium of thickness z, where the flow is in the positive x direction only (i.e. there is an impenetrable membrane at $x = 0$, preventing flow in the negative x direction), is given by (6),

$$C_e(x,t) = 2M \sum_{m=-\infty}^{\infty} (-1)^m \frac{e^{-\frac{(2mz+x)^2}{4Et}}}{\sqrt{4\pi Et}} \qquad (8)$$

where E is the diffusion constant and the concentration is zero at $x = z$. Note that the same value of the diffusion constant has been taken for effusion through the ioniser and the target. The length of the ioniser is assumed a variable and is adjusted to obtain the "true" value. The flow is given by,

$$F_e(x,t) = 4ME \sum_{m=-\infty}^{\infty} (-1)^m \frac{(2mz+x) e^{-\frac{(2mz+x)^2}{4Et}}}{4Et\sqrt{4\pi Et}} \qquad (9)$$

At $x = z$ the flow becomes,

$$F_e(t) = 4ME \sum_{m=-\infty}^{\infty} (-1)^n \frac{(2m+1)z e^{-\frac{(2n+1)^2 z^2}{4Et}}}{4Et\sqrt{4\pi Et}} \qquad (10)$$

The target consists of a large number of foils regularly spaced along the axis of the target. Therefore, there are a large number of surfaces or sheets of sources, uniformly spaced along the target. The target length is $2h$ with an ioniser of length s. Because of the symmetry about the centre of the target through the axis of the ioniser tube, it is possible to consider only one half of the target. The total flow from the ioniser will be twice that for each half.

Putting,

$$z = h + s \qquad (11)$$

Inserting (11) into (9) and integrating with respect to h from 0 to h, gives the flow from all the foils (in half the target),

$$F_{eh}(t) = \frac{\int_0^h F_e(t) dh}{\int_0^h dh} \qquad (12)$$

$$F_{eh}(t) = \frac{ME}{h} \sum_{m=-\infty}^{\infty} (-1)^m \frac{e^{-\frac{(2m+1)^2 s^2}{4Et}} - e^{-\frac{(2m+1)^2 (h+s)^2}{4Et}}}{(2m+1)\sqrt{4\pi Et}} \qquad (13)$$

At time $t = T$, M may be replaced by (7). The effusion from this sheet at $t = T$ is just starting and hence t in (12) must be replaced by $(t - T)$, so the flow becomes,

$$F_{de}(t) = F_d(T)dT \cdot F_{eh}(t - T) \qquad (14)$$

Integrating with respect to T from 0 to t, gives the resultant flow over all the time up to t,

$$F_{de}(t) = \int_0^t F_d(T) \cdot F_{eh}(t - T) dT \qquad (15)$$

$$F_{de}(t) = \frac{2C_{d0}DE}{h} \int_0^t \sum_{n=-\infty}^{\infty} (-1)^n \frac{e^{-\frac{(nd)^2}{DT}}}{\sqrt{\pi DT}} \sum_{m=-\infty}^{\infty} (-1)^m \frac{e^{-\frac{(2m+1)^2 s^2}{4E(t-T)}} - e^{-\frac{(2m+1)^2 (h+s)^2}{4E(t-T)}}}{(2m+1)\sqrt{4\pi E(t-T)}} dT \qquad (16)$$

This can not be integrated analytically, but may be accomplished numerically.

The flow rate probability per particle produced in the foils is,

$$p_{de}(t) = \frac{\delta \varepsilon_h}{\pi} \int_0^t \sum_{n=-\infty}^{\infty} (-1)^n \frac{e^{-\frac{n^2}{\delta^2 T}}}{\sqrt{T}} \sum_{m=-\infty}^{\infty} (-1)^m \frac{e^{-\frac{(2m+1)^2}{4(t-T)}\left(\frac{1}{\varepsilon_s}\right)^2} - e^{-\frac{(2m+1)^2}{4E(t-T)}\left(\frac{1}{\varepsilon}+\frac{1}{\varepsilon_h}\right)^2}}{(2m+1)\sqrt{(t-T)}} dT \qquad (17)$$

where.

$$\delta = \sqrt{\frac{D}{d^2}} \qquad \varepsilon_s = \sqrt{\frac{E}{s^2}} \qquad \varepsilon_h = \sqrt{\frac{E}{h^2}} \qquad (18)$$

Equation 17 is plotted in Figures 4a - 4d, for various values of δ, ε_s, ε_h, as a function of t.

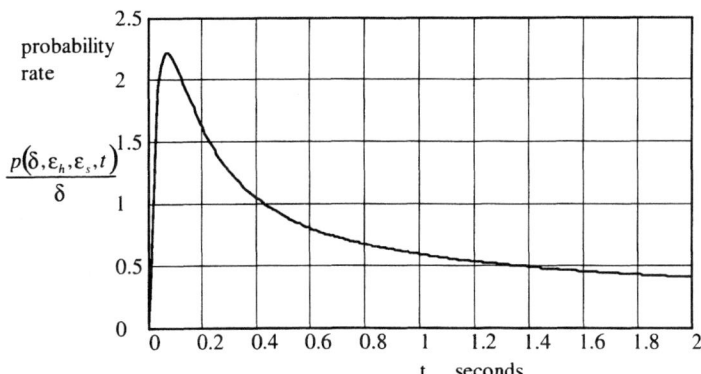

FIGURE 4a. The probability rate, $\frac{p_{de}(\delta,\varepsilon_h,\varepsilon_s,t)}{\delta}$, plotted versus time, for δ = 0.001, 0.03 and 0.01, ε_h = 3 and ε_s = 4. All 3 curves lie on top of each other.

It will be seen that in Figure 4a, where only the diffusion in the foil is altered, that the shapes of the curves are identical and only their amplitudes vary. The three curves of $p_{de}(t)/\delta$ all lie on top of each other. In the cases where DT/d^2 is small, it is possible to approximate the first sum term using the relation,

$$\sum_{n=-\infty}^{\infty}(-1)^n e^{-\frac{n^2}{x^2}} = 1 \qquad \text{for } x < \sim 0.4, \tag{19}$$

then (17) becomes,

$$p_{de}(t) = \frac{\delta\varepsilon_h}{\pi}\int_0^t \sum_{m=-\infty}^{\infty}(-1)^m \frac{e^{-\frac{(2m+1)^2}{4(t-T)}\left(\frac{1}{\varepsilon_s}\right)^2} - e^{-\frac{(2m+1)^2}{4E(t-T)}\left(\frac{1}{\varepsilon}+\frac{1}{\varepsilon_h}\right)^2}}{(2m+1)\sqrt{(t-T)T}} dT \tag{20}$$

which is a simple function of δ. As t (and hence T) increases, then $DT/d^2 \geq 0.4$ and equation (19) does not hold, in which case equation (17) must be used. In practice this will only be important at times below the radioactive decay time, $1/\lambda$, since the release curve will be dominated by the radioactive decay at greater times. Thus, for $D/d^2 < \sim 0.4 \cdot \lambda$ equation (20) is sufficiently accurate. Hence, it is important to have thin foils to increase the yield when the radioactive decay time is relatively short.

It is interesting to note that since equation (20) is usually applicable for the targets considered here, the diffusion terms $\delta \cdot \varepsilon_h$ multiply the whole equation but δ does not alter the shape of the curve while ε_h does. The effect of δ is only observed on the shape of the curve at relatively long times. Figure 4b illustrates the effect on the shape of the curves at longer times - i.e. at larger values of Dt/d^2.

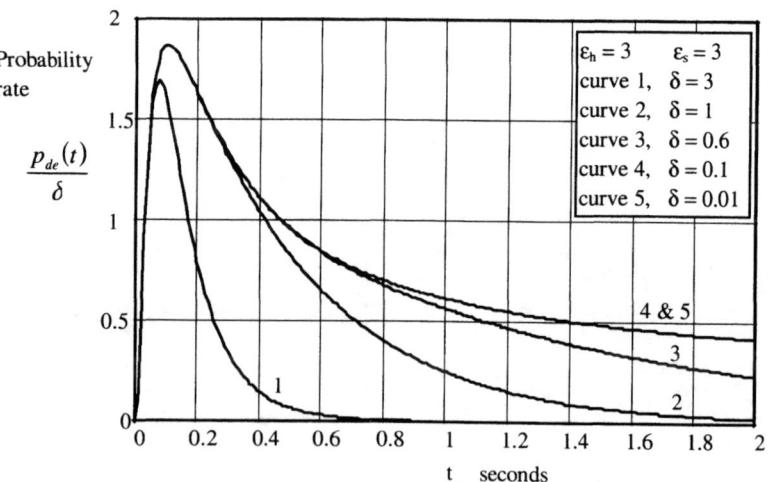

FIGURE 4b. Plots of $\dfrac{p_{de}(t)}{\delta}$ showing the effect of varying δ. As δ becomes larger the curves only coincide at smaller and smaller times. The values of the other parameters are shown on the graph.

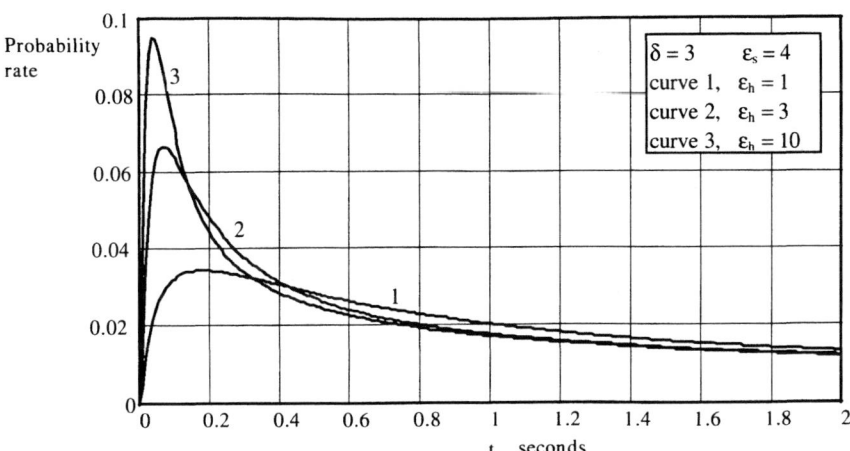

FIGURE 4c. $p_{de}(\delta, \varepsilon_h, \varepsilon_s, t)$ plotted versus t for ε_h = 1, 3 and 10. The values of the other parameters are shown on the graph.

Figure 4c shows the effect of changing ε_h; decreasing ε_h speeds up the release both on the rise and fall of the curve.

Figure 4d shows the effect of changing ε_s; decreasing ε_s shifts the peak to earlier times until the peak is at time zero and increases the rate of rise and, to a lesser extent, the rate of fall.

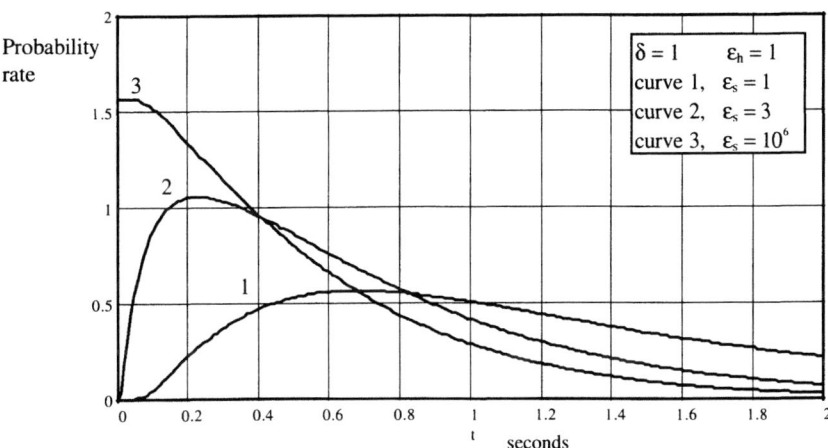

FIGURE 4d. $p_{de}(\delta, \varepsilon_h, \varepsilon_s, t)$ plotted versus t for ε_s = 1, 3 and ∞. The values of the other parameters are shown on the graph.

Both Figures 4c and 4d show the importance of having a high conductance ioniser (short, with a large cross section) and a high conductance through the target. Making the foils thin will not improve the yield if the effusion through the target and ioniser is slow relative to the radioactive decay time.

Note that equation (17) is suitable for relatively short times if the series sums are to converge for reasonably small values of n and m. Figure 4 was calculated using $n = m = 10$. At longer times more sum terms must be used or the alternative forms of the solution to the diffusion equation, which are better at long times.

FITTING TO THE MEASUREMENTS

Measurements of the release curves of a number of radioactive species from a RIST geometry target with foils of 0.0025 cm (target number Ta050) and 2 targets with foils of 0.01 cm thickness (target numbers Ta075 and Ta126) have been taken (7) at ISOLDE. Unfortunately, the measurements on Ta075 and particularly Ta126 were unreliable due to unknown problems in the detectors or electronics. This gave ragged peaks in the release curve, probably because the problems were exacerbated at high count rates. Measurements were taken on ^{8}Li for all three targets and ^{9}Li for target Ta126. The calculated release curve has been fitted to the measured points of ^{8}Li and ^{9}Li by optimising (minimising chi squared) the values of D, ε_h and ε_s. Fits to other isotopes have yet to be made. Figures 5 shows the results for ^{8}Li from target Ta050. The fit of the calculated points to the measurements is very good using the values shown. The vertical axis is actually the measured count in a given time, so is proportional to the count rate. Also, note that the measured counts are reduced by the

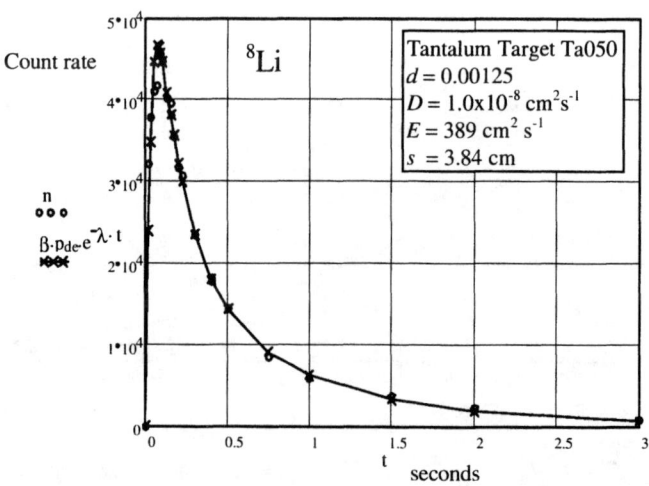

FIGURE 5. ^{8}Li results. Measured data (circles) and calculated fit (crosses and line) for target Ta050.

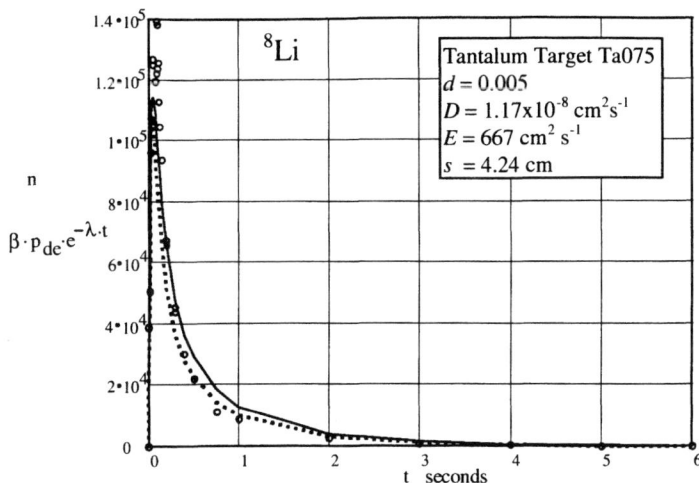

FIGURE 6. ^{8}Li results. Measured data (circles) and the best calculated fit (line) for target Ta075. A better fit (dotted line) on the falling part of the curve is obtained by using the values from Figure 5, scaled by the increased foil thickness.

decay of the radioactive particles so the calculated values have been multiplied by the decay factor, $e^{-\lambda t}$ ($1/\lambda = 1.209$ s for ^{8}Li and 0.257 s for ^{9}Li). Since the calculated relationship is a probability rate, it has been multiplied by an arbitrary constant β to scale to the data points.

Figure 6 shows the measured data for ^{8}Li from target Ta075 and the best fit to all the data points. The measured count rates have a very large scatter around the peak, indicating the typical scale of the error in the measurements on this target and Ta126. The fit is not as good as that for target Ta050 (Figure 5). If the values found from Ta050 are used, but scaled according to the conductance theory (8), then a better fit is obtained on the falling part of the curve. If the data points around the peak are in error, as suspected, then fitting to the falling part of the curve is better. That the scaled values obtained from Ta050 give a good fit supports the premise that the values obtained from Ta050 are reliable.

TABLE 1. Calculated Values of the Diffusion Constants for Lithium

Target	d cm	h cm	Particle	Calculated values		
				D cm^2 s^{-1}	E cm^2 s^{-1}	s cm
Ta050	0.0025	10	^{8}Li	1×10^{-8}	400	3
Ta075	0.0100	10	^{8}Li	1×10^{-8}	700	4
Ta126	0.0100	10	^{8}Li	1×10^{-8}	2000	8
Ta126	0.0100	10	^{9}Li	0.3×10^{-8}	400	2

Table 1 shows the best fits to the data for the three targets. Because of the problems of the measurements with targets Ta075 and Ta126, less reliance should be placed on these values.

DISCUSSION

Figure 5 shows a very good fit of the analytical curve to the measured data points for target Ta050 using a value of diffusion constant of 1×10^{-8} cm^2 s^{-1} for lithium in tantalum at ~2300 K. This value is reasonable compared to other measurements (9). While the data and the fit of the analytical curve with the other targets is not so good, the diffusion constant is still calculated to be 1×10^{-8} cm^2 s^{-1}, as is expected. The value of E should increase for the targets with the bigger gaps between foils and this is found to be true, although the values are not the same for Ta075 and Ta126; the values should be should be since they have the same internal geometry.

It would be expected that the values of D, E and s would be almost identical for ^{8}Li and ^{9}Li (these values should vary as the inverse of the square root of their masses) for the same target. The discrepancy may be due to the measurement problems or variations in temperature between measurements. The value of s is close to the actual ioniser length of ~5 cm.

Although the results to date are promising, the equation needs further verification by comparison to experiment and it should prove a useful tool in understanding the working of foil targets. The calculation can be extended to other target geometries and used to improve the performance of targets, optimising the geometry for different isotopes, and in particular, for the rare short lived nuclei.

REFERENCES

1. Lettry, J., R., Cathrall, Drumm, P., van Duppen, P., Evenson, A. H. M., Focker, G. J., Jokin, A., Jonsson, O. C., Kugler, E., Ravn, H. L., ISOLDE Collaboration, *Nucl. Instr. and Meth. B*, **126**, 130 (1997).
2. Kugler, E., Fiander, D., Jonson, B., Haas, H., Prewloka, A., Ravn, H. L., Simon, D. J. and Zimmer, K., *Nucl. Instr. and Meth. B*, **70**, 41 (1992).
3. Kirchner, R., *Nucl. Instr. and Meth. B*, **70**, 186 (1992).
4. Fick, A., *Ann. Phys. Lpz.* **170**, 59 (1855).
5. Bennett, J. R. J., Densham, C. J., Drumm, P. V., Evans, W. R., Holding, M., Murdoch, G. R. and Panteleev, V., *Nucl. Instr. and Meth. B*, **126**, 117 (1997).
6. Crank, J., *The Mathematics of Diffusion*, Clarendon Press, Oxford, 1967.
7. Drumm, P.V. et al., *Nucl. Instr. and Meth. B*, **126**, 121 (1997).
8. Bennett, J. R. J., *Nucl. Instr. and Meth. B*, **126**, 146 (1997).
9. Densham, C. J., Thwaites, C. and Bennett, J. R. J., *Nucl. Instr. and Meth. B*, **126**, 154 (1997).

A Single Accelerator RIB Facility using the Recoil Mass Spectrometer HIRA at NSC, New Delhi

J J Das*, P Sugathan, N Madhavan, T Varughese, B Kumar,
P V Madhusudhana Rao[1], A K Sinha

Nuclear Science Centre, Post Box # 10502, New Delhi - 110067, India.
[1] *Dept. of Nuclear Physics, Andhra University, Visakhapatnam - 530003, India.*
* e-mail : jjdas@nsc.ernet.in

Abstract. An on-line Radioactive Ion Beam (RIB) facility is being developed at NSC, New Delhi utilizing the existing 15 UD Pelletron accelerator and the recoil mass separator, HIRA in a new ion-optical mode. The primary beams from the accelerator will be used in inverse kinematics for the production of light RIB species such as ^{7}Be, ^{8}Li, ^{13}N, ^{17}F, ^{18}Ne, etc. These products are separated and transported by HIRA which will be operated in a new ion-optical mode specialy tuned for excellent beam rejection and better transportation efficiency. The RIB particles are collected and focussed at the secondary target site with better intensity and optimal beam qualities(an intensity of about 10^4 - 10^6 pps and size ~ 3 mm dia.). The facility is expected to be operational by the end of 1998.

Introduction

With the availability of Radioactive ion beams from many accelerator facilities, physics with RIBs has become one of the most exciting topics of the present day nuclear physics. Experiments with RIBs have been done in many fields like, creation of exotic nuclear species, study of the scattering of halo nuclei, and, nuclear astrophysics. The main challenge in this emerging field is the experimental methods to produce various RIBs of sufficient energy, intensity and beam quality for performing precise nuclear reactions.

An experimental facility for the production and separation of secondary Radioactive Ion Beams (RIBs) is in progress at Nuclear Science Centre, New Delhi. The facility utilizes the existing 15-UD Pelletron accelerator(1) and the on-line Recoil Mass Spectrometer (RMS) HIRA(2) specially tuned for providing high quality secondary ion beams from inverse kinematic reactions. In this paper we report the ion optical design and capability of the setup at NSC.

Production of RIB species

The NSC RIB facility will be suitable for producing secondary beams of medium mass particles having energy of few MeV/amu and life-times of the order of few microsecond and more. Primary beams from the Pelletron accelerator, will be used in inverse reaction to produce radioactive ion beams (RIBs) from (p,n), (d,n), (^{3}He,n) type of reactions. The inverse kinematics will provide better kinematic focusing of the secondary ion beams and sufficiently large energy suitable for nuclear reactions with unstable beams. The radio-active reaction products will be separated and transported on-line using the recoil mass separator HIRA, which will be operated in new ion optical mode tuned for inverse kinematics reaction and thereby increasing collection efficiency. In a typical experiment the primary beam current from the accelerator will be of few tens of pna and we will be using thick target foils of $(CH_2)_n$, $(CD_2)_n$ materials. To prevent any beam induced damage to the target, a rotating target wheel assembly will be used in our facility. This will provide an active area of ~ 60 cm^2 in the target.

Separation / Transport of RIBs :

Many of the existing facilities for separation of the RIBs uses mainly magnetic devices and they have the limitations of large beam spot size, low transport efficiency and somewhat lower beam rejection capability. AT NSC, we try to improve upon these limitations using our on-line recoil mass spectrometer (RMS) HIRA for separating and transporting the RIBs. The device will be operated in a new focusing mode which will improve the RIB beam spot size, and also provides better collection efficiency. A brief description of the new ion optical mode is given below.

Beam optics of HIRA : The recoil mass spectrometer HIRA and its normal mode of operation is described in ref (2). The electromagnetic elements of HIRA consists of two magnetic quadrupole doublets. two electric dipoles (EDs), one magnetic dipole (MD) and a magnetic multipole(M). In the new mode of optics to be used for RIB separation, the configuration will be $Q1_y Q2_x$-ED1-M-MD-ED2-$Q3_x Q4_y$ (3), where the two quadrupole doublets are excited in reverse direction compared to the normal mode of operation. The ion-optics calculation were carried out using the code GIOS(4). Figure1 shows the focusing characteristic of the new mode for x and y directions. The new mode allows a point-to-point focusing of the particles from primary target to the MD-centre and from MD centre to secondary target position. There is large momentum dispersion at the MD centre where proper slit can be installed to filter the primary beam of different charge states. This intermediate momentum focal plane at the centre of MD will have energy resolution of $\sim 1\%$ and the second focal plane where secondary target will be placed will have zero mass/energy dispersion with unit magnification in both x and y direction. Figure 2 shows the phase space contour plot (x Vs y) of typical RIB ^{18}Ne beam along with the beam at the intermediate focal plane and the RIB spot size at the final focal plane.

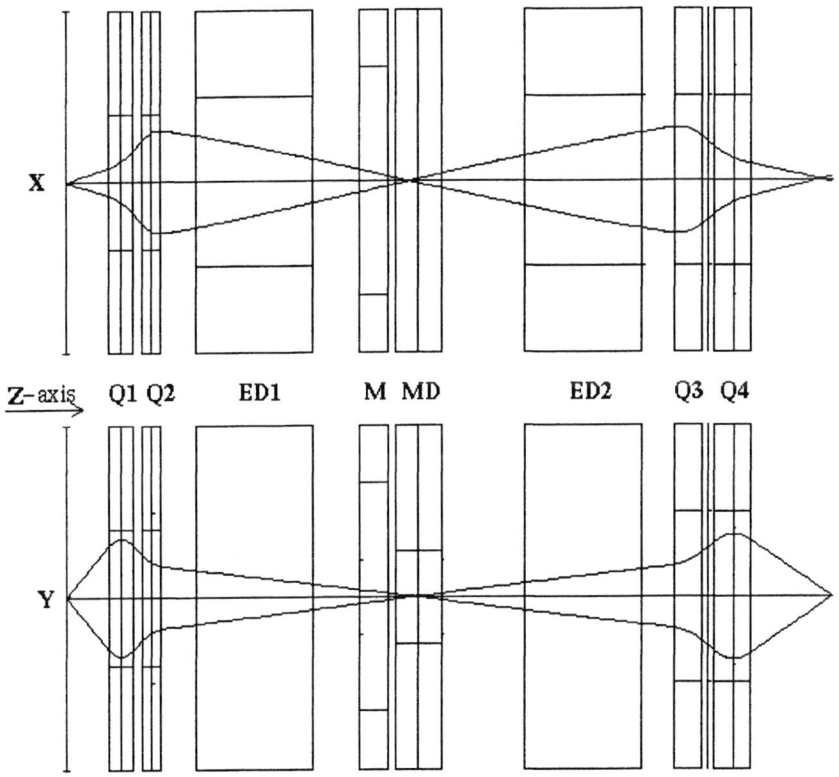

FIGURE 1. Shows the trajectories for the central particle along horizontal and vertical plane for 10 mSr solid angle.

This mode of optical arrangement collect RIBs of different charge states (within 10% of m/q) and focus them into single spot at the secondary target position with beam size as good as the beam spot produced from the pelletron accelerator. Also since the collection efficiency is more, large number of particles are collected by the system. A remote controlled slit system installed at the MD centre will filter the beam particles thereby eliminating the intense beam background.

Features of NSC RIB facility.

As compared to other facilities using only magnetic separators, our method of extracting the RIBs have the following added advantage.

(a) Energy Selection: As the first dispersive element in the NSC facility is electric dipole (ED1) which disperses in energy, the new optics ensures not only good separation between primary beam and the RIBs but also good energy selectivity (resolution of 1% or better) of RIBs. Even in those reactions where the primary beam and the RIBs have nearly the same p/q (momentum to charge state ratio) such as in the production of ^{18}Ne from ^{3}He(^{16}O,^{18}Ne)n reaction at $(E_{lab}(^{16}O) = 60$ McV), the difference in the E/q (energy to charge state ratio) between ^{16}O and ^{18}Ne is very large to allow good separation. A pure magnetic separator would fail to separate the primary

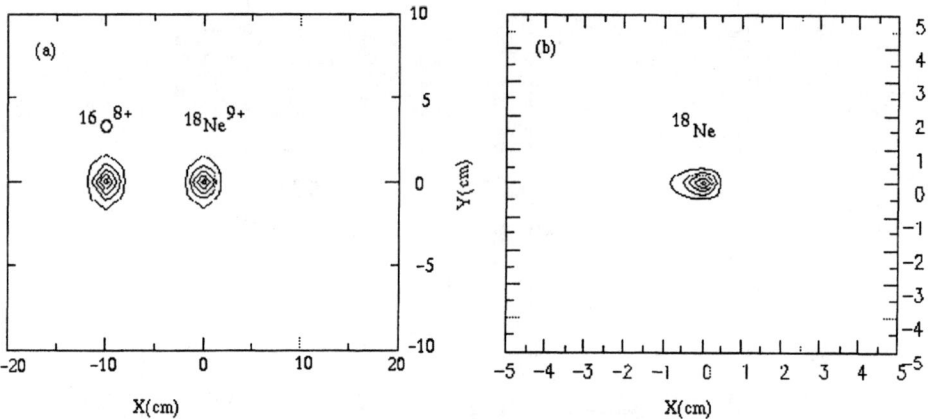

FIGURE 2. (a): Phase space(x Vs y) plot for the particles ^{18}Ne^{9+} and ^{16}O^{8+} at the intermediate focal plane and (b) the beam spot size for ^{18}Ne particles at the final focal plane. The spectrometer is tuned for ^{18}Ne beam produced from the reaction ^{3}He(^{16}O,^{18}Ne)n at 60 MeV energy.

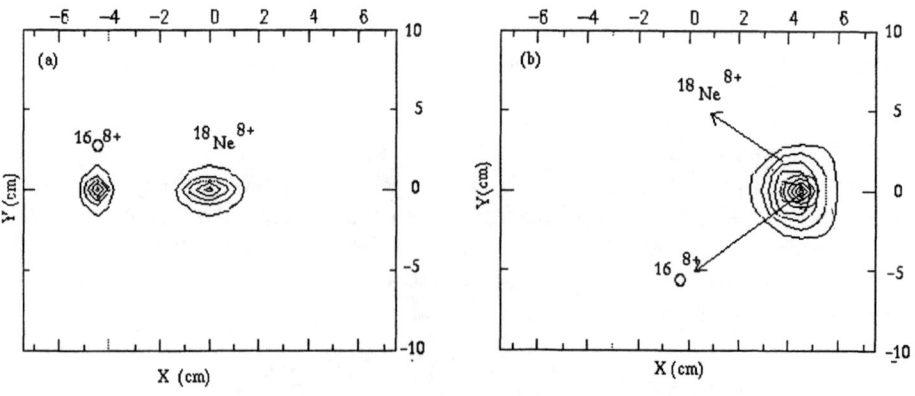

FIGURE 3. Phase space plot from (a) RMS HIRA and (b) a pure magnetic separator for particles ^{18}Ne^{8+} and ^{16}O^{8+}. The two beam particles are well separated in the case of RMS. The momentum resolution of magnetic device is taken to be 1/1000.

and secondary beams in such a case. Fig. 3 shows the phase space plot for a typical case where a magnetic separator is compared with RMS like HIRA for inverse kinematic reaction.

(b) High primary beam rejection in inverse kinematic reactions: The new optics and the related hardware (remotely controlled slit system at MD centre), the longitudinal slot in the anode plate of electric dipole (ED1) of HIRA and the rotation option of HIRA independently help in effective primary beam flux and high throughput of the secondary RIBs.

(c) Selection of more than one charge state of RIBs: The overall dispersion cancellation of HIRA and the choice of electric dipole as the first dispersive element allows more than one charge state of RIBs to be effectively separated from the primary beam and to cause them to arrive at a single spot at the secondary target location.

(d) Other key features of NSC RIB facility

Solid angle of acceptance for RIBs	: ~ 10 mSr
Energy acceptance for RIBs	: ~ +/- 10%
Mass (m/q) acceptance for RIBs	: ~ +/- 7%
Energy resolution at intermediate focal plane (i.e. at MD centre)	: ~ 1%
Horizontal (x) and Vertical (y) spatial and angular magnifications	: Unity
Final RIB spot size	: ~3 mm dia.
Final dispersion ((E/q) and (m/q))	: Zero
Types of RIBs	: ^{7}Be, ^{8}Li, ^{13}N, ^{17}F, ^{18}Ne, etc.
RIB intensity (with additional rotating target foil assembly)	: ~ 10^4 - 10^6 pps
Energy of RIBs	: few MeV/amu
Energy spread of RIBs	: ~ 1 MeV
Maximum value of (E/q) of RIBs	: 7.0 MV
Maximum value of (ME/q^2) of RIBs	: 48.5 MeV-amu

References :

1. G. K. Mehta et. al. *Nucl. Instr. and Meth.* **A268** 334(1988) and
   D Kanjilal et. al., *Nucl. Instr. and Meth.* **A328** 97(1993).
2. A K Sinha et. al., *Nucl. Instr. and Meth.* **A339** 543(1994).
3. J.J Das et. al. *J. Phys. G: Nucl. & Part. Phys.* **24** 1371(1998).
4. H. Wollnik et. al., *Nucl. Instr. and Meth.* **A258** 408(1987).

Beam Tests of the 12 MHz RFQ RIB Injector for ATLAS

R.A. Kaye, K.W. Shepard, B.E. Clifft, and M. Kedzie

Argonne National Laboratory
Argonne, Illinois 60439

Abstract. In recent tests without beam, the Argonne 12 MHz split-coaxial radio-frequency quadrupole (RFQ) achieved a cw intervane voltage of more than 100 kV, the design operating voltage for the device. This voltage is sufficient for the RFQ to function as the first stage of a RIB injector for the Argonne Tandem Linear Accelerator System (ATLAS). Previously reported beam dynamics calculations for the structure predict longitudinal emittance growth of only a few keV·ns for beams of mass 132 and above with transverse emittance of 0.27 π mm·mrad (normalized). Such beam quality is not typical of RFQ devices. The work reported here is preparation for tests with beams of mass up to 132. Beam diagnostic stations are being developed to measure the energy gain and beam quality of heavy ions accelerated by the RFQ using the Dynamitron accelerator facility at the ANL Physics Division as the injector. Beam diagnostic development includes provisions for performing the measurements with both a Si charged-particle detector and an electrostatic energy spectrometer system.

INTRODUCTION

A prototype RFQ accelerator [1] suitable for the initial stage of a secondary-beam accelerator system for an ISOL-type source of radioactive ions has recently been completed. Initial tests of this RFQ [2] have already shown that the RFQ is capable of running cw at an intervane voltage of 102 kV, a value above the design operational level of 100 kV, with an rf input power of 17 kW.

The RFQ is planned to be the initial element of a preaccelerator injector system for radioactive ions into the existing ATLAS accelerators [3,4] and, because of the excellent beam quality of ATLAS, the RFQ will determine the beam quality of the facility. In particular, the longitudinal emittance growth needs to be much smaller than that of typical RFQ implementations.

The ISOL-type ion sources appropriate for RIB produce ions having a low charge state ($q = 1$). The benchmark beam used for the RFQ design was radioactive ^{132}Sn both because of interest in this beam in the nuclear physics research community, and also because mass $A = 132$ represented a considerable increase in mass beyond previously developed RIB injectors for singly-charged ions ($A = 30$).

Numerical simulation of the RFQ beam dynamics for $A/q = 132$ ions [1] indicate a longitudinal emittance growth of only 1.4 π keV·ns through the RFQ structure. Experimental tests are needed to verify these calculations. These tests are to be made with stable beams, using ions of similar mass-to-charge ratio $A/q \leq 132$.

EXPERIMENTAL SETUP

Beam tests of the RFQ are being set up at the 4 MV Dynamitron facility at the ANL Physics Division. The Dynamitron is capable of operating at voltages in the range of about 0.2–4 MV and can deliver high-current positive-ion beams of a variety of species and charge states. A dedicated beam line is under construction which will couple the post-Dynamitron beam transport system to the RFQ. A schematic view of the Dynamitron and existing beam transport system relative to the position of the RFQ is shown in Fig. 1. To inject beam into the RFQ, ions exiting the Dynamitron are first deflected $-8°$ (relative to the initial ion trajectory) by a magnetic dipole ($M1$) magnet and then by $-25°$ with a second $M1$ magnet. Horizontal and vertical steering can be obtained using $M1$ deflectors internal to the Dynamitron, but additional steering is available from a set of three $M1$ magnets located along the beam line (see Fig. 1). Transverse focusing in this beam line section is provided by a single magnetic quadrupole doublet lens located about 3.9 m from the RFQ entrance. Electric quadrupole triplet lenses will eventually be added near the RFQ entrance and exit to provide the proper radial matching needed for $A = 132$, low-velocity beams.

Beam diagnostics systems for both the pre- and post-RFQ beam lines are currently under development. A planned layout of the initial systems and their relation to the RFQ is shown in Fig. 2. As shown in the figure, there will be three beam diagnostics and control systems before beam is injected into the RFQ. A Faraday cup will be used to measure beam current, and a wire scanner will be used to measure the transverse profiles of the beam. Adjustable slits will be used to control the intensity and position of the beam entering the RFQ. Ions accelerated by the RFQ will encounter similar diagnostic systems, as shown in Fig. 2. At this point, however, we require additional diagnostics capable of measuring ion energy and arrival times. For this purpose, two diagnostic systems are under development which will be discussed in detail in the next section.

DIAGNOSTICS FOR ACCELERATED BEAM

Beam diagnostic stations are being designed to measure the energy and beam quality of heavy ions ($A/q \leq 132$) accelerated by the RFQ. Energy measurements of heavy ions in the relatively low-energy regime such as that expected from the initial acceleration of exotic beams pose particular challenges which need to be addressed during the diagnostic design process. Therefore, a system of two complementary diagnostic stations is being developed (see Fig. 2).

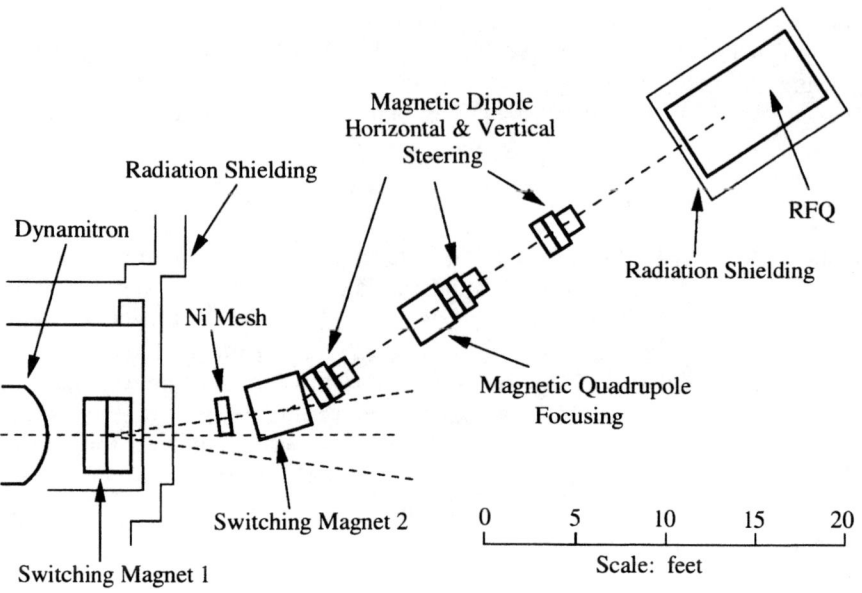

FIGURE 1. Schematic showing an overhead view of the current beam transport system between the Dynamitron injector accelerator and the RFQ.

The first diagnostic station (labeled Diag 1 in Fig. 2), which will be placed approximately 74 cm from the RFQ exit, consists of a Faraday cup to measure beam transmission through the RFQ, and a silicon (Si) charged-particle detector to measure the beam energy and arrival times. Both devices are retractable to both intercept the beam and to allow the beam to pass through when desired. The choice of including a Si detector was based on the relative ease of performing pulse-height (energy) spectrometry on a beam of incident heavy ions with the option of deriving a high resolution (< 1 ns), fast timing signal which would be useful for longitudinal emittance measurements. However, difficulties arise from the beam characteristics of this particular application and from constraints inherent to the Si detector itself. Considering the relatively low beam energies expected following RFQ acceleration, there will be a significant pulse-height signal defect from the output of the Si detector due to energy losses occurring in the front surface electrode and Si crystal dead layer. Detectors using an ion-implanted, ≈ 500 Å thick boron front-surface electrode contact should suffice for this application. Moreover, the large amount of charge carriers liberated by the heavy ion beam may degrade the energy resolution, but utilizing a detector with a minimal Si depletion depth and increasing the bias voltage will enhance the electric field strength and hence the charge collection capabilities.

The beam flux incident on the Si detector must be kept to a minimum, such that

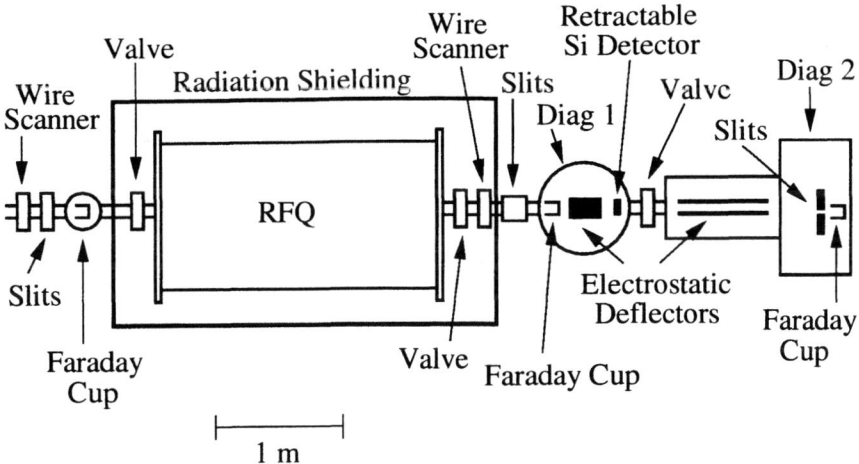

FIGURE 2. Schematic showing an overhead view of the proposed beam diagnostic stations used at the entrance and exit sections of the RFQ.

less than about 1000 heavy ions are incident on the detector each second to avoid Si crystal damage. Initially, the detector will be placed at 0° relative to the beam direction since very low-energy ions will be used and since proper beam current reduction can be achieved with the inclusion of beam degraders, apertures, and slits. One such set of degraders includes the overlap of five electroformed nickel meshes, each having wires that are 0.00020 in. thick spaced 0.00020 in. apart, forming a grid of 2000 wires per inch. The meshes are placed between the two switching magnets (see Fig. 1) to ensure that only ions not suffering from an energy loss upon passage through the meshes will be injected into the RFQ. Preliminary tests with a ^{132}Xe beam have shown that these meshes reduce beam intensity by a factor of 10^3. Overall, intensity reductions on the order of 10^8 were achieved by adding a $\frac{1}{16}$ in. diameter aperture and a pair of adjustable jaw slits. Future experiments could reduce beam intensities by performing elastic scattering from a very thin ($\sim 30\mu g/cm^2$) gold foil placed near the Si detector.

A separate but complementary diagnostic system, shown as Diag 2 in Fig. 2, is also under development. It comprises an electrostatic energy spectrometer system in which a set of parallel plates, with a given potential difference between them, will deflect a charged-particle beam according to the energy and charge of the ions. The location and intensity of deflected beam is then quantified by a Faraday cup positioned at a point offset from the undeflected beam axis. Slits are placed in front of the Faraday cup to limit the uncertainty in the beam position. If the Faraday cup position is held fixed, then the beam energy may be deduced by measuring the beam current on the Faraday cup as a function of a varying potential difference between the plates and comparing the results with that expected from theoretical predictions. Such a system has the advantage of performing an energy measurement rather easily regardless of the incoming beam intensity, however it lacks timing information unless a time-of-flight system is coupled to the fore mentioned diagnostics. Use of this system is also motivated from the fact that the electrostatic plates and their vacuum housing have already been fabricated, installed, and optically aligned from use in past atomic physics experiments.

EXPERIMENTS WITH BEAM

As mentioned in the Introduction, a wide variety of beams which satisfy the condition of $A/q \leq 132$ may be accelerated by the RFQ. Ions having A/q values less than 132 are accommodated by simply scaling the RFQ vane voltage and by maintaining the appropriate ion injection velocity profile. However, since it is the goal of the prototype RFQ to successfully accelerate $A/q = 132$ radioactive ions, this discussion will focus primarily on experiments intended to test this capability. In order to simulate the conditions expected from radioactive ^{132}Sn beams, the RFQ will need to operate at an intervane voltage of 100 kV, and in cw mode with the intent to preserve beam intensity. Only routine high-voltage conditioning of the RFQ appears to be needed for this based on the results of earlier tests [2].

Stable beams of singly-charged ^{132}Xe ions may be used to simulate the conditions expected for ^{132}Sn beam extracted from an ISOL-type ion source. Beam dynamics calculations which have investigated the acceleration of ^{132}Sn [1] have shown that ions with $A/q = 132$ will be accelerated by the RFQ to a velocity of $0.0049c$ (corresponding to an energy of 1508 keV) if they are injected with an initial velocity of $0.0025c$ (corresponding to an energy of 378 keV). Identical injection velocities can be obtained for ^{132}Xe ions using a Dynamitron terminal potential of 378 kV.

The RFQ was designed with external bunching primarily to enhance beam quality, but also to increase efficiency by using the full length of the RFQ for acceleration. Initial beam tests will inject dc (unbunched) beams into the RFQ and hence result in reduced beam transmission. However, beam simulations predict that about 30% of the incoming ions will lie in the proper rf phase bucket to emerge from the RFQ having the maximal energy gain. Longitudinal matching will be provided in the future from a bunching system similar to the 12 MHz gridded-gap, four-harmonic system presently in use for the ATLAS accelerators [5].

Preliminary measurements of accelerated beam will focus on verifying the energy gain expected for $A/q = 132$ ions from theoretical calculations. Both diagnostic systems mentioned earlier could be used to independently check this result. Standard spectroscopic techniques could be used to obtain an ion energy spectrum using the Si charged-particle detector. The electrostatic energy spectrometer system would also be effective for measuring the energy of these $A/q = 132$ ions, as demonstrated by the results of a simulation using the SIMION code shown in Fig. 3. The calculations utilized the existing dimensions of the pair of plates, as shown in the figure, and assumed a drift space of 46 cm based on an existing vacuum chamber coupled

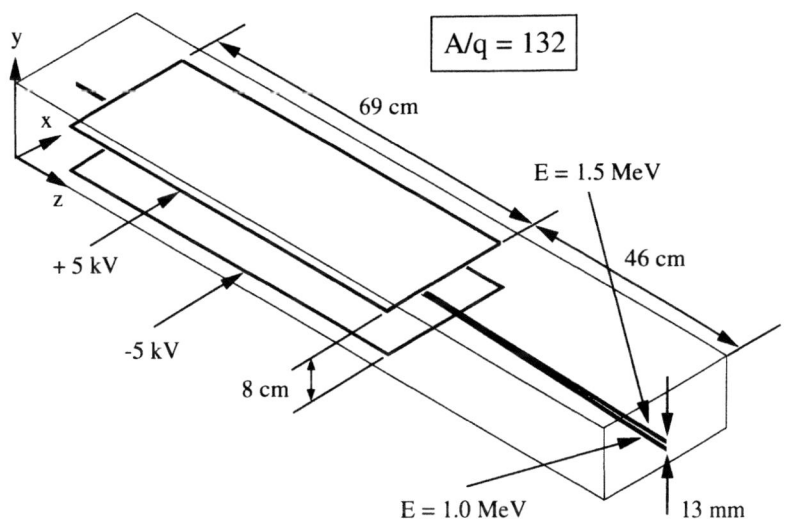

FIGURE 3. Results of a SIMION simulation indicating the flight paths of 1.0 MeV and 1.5 MeV beams, each having $A/q = 132$ and diameters of 6 mm, through a potential difference of 10 kV between a pair of electrostatic deflector plates having the dimensions as indicated. The y-direction deflection difference of 13 mm refers to the centroid position of each beam.

to the chamber which houses the plates. Monoenergetic 1.0 MeV and a 1.5 MeV beams, each with $A/q = 132$ and 6 mm in diameter, were allowed to travel through the electric field generated by the plates and the field-free drift space. When a potential difference of 10 kV is applied between the plates, the calculations show that the centroid positions of the two beams are separated by 13 mm at the termination of the drift space, indicating that energy resolutions on the order of 500 keV can be expected from this system. This is comparable to the typical energy resolutions of Si detectors performing heavy-ion spectroscopy.

Further tests will probe the transverse and longitudinal emittances of the beam and compare them to the predicted results [1]. Of particular interest is the quality of the longitudinal emittance, since calculations with a bunched ^{132}Sn beam show that a value of 6.1 π keV·ns is predicted at the RFQ exit, assuming an input longitudinal emittance of 4.7 π keV·ns. These values are substantially smaller than typical RFQ implementations, but are required to maintain the beam quality of the existing ATLAS accelerators. Figure 4 shows the numerically simulated longitudinal emittance of 200 ^{132}Sn ions which were accelerated by the RFQ after being injected in 1 ns bunches. From the figure, it is clear that the energy spread $\Delta E \approx 20$ keV of the bunch is much smaller than the intrinsic resolutions of both the Si detector and electrostatic energy spectrometer systems mentioned above. In fact, for the energy spectrometer to achieve the same separation as shown in Fig. 3

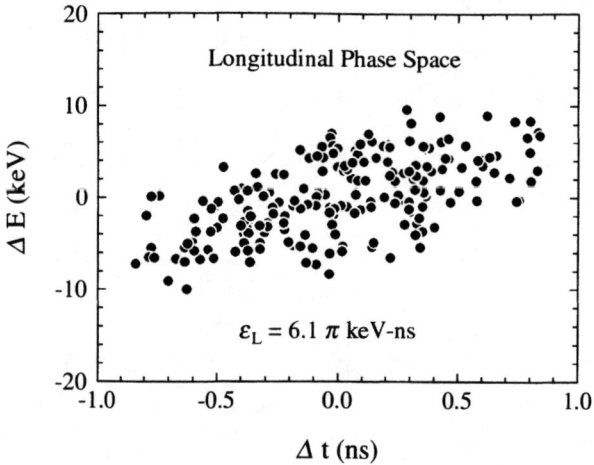

FIGURE 4. Numerically simulated longitudinal phase space of 200 ^{132}Sn ions at the exit of the RFQ [1].

(13 mm) of beams differing in energy by only 20 keV, then calculations show that a plate potential difference of 30 kV and a drift space of 6.8 m is required. A more physically reasonable way to measure the longitudinal emittance would make use of the timing information of the beam. A time width of $\Delta t \approx 2$ ns is predicted for each beam bunch at the RFQ exit, as seen from Fig. 4, which is within the timing resolution of many Si detectors. Therefore, precise timing correlations can be measured between beam bursts and rf pulses from the RFQ. Moreover, this same timing technique may be used to accurately determine the velocity and hence the energy of the beam. A comparison of the time differences between beam and rf pulses as measured by the Si detector at two different positions along the beam axis could give a very accurate determination of ion velocities. Since the exit velocity of the mass 132 ions of interest are about 1.47×10^6 m/s, or 0.147 cm/ns, changes on the order of nanoseconds can be seen in the time correlation between beam and rf pulses for ion flight path differences on the order of millimeters, making this technique quite reasonable to implement. The energy spectrometer may also be used for this measurement if a time-of-flight system is incorporated. One possibility involves the installation of a fast Faraday cup in the Diag 2 chamber of Fig. 2.

SUMMARY

Preparations for beam tests of the Argonne 12 MHz RFQ are in progress. Since the RFQ is intended to be the initial component of a pre-accelerator injector system of radioactive beams into the existing ATLAS accelerators, the tests will proceed with the intent to simulate the conditions of the benchmark beam of exotic ^{132}Sn.

Tests without beam demonstrated that the RFQ is capable of sustaining an operating intervane voltage of 100 kV cw and is thus capable of accelerating $A/q = 132$ ions. Stable beams of singly charged ^{132}Xe ions will eventually be used as the ultimate test of the ability to accelerate ^{132}Sn beam. For this purpose, a dedicated beam line is under construction at the 4 MV Dynamitron facility at the ANL Physics Division consisting of beam transport systems and diagnostics.

In order to measure the energy of beam accelerated by the RFQ, two separate diagnostic systems are under development. One system consists of a Si charged-particle detector from which energy resolutions of about 500 keV are typical for heavy ions. Although this intrinsic resolution is larger than that expected for the energy width of bunched $A/q = 132$ beams, the subnanosecond timing resolutions of these detectors makes it possible to directly and accurately measure the time widths and, as a result, infer energy spreads as well. This will be particularly useful in the determination of the longitudinal emittance of the beams. Another diagnostic system consisting of a pair of electrostatic deflector plates and a Faraday cup forms an energy spectrometer that has an energy resolution similar to that of the Si detector system. Adding a time-of-flight system to these diagnostics would be necessary to determine longitudinal emittances, however.

ACKNOWLEDGMENTS

The authors gratefully acknowledge the technical assistance of M. Portillo, J. Joswick, R. Kickert, C. Batson, S. Davis, and the ANL Physics Division machine shop staff. Helpful conversations with J. Nolen, R. Pardo, A. Woosmaa, and R. Laxdal are also greatly appreciated. This work is supported by the U.S. Department of Energy, Nuclear Physics Division, under Contract W-31-109-ENG-38.

REFERENCES

1. K.W. Shepard and W.C. Sellyey, Proc. of the 1996 Linear Accelerator Conference, August 26-30, Geneva, Switzerland, CERN 96-07, 68 (1996).
2. K.W. Shepard, M. Kedzie, and R.A. Kaye, to be published in the Proc. of the 1998 Linear Accelerator Conference, August 23-28, Chicago, IL.
3. J.A. Nolen, Proc. of the 1995 IEEE Particle Accelerator Conference, May 1-5, Dallas, Texas, 95CH35843, 354 (1996).
4. K.W. Shepard and J.W. Kim, Proc. of the 1995 IEEE Particle Accelerator Conference, May 1-5, Dallas, Texas, 95CH35843, 1128 (1996).
5. F.J. Lynch, R.N. Lewis, L.M. Bollinger, W. Henning, and O.D. Despe, Nucl. Instrum. and Meth. **159**, 245 (1979).

The Munich Accelerator for Fission Fragments (MAFF)

Oliver Kester*, Dietrich Habs*, Martin Groß*, Thomas Sieber*, Henning Bongers*, Alfred Kolbe*, Hans Jürgen Maier*, Peter Thirolf*, Till von Egidy†, Ulli Köster†, Erich Steichele†, Paul Kienle†, Hans Joachim Körner†, Alwin Schempp+, Ulrich Ratzinger°

*Ludwig-Maximilians Universität München, 85748 Garching, Germany
†Technische Universität München, 85748 Garching, Germany
+J.W. Goethe Universität Frankfurt, 60325 Frankfurt, Germany
°GSI, Postfach 110552, 64220 Darmstadt, Germany

Abstract. An accelerator (MAFF) for the new Munich high flux reactor FRM-II is under design [1,2] in order to deliver intense beams of very neutron-rich fission fragments of up to 10^{12} particles per second with final energies between 3.7 and 5.9 MeV/u to perform experiments for the production of heavy elements [3]. To obtain an efficient acceleration in a short LINAC, charge breeding of the 1^+-ion beam from the reactor to a q/A $\geq$ 0.16 is required. New measurements with an electron cyclotron resonance ion source (ECRIS) [4] have shown that the requirements for a low duty cycle LINAC (10%) can be fulfilled by the ECRIS. To reach a high flexibility in the final energy with a small number of structures, new kinds of IH-structures are under development at the Munich tandem laboratory.

INTRODUCTION

Advanced radioactive nuclear beam facilities have a strong scientific case [5,6]. Three areas of physics benefit from experiments in such facilities: Nuclear Physics (exotic compositions of nuclei in terms of protons and neutrons, exotic shapes, superheavy elements), Nuclear Astrophysics (stellar energy generation, origin of the elements) and the limits of the Standard Model. This is combined with a broad range of interdisciplinary and practical applications. The physics and the applications of neutron-rich fission fragment beams were discussed also in detail on a workshop in 1996 at Benedkitbeuern [7]. Several different projects of radioactive beam facilities are under way throughout the world [8]. With the ISOL (Isotope Separation On-line) technique intense high quality beams of short-lived, unstable isotopes can be supplied using very different reactions for the production: i) high energy protons, ii) fast neutrons and iii) thermal neutrons. While the production with proton beams is limited by the deposited energy and radiation cooling of the target, neutrons avoid the additional energy loss of charged particle beams. The small fission cross sections for fast neutrons ($\sigma_{fast}(^{238}U)$=43 mb) result in rather large amounts of target material (e.g. 0.5 kg U) and lead to the development of liquid U-targets. On the other hand the cross sections for thermal fission ($\sigma_{therm}(^{235}U)$=580 b) are several orders of magnitude larger than other production cross sections and thus require much smaller amounts of target material (typically 1-2 g of ^{235}U). The high neutron flux of a high flux reactor ($1.5 \cdot 10^{14}/(cm^2 s)$) allows to produce very large intensities of neutron-rich beams (e.g. $^{91}Kr : 4 \cdot 10^{11}/s$), where finally the removal of the fission energy from the target (several kW) and the activity of the target (several 10 kCi) limit the amount of used ^{235}U.

A first realization of this concept was studied in the PIAFE-project (**P**roduction, **I**onisation et **A**ccélération de **F**aisceaux **E**xotiques) [9,10] at the ILL- (Institut Laue-Langevin) high flux reactor in Grenoble. The PIAFE project was stopped in July 1998 for different political reasons. The MAFF-Project (**M**unich **A**ccelerator for **F**ission **F**ragments) will be the only facility in the world, which uses the thermal neutrons of a high flux reactor for the production of radioactive ions. While for PIAFE a single ended beam tube was foreseen, which reached only partially into the heavy water tank of the reactor, the MAFF project has a through-going beam tube. One side is used for the operation and the change of the ion source; the other side for the extraction of the radioactive ion beam. This design, with access from both sides has many technical and safety relevant advantages. Thus the small ion source can readily be changed from the one side without demounting the electrostatic lens system for ion extraction. Compared to other radioactive beam accelerators under construction the MAFF project expects beam intensities of fission fragments which are typically a factor of 1000 larger.

PHYSICS MOTIVATION

Our knowledge of exotic nuclei close to the drip line on the neutron rich side of the nuclear chart is rather small except for light nuclei, due to the fact that nuclides at the neutron drip line are far beyond the scope of the present experimental techniques. Intense beams of neutron rich nuclei open up a wide field of research and applications in nuclear physics, astrophysics, solid state physics, atomic physics and chemistry.

Heaviest nuclei might be produced by fusion of neutron rich nuclei. Heavy element studies with neutron rich beams will be important for the study of complex nuclear systems, the behaviour of shells with large spin, the properties of bulk nuclear matter, and the effect of large Coulomb forces on the nucleus. High spin physics with neutron rich heavy nuclei can investigate new regions of superdeformed and hyperdeformed nuclei. Information on neutron rich nuclei through in-beam γ-spectroscopy can elucidate the astrophysical r-process. The examination of the nuclear shell structure for neutron rich isotopes becomes possible. Of particular interest are nuclei around the doubly magic ^{132}Sn and ^{78}Ni the N=50 and 82 shell closures and neutron-rich $A \approx 100$ nuclei. More nuclear spectroscopy data far from beta-stability are needed to verify the different calculations predicting "shell-quenching", i.e. the reduction or even vanishing of shell effects when approaching the neutron drip-line [11]. The study of nuclei close to the drip lines is of interest due to the combination of weak binding and the proximity of the particle continuum. Predictions for heavier systems are strongly model dependent due to their high sensitivity to various details of the theoretical treatments [12]. Precision mass measurements will be possible with Penning ion traps. With those tools a stringent test and improvement of the nuclear theories for the most complex nuclear systems becomes possible. Laser experiments on trapped atoms would reveal nuclear deformations, ionization potentials and electronic level shifts due to relativistic effects of the inner electron orbits. Of strong interest is the study of the electron shell under the influence of a large central Coulomb field, in particular the inner electron orbits, the modification of atomic level structure by relativistic effects and its influence on the chemical properties. This can be deduced by laser spectroscopy of the long living fusion products. A fundamental question is the validity of the Mendelejev systematics towards the heaviest elements at the end of the periodic table. The neutron rich nuclides with sufficiently long half-lives are valuable objects for systematic chemical studies. During the last decade the application of radioactive isotopes in solid state physics and particularly in semiconductor physics experienced an immense boom. This development was mainly triggered by combining classical nuclear methods and newly developed tracer techniques with the wide range of radioactive ion beams available at ISOLDE. Deep implantation of radioactive nuclei for the analysis of the structure of solid state systems and migration studies of Hydrogen in solid states are interesting topics for the solid state physics. The examination of the migration of fission products in solid states is important for the final storage of nuclear waste. Nuclear medicine means radioactive tracer application in vivo. Compared to the nuclear medical diagnosis the radiopharmaceutical development for therapy is considerably delayed. This field is now intensively progressing and may be significantly supported by a reactor based isotope separator on-line (ISOL) facility. The provided β^- emitters have a great potential in the field of radionuclide therapy, while the β^+ emitters, needed for positron emission tomography can be produced with spallation reactions at accelerator based ISOL facilities.

OVERVIEW OF THE PROJECT

The overall layout of the facility is shown in fig.1 with the reactor and the new eastern experimental hall. Within the reactor the ion source is mounted from the western side into the evacuated beam tube SR6 while the singly charged ion beam is extracted on the opposite side.

After a mass separation several ion beams can be transfered into the eastern experimental hall. There the singly charged ions are converted into highly charged ions by breeding in an ECR source. Thus they can be accelerated in a rather compact linear accelerator to energies up to 6 MeV·A. The beam can be switched to several experimental areas. In the main beam line the fusion target for heavy element production and the Munich On-line Recoil Residue Ion Separator (MORRIS) [3] will be installed. A system similar to the SHIPTRAP system being set up at GSI allows the identification of the fusion products.

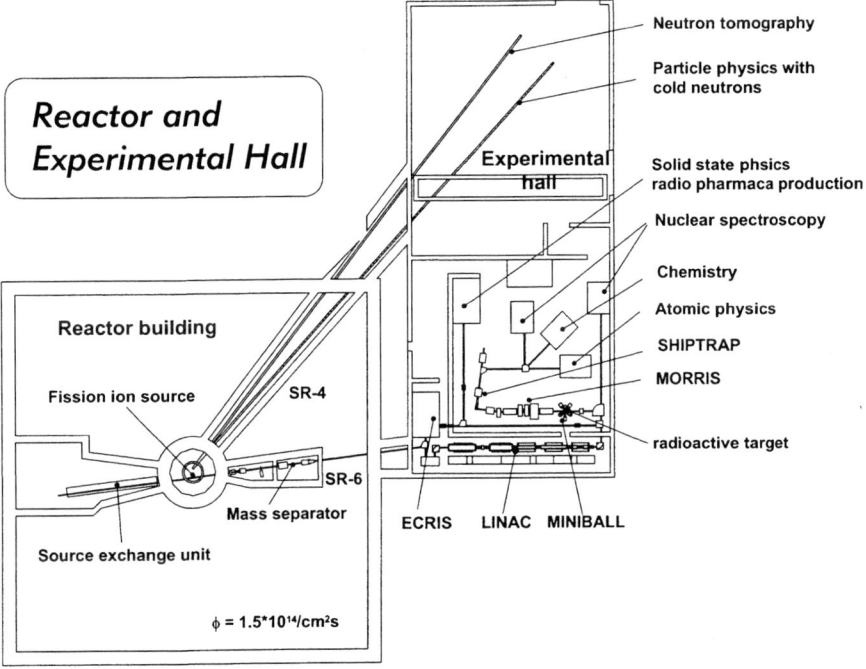

FIGURE 1. Overview of the MAFF accelerator and experimental facility

The in-pile part and the source handling

Figure 2 shows a schematic view of the reactor tube and the source exchange unit. Inside the through-going evacuated beam tube SR6 of the FRM-II reactor the fission fragment ion source is moved with a trolley. In operation the source is 60 cm off the center of the fuel element. The fission fragments are extracted as singly charged ions by different techniques of ionization. The fission fragment ion source is on 30 kV high voltage with respect to the beam tube and the extraction electrode. After acceleration the 30 keV ion beam then is guided by electrostatic lenses to the outside. From both ends of the beam-tube a Helium cryo-pumping system is introduced. Gaseous Helium at about 20 K and 3 bar is circulated through aluminium tubes which also carry cooling panels. In this way the central region of the beam-tube is evacuated with high pumping power. A second aim of this cryo-pumping system is to localize the radioactive gases from the hot ion source on the cryo-panels.

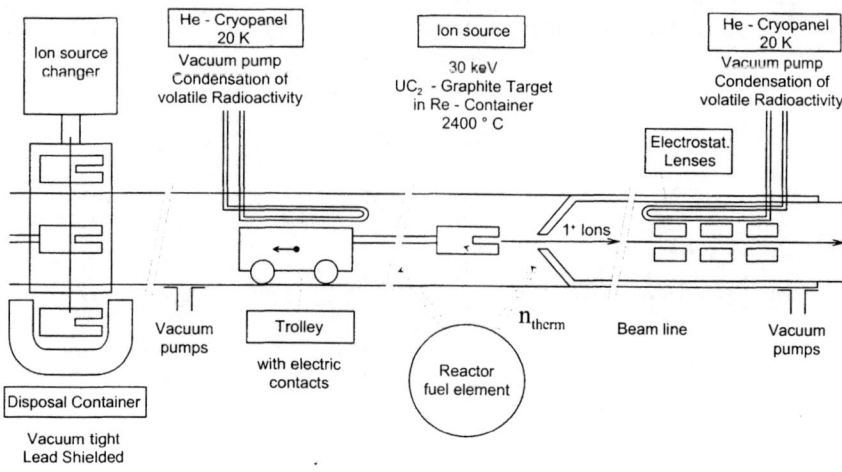

FIGURE 2. Scheme of the in-pile part of MAFF

The source has to be changed after one reactor cycle (50 days each). It will be pulled out of the reactor beam tube by a trolley to the "station", where it will be disconnected from the carriage, and then put in a waste disposal container. After the source has been moved back into the source exchanger the source is lifted up by a manipulator. The ion source wagon is moved out of the way and the radioactive source is transfered vertically down to the carrier on the lower level. Here a horizontally moving manipulator transfers the carrier and the source into the shielded transport container. After removing the manipulator the transport container is sealed by a vacuum tight valve. After removing the used ion source a new source is transfered to the trolley. It has been backed out in the preparatory chamber to reduce the gas load for the vacuum system of the fission fragment accelerator.

The in-pile target ion source

The main new element of the fission fragment accelerator is the ion source for radioactive fission fragments shown in fig.3. The ion source design is similar to the source in Studsvik [13,14]. It contains up to 1g of 93% enriched ^{235}U. The ^{235}U is dispersed homogeneously in the form of UC_2 in a matrix of graphite. This graphite rod of 6 cm length and 2.5 cm diameter is heated up to a temperature of 2400^0C allowing for a fast diffusion of the fission fragments. This graphite rod is encapsulated in a Rhenium cylinder, which at the one end has a 2 mm diameter hole, where the fission fragments exit. The source is placed into an unperturbed neutron flux of $1.5 \cdot 10^{14}/(cm^2 s)$. With 1g of ^{235}U typically 10^{14} fissions per second occur. The maximum produced fission heat of 3 kW is cooled away by radiation. Additionally the source may be heated externally with up to 600 W. In this way the source temperature can be kept constant though 18% of the ^{235}U are burnt during a reactor cycle of 52 days. During first tests of the source the uranium content of the target will be much smaller, requiring a total electric heating of the source. By the use of several heat shields the heating power of 600 W is sufficient to reach the required temperature. Presently we setup a test experiment, where we heat an ion source externally to 2400^0C to study the long term behaviour in detail.

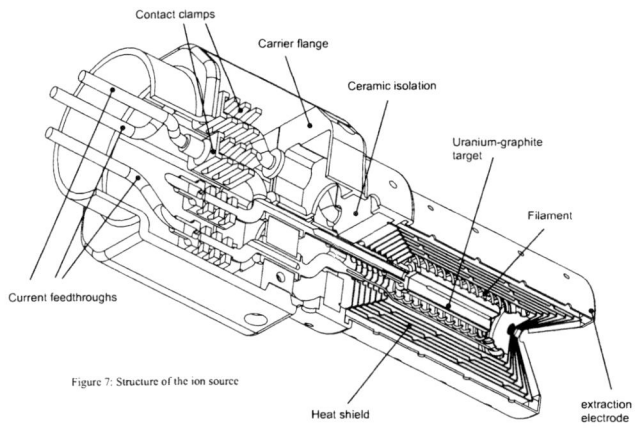

FIGURE 3. Drawing of the target ion source using surface ionization

The in-pile ion sources of MAFF have to fulfill the main requirements of any ISOL source [15]: high efficiency and if possible element selectivity, plus some additional constraints due to their position in a reactor beam tube: reliable long term operation and compact design. It is evident, that there is no general purpose ion source, which can fulfill all these requirements for all elements which are produced abundantly in thermal neutron induced fission, i.e. Z=28 to 63 except refractory elements which are barely released from the target. Therefore three different types of ion source will be used for dedicated elements. All of them are in routine operation at different ISOL facilities since long time [16]: The thermal ionization source, the laser ion source and the plasma source.

THE ACCELERATOR

In order to perform experiments with fission products at the Coulomb barrier post acceleration of the radioactive ions in a low β heavy ion LINAC is required. Several schemes of post acceleration can be considered [15]. The brute force and somehow expensive way is the continuous acceleration of the singly charged fission products by a low frequency LINAC to stripping energies in the range of several MeV and subsequent postacceleration of the highly charged ions in either a cyclotron or a LINAC [17]. A more economical scheme is to use charge multiplication of the fission products in an ion source delivering highly charged ions like the Electron Beam Ion Source (EBIS) [18,19] and the Electron Cyclotron Resonance Ion Source (ECRIS) [19–21] for efficient acceleration in a short LINAC.

Charge breeding

For high charge state acceleration by compact LINACs a typical $q/A = 1/9...1/4$ is required. In addition to reduce the duty cycle down to 10% accumulation and bunching of the cw-ion beam from the fission ion source two main scenarios can be taken into account.

- The slow EBIS injection or "accu" mode, where the 1+ ions are continuously injected into the EBIS during the confinement period [22]. The accu-EBIS mode has the advantage, that the full trapping capacity of the electron beam can be used to confine the ions while charge breeding takes place. Losses will occur in the extraction and the cleaning cycle and the charge state distribution will be broader due to the continuous injection. A good overlap of the injected beam and the electron beam is essential, i.e. the emittance of the 1+ sources has to fit into the EBIS acceptance. To improve the injection efficiency a single pass emittance improver like a gas filled Radio-Frequency-Quadrupole (RFQ)-ion guide is under development [23,24].

- The pulsed extraction in the afterglow mode while continuous injection into the ECRIS [4]. This scheme deals with a continuous injection and pulsed extraction in the afterglow mode. This breeding scheme is the accu-mode in case of using the ECRIS. In order to use a LINAC with a duty cycle below 25% at repetition rates between 25 and 100 Hz, such a pulsing is required. This mode uses the ECRIS as an ion trap in the so called ECRIT mode. It has been shown in measurements that the ions in the main volume will be stored also in the afterglow mode and can be catched again, when the plasma is recycled. It is possible to achieve repetition rates between 20-40 Hz.

To compare the different schemes one has to take into account the maximum intensities, the reachable charge states, the efficiencies and different considerations concerning the practical application. In general the mean charge state of an ECRIS is lower compared to the EBIS or it needs more time to reach the same charge state for the ECRIS and the charge state distribution of the ECRIS is broader. On the other, hand only an optimal EBIS (higher electron current, shorter confinement times to reach the required charge state) can reach the required ion beam intensity ($> 10^{10}$ s^{-1}). Such an EBIS has to be developed, which coincides with the recent development plans of the whole EBIS community [25]. Concerning the ion beam intensities, until now the EBIS tests have been run with a 25 Hz repetition rate, giving a total extracted current of 3.7 pnA [26], whereas the ECRIS can deliver 2 pμA without saturation effects. Recent experiments at CRYSIS show that relatively good efficiencies may be achieved in the accu-EBIS mode for injected currents up to 1 μA of N$^+$. The peak efficiency [n+]/[1+] for the ECRIS was 7.9 % for Ar^{8+} (with only 20 W of RF power at 10 GHz) [27], for the EBIS 9.4% for Ar^{14+} (with a 400 A/cm^2, 9 keV electron beam and 40 ms confinement time) [26] were reached. So the charge breeding efficiencies are comparable.

The accu-EBIS showed for nitrogen efficiencies far below 0.1% [26], but the beam parameters were not optimized. Better results with up to two orders of magnitude higher efficiencies might be possible with careful tuning. The use of an ECRIS as a trap provides a cumulated efficiency of about about 2.2% for Rb^{15+} during the first 20 ms of the extraction.

The EBIS has to work under ultra high vacuum conditions to achieve long lifetimes of the cathode and low ion intensities from residual gas. The ECRIS runs with support gas (pressures around 10^{-6} mbar), which might cause more contaminations of the extracted beam. The emittance of an ion beam from an EBIS is much smaller than the emittance of a beam from the ECRIS. This is an advantage concerning emittance growth in the LINAC and in the deceleration mode in order to reach the lower energy region without major losses.

The LINAC

The radioactive ions will be accelerated to a tunable final energy between 3.7 and 5.9 MeV/u by a LINAC similar to the accelerator of REX-ISOLDE [28,29]. The LINAC will consist of several structures: a Radio Frequency Quadrupole (RFQ) accelerator, three IH-cavities similar to the CERN "lead-LINAC" and two seven-gap (IH)-resonators which are used to vary the final energy. The proposed LINAC is shown in fig.4. The LINAC is considered to accelerate particles with a A/q < 6.3 to energies at the Coulomb barrier for the mass range of 75-150 and elements between Ni and Eu. The maximum duty cycle should not exceed 10% which means 10 ms rf pulse for 10 Hz repetition rate or 20 ms rf power at 5 Hz repetition rate. The energy width of the delivered beams has to be rather small ($\Delta E/E \leq 0.2\%$), because the excitation functions for the production of heavy elements via fusion reactions may be very narrow [30]. The jump in the resonance frequency from 101.28 MHz to the second harmonic after the first IH-tank is required to reduce the resonator length significantly, which otherwise increases due to the increasing cell length.

The injection energy into the RFQ should be 2.5 keV/u to get a high bunching efficiency in the first RFQ section. With an $q/A = 0.16$ and a final energy of 300 keV/u, a rod voltage of 59 kV is required. For the preliminary design the calculations show a normalized acceptance of 0.5 π mm mrad and a transmission

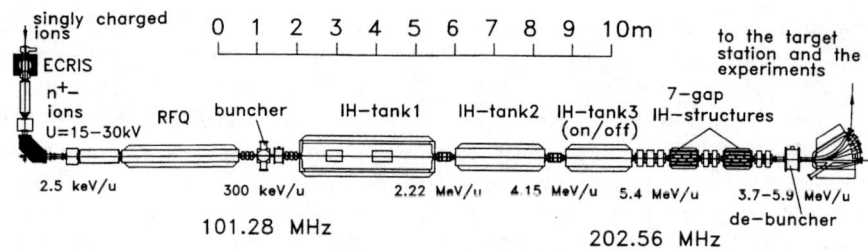

FIGURE 4. LINAC and charge breeding lay-out for MAFF.

of about 98.5%. In order to reach energies close to the Coulomb barrier a booster LINAC is required with a maximum final energy of 5.4 MeV/u. According to the injection energy of 300 keV/u an effective voltage gain of 32.1 MV is required for particles with A/q=6.3. Therefore three IH-cavities are taken into account. The variation of the large range in final energy with only two seven-gap resonators is possible by switching tank 3 on and off depending on the energy range which has to be covered. Thus the two 7-gap resonators can accelerate or decelerate at different injection energies namely at 4.15 MeV/u or at 5.4 MeV/u which corresponds to the final energies of tank 2 and tank 3 respectively. Thus tank 3 of the booster LINAC needs twice the effective voltage of both 7-gap resonators. The booster LINAC is described in more detail in [1,31].

The 7-gap IH-cavities will operate at 202.56 MHz eigenfrequency with a $0°$-synchronous particle structure. A total resonator voltage of 2.1 MV at 80 kW incoupled rf-power is expected, which requires a shunt impedance of 110 MΩ/m. The drift tube structure of a 7-gap structure is optimized to a certain particle velocity. Here the particle velocity varies between $\beta = 8.9\%$ and 11.3%, hence a design value of $\beta = 10\%$ or 4.6 MeV/u is chosen for both resonators in order to get the highest possible transit time factors in both energy ranges. The cell length is 7.4 cm. The gap length should be 2.5 cm in order to stay below 150 kV/cm which corresponds to the maximum Kilpatrick field strength [32] at 202.56 MHz. Thus a safety margin of 400 kV per gap is realistic. Particle dynamics calculations show that a continuous variation of the output energy from 3.64 to 5.94 MeV/u is possible. A higher resonator voltage may even increase the range. From MAFIA calculations a much higher shunt impedance of the 7-gap IH-resonators of about 300 MΩ/m is expected, which could reduce either the required rf-power or could increase the total resonator voltage.

REFERENCES

1. Kester, O., et al., *Nucl. Instr. and Meth.* **B139** 28 (1998).
2. Groß, M., et al., "The Status of the Munich Fission Fragment Accelerator Project", presented at the 2nd international workshop on nuclear fission and fission-product spectroscopy, Grenoble, France, April 1998.

3. Thirolf, P., et al.,*Nucl. Instr. and Meth.* **B126** 242 (1997).
4. Chauvin, N., et al., "ECRIT: Electron Cyclotron Resonance Ion Trap, a multicharged ion breeder/buncher.", to be published in *Nucl. Instr. and Meth.* **A** (1998).
5. Nuclear Physics in Europe: Highlights and Opportunities, Dec. 1997, NuPECC Report, ed. J. Vervier, J. Äystö, H. Doubre, S. Gales, G. Morrison, G. Ricco, D. Schwalm and G.-E. Körner.
6. Proceedings of the Workshop on the science for an advanced ISOL facility, Ohio State University, Columbus, Ohio, July 30 - August 1, 1997.
7. Proceedings of the International Workshop on Research with Fission Fragments, Benediktbeuern, Germany, ed. by T. von Egidy, D. Habs, F.J. Hartmann, K.E.G. Löbner and H. Nifenecker, World Scientific, 1997.
8. European radioactive beam facilities, Statement by NuPECC (1993), Report by study group, ed. R.H. Siemssen.
9. PIAFE Project report, march 1998, ed. by U. Köster and J.-A. Pinston.
10. PIAFE Collaboration, Technical report of the project covering the period 1993-1996, ISN report ISN 97-52.
11. Pearson, J.M., Nayak, R.C., and Goriely, S., *Phys. Lett.* **B387** 455 (1996).
12. Dobaczewski, J., et al., *Phys. Rev.* **C53** 2809 (1996).
13. Jacobsson, L., Fogelberg, B., Ekstrom, B. and Rudstam, G., *Nucl. Instr. Meth.* **B26** 223 (1987).
14. Vogelberg, B., et al. *Nucl. Instr. Meth.* B70 137 (1992).
15. Köster, U., Kester, O. and Habs, D., *Rev. Sci. Instr.* **69** 1316 (1998).
16. Ravn, H., in *Proc. Int. Workshop on Research with Fission Fragments*, Benediktbeuern, ed. by T. von Egidy et al., World Scientific, Singapore, 1997, p. 62.
17. Nolen, J. A. et al., *Rev. Sci. Instr.* **69** 742 (1998).
18. E. D. Donets, E.D., *The Physics and Technology of Ion Sources*, ed. by I.G. Brown, Wiley, New York, 1989, p.245.
19. 'Handbook of Ion Sources', ed. by B. Wolf, CRC Press, Boca Raton, 1995.
20. Geller, R., *Ann. Rev. Nucl. Part. Sci.* **40** 15 (1990).
21. Geller, R., Tamburella, C., and Belmont, J. L., *Rev. Sci. Instrum.* **67** 128 (1996).
22. Becker, R., et al., Proc. of EPAC'92, Edition Frontières 1992, p.981.
23. Lunney, M. D., et al., report CSNSM 97-02, Orsay, 1997.
24. H. J. Xu, H. J., et al., *Nucl. Instr. and Meth.* **A333** 274 (1993).
25. Proc. of the EBIS Workshop, Gelnhausen, 1997, parts are published in *Rev. Sci. Instr.* **69**(1998).
26. Visentin, B., et al., *Nucl. Instrum. and Meth.* **B101** 275 (1995).
27. Tamburella, C., et al., *Rev. Sci. Instr.* **68** 2319 (1997).
28. Kester, O., et al., "The REX-ISOLDE Project" in *Proceedings of the CAARI'96*, AIP conf. proc. 392, 1997, p.417.
29. Kester, O., et al., "The REX-ISOLDE LINAC", presented at the EPAC98, Stockholm, June 22-26, 1998.
30. Hofmann, S., et al., *Nucl. Instr. and Meth.* **B126** 310 (1997).
31. Kester, O., et al., "The LINAC of the Munich Accelerator for Fission Fragments (MAFF)", presented at the LINAC98, Chicago, August 23-28, 1998.
32. Kilpatrick, W.D., *Rev. Sci. Instr.* **28** 824 (1957).

RFQ - IH Radioactive Beam Linac for ISAC

Robert E. Laxdal

TRIUMF, 4004 Wesbrook Mall, Vancouver, Canada, V6T 2A3

Abstract. The ISAC radioactive beam facility under construction at TRIUMF includes a 500 MeV proton beam ($I \leq 100\mu A$) impinging on a thick target, an on-line source, a mass-separator, an accelerator complex, and experimental areas. The accelerator chain includes a 35 MHz RF Quadrupole (RFQ) to accelerate beams of $A/q \leq 30$ from 2 keV/u to 150 keV/u and a post-stripper, 105 MHz variable energy drift tube linac (DTL) to accelerate ions of $3 \leq A/q \leq 6$ to a final energy from 0.15 to 1.5 MeV/u. The design concept and present status of the ISAC linear accelerator complex will be summarized. In particular, first rf and beam tests with the RFQ and the status of fabrication and testing of DTL components will be reported.

INTRODUCTION

A radioactive ion beam facility with on-line source and linear post-accelerator is being built at TRIUMF [1]. [1] In brief, the facility includes a 500 MeV proton beam ($I \leq 100\,\mu A$) from the TRIUMF cyclotron impinging on a thick target, an on-line source to ionize the radioactive products, a ~1/10000 mass-separator for mass selection, an accelerator complex and experimental areas. Beams of $E \leq 60$ keV and $A \leq 238$ will be delivered to the low energy experimental area. A post-accelerator is being installed to supply radioactive beams to the high energy experimental area. The accelerator chain includes a 35 MHz RFQ to accelerate beams of $A/q \leq 30$ from 2 keV/u to 150 keV/u and a post stripper, 105 MHz variable energy drift tube linac (DTL) to accelerate ions of $3 \leq A/q \leq 6$ to a final energy between 0.15 MeV/u to 1.5 MeV/u. The accelerators have several noteworthy features. The RFQ, a four vane split-ring structure, has no bunching section; instead the beam is pre-bunched at 11.7 MHz with a single-gap, pseudo saw-tooth buncher. The variable energy DTL is based on a unique separated function approach with five independent inter-digital H-mode (IH) structures providing the acceleration and quadrupole triplets and three-gap bunching cavities between tanks providing transverse and longitudinal focussing respectively. Both linacs are required to operate *cw* to preserve beam intensity. A layout of the ISAC accelerator chain is shown in Figure 1.

---

[1] http://www.triumf.ca/isac/lothar/isac.html

Occupancy of the accelerator floor of the new ISAC building began in July 1997. In less than a year an off-line source and injection line have been commissioned and initial rf and beam tests with the RFQ in an intermediate configuration have been completed. In addition the first DTL buncher has been delivered and signal level tests completed in preparation for high power tests this month. The first DTL IH tank is being copper plated for rf tests this fall.

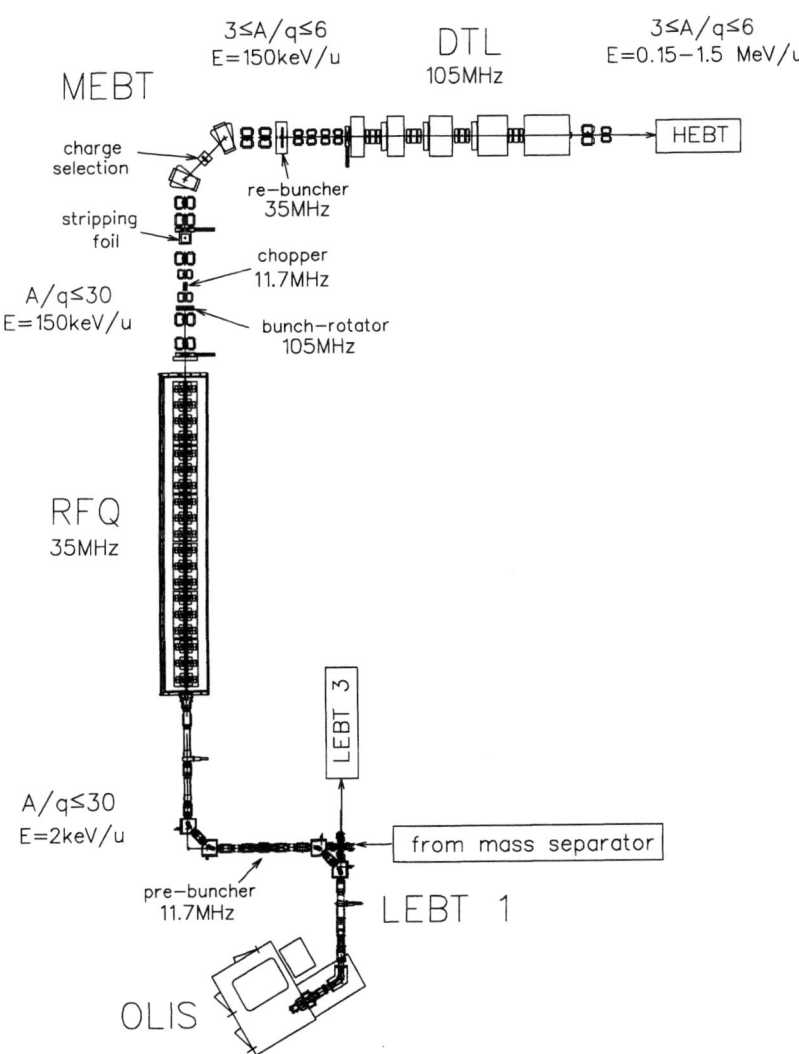

FIGURE 1. The ISAC linear accelerator.

LEBT

The low energy beam transport [2] (LEBT) consists of electrostatic elements; quadrupoles, steering plates and spherical bends, that transport the exotic beams from the mass-separator located on the target floor and stable beams from the off-line ion source (OLIS) located on the accelerator floor. An electrostatic switchyard allows selection of either the stable or unstable beam for acceleration while the other beam can be sent to the low energy experimental area.

The beam to be accelerated is pre-bunched at 11.7 MHz, the third sub-harmonic of the RFQ frequency, in a single gap, multi-harmonic pseudo-sawtooth pre-buncher. The pre-buncher frequency was selected at the request of experimenters to give a longer bunch spacing (86 ns), a useful feature for certain TOF and coincidence rejection techniques. The fundamental frequency and the first three harmonics are individually phase and amplitude controlled and combined at signal level. The signal is amplified by an 800 W broad-band amplifier that drives the two plates in push-pull mode with a peak voltage of about 200 V (400 V between plates). Optimization of amplitude and phase of each harmonic results in an al-

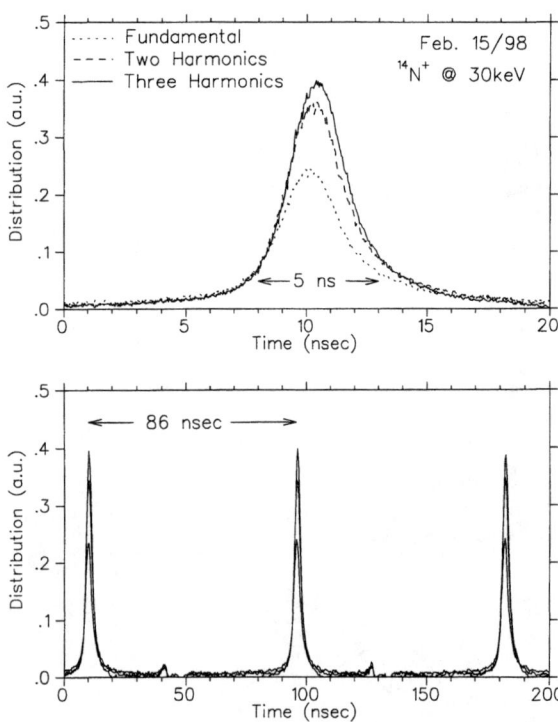

FIGURE 2. Beam bunches measured on the fast Faraday cup in LEBT for one, two or three harmonic bunching.

most saw-tooth modulation on the beam velocity. The variation in the gap-crossing efficiency for each harmonic means that the driving voltage is far from a sawtooth, being dominated by the higher, less efficient harmonics. In fact the present amplifier band width rolls off after 35 MHz and so initial testing was done with only three harmonics.

The pre-buncher is positioned ~5 m upstream of the RFQ. The last four quadrupoles upstream of the RFQ match the beam to the RFQ acceptance.

OLIS and LEBT Commissioning

Installation of the off-line ion source (OLIS) began in July 97 with first beam extracted in November 97. Commissioning of the LEBT from OLIS to the RFQ followed soon after. A 2.45 GHz micro-wave source with a magnetic cusp plasma confinement is installed. The source efficiently produces positive ions of gaseous species from Hydrogen to Argon. The ions for the initial RFQ tests are N^+ and N_2^+ with measured transverse emittances of $20\pi\mu$m and $10\pi\mu$m respectively. The saw-tooth prebuncher was installed and commissioned with three harmonics in February 98. The fourth harmonic will be added following an upgrade to the wide band amplifier. During commissioning the bunched beam structure was measured with a 50Ω cone-type fast Faraday cup [3](Figure 2). Tuning proved relatively straight forward, the phase of each harmonic was determined with the beam and the amplitudes were set to pre-determined values followed by empirical optimization.

RF QUADRUPOLE

The ISAC RFQ (Figure 3) is a 4-vane split ring structure [4]. A total of 19 split rings, each feeding a 40 cm length of modulated electrodes, are housed in a square 1m×1m tank with a total length of almost 8 m. The gross specifications include a bore radius of $r_0 = 7.4$ mm, and a maximum inter-vane voltage of 74 kV corresponding to a power of 85 kW. Rings and electrodes are water cooled. A unique feature of the design is the constant synchronous phase of -25° [5]. The buncher and shaper sections of the RFQ have been eliminated in favour of the pre-buncher previously described. This shortens the RFQ but in addition, injecting a pre-bunched beam yields a smaller longitudinal emittance at the expense of a slightly lower beam capture. When operating with four harmonics on the prebuncher we expect 81% of the beam to be accelerated in the 11.7 MHz bunches, ~4% accelerated in the two neighbouring 35 MHz buckets, with 15% of the beam unaccelerated and lost in the first few quadrupoles of the MEBT. (With the existing three harmonics we expect a distribution of 76%, 4% and 20% respectively.)

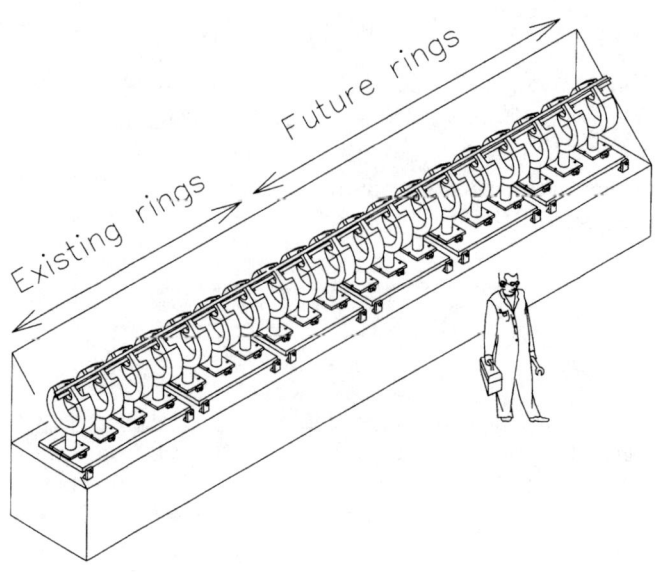

FIGURE 3. The ISAC 35 MHz RFQ.

Interim Test

The RFQ is being installed in two stages. The first seven split rings have been installed in the final RFQ tank for an interim beam test subsequent to installation of the remaining twelve rings. The test allows us to perform rf measurements, establish alignment procedures, commission the injection line, determine matching conditions and establish capture efficiencies all well in advance of the RFQ completion in 1999.

A schematic of the test set-up is shown in Figure 4. A copper wall installed downstream of the seven ring section provides rf containment. Beams produced in OLIS are transported through a ∼15 m long LEBT section to arrive at the RFQ (Figure 1). The beam is accelerated to 53 keV/u, then a series of eight electrostatic quadrupoles transports the beam to a diagnostic station just downstream of the RFQ tank. The diagnostic station includes a fast Faraday cup, transverse emittance rig and energy analyzing magnet.

Signal Level Tests

RF signal level tests have been performed on the 7-ring volume using both unmodulated *dummy* electrodes and the final copper electrodes. The *dummy* electrodes were installed to get an early frequency measurement and to check the mechanical alignment of the electrodes. The results of the tests are a frequency of 35.7 MHz, a quality factor of $Q = 8700$, and a shunt impedance of 292 kΩ·m. The frequency will be slightly lower for the full 19 rings and tuners can be used

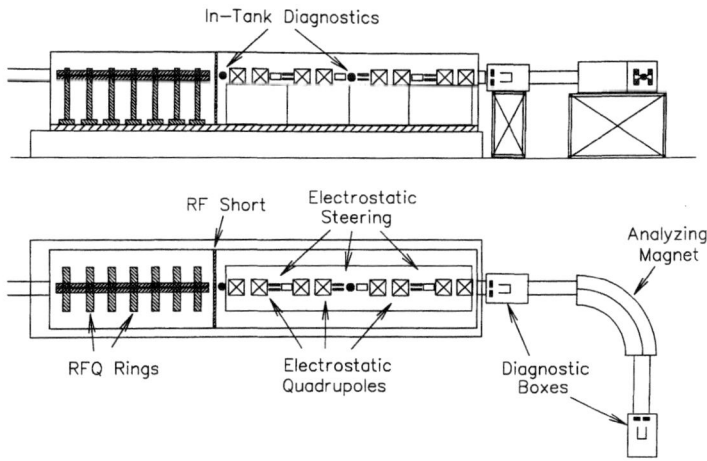

FIGURE 4. A schematic view of the RFQ and test station.

to reach the design frequency. Bead pull measurements show that the quadrupole fields are within the prescribed tolerances; that is that the average peak field along the electrode length varies by no more than ±1% and that the asymmetry in the quadrupole field is less than ±1%. The electrodes themselves are aligned to the ideal beam axis to within the tolerance of ±80μm.

RFQ Beam Tests

The initial set of rf and beam tests have been completed. The RFQ was operated in cw mode for all beam tests. The operation of the RFQ at peak voltage (74 kV) is stable [6]. Beams of both N^+ and N_2^+ have been accelerated to test the RFQ at both low and high power operation.

Beam capture as a function of RFQ vane voltage measurements have been completed for each ion and for both unbunched and bunched input beams. The results are given in Figure 5 (solid lines) along with predicted efficiencies based on PARMTEQ calculations (dashed lines). The N_2^+ capture efficiency at the nominal voltage is 80% in the bunched case (three harmonics) and 30% for the unbunched case in agreement with predictions. The capture for one harmonic and two harmonic pre-bunching are 63% and 74% respectively also in good agreement with predictions. The results are a strong confirmation of the pre-buncher performance and RFQ longitudinal acceptance as well as of the accuracy of the design and simulation codes.

The results for N^+ are somewhat lower in the bunched case (74%). The experimental evidence suggests that a transverse centering error either generated at injection and/or during acceleration has reduced the transverse acceptance such that the larger N^+ beam is not transmitted cleanly. Investigations are continuing.

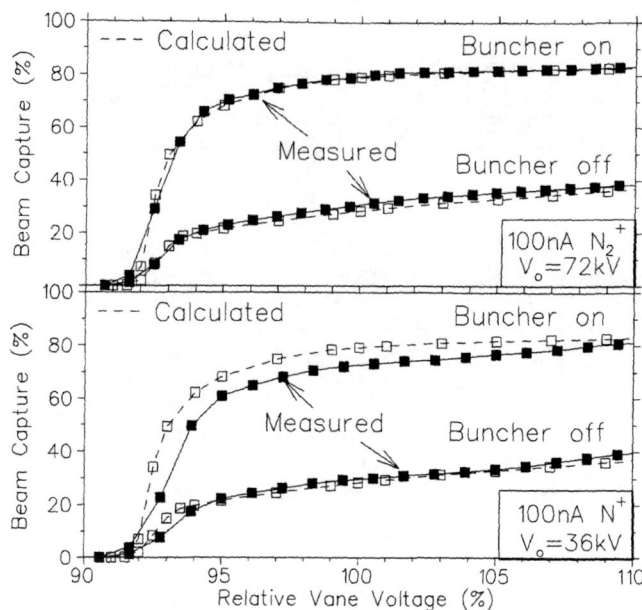

FIGURE 5. RFQ beam test results showing capture efficiency as a function of relative vane voltage for N_2^+ (top) and N^+ (bottom). The nominal vane voltage in each case is 72 kV and 36 kV respectively. The beam capture for both bunched and unbunched initial beams are recorded (solid line and solid squares) and are compared with PARMTEQ calculations (dashed lines and open squares).

The energy of the beam as measured with the analyzing magnet is 55 keV/u. The beam energy is higher than originally quoted since the RFQ test frequency is 2% higher than the design frequency. (The actual frequency is expected to drop with the addition of the remaining 12 rings.) The energy spread for the bunched and unbunched cases were measured at $\pm 0.4\%$ and $\pm 0.7\%$ respectively and compare well with PARMTEQ predictions (Figure 6).

We are presently conducting a further set of tests [7] including measurements of longitudinal and transverse emittance and acceptance. These measurements will proceed through October 98 before we commence with the installation of the rest of the rings. Commissioning of the RFQ in the final configuration is expected in Fall 1999.

MEBT

The beam is stripped in the medium energy beam transport (MEBT) with a thin carbon foil (3 μgm/cm^2) to boost the charge state before acceleration in the DTL. The beam from the RFQ is focussed in three dimensions onto the stripping foil

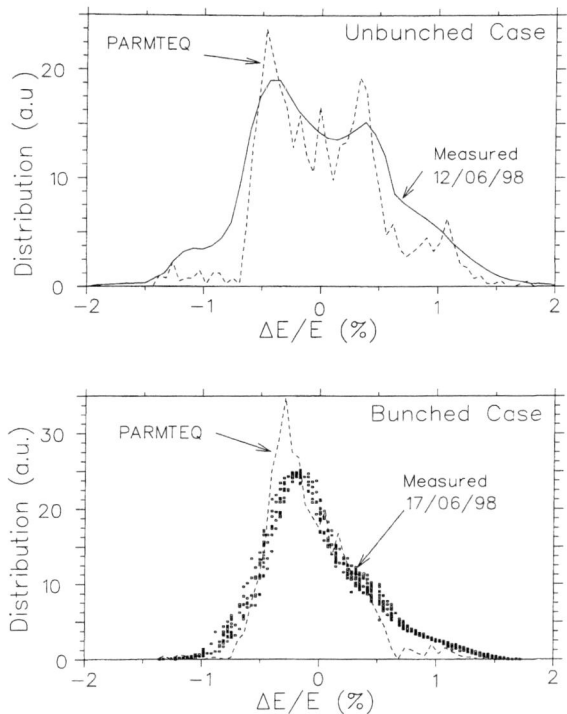

FIGURE 6. Results of energy spread measurements of accelerated N$^+$ beams for both (a) unbunched and (b) bunched cases. PARMTEQ simulation results are plotted for comparison.

with quadrupoles and a 105 MHz double gap bunch rotator to minimize emittance growth due to multiple scattering and energy straggling. A chopper eliminates the small quantity of beam (~4%) accelerated in the two 35 MHz buckets neighbouring the main pulse. After charge selection the beam is matched into the DTL with quadrupoles and a 35 MHz re-buncher.

All quadrupoles have been received. A two-gap spiral 35 MHz re-buncher is presently in development. A model has been completed to study the mechanical rigidity. The MEBT, initially without bunch rotator and chopper, will be installed in the summer of 1999 in time to commission the full energy RFQ followed by the DTL.

DRIFT TUBE LINAC

The design concept for the DTL has been reported previously [8]. The drift tube linac (DTL) is required to accelerate, in cw mode, ions with $3 \leq A \leq 6$ from an injection energy of 0.15 MeV/u to a final energy variable from 0.15 to

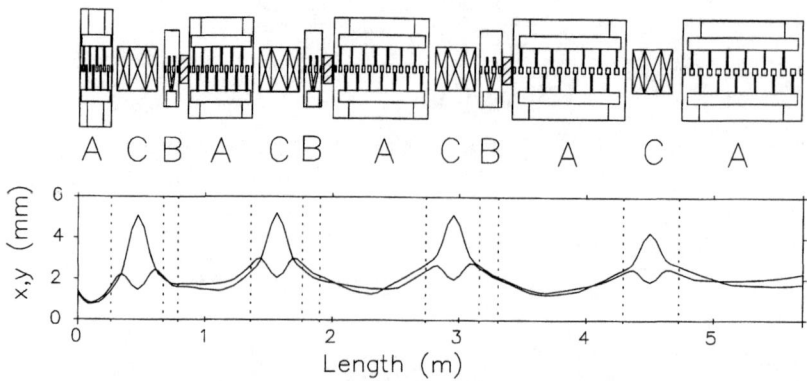

FIGURE 7. Schematic drawing of the ISAC variable energy 105 MHz DTL (upper figure) consisting of five IH tanks (A), three split-ring resonators (B) and quadrupole triplets (C). Beam envelopes (lower figure) define the x and y maximum half sizes as a function of linac length. The calculations are for a beam with normalized emittances of $0.15\pi\,\mu$m and 1.6π keV/u·ns longitudinally.

1.5 MeV/u. The aim is to build an efficient yet flexible room temperature structure. A *separated function* DTL concept has been developed. Five independently phased Interdigital H-mode (IH) tanks operating at $\phi_s = 0°$ provide the main acceleration. These structures are characterized by their simple construction and very high shunt impedance. Longitudinal focussing is provided by triple gap, split-ring resonator structures [9] positioned before the second, third and fourth IH tanks. Quadrupole triplets placed after each IH tank maintain transverse focussing. To achieve a reduced final energy, the higher energy IH tanks are turned off sequentially and the voltage and phase in the last operating tank is varied. The split-ring resonator cavities are adjusted to maintain longitudinal bunching. A schematic drawing of the DTL is shown in Figure 7. Diagnostic boxes are placed between the bunchers and the IH tanks.

At full voltage the beam dynamics are typical for a 0° accelerating structure [10] with the benefits that the acceleration efficiency is optimum and rf defocusing is reduced. The calculated transverse beam envelopes for the full energy case are shown in Figure 7 for matched, normalized emittances of $0.15\pi\,\mu$m and 1.6π keV/u·ns. (For design purposes the longitudinal emittance is larger than expected by a factor of two.) The beam enters each accelerating section converging in all three dimensions, goes through a waist during acceleration and is refocussed both longitudinally and transversely in the intertank space. The strong periodic focussing yields small beam sizes and an increased acceptance.

The gross specifications of the five IH tanks and the three split-ring resonators for the design particle of $q/A = 1/6$ are given in Table 1. The chief design considerations are the variable energy requirement which limits tank and intertank lengths and the *cw* operation which limits the accelerating gradient. The high

TABLE 1. Parameter specifications for each IH tank and buncher (B1-B3) for the design particle of $q/A = 1/6$ in full energy mode. All cavities operate at 105 MHz. Here L is the length, R is the tank radius, $\overline{E_o \cdot T}$ is the effective field gradient, E_s is the peak surface field and Z is the effective shunt impedance. The quoted shunt impedance values are from **MAFIA**. The power calculations assume a shunt impedance 75% of the value quoted.

Tank	No. Cells	L (cm)	R (cm)	$\overline{E_o \cdot T}$ (MV/m)	E_s (MV/m)	V_{eff} (MV)	Z (MΩ/m)	P (kW)	E_{out} (MeV/u)
1	9	26	46	2.1	10	0.5	480	3.3	0.23
2	13	50	38	2.4	12	1.2	495	7.8	0.44
3	15	77	38	2.5	14	2.0	464	14	0.78
4	14	90	38	2.4	14	2.2	400	17	1.14
5	13	98	38	2.3	14	2.2	365	19	1.50
B1	3	10	28	1.9	9.8	0.19	77	6.4	0.23
B2	3	12	28	2.3	11	0.26	75	11	0.44
B3	3	14	28	2.3	11	0.32	72	14	0.78

shunt impedance of the IH structure yields a relatively low power consumption (∼90 kW) for the whole DTL. Each triplet unit has a bore aperture of 24 mm and a maximum gradient of ∼65 T/m. They will occupy a 40 cm space between tanks.

A plot of the tank voltage and phase required for a given final energy is shown in Figure 8. Each point represents a tune that has been simulated. In all cases, for input beams of 1.6π keV/u·ns (to be conservative this is a factor of two larger than expected), the final emittance was $\epsilon_z \leq 1.7\pi$ keV/u·ns. For a reduced voltage the particle bunch is phased negatively with respect to the synchronous phase so that as the particles lose step with the synchronous particle and drift to more positive phases they gain the required energy. Below some minimum voltage (set by multi-pactoring criteria) the phase alone is used to fine tune the output energy. For the lower energies the upstream buncher is used to match the beam to the detuned tank. The buncher following this tank is then used to capture the diverging beam. The three gap split-ring structure is chosen for its large velocity acceptance and large multipactor-free voltage range. The three bunchers must operate over β regimes given by 1.8%→2.2%, 1.8%→3.1%, 1.8%→4.1% respectively and over voltage ranges from 15% to the tabulated value. Three resonators with gap structures synchronized to beam velocities of $\beta = 2.3\%$, 2.7% and 3.3% have been specified.

Fabrication of the first IH tank and the first split-ring buncher has proceeded in advance of the bulk of the DTL in order to get experience with the fabrication techniques. The first DTL buncher, a split-ring three gap structure operating at 105 MHz, has been designed and fabricated at INR-RAS Troitsk and has been delivered to TRIUMF. Signal level tests on the buncher are now successfully completed. The rf frequency is 105.21 MHz with a measured Q of 4200 and shunt impedance of $R_s = (U_{eff}/T)^2/P = 9.3$ MΩ. The mechanical vibration of the tubes was measured at ∼1 μm. The field distribution as calculated from a bead-pull measurement is

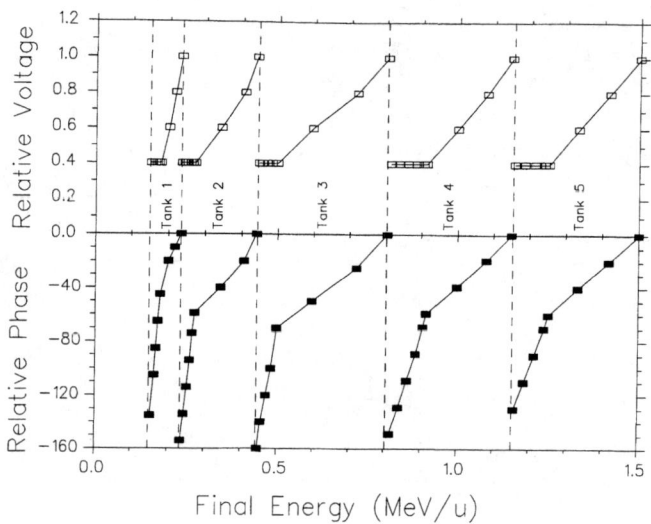

FIGURE 8. Tank voltage and phase required for a certain final energy. Upstream tanks are turned off. The full energy case corresponds to tank voltages of 1.0 and phases of 0°.

shown in Figure 9 (squares) and compares well to a MAFIA simulation (dashed line). Power level tests are due to begin this month. The fabrication of the remainder of the cavities will proceed after the acceptance tests are complete.

The stems and ridges of the first tank have been received. They are both fabricated from solid copper. Fabrication of the mild steel tank is now complete. Assembly of the first DTL tank will commence after the completion of the copper-plating. Power tests are scheduled for late fall.

The quadrupole triplets have been specified and are presently in design. Detailed design and fabrication of the remainder of the tanks and bunchers is scheduled for 1999. Commissioning of the full DTL is expected in the latter half of 2000.

BEAM QUALITY

The beam from the source is expected to have a transverse emittance no larger than $\epsilon_{x,y} = 50\pi\mu m$ corresponding at 2 keV/u to a normalized emittance of $\epsilon_n = \beta\gamma\epsilon = 0.1\pi\mu m$. The expected transverse and longitudinal emittances at various locations in the accelerator chain are shown in Table 2. The values quoted enclose 98% of the particles. Note the effect of multiple scattering and energy straggling in the stripping foil and the small emittance growth in the MEBT rebuncher. The transverse acceptance of the RFQ and DTL are large enough that no emittance increase is expected during acceleration. The technique of pre-bunching the beam entering the RFQ and eliminating the bunching and shaping section in the RFQ gives a very compact beam in longitudinal phase space. The longitudinal

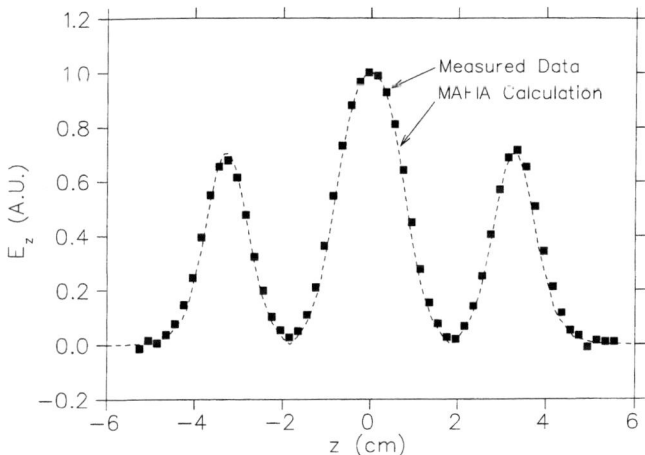

FIGURE 9. Measured electric field distribution (squares) along the beam axis. `MAFIA` calculation results are plotted for comparison.

acceptance of the DTL and flexibility of the design are such that not more than 10% emittance growth is expected over the whole energy range.

CONCLUSIONS

The successful RFQ beam tests demonstrate a strong confirmation of both the beam dynamics design and the engineering concept and realization. The ISAC linac provides efficient cw acceleration at room temperature with a beam quality and flexibility approaching that of a superconducting linac but at a reduced cost and complexity. The ISAC facility is scheduled to supply high energy beam to experimenters by the end of 2000.

TABLE 2. Simulation results showing the beam emittance at various locations in the ISAC accelerator chain. The values quoted enclose 98% of the particles.

Position	Transverse		Longitudinal	
	$\epsilon_{x,y}$	$\beta\epsilon_{x,y}$	ϵ_z	ϵ_z
	$(\pi\mu m)$	$(\pi\mu m)$	$(\pi\%\ ns)$	$(\pi keV/u\ ns)$
LEBT	50	0.1	DC	DC
After RFQ	5	0.1	0.33	0.5
After Foil	10	0.2	0.45	0.67
Before DTL	11	0.22	0.50	0.74
After DTL	$11\cdot(0.018/\beta_{final})$	0.22	$0.5\cdot(0.15/E_{final})$	0.74

ACKNOWLEDGMENTS

TRIUMF is first and foremost a cyclotron laboratory and has entered the linear accelerator field only very recently. We are indebted, therefore, to various members of this community for their very real assistance in the design and implementation of our new facility. It is also a pleasure to acknowledge the dedicated collective efforts of the TRIUMF-ISAC staff whose many contributions have been recorded here.

REFERENCES

1. P. Schmor, "Overview of ISAC", this conference.
2. R. Baartman, J. Welz, "60 keV Beam Transport Line and Switchyard for ISAC", Proceedings of the 1997 Part. Acc. Conf., Vancouver, 1997.
3. J. Bogaty, et al, "A Very Wide Band-Width Faraday Cup...", Proceedings of the 1990 Linear Accelerator Conference, 1990.
4. R. Poirier, et al, "Construction Criteria and Prototyping for the ISAC RFQ Accelerator at TRIUMF", Proceedings of the 1997 Part. Acc. Conf., Vancouver, 1997.
5. S. Koscielniak, et al, "Beam Dynamics Studies on the ISAC RFQ at TRIUMF", Proceedings of the 1997 Part. Acc. Conf., Vancouver, 1997.
6. R. Poirier, et al, "RF Tests on the Initial 2.8m of the 8m Long ISAC RFQ at TRIUMF", Proceedings of the 1998 Linac Conference, Chicago, to be published.
7. R.E. Laxdal, "Testing the RFQ at 53 keV/u", TRIUMF-ISAC Design Note, July 1997.
8. R.E. Laxdal, et al, "A Separated Function Drift Tube Linac for the ISAC Project at TRIUMF", Proceedings of the 1997 Part. Acc. Conf., Vancouver, 1997.
9. Y. Bylinsky, et al, "A Triple Gap Resonator Design for the Separated Function DTL at TRIUMF", Proceedings of the 1997 Part. Acc. Conf., Vancouver, 1997.
10. U. Ratzinger, "Interdigital IH Structures", Proc. 1990 Linear Accelerator Conf., Los Alamos, p.525, (1990).

The Radioactive Ion Beams Facility Project for the Legnaro Laboratories

Luigi B. Tecchio on behalf of the SPES Study Group

Laboratori Nazionali di Legnaro
Via Romea 4, 35020 Legnaro (PD), Italy

Abstract. In the frame work of the Italian participation to the project of a high intensity proton facility for the energy amplifier and nuclear waste transmutations, LNL is involving in the design and construction of prototypes of the injection system of the 1 GeV linac that consists of a RFQ (5 MeV, 30 mA) followed by a 100 MeV linac. This program has been already financially supported and the work is actually in progress. In this context, the LNL has been proposed a project for the construction of a second generation facility for the production of radioactive ion beams (RIBs) by using the ISOL method. The final goal consists in the production of neutron rich RIBs with masses ranging from 80 to 160 by using primary beams of protons, deuterons and light ions with energy of 100 MeV and 100 kW power. This project is proposed to be developed in about 10 years from now and intermediate milestones and experiments are foreseen and under consideration for the next INFN five year plan (1999 - 2003). In such period of time is proposed the construction of a proton/deuteron accelerator of 10 MeV energy and 10 mA current, consisting of a RFQ (5 MeV, 30 mA) and a linac (10 MeV, 10 mA), and of a neutron area dedicated to the RIBs production, to the BNCT applications and to the neutron physics. Some remarks on the production methods will be presented. The possibility of producing radioisotopes by means of the fission induced by neutrons will be investigated and the methods of production of neutrons will be discussed.

INTRODUCTION

The international community is showing growing interest in high intensity linacs for scientific, industrial and social applications. Proton linacs with final energies of about 1 GeV and CW operation are proposed for secondary beams production, tritium production, nuclear waste transmutation or energy production in sub-critical accelerator driven reactors. The beam intensities vary for different proposed application and are ranging from 10 to 100 mA.

In the frame work of the Italian participation to the project of a high intensity proton facility for the energy amplifier and nuclear waste transmutations, LNL is involving in the design and construction of prototypes of the injection system of the 1 GeV linac that consists of a RFQ (5 MeV, 30 mA) followed by a 100 MeV linac. This program has been already financially supported and the work is actually in progress. In this context, the LNL has been proposed a project for the construction of a second generation facility for the production of radioactive ion beams (RIBs) by using the ISOL method. The final goal consists in the production of neutron rich RIBs with masses ranging from 80 to 160 by using primary beams of protons, deuterons and light ions with energy of 100 MeV and 100 kW power (see Fig. 1). This project is proposed to be developed in about 10 years from now and intermediate milestones and experiments are foreseen and under consideration for the next INFN five year plan (1999 - 2003). In such period of time is proposed the construction of a proton/deuteron accelerator of 10 MeV energy and 10 mA current, consisting of a RFQ (5 MeV, 30 mA) and a linac (10 MeV, 10 mA), and of a neutron area dedicated to the RIBs production, to the BNCT applications and to the neutron physics. The RFQ it will be of the same type of that designed for the high intensity project mentioned above. An intense R&D program on high intensity accelerator techniques and targetry is already in progress.

THE ACCELERATOR

The sequence RFQ-DTL (Drift Tube Linac) is, by far, the most used scheme for proton linacs in the energy range of 10 - 100 MeV. In our design both DTL and RFQ operates at the main linac frequency of 352 Mhz; in this way we avoid any frequency jump, and the bore hole inside the DTL structure can be kept large enough to have a good margin between beam dimensions and machine acceptance.

The RFQ structure is, nowadays, the natural choice for the low energy part of any linear accelerator. It is very efficient up to the energy of few MeV giving a transmission in excess of 90% of the continuos beam coming from the source at energies of few tens of keV. The acceleration efficiency of the RFQ falls down very rapidly in the range of 1 to 10 MeV and it is mandatory to change structure. As usual we consider a DTL as following accelerating segment and the transition has been put at 5 MeV trading off the RFQ low efficiency at the end of the structure with the higher DTL shunt impedance at its beginnings. The DTL shows a good efficiency up to hundreds of MeV. In Table 1 the main accelerator parameters are summarized.

An intermediate milestone of this project consists on the construction of a proton/deuteron accelerator of 10 MeV energy and 10 mA current and of a neutron area dedicated mainly to the RIBs production, to the BNCT applications and to the neutron physics.

In the frame work of the high intensity proton linac project a first prototype (in alluminum) of the RFQ accelerator (5 MeV, 30 mA) has been designed and constructed. The prototype has been recently delivered and actually is under RF

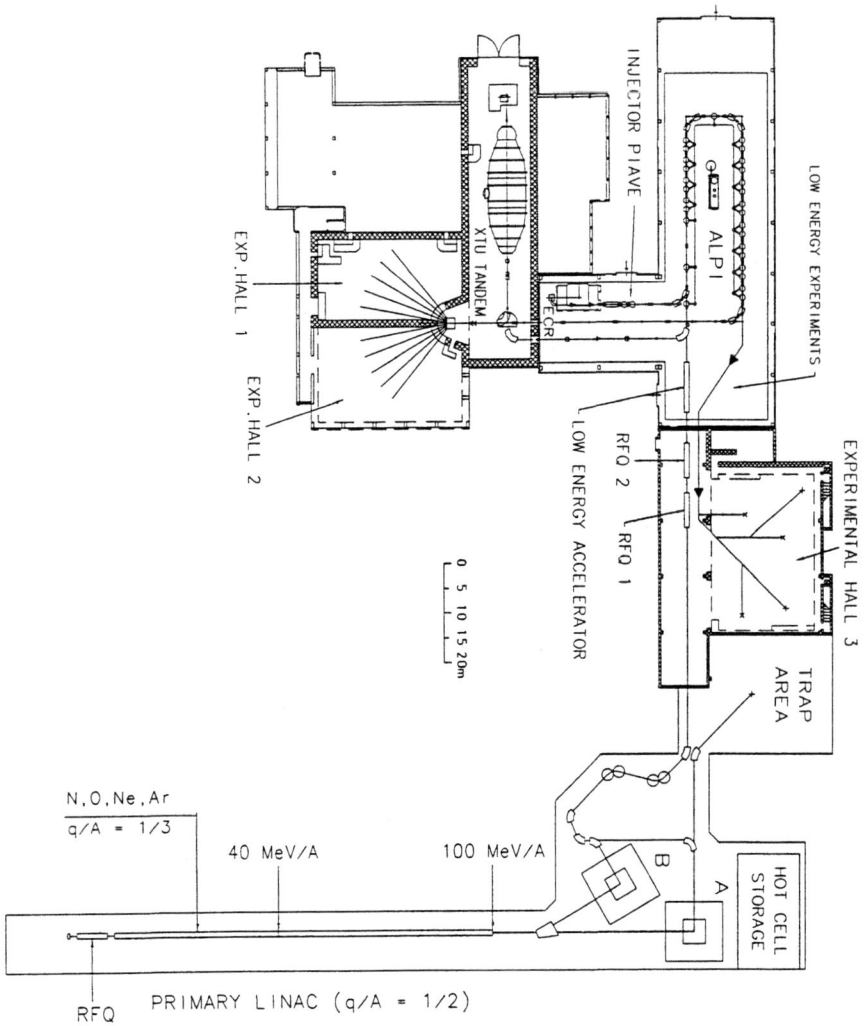

FIGURE 1. Layout of the RIBs Facility

TABLE 1. Main Parameters of the Accelerator

Parameters	Unit	RFQ	DTL
Input energy	MeV	0.05	5
Output energy	MeV	5	100
Beam current	mA	30	10
RF frequency	Mhz	352.2	352.2
Duty Cycle	%	100	100
Total length	m	5.3	80
Transmission	%	94.6	100
RF power diss.	MW	0.6	8.3
Beam loading	MW	0.15	2.8
Quad. diss.	MW	-	0.6

measurements in the laboratory. After measurements shall we proceed to the construction of the final RFQ that is planned to be ready in two years. At the same time, in collaboration with the ARGONNE National Laboratory, an intense R&D program dedicated to the study of superconducting cavities for the DTL linac has been initiated. In any case, a low energy proton accelerator (up to 5 MeV) like the RFQ above described is a very suitable accelerator to produce high intensity neutron beams. In fact, the reaction ^{7}Li (p,n)^{7}Be has been proposed as an accelerator-based source of neutrons. This reaction displays a large resonance in the forward direction around 2.3 MeV which extends to about 2.5 MeV. The angular distribution of the produced neutrons shows a pronounced peak at zero degree. The neutron yield (per incident proton) between 0^0 and 30^0 is about 4×10^{-3} (n/p) so that, in our case, an intensity of the order of 2.5×10^{14} (n/s) is expected. The neutrons induced fission presents remarkable advantages for the production of neutron rich RIBs with respect to the direct production (proton beam directly on the production target), and exotic beams with intensities of the order of 10^9 - 10^{12} ions/s are possible.

RIBs PRODUCTION METHODS

One of the most used methods for the production of RIBs is the combined process of fission and spallation by means of protons on a target of different chemical-physical nature. In particular, the results of the experimentation at ISOLDE of CERN (1), where the primary beam consists of 600 MeV protons, are well known. The experimental results make us believe that the fission process becomes prevalent at low energies, namely some tens of MeV, whereas the spallation process is dominant at energies higher than 100 MeV. In the 100 MeV region the two processes are in conflict. The simulations carried out by the Monte Carlo codes LAHET (2) and FLUKA (3) provide discordant results on the percentage of the spallation processes for 100 MeV protons on a U238 target, 11% and 62% respectively. On the other hand experimentation provides unambiguous data on the production of RIBs, when using protons of different energies as primary beam or neutrons (4). Increasing the proton beam energy, the production region extends more and more towards the direction of the proton-rich isotopes. On the contrary, a great increase of the cross section in the neutron-rich region may be obtained by means of the fission process induced by neutrons on fissile targets, like Uranium and Thorium.

The use of the fission reaction induced by neutrons of high energy (100 MeV) appears to be very convenient in order to produce exotic beams of high intensity. Moreover, the conversion of the primary proton/deuteron beam to a neutron flux allows to simplify and partially to solve the problems related to the power dissipation (100 kW) in the production target.

Neutrons production

High neutrons fluxes may generally be obtained by the conversion of primary beams of protons and deuterons to targets of different materials (deuterium, tritium, beryllium,...). The most fruitful reactions are: D(d,xn), Be(d,xn), Th(d,xn), U(d,xn). For the sake of convenience we analyze the reactions that are generally most employed, i.e. Be(d,xn) and U(d,xn).

Independently from the conversion target, the deuterons of 100 MeV/u are the most appropriate projectiles for the production of neutrons. Conversion targets as uranium and beryllium both present characteristics which are suitable for the task, even if they have different angular and energetic distributions. Table 2 show the mean values of the neutron multiplicity for the different regions of the neutron energy spectrum.

Moreover, the cascade of secondary neutrons produced in the target, which contribute to a great extent to the fission process, must be taken into account. The neutrons generation in the Be(d,xn) reaction is the most efficient (secondary neutron multiplicity $<M_{n2}>$ = 3) in comparison with the generation in the U(d,xn) reaction ($<M_{n2}>$ = 1) and seems to be the most favorable for the production of exotic beams.

R&D PROGRAM AT THE LEGNARO LABORATORIES

The R&D program on the radioactive beams at the LNL consists into study the production methods of RIBs through the fission induced by thermal and non-thermal neutrons on targets of Th232, U235 and U238. In order to minimize the release time from the target and to maximize the ionization efficiency in the ion source the technology of the target/ion source system will be study. This R&D program has been funded by INFN and will take the next two financial years (1999-2000). The experiments will be performed at the Van de Graaff accelerator by using the 7 MeV,

Table 2. Mean value $\langle M_n \rangle$ of the multiplicity of neutrons, in units of emitted nucleons per incident deuteron (100 MeV/u) in the different regions of the energetic spectrum.

Neutron Energy range (MeV)	$\langle M_n \rangle$ for Be	$\langle M_n \rangle$ for U238
≥50	0.245	0.080
≥25	0.546	0.224
≥10	0.898	0.385
≥5	1.111	0.554
≥0	1.556	2.210

3 µA deuteron beam for the generation of neutrons subsequently employed in the fission processes. The neutrons will be produced by the following reactions: Be(d,n), D(d,n) and t(d,n).

By using the Be(d,n) reaction at 7 MeV, it is possible to obtain about 1.5×10^{10} n s^{-1} sr^{-1}, at zero degree, of 3.2 MeV average energy. By thermalizing adequately the produced neutrons, a flux of 2×10^{8} cm^{-2} s^{-1} thermal neutrons may be obtained (5). The high energy (14 MeV) neutrons will be produced by the reaction t(d,n); about 10^9 n/s may be obtained with primary deuterons of 0.5 - 1 MeV energy and 3 µA current (6).

The experimental set up has already been installed and consists of a bunker housing the beryllium foil for neutron generation and the target/ion source system, followed by a magnetic spectrometer (M/ΔM~800) for the isotopic mass separation and of a detector for isotopes identification. Both the production target and the ion source operates at high temperature (<2500 °C); the source is a conventional surface ionization source charge 1+; the operation voltage is 20 kV.

A first prototype of the target/ion source system has been succesfully tested and the on-line separator has been calibrated with stable ions.

CONCLUSIONS

The possibility of disposing of a neutron facility for the RIBs production at LNL within the next five/seven years become rather realistic, but still depend upon the decision concerning the next five years plan of INFN. A part such a considerations, in the frame work of the high intensity proton linac program a first prototype (in alluminum) of the RFQ accelerator (5 MeV, 30 mA) has been designed and constructed. The prototype has been recently delivered and actually is under RF measurements in the laboratory. After measurements shall we proceed to the construction of the final RFQ that is planned to be ready in two years from now. Its features it self will fulfill completely the requirements for a neutron facility of average intensity. In parallel, an R&D program to investigate the feasibility of RIBs production through the fission induced by neutrons on fissile targets has been financially supported and the experimental set up has been already installed at the CN accelerator of the LNL. Preliminary tests of the on-line separator are actually on progress.

ACKNOWLEDGMENTS

I would like to thanks all my colleagues of the Accelerator Division and of the SPES study group for they support during the preparation of this paper.

REFERENCES

1. Ravn, H.L., *Physics Reports* **54**,201 (1979).
2. Prael, R.E., Bozoian, M., *LA-RU-88-3238*, Los Alamos 1998.
3. Ferrari, A., Sala,P., *Proceedings of MC93*, Tallahassee, Florida 1993, Ed. World Scientific, Singapore 1994.
4. Ravn, H.L., et al., *Nucl. Instrum. Meth.* **B88,** 441 (1994).
5. Agosteo, S., et al., *Rad. Prot. Dos.* **70**, 559 (1997).
6. Lee, W.C., et al., *Nucl. Instrum. Meth.* **B99**, 739 (1995).

Development of a Radioactive Ion Beam Test Stand at LBNL

D. Wutte, J. Burke, B. Fujikawa, P. Vetter, S.J. Freedman, R.A. Gough, C. M. Lyneis, Z. Q. Xie

Ernest Orlando Berkeley National Laboratory, University of California at Berkeley Berkeley, California 94720, USA

Abstract. For the on-line production of a $^{14}O^+$ ion beam, an integrated target - transfer line ion source system is now under development at LBNL. ^{14}O is produced in the form of CO in a high temperature carbon target using a 20 MeV ^{3}He beam from the LBNL 88" Cyclotron via the reaction $^{12}C(^3He,n)^{14}O$. The neutral radioactive CO molecules diffuse through an 8 m room temperature stainless steel line from the target chamber into a cusp ion source. The molecules are dissociated, ionized and extracted at energies of 20 to 30 keV and mass separated with a double focusing bending magnet.
The different components of the setup are described. The release and transport efficiency for the CO molecules from the target through the transfer line was measured for various target temperatures. The ion beam transport efficiencies and the off-line ion source efficiencies for Ar, O_2 and CO are presented. Ionization efficiencies of 28% for Ar^+, 1% for CO, 0.7% for O^+, 0.33 for C^+ have been measured.

INTRODUCTION

At the Lawrence Berkeley National Laboratory we have commissioned an ion source test stand for radioactive ion beam development. The primary goal of this test stand is the on-line production of an $^{14}O^+$ ion beam. We are interested in measuring the shape of the decay spectrum for the Gamow-Teller branch in the ^{14}O beta-decay (Figure 1) in order to test the conserved vector current hypothesis (CVC) proposed by Feynman and Gell-Mann(1). The ^{14}O half-life of 70 seconds requires producing the isotope on-line at the 88" Cyclotron. ^{14}O is generated in the form of CO in a high temperature carbon target using a 20 MeV $^3He^+$ beam from the LBNL 88" Cyclotron via the reaction $^{12}C(^3He,n)^{14}O$.

The ^{14}O atoms must be then separated from the other radioactive isotopes produced in the carbon target and implanted into a thin carbon foil in order to:

(i) minimize the radiation background
(ii) maximize the signal in the beta spectrometer by concentrating the ^{14}O sample size.

For this purpose an 8 m stainless steel transfer line connects the target chamber to a cusp ion source (2) through a turbo molecular pumping stage. Thus, the turbo pump separates the target vacuum chamber from the ion source (Figure 2).

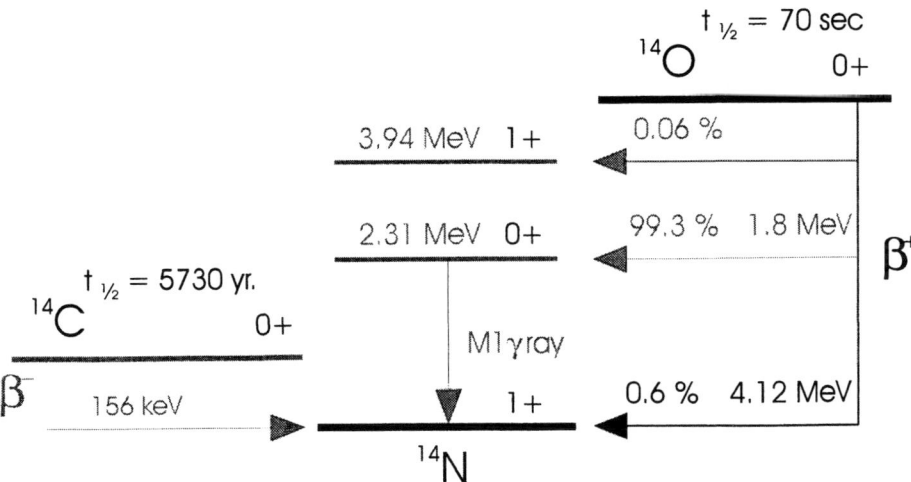

FIGURE 1. Mass-14 isospin triplet system. The transition of interest (4.12 MeV β^+, ^{14}O → ^{14}N) is indicated in black

The gas coming from the turbo pump is fed into the ion source and ionized, extracted to energies of 20 to 30 keV and mass separated. To achieve a small sample size for the beta spectrometer, it is planned to implant the ^{14}O$^+$ ions on a 2 mm spot into a thin carbon foil. This sample will then be transferred to the beta-spectrometer. At an implantation rate of $2 \cdot 10^7$ pps the required counting time in the spectrometer will be about 150 hours.

The first part of the paper describes the target setup and first experimental results of the high-temperature graphite target. The second part presents the design and commissioning of the new radioactive ion beam test stand. The third part discusses the off-line performance of the RF driven multi cusp ion source.

TARGET SETUP

The half-life of ^{14}O is approximately 70 seconds and oxygen is chemically reactive. Given these constraints we decided to develop a high porosity heated solid carbon target. An ^{14}O atom produced in the carbon chemically bonds with a carbon atom to form a ^{12}C^{14}O molecule. This molecule has to diffuse to the surface of the local carbon grain and out of the bulk material. The diffusion time of the ^{12}C^{14}O out of the target decreases with target temperature. The molecules then diffuse through 8 m of 5 cm inner diameter stainless steel tubing via molecular flow. On the other end of the transport line a 220 l/s turbo molecular pump maintains a constant pressure differential between the target chamber and the far end of the transport line. The output of the turbo molecular pump can be periodically trapped in a small volume, which we then monitor for ^{14}O activity. Two sodium-iodide detectors placed on either side of the sample volume operate in coincidence to detect the back-to-back 511 keV gamma rays from the beta particle annihilation. A germanium detector is used to detect the 2.3 MeV gamma ray from the ^{14}O decay simultaneously with the sodium-iodide detectors.

EXOTIC ION BEAM TEST STAND

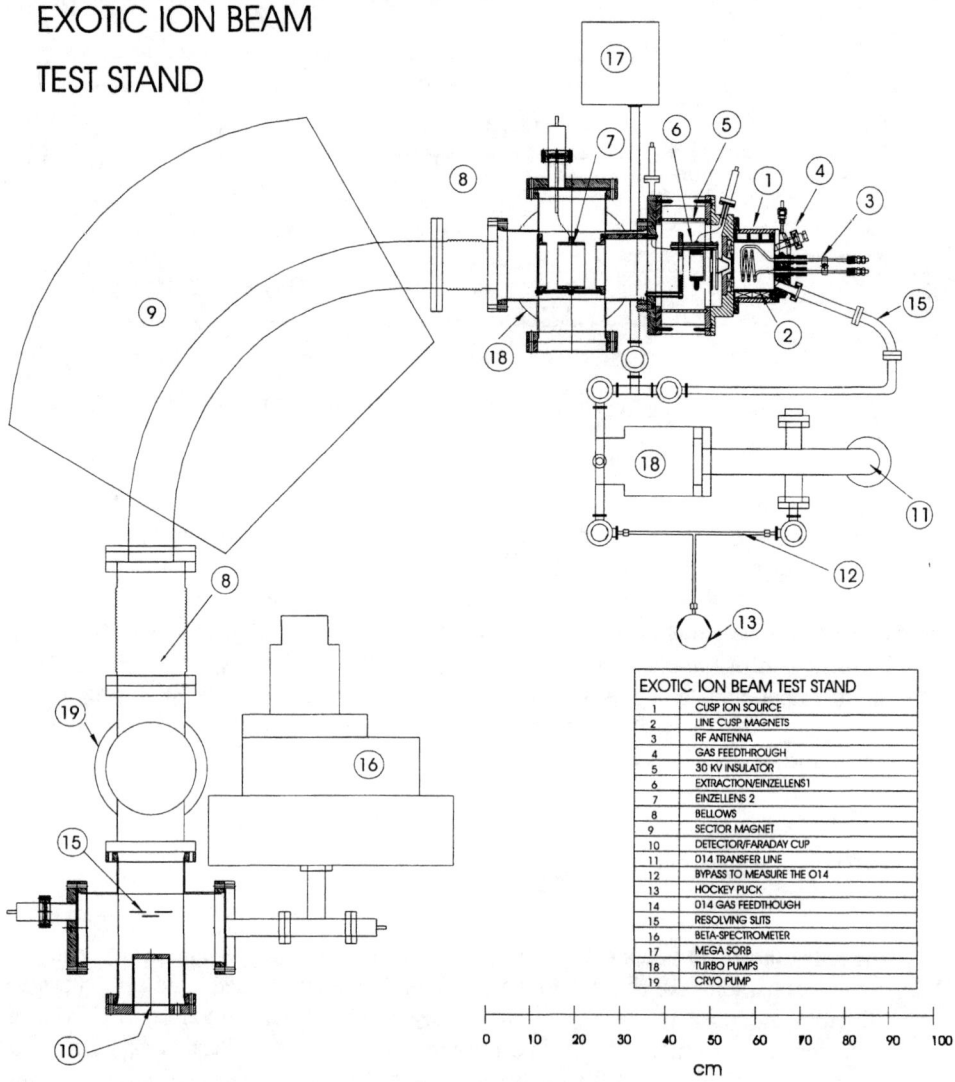

FIGURE 2. Setup for the Exotic Ion Beam Test Stand and the O^{14} experiment.

The prototype carbon target was made of carbon felt with a density of 0.082 g/cm^3 and an average fiber diameter of a few microns. It was indirectly heated by a tantalum heating filament. With this setup we produced about $2 \cdot 10^6$ atoms of ^{14}O per second as measured at the outlet of the turbo molecular pump at a beam current of 5 μA on target and a target temperature of 1500 degrees Celsius. This corresponds to an efficiency of only 0.3 %, of the assumed thick target production rate of $2 \cdot 10^8$ pps/μA. We believe that this low number can be attributed to the fact that the hot tantalum surface acts as a getter for CO gas.

Therefore, an all-carbon target was designed with no metals near the heated section of the target. The all carbon target was constructed using reticulated vitreous carbon (RVC, available from ERG Materials and Aerospace). We used the lowest density (0.048 g/cm^3) and highest surface to volume ratio. The electrical properties are suitable for direct resistive heating. The target (length of 5 cm, a width of 3.8 cm, and a thickness of 1 cm) has a resistance of 0.20 ohms at 1750 degrees Celsius in a 10^{-5} Torr vacuum. Electrical contacts are made by bolting the target to water-cooled electrodes using molybdenum threaded rod. Molybdenum was used because of its high melting point (2600 degrees Celsius). Graphite and boron nitride heat shields were then added concentrically around the target.

With the new target heated to 1720 °C and 2 μA cyclotron beam current on the target, we measured $3 \cdot 10^7$ pps of ^{14}O at the output of the turbo pump. Figure 3 shows the production rate as a function of target temperature. Therefore at 20 μA primary beam current a production rate of $3 \cdot 10^8$ pps of ^{14}O can be expected.

Using the assumed thick target production rate of $2 \cdot 10^8$ pps/μA, an efficiency of 7.5 % for the new target has been achieved.

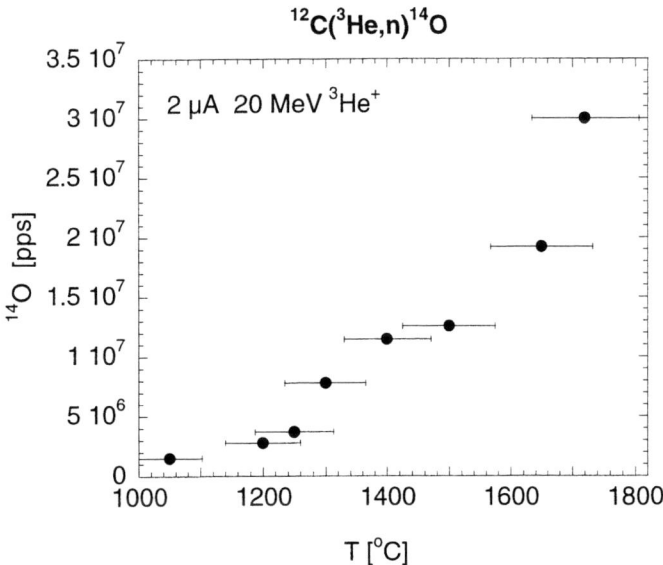

FIGURE 3. Production rate as a function of target temperature as measured at the end of the transport line.

Beam Transport Line

Currently the ion beam test stand consists of a radioactivity transfer line, a multicusp ion source, an accel-decel type extraction system with a 2 mm plasma outlet aperture, two electrostatic einzel lenses, and a mass analyzing magnet.

The ion beam transport line has been designed around an existing double focusing 90° sector magnet from the former HILAC injector line at LBNL. It has a bending radius of 54 cm, edge angles of 30 degrees, and a gap with of 3.8 cm. The horizontal waist is located about 43 cm downstream from the vertical waist. Therefore, the ion beam has an elliptical shape after the sector magnet. An additional focusing element will be needed to achieve the required beam spot size of 2mm diameter at the implant foil. The ion source extraction system has been optimized to match the extracted ion beam from the 10 cm RF driven cusp ion source to the acceptance of the analyzing magnet.

The ion beam extraction from a cusp ion source is highly dominated by transverse space charge effects. Since the ion source has to run at a neutral pressure of at least 1 mTorr to sustain the plasma, a high current of ions from the plasma sustaining gas must be unavoidably extracted together with the low current radioactive ion. For instance together with projected $3 \cdot 10^7$ $^{14}O^+$ an additional 1 mA of Argon ions (corresponding to a ion beam density of 30 mA/cm^2) have to be extracted at a relative low extraction voltage of 30 kV (required as implant energy of the ^{14}O experiment).

The 30 kV extraction system and the following transport line have been optimized with the ion trajectory code IGUN (3). The use of two einzel lenses allows limited independent control over both beam size and divergence at the magnet entrance. Therefore, the ion optics can be adjusted over a wide range of extraction voltages and current densities (3 mA/cm^2 to 60 mA/cm^2, corresponding to a total extracted current of 100 eµA to 2 emA), as verified experimentally. The extraction system and the two einzel lenses are mounted on a single flange to ensure a proper alignment.

Figure 4 shows an IGUN simulation for a 30 keV 1 mA Ar$^+$ beam (30 mA/cm^2) through the extraction system and the first einzel lens. Figure 5 shows the continuation of the ion beam through the second einzel lens to the entrance of the sector magnet. Figure 6 demonstrates the rotation of the RR' emittance figure as the ion beam proceeds through the optic system for an 18 keV 0.4 mA and a 30 keV 1 mA Ar$^+$ beam. At the entrance of the magnet the ion beam is slightly divergent and provides a virtual image point about 3 m upstream. The ion beam through the electrostatic einzel lens system was simulated without space charge compensation.

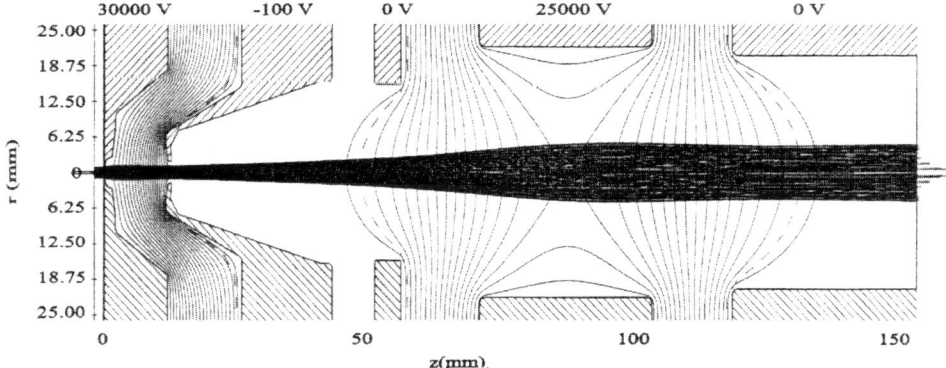

FIGURE 4. IGUN simulation run for a 30mA/cm^2 (1 mA) Ar$^+$ beam showing an overview of the extraction system including the 1st einzel lens.

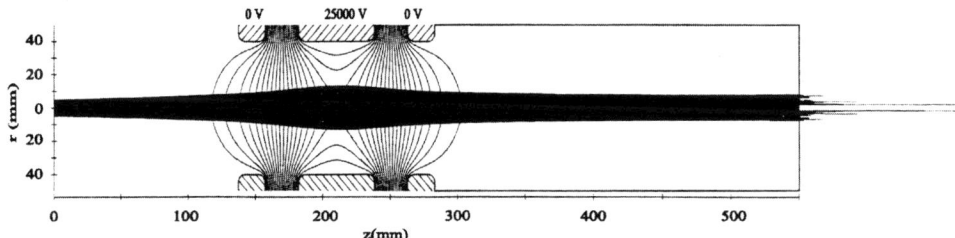

FIGURE 5. Continuation of the ion beam trajectories from figure 4 through the second set of einzel lenses to the entrance of the sector magnet.

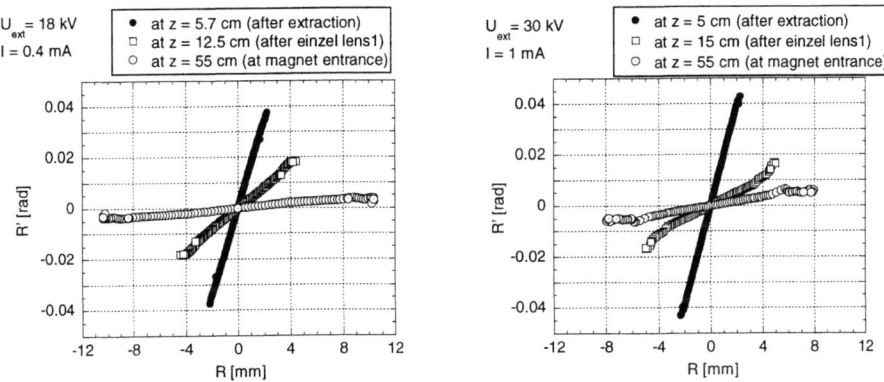

FIGURE 6. Change of the RR' emittance as the beam proceeds through the optics system for a 12 mA/cm^2 Ar$^+$ beam and a 30 mA/cm^2 Ar$^+$.

TABLE 1. Predicted beam size and measured transmissions for various ion beam densities and open resolving slits (30 x 30 mm) after the sector magnet.

current	Argon			Oxygen		
	max. experim. transmission for 30x30 mm resolving slits	TRACE 2D beam ellipse at resolv. slits x[mm]	y[mm]	max. experim. transmission for 30x30 mm resolving slits	TRACE 2D beam ellipse at resolv. slits x[mm]	y[mm]
0 mA		10	2		9	2
0.15 mA	95 %	16	8			
0.2 mA	85 %	18	10			
0.4 mA	67 %	27	17	100 %	20	12
1 mA	50 %	50	35	82 %	35	25
1.8 mA				63 %	55	40
2.2 mA				50 %	60	45

The beam transport through the analyzing magnet to the Faraday Cup was simulated with the first order matrix code TRACE2D, which includes space charge effects. The output beam parameters of IGUN have been used as input parameters for TRACE2D, providing a full simulation from the plasma meniscus into the Faraday Cup.

We have found a good agreement between the simulated and experimental lens voltages for optimal transmission of the ion beam through the sector magnet. As predicted in the simulations, there are no beam losses in the electrostatic lens system.

The measured ion beam transmission for different plasma parameters for fully opened resolving slits (30 mm x 30 mm) after the sector magnet is summarized in table1. For comparison, the TRACE2D calculated beam spot size at the resolving slits is also listed. Depending on the discharge pressure, the beam line pressure is in the pressure range of $1 \cdot 10^{-7}$ and $7 \cdot 10^{-6}$ Torr. At these low pressures, the degree of space charge neutralization will be small and we have therefore neglected the effect in the calculation.

Experimental ion beam transmissions of 85 %-100 % for the lower ion beam densities and 50-65 % for the higher current densities have been achieved.

OFF-LINE ION SOURCE TESTING

A detailed description of the RF driven multi cusp ion source used in this study, together with its basic characteristics can be found elsewhere (2) The magnetic plasma confinement is achieved by 14 columns of samarium-cobalt permanent magnets, which form a longitudinal line cusp configuration. The plasma is inductively heated by up to 2kW of RF power (13.56 MHz).

With this type of ion source, the main concern with respect to the production of radioactive ion beams is the minimum discharge pressure required to sustain a plasma, which complicates the beam transport with a high extracted current. The other concern is the lifetime of the rf-antenna, which is reduced due to the continuous bombardment of heavy ions in the plasma (4,5).

Therefore the performance of the cusp ion source has been evaluated off-line with respect to the three major requirements of the planned experiment:

(i) To achieve the necessary ^{14}O particle current of $1\text{-}2 \cdot 10^7$ pps at the implant target, the ion source should be able to provide 10 % ionization efficiency for $^{14}O^+$.

(ii) The gas hold up time in the ion source must be less than one ^{14}O half-life.

(iii) At the above mentioned implantation rate of $1\text{-}2 \cdot 10^7$ pps the expected continuous run of the experiment will be at least 150 hours. Therefore, the ion source should continuously operate at least 200 hours.

Ionization efficiencies were measured off-line for singly charged argon, oxygen, carbon, and carbon monoxide ions with calibrated leaks. The presented efficiencies quote the overall system efficiencies (ion source and transport line). The maximum ionization efficiencies for all measured species are summarized in Table2. The experimental gas hold up time is described by the exponential fit $A \cdot \exp(-t/\tau_{fast}) + B \cdot \exp(-t/\tau_{slow})$. The fast component describes the holdup time of the ions in the plasma, the slow component is related to the wall sticking time. About 70 % of the signal is dropped within τ_{fast}.

A promising ionization efficiency of up to 28% for a calibrated argon leak (leak rate $1.9 \cdot 10^{-5}$ scc/sec) has been measured with the RF driven multicusp ion source (see Figure 7). O_2 was used as support gas. The maximum ionization efficiency has been achieved at an oxygen discharge pressure of 2 to 3 mTorr. At lower pressure, the RF coupling was less effective, thus resulting in high reflected power and a reduction in the ionization efficiency.

Nevertheless, for CO^+ and O^+ the ionization efficiencies are much lower than with noble gases. Efficiencies have been measured with a calibrated CO leak ($2.3 \cdot 10^{-6}$ scc/sec) with argon as support gas for C^+, O^+ and CO^+. With the cusp source the best efficiencies achieved have been 1 % for CO^+, 0.7 % for O^+ and 0.33 % for C^+. In general rather long ion source hold up times have been observed, the longest has been measured for O^+.

TABLE 2. Ionization efficiencies and hold up times of the RF driven cusp ion source.

ion	cal. leak	leak rate (scc/sec)	equival. pµA	support	efficiency	holdup time	
						τ_{fast}	τ_{slow}
C^+	CO	$2.3 \cdot 10^{-6}$	9.5	Ar	0.33 %	13 sec	86 scc
CO^+	CO	$2.3 \cdot 10^{-6}$	9.5	Ar	1 %	17 sec	84 sec
O^+	CO	$2.3 \cdot 10^{-6}$	9.5	Ar	0.7 %	42 sec	77 sec
O^+	O_2	$2.1 \cdot 10^{-5}$	154	Ar	0.7 %	36 sec*	—
O^+	O_2	$2.1 \cdot 10^{-5}$	154	He	0.45 %	154 sec*	—
Ar^+	Ar	$1.9 \cdot 10^{-5}$	70	O_2	28 %	6 sec	32 sec

*Fast component fitted only.

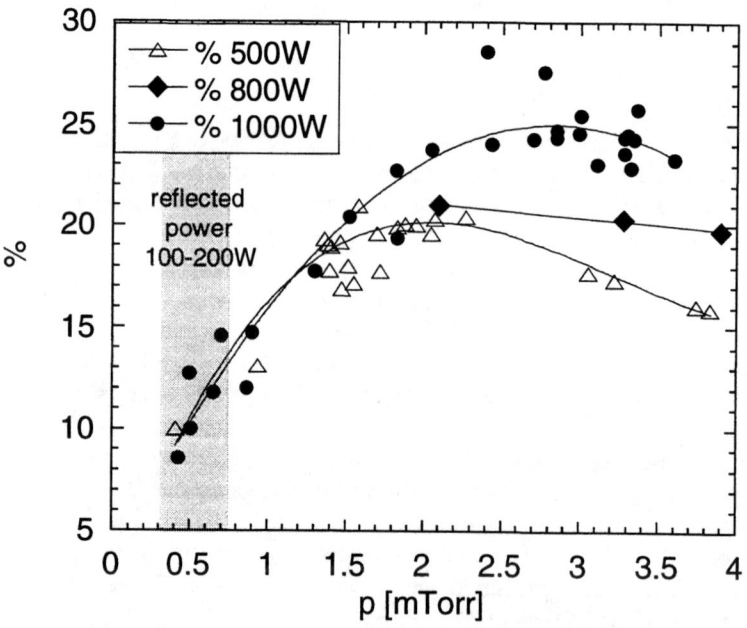

FIGURE 7. Ionization efficiency for Ar^+ (100% equivalent to 70µA Ar^+) with O_2 as support gas for 500W, 800W and 1000W RF input power and various O_2 discharge pressures.

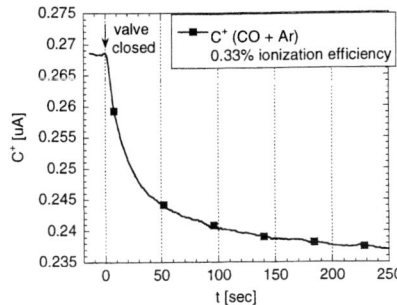
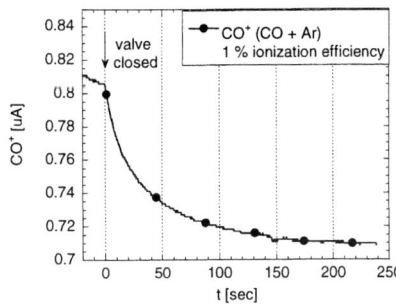

FIGURE 8. CO^+ and C^+ ionization efficiency and hold up time in the cusp.

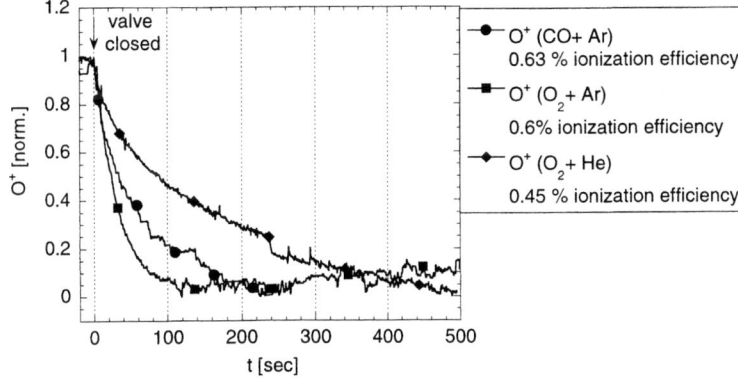

FIGURE 9. O^+ ionization efficiencies and hold up times in the cusp source for the CO leak (Ar as support gas), the O_2 leak (Ar as support gas), and the O_2 leak (He as support gas).

With an O_2 leak ($2.1 \cdot 10^{-5}$ scc/sec) argon and helium have been used as support gases to measure the O^+ ionization efficiencies. With argon as support gas an ionization efficiency of 0.7 % has been measured and with helium 0.45 %. For He as discharge gas the ion source had to be operated at higher gas pressure to sustain a plasma, thus explaining the lower gas efficiency and longer hold up time.

The hold up times for O^+ varied for all three cases (CO leak with Ar, O_2 leak with Ar, O_2 leak with He). The shortest hold up time was measured with argon as support gas and the O_2 leak, the longest with helium and the O_2 leak.

The results for the CO and O_2 leak are summarized in Figure 8 and Figure 9.

The discrepancy between the argon efficiencies and the carbon or oxygen efficiencies may be explained by the differences in the plasma wall sticking probabilities. Noble gases can be recycled into the plasma, explaining the high efficiency for argon. On the contrary, carbon and oxygen tend to stick at the plasma chamber wall, leading to low source efficiencies in the cusp ion source. For carbon, this effect was demonstrated by the deposition of a carbon film on the plasma chamber walls. The plasma potential for the RF driven cusp source has been measured to be in the order of 5 V to 10 V. Therefore, the ions can not gain enough energy to sputter the adsorbed atoms and molecules from the plasma chamber walls.

The average ion source lifetime for the above mentioned performance tests was about 15 hours, limited by the failure of the porcelain-coated copper antenna. We believe that the performance of the antenna is limited by micro cracks in the porcelain coating. The RF voltage (in the order of a few 100 V) can penetrate through these cracks and ion sputtering of the coating material becomes the lifetime limitation. This explanation is confirmed by the facts that

i) copper appears in the ion beam after 6-10 hours of operation, as soon as the antenna begins to fail
ii) the antenna lifetime is longer for lower masses.

To reduce the sputtering problem, the porcelain-coated antenna was replaced by a quartz antenna. Tin-coated copper threaded wire strands were placed as RF conductors inside a water-cooled quartz tube. The ion source performance of both antenna types was similar. The average lifetime was about 20 h for the quartz antenna, limited by sudden failures of the glass tubes. Since the quartz antenna showed a faint opacity (initially fully transparent), we believe that again plasma sputtering was the lifetime limitation.

CONCLUSION AND OUTLOOK

A radioactive ion beam test stand has been commissioned with an RF cusp source at the 88" Cyclotron at LBNL. The beam transport line consists of an accel-decel extraction system, two electrostatic einzel lenses, and a double focusing sector magnet. An ion beam transmission of up to 100% has been measured.

A high temperature all-carbon target has been developed. O^{14} intensities of up to $3 \cdot 10^7$ pps have been measured at the entrance to the ion source with 2μA primary beam intensity from the cyclotron at a target temperature of 1720° C.

The RF cusp source performance has been tested off-line with respect to the O^{14} experimental requirements. The cusp ion source can not fulfill the three major experimental requirements:

(i) The highest O^+ efficiency was only 0.7 %.
(ii) The gas hold up time in the ion source is in the order of one half-life of O^{14}.
(iii) The average source lifetime with the porcelain-coated antenna is only about 15 hours, and for the quartz antenna about 20 hours.

As a comparison the off-line gas efficiency has been measured on the AECR-U at the 88" Cyclotron for various gases. In Figure 10 the holdup times and measured ionization efficiencies in the cusp source for O^+ are compared with the results for O^{6+} in the AECR-U, which has the highest ionization efficiency of all oxygen charge states. Because the experimental requirement can not be met with the cusp source without further development, and because of the promising results measured on the AECR-U, the cusp ion source will be replaced by a small ECR ion source.

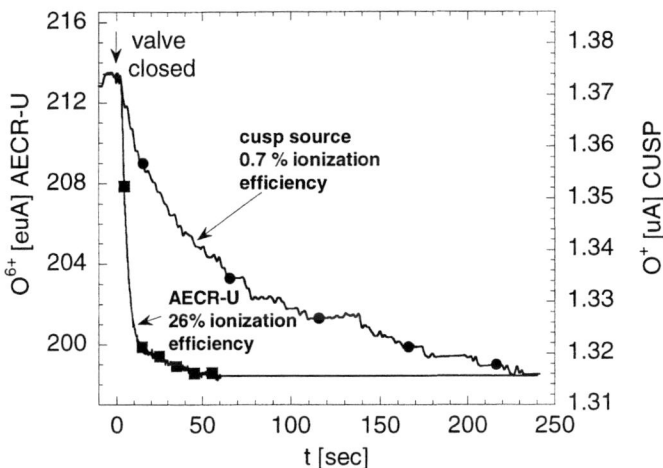

FIGURE 10. Comparison of the hold up times for O^+(cusp) and for O^{6+} (AECR-U) ion source for a calibrated CO leak. The ion ionization efficiency is indicated in the graph also.

ACKNOWLEDGEMENT

We would like to thank K.N. Leung for providing the cusp source and M. Leitner for valuable discussions. Furthermore, we would like to thank D. Garfield, L. Mills, S. Peterson, Jim Rice, D. Syversrud, V. Uno, S. Wilde, and M. D. Williams for their technical assistance. This work was supported by the Director, Office of Energy Research, Office of High Energy Physics and Nuclear Physics Division of the U.S. Department of Energy under contract No. DE-AC03-76SFF00098.

REFERENCES

1. Feynman R. P., Gell-Mann M., Phys. Rev. **109**, 193 (1958)
2. Wutte D.C., S. Freedman, R.A. Gough, Y. Lee, M. Leitner, K. N. Leung, C.M. Lyneis, D.S. Pickard, M. D. Williams, Z. Q. Xie, NIMB **42** (1998)
3. R. Becker, W. B. Hermannsfeldt, RSI **63**, 2756 (1991)
4. Lee Y., R.A. Gough, K.N. Leung, L.T. Perkins, D.S. Pickard, J. Vujic, L.K. Wu, M. Olivio, H. Einenkel, RSI **69** 1023 (1998)
5. Peters J., to be published in the Proceedings of the international Linac98 conference, Chicago (1998)

List of Participants

Nemitala Added
DPTO Fisica Nuclear - IFUSP
Laboratório Pelletron
Gaixa Postal 66318
CEP 05315-970
Sao Paulo
BRAZIL
Phone: 55-11-8186939/8186942
Fax: 55-11-2112742
EMail: nemitala@uspif1.if.usp.br

Gerald D. Alton
Physics Division
Oak Ridge National Laboratory
Building 6000, MS-6368
P.O. Box 2008
Oak Ridge, TN 37831-6368
Phone: 423/574-4751
Fax: 423/574-1268
EMail: gda@ornl.gov

Per Arndt
Department of Physics
Hahn-Meitner-Institut
Glienicker Strasse 100
D-14109 Berlin
GERMANY
Phone: 0049-30-8062-2427
Fax: 0049-30-8062-2293
EMail:

Jeffery Ashenfelter
Wright Nuclear Structure Laboratory
Yale University
272 Whitney Avenue
New Haven, CT 06443
Phone: 203/432-3090
Fax: 203/432-3522
EMail: ash@mirage.physics.yale.edu

Thomas A. Barker
Physics Department
Yale University
272 Whitney Avenue
New Haven, CT 06520
Phone: 203/432-3096
Fax: 203/432-3522
EMail: tom@mirage.physics.yale.edu

J. Roger J. Bennett
ISIS Department
CLRC
Rutherford Appleton Laboratory
Chilton, Didcot Oxon. 0X11 0QX
UNITED KINGDOM
Phone: 44-1235-44-6379
Fax: 44-1235-44-5607
EMail: jrjb@isise.rl.ac.uk

Yang Bingfan
Department of Nuclear Physics
China Institute of Atomic Energy
P.O. Box 275 (62)
Beijing 102413
PRC
Phone: 86-10-69357817
Fax: 86-10-69357787
EMail: bingfan@mipsa.ciae.ac.cn

Giovanni M. Bisoffi
Divisione Acceleratori
Laboratori Nazionali di Legnaro
Via Romea 4
I -35020 Legnaro, Padova
ITALY
Phone: 39-49-8068-349
Fax: 39-49-641925
EMail: bisoffi@lnl.infn.it
Phone: 39-49-8068-349

Marco Cavenago
Divisione Acceleratori
Laboratori Nazionali di Legnaro
Via Romea 4
I -35020 Legnaro, Padova
ITALY
Phone: 39-49-8068-503
Fax: 39-49-641925
EMail: cavenago@lnl.infn.it

Anup Choudhury
National Superconducting Cyclotron Laboratory
Michigan State University
South Shaw Lane
East Lansing, MI 48824-1321
Phone: 517/353-5967
Fax: 517/353-9671
EMail: choudhury@maya.nscl.msu.edu

Giovanni Ciavola
INFN-LNS
Via S. Sofia 44
95123 Catania
ITALY
Phone: 39-095-542-262
Fax: 39-095-542-302
EMail: ciavola@lns.infn.it

Benny E. Clifft
Physics Division
Argonne National Laboratory
Building 203
9700 South Cass Avenue
Argonne, IL 60439
Phone: 630/252-9959
Fax: 630/252-4118
EMail: clifft@anlphy.phy.anl.gov

Antonio Dainelli
Divisione Acceleratori
Laboratori Nazionali di Legnaro
Via Romea 4
I-35020 Legnaro, Padova
ITALY
Phone: 39-49-8068-352
Fax: 39-49-641925
EMail: dainelli@lnl.infn.it

Scott Dix
Leybold Vacuum/Vacuum One
2502 North Clark Street
Chicago, IL 60614
Phone: 773/244-3102
Fax: 773/244-3975
EMail: sdix@vacuumone.com

Alberto Facco
Accelerator Division
INFN-LNL
Via Romea 4
35020 Legnaro, Padova
ITALY
Phone: 39-49-8068-352
Fax: 39-49-641925
EMail: facco@lnl.infn.it

Graziano Fortuna
Divisione Acceleratori
Laboratori Nazionali di Legnaro
Via Romea 4
I-35020 Legnaro, Padova
ITALY
Phone: 39-49-8068-442
Fax: 39-49-641925
EMail: fortuna@lnl.infn.it

Santo Gammino
INFN-LNS
Via S. Sofia 44
95123 Catania
ITALY
Phone: 39-95-542270
Fax: 39-95-542302
EMail: gammino@lns.infn.it

Jerry D. Garrett
Physics Division
Oak Ridge National Laboratory
Mail Stop 6368
P.O. Box 2008
Oak Ridge, TN 37831-6368
Phone: 423/576-5481
Fax: 423/574-1268
EMail: garrettjd@ornl.gov

Richard Geller
I.S.N. Grenoble
Av. des Martyrs 53
38026 Grenoble
FRANCE
Phone: 334-76-284162
Fax: 334-76-284004
EMail:

Subhendu Ghosh
Nuclear Science Centre-New Delhi
c/o Argonne National Laboratory
9700 South Cass Avenue, Building 203
Argonne, IL 60439
Phone: 630/252-6120
Fax: 630/252-9647
EMail: ghosh@anlphy.phy.anl.gov

Geroge Gillespie
G. H. Gillespie Associates
10855 Sorrento Valley Road, Ste. 201
San Diego, CA 92121
Phone: 619/677-0076
Fax: 619/677-0079
EMail:

Tom J. Gray
Physics Department -
James R. Macdonald Laboratory
Kansas State University
Manhattan, KS 66506-2601
Phone: 785/532-2663
Fax: 785/532-6806
EMail: tgray@phys.ksu.edu

Van Griffin
Physics Department
Florida State University
Tallahassee, FL 32306-4350
Phone: 850/644-5814
Fax: 850/644-9848
EMail: van@nucmar.physics.fsu.edu

Willy Haeberli
Department of Physics
University of Wisconsin
1150 University Avenue
Madison, WI 53706
Phone: 608/262-0009
Fax: 608/202-3598
EMail: whaeberli@uwnuc0.physics.wisc.edu

Kjell Håkansson
Department of Physics
University of Lund
Sölvegatan 14
SE-22362 Lund
SWEDEN
Phone: 46-046-2227642
Fax: 46-046-2224709
EMail: kjell.hakansson@nuclear.lu.se

Ragnar Hellborg
Department of Physics
University of Lund
Sölvegatan 14
SE-22362 Lund
SWEDEN
Phone: 46-046-2227644
Fax: 46-046-2224709
EMail: ragnar.hellborg@nuclear.lu.se

Walter F. Henning
Physics Division
Argonne National Laboratory
Building 203
9700 South Cass Avenue
Argonne, IL 60439
Phone: 630/252-4004
Fax: 630/252-3903
EMail: henning@anl.gov

Dale K. Hensley
Solid State Division
Oak Ridge National Laboratory
Building 3003, MS 6048
1 Bethel Valley Road
Oak Ridge, TN 37831
Phone: 423/576-6697
Fax: 423/576-6720
EMail: dkz@ornl.gov

Rainer Hölzle
Institut fuer Festkörperforschung
Forschungszentrum Jülich
Leo-Brandt-Strasse
52428 Jülich
GERMANY
Phone: 0-2461-61-3151
Fax: 0-2461-61-2410
EMail: r.hoelzle@fz-juelich.de

Richard Hyder
Manor Farm House
12 Manor Road
Woodstock, OXON OX20 1XJ
UNITED KINGDOM
Phone: 44-1993-810-147
Fax:
EMail:

Nathan L. Jones
Physics Division
Oak Ridge National Laboratory
Building 5500
P.O. Box 2008
Oak Ridge, TN 37831-6377
Phone: 423/574-3104
Fax: 423/574-1118
EMail: nnj@ornl.gov

Robert A. Kaye
Physics Department
Argonne National Laboratory
Building 203
9700 South Cass Avenue
Argonne, IL 60439
Phone: 630/252-4039
Fax: 630/252-9647
EMail: kaye@anlphy.phy.anl.gov

Mark J. Kedzie
Physics Division
Argonne National Laboratory
Building 203
9700 South Cass Avenue
Argonne, IL 60439
Phone: 630/252-1635
Fax: 630/252-4118
EMail: kedzie@anl.gov

Allan Kern
Physics Department
Western Michigan University
Kalamazoo, MI 49008-5151
Phone: 616/387-4957
Fax: 616/387-4939
EMail: allan.kern@umich.edu

Oliver Kester
Sektion Physik, LS Habs
LMU München
Am Coulombwall 1
85748 Gardhing
GERMANY
Phone: 49-89-283-14076
Fax: 49-89-283-14072
EMail: oliver.kester@physik.uni-muenchen.de

Walter Kutschera
Institut für Radiumforschung und Kernphysik
University of Vienna
Währinger Strasse 17
A-1090 Vienna
AUSTRIA
Phone: 43-1-4277-51700
Fax: 43-1-4277-9517
EMail: walter.kutschera@univie.ac.at

Nicolas Lahera
IReS
IRES (CNRS-IN2P3)
S3, rue du Loess
BP 28
67037 Strasbourg
FRANCE
Phone: 03-88-10-66-66
Fax:
EMail: nicolas.lahera@ires.in2p3.fr

James D. Larson
10011 East 35th Street Terrace South
Independence, MO 64052-1107
Phone: 816/353-1527
Fax:
EMail:

Robert Laxdal
TRIUMF
4004 Wesbrook Mall
Vancouver, BC V6T 1K8
CANADA
Phone: 604/222-7322
Fax: 604/222-1074
EMail: lax@triumf.ca

Michel Letournel
VIVIRAD
23 rue Principale
67117 Handschuheim
FRANCE
Phone: 33-3-88-69-13-25
Fax: 33-3-88-69-16-18
EMail: vivirad@aol.com

Yuan Liu
Physics Division
Oak Ridge National Laboratory
Building 6000, MS-6368
P.O. Box 2008
Oak Ridge, TN 37831-6368
Phone: 423/574-4761
Fax: 423/574-1268
EMail: liuy@ornl.gov

Martha J. Meigs
Physics Division
Oak Ridge National Laboratory
MS 6368
P.O. Box 2008
Oak Ridge, TN 37831
Phone: 423/574-4950
Fax: 423/574-1268
EMail: meigs@mail.phy.ornl.gov

Claude M. Lyneis
NSD
Lawrence Berkeley National Laboratory
Building 88-219
One Cyclotron Road
Berkeley, CA 94720
Phone: 510/486-7815
Fax: 510/486-7983
EMail: cmlyneis@lbl.gov

Bruce Milton
TRIUMF
4004 Wesbrook Mall
Vancouver, BC V6T 2A3
CANADA
Phone:
Fax:
EMail: milton@triumf.ca

Bonnie Maratea
AccelSoft, Inc.
10855 Sorrento Valley Road, Ste. 202A
San Diego, CA 92121
Phone: 619/677-0133
Fax: 619/784-3736
EMail: bonnie@ghga.com

Radwan Mourad
Sciaky, Inc.
4915 West 67th Street
Chicago, IL 60638-6493
Phone: 708/594-3800
Fax: 708/594-9213
EMail: www.sciaky.com

Donald May
Cyclotron Institute
Texas A&M University
College Station, TX 77840-3366
Phone: 409/845-1411
Fax: 409/845-1899
EMail: may@comp.tamu.edu

Floyd H. Munson
Physics Division
Argonne National Laboratory
Building 203
9700 South Cass Avenue
Argonne, IL 60439
Phone: 630/252-8749
Fax: 630/252-2864
EMail: munson@anlphy.phy.anl.gov

John McKay
Box 463
Deep River, ON K0J 1P0
CANADA
Phone: 613/584-4975
Fax:
EMail: mckayj@intranet.ca

John Noe
Department of Physics & Astronomy
SUNY-Stony Brook
Stony Brook, NY 11794-3800
Phone: 516/632-8146
Fax: 516/632-8573
EMail: john.noe@sunysb.edu

Jerry S. Nolen
Physics Division
Argonne National Laboratory
Building 203
9700 South Cass Avenue
Argonne, IL 60439
Phone: 630/252-6418
Fax: 630/252-9647
EMail: nolen@anl.gov

Gregory A. Norton
National Electrostatics Corp.
P.O. Box 620310
7540 Graber Road
Middleton, WI 53562-0310
Phone: 608/831-7600
Fax: 608/256-4103
EMail: nec@pelletron.com

Francis Osswald
Accelerators Department
IRES (CNRS-IN2P3)
BP 28
Cedex 2
67037 Strasbourg
FRANCE
Phone: 0033-38810-6299
Fax: 0033-38810-6273
EMail: francis.osswald@ires.in2p3.fr

Richard Pardo
Physics Division
Argonne National Laboratory
Building 203
9700 South Cass Avenue
Argonne, IL 60439
Phone: 630/252-4029
Fax: 630/252-9647
EMail: pardo@anlphy.phy.anl.gov

Howard Pavasko
Leybold Vacuum Products
5700 Mellon Road
Export, PA 15632
Phone: 724/325-6566
Fax:
EMail:

Maria Petra
Physics Division
Argonne National Laboratory
Building 203
9700 South Cass Avenue
Argonne, IL 60439
Phone: 630/252-4039
Fax: 630/252-9647
EMail: petra@anlphy.phy.anl.gov

Andrea Pisent
Divisione Acceleratori
Laboratori Nazionali di Legnaro
Via Romea 4
I-35020 Legnaro, Padova
ITALY
Phone: 39-49-8068-358
Fax: 39-49-641925
EMail: pisent@lnl.infn.it

Prakash Potukuchi
Nuclear Science Centre-New Delhi
c/o Argonne National Laboratory
9700 South Cass Avenue, Building 203
Argonne, IL 60439
Phone: 630/252-6120
Fax: 630/252-9647
EMail: prakash@anlphy.phy.anl.gov

Kenneth H. Purser
Southern Cross Corporation
2 Centennial Drive
Peabody, MA 01960
Phone: 978/531-4526
Fax: 978/531-4580
EMail: kpurser@shore.net

Robert Rebmeister
Service des Accélérateurs
Institut de Recherches Subatomiques
23, rue du Loess
B.P. 28
F-67037 Strasbourg Cedex 2
FRANCE
Phone: 33-03-88-10-6564
Fax: 33-03-88-10-6664
EMail: robert.rebmeister@ires.in2p3.fr

Jani Reijonen
Accelerator and Fusion Research Division
Lawrence Berkeley National Laboratory
MS 5-121
1 Cyclotron Road
Berkeley, CA 94720
Phone: 510/495-2774
Fax: 510/486-5105
EMail: reijonen@mh1.lbl.gov

Roland Repnow
Max-Planck-Institut für Kernphysik
Postfach 10 39 80
D-69029 Heidelberg
GERMANY
Phone: 49-6221-516-364
Fax: 49-6221-516-234
EMail: rep@mpi-hd.mpg.de

Ludwig F. Rohrer
Beschleunigerlabor der LMU and TU München
Forschungsgelände
85748 Garching
GERMANY
Phone: 89-289-14272
Fax: 89-289-14280
EMail: ludwig.rohrer@physik.uni-muenchen.de

John Schiffer
Physics Division
Argonne National Laboratory
Building 203
9700 South Cass Avenue
Argonne, IL 60439
Phone: 630/252-4066
Fax: 630/252-2864
EMail: schiffer@anl.gov

Paul Schmor
TRIUMF
4004 Wesbrook Mall
Vancouver, BC V6T 2A3
CANADA
Phone: 640/222-7415
Fax: 640/222-1074
EMail: schmor@triumf.ca

Robert J. Schneider
Nat'l Ocean Sciences Accelerator Mass Spectrometry Fa
Woods Hole Oceanographic Institution
MS 8
Woods Hole, MA 02543
Phone: 508/289-2756
Fax: 508/457-2183
EMail: rschneider@whoi.edu

Kenneth Shepard
Physics Division
Argonne National Laboratory
Building 203
9700 South Cass Avenue
Argonne, IL 60439
Phone: 630/252-4899
Fax: 630/252-9647
EMail: kwshepard@anl.gov

David M. Shepherd
Megavolt Ltd.
Unit 4B Westcombe Trading Estate
Ilminster, Somerset TA19 9DW
UNITED KINGDOM
Phone: 44-146-57458
Fax: 44-146-57458
EMail:

Jim Specht
Physics Department
Argonne National Laboratory
Building 203
9700 South Cass Avenue
Argonne, IL 60439
Phone: 630/252-3610
Fax: 630/252-4118
EMail: jrspecht@anl.gov

David Spingler
Physics Department
Florida State University
Tallahassee, FL 32306-4350
Phone: 850/644-4163
Fax: 850/644-9848
EMail: spingler@nucmar.physics.fsu.edu

Dannie B. Steski
AGS Department
Brookhaven National Laboratory
Building 901A
Upton, NY 11973
Phone: 516/344-4581
Fax: 516/344-4583
EMail: steski@bnl.gov

Derek W. Storm
Nuclear Physics Laboratory
University of Washington
P.O. Box 354290
Seattle, WA 98195
Phone: 206/543-4085
Fax: 206/685-4634
EMail: storm@npl.washington.edu

Suehiro Takeuchi
Department of Material Science
Japan Atomic Energy Research Institute
Accelerators Division
Tokai, Ibaraki 319-1195
JAPAN
Phone: 81-29-282-5860
Fax: 81-29-282-6321
EMail: takeuchi@tdmalph0.tokai.jaeri.gc.jp

Willard L. Talbert
Amparo Corporation
1 East Sunrise Drive
Santa Fe, NM 87501
Phone: 505/983-0467
Fax:
EMail: willtalb@aol.com

Peter Thieberger
Alternating Gradient Synchrotron
Brookhaven National Laboratory
59 Cornell Avenue
Building 901A
Upton, NY 11973-5000
Phone: 516/344-4004/4581
Fax: 516/344-4583
EMail: pt@bnl.gov

Masahito Tomizawa
Tanashi Branch
High Energy Accelerator Research Organization
3-2-1 Midori-Cho
Tanashi, Tokyo 188-8501
JAPAN
Phone: 81-424-69-9541
Fax: 81-424-68-5543
EMail: tomizawa@tanashi.kek.jp

Karl F. von Reden
Department of Geology & Geophysics
Woods Hole Oceanographic Institution
McLean Laboratory, MS 8
Woods Hole, MA 02543
Phone: 508/289-3384
Fax: 508/457-2183
EMail: kvonreden@whoi.edu

Richard Vondrasek
Physics Division
Argonne National Laboratory
Building 203
9700 South Cass Avenue
Argonne, IL 60439
Phone: 630/252-5972
Fax: 630/252-4118
EMail: vondrasek@anlphy.phy.anl.gov

Bernard Waast
Acclerator Tandem
Institut de Physique Nucleaire
91406 Orsay - Cedex
FRANCE
Phone: 33-1-6915-5291
Fax: 33-1-6915-5108
EMail:

David C. Weisser
Department of Nuclear Physics
Australian National University
Research School of Physical Science and Engineering
Canberra ACT 0200
AUSTRALIA
Phone: 61-6249-2080
Fax: 61-6249-0748
EMail: david.weisser@anu.edu.au

Doug Will
Nuclear Physics Laboratory
University of Washington
Box 354290
Seattle, WA 98195
Phone: 206/543-4026
Fax: 206/685-4634
EMail: will@npl.washington.edu

Kenneth F. Wilson
Sales and Marketing
Meyer Tool & Manufacturing, Inc.
4601 Southwest Highway
Oak Lawn, IL 60453
Phone: 708/425-9080
Fax: 708/425-2612
EMail: kfw@megsinet.net

Hermann Wollnik
2. Physik Institut
University of Giessen
Heinrich-Buff-Ring 16
D-35392 Giessen
GERMANY
Phone: 011-49-641-9933-240
Fax: 011-49-641-9933-239
EMail: wollnik@uni-giessen.de

Daniela Wutte
Nuclear Science Division
Lawrence Berkeley National Laboratory
MS 88
1 Cyclotron Road
Berkeley, CA 94720
Phone: 510/486-2918
Fax: 510/486-7983
EMail: daniela_wutte@lbl.gov

Zu Qi Xie
Nuclear Science Division
Lawrence Berkeley National Laboratory
Building 88, Rm. 113
Berkeley, CA 94720
Phone: 510/486-7814
Fax: 510/486-7983
EMail: zqxie@lbl.gov

Gary P. Zinkann
Physics Division
Argonne National Laboratory
Building 203
9700 South Cass Avenue
Argonne, IL 60439
Phone: 630/252-4892
Fax: 630/252-9647
EMail: zinkann@anlphy.phy.anl.gov

Author Index

A

Abe, S., 152
Abrioni, G., 360, 366
Added, N., 168
Ahmad, I., 477
Ahuja, R., 272
Ajithkumar, B. P., 267
Alton, G. D., 321, 330, 341, 352, 466
Andreev, V., 173
Anthony, J., 267
Arai, S., 451
Arakaki, Y., 451
Auble, R. L., 466

B

Back, B. B., 477
Badan, L., 228
Baktash, C., 466
Ball, J. A., 352
Bao, Y., 341, 352
Barua, P., 272
Beene, J. R., 466
Beis, J., 56
Bellino, M., 410
Beltramin, A., 228
Bennett, J. R. J., 490, 512
Berners, E. D., 92
Bertazzo, L., 228
Bertrand, F. E., 466
Bhowmik, R. K., 267, 272
Bingfan, Y., 80
Bisoffi, G., 173, 228
Bissiato, E., 173
Bongers, H., 536
Boscagli, L., 228
Bouly, J. L., 287
Bruandet, J. F., 287
Burke, J., 566

C

Canella, S., 360, 366
Canzhe, Z., 80
Carlucci, D., 228
Cavenago, M., 360, 366
Cerny, J., 312
Cervellera, F., 366
Chacko, J., 267
Changrani, T., 267
Chauvin, N., 287
ChauXin, K., 80
Chiurlotto, F., 173, 228
Chopra, S., 39, 272
Choudhury, A., 267
Ciavola, G., 300
Clifft, B. E., 272, 279, 477, 528
Comunian, M., 173
Contran, S., 228
Contran, T., 228
Corradin, E., 173
Cui, B., 341, 352
Curdy, J. C., 287

D

Dainelli, A., 228
Dale, D., 56
Daniel, R. E., 3
Das, J. J., 523
Datta, S. K., 39, 272
Debiak, T., 56
De Lazzari, M., 228
Delfitto, G., 360
Dezhong, L., 80
Donoghue, J., 410
Drake, D. M., 495

E

Elder, K. L., 410

F

Facco, A., 138, 185
Ferguson, S. M., 92
Ferry, J. A., 3
Fox, J. D., 466

Freedman, S. J., 566
Fujikawa, B., 566

G

Gagnon, A. R., 410
Gammino, S., 300
Gargari, S., 39, 272
Garrett, J. D., 466, 505
Geller, R., 287
Ghosh, S., 236, 267, 272
Gomes, I., 477
Gough, R. A., 466, 566
Gove, H. E., 426
Greene, J. P., 477
Groß, M., 536
Guilian, Z., 80
Guo, F. Q., 312

H

Habs, D., 536
Håkansson, K., 94, 100
Halbert, M. L., 466
Hanashima, S., 152
Hassanein, A. M., 477
Haustein, P., 312
Hayes, J. M., 410, 422
Hellborg, R., 94, 100
Henning, W. F., 477
Heugel, J., 104
Horie, K., 152

I

Imanishi, A., 451
Ishizaki, N., 152

J

Janssens, R. V. F., 477
Jegham, E., 104
Jiang, C.-L., 477
Jiazheng, Z., 80
Jiuchang, Q., 80
Joosten, R., 312

Joshi, R., 39, 272

K

Kalnins, J. G., 466
Kamykowski, E., 56
Kanda, S., 152
Kanjilal, D., 39, 272
Kaye, R. A., 477, 528
Kedzie, M., 528
Kester, O., 536
Key, R. M., 410
Kienle, P., 536
Klody, G. M., 3
Kobayashi, C., 65
Kolbe, A., 536
Körner, H. J., 536
Köster, U., 536
Krishnan, S. A., 267, 272
Kulevoy, T., 366
Kumar, B., 523
Kumar, M., 267
Kumar, R., 267, 272
Kutschera, W., 399

L

Lahera, N., 104
Lamy, T., 287
Larimer, R. M., 312
Laxdal, R. E., 546
Leitner, M. A., 384
Lenz, J. W., 495
Litherland, A. E., 426
Liu, Y., 321, 330, 466
Liyong, W., 80
Lobanov, N. R., 117
Lohwasser, R., 341
Lollo, M., 173
Lombardi, A., 173
Long, P., 410
Lyneis, C. M., 312, 384, 566

M

Madhavan, N., 523
Madhusudhana Rao, P. V., 523
Maier, H. J., 536

Malyadri, A. J., 272
Mandal, A., 267
Matsuda, M., 65, 152
McK. Hyder, H. R., 47
McKay, J. W., 74
McMahan, P., 312
McNichol, A. P., 410
Mehta, G. K., 39, 267
Mehta, R., 267
Melnychuk, S., 56
Meyer, F. W., 321
Milton, B. F., 56
Morin, T., 410
Munson, F. H., 258
Murray, S. N., 330

N

Narayanan, M. M., 39
Nifenecker, H., 287
Niki, K., 451
Noé, J. W., 192
Nolen, J. A., 279, 477
Norman, E. B., 312
Norton, G. A., 3

O

Ogan, M. W., 466
Ohuchi, I., 152
Okada, M., 451
O'Neil, J. P., 312
Ophel, T. R., 24
Osswald, F., 104

P

Palmieri, V., 228
Pardo, R. C., 279, 477
Peden, J. C., 410
Petra, M., 477
Pisent, A., 173, 214
Plasil, F., 466
Poggi, M., 228
Porcellato, A. M., 173, 228
Portillo, M., 477
Potukuchi, P. N., 236

Powell, J., 312
Prakash, P. N., 267
Purser, K. H., 426

R

Rao, S., 272
Rathke, J., 56
Ratzinger, U., 536
Rebmeister, R., 104
Reed, C. A., 341, 352
Reed, C. B., 477
Rehm, C. E., 279
Rehm, K. E., 477
Rodrigues, G. O., 267
Rogers, J., 56
Rohrer, L., 88, 252
Rowe, M. W., 312
Roy, A., 267, 272
Ruegg, R., 56

S

Sahu, B. K., 267
Sarkar, A., 267, 272
Savard, G., 477
Scarpa, F., 185
Schempp, A., 536
Schiffer, J. P., 477
Schmor, P. W., 439
Schneider, R. J., 410, 422
Schnitter, H., 252
Shapira, D., 466
Shen, W. Q., 279
Shengyong, S., 80
Shepard, K. W., 236, 267, 477, 528
Shirai, T., 173
Sieber, T., 536
Siemssen, R. H., 477
Sinha, A. K., 523
Sole, P., 287
Sortais, P., 287
Spampinato, P. T., 466
Specht, J. R., 477
Sredniawski, J., 56
Staples, J. W., 466
Stark, S. Y., 228
Steichele, E., 536

Steski, D. B., 375
Stivanello, F., 228
Stuart, D., 410
Sugathan, P., 523
Suresh Babu, M. V., 267

T

Takeda, Y., 451
Takeuchi, S., 65, 152, 244
Talbert, W. L., 495
Tayama, H., 152
Taylor, C. E., 384
Tecchio, L. B., 559
Thirolf, P., 536
Tilbrook, I. R., 272
Tojyo, E., 451
Tomizawa, M., 451
Tovo, E., 173
Tovo, R., 173
Tsukihashi, Y., 152

U

Uthas, S., 100

V

VanBrocklin, H. F., 312
Varughese, T., 523
Vetter, P., 566
Vieux-Rochaz, J. L., 287
von Egidy, T., 536
von Reden, K. F., 410, 422

W

Walker, J. J., 495
Weimin, Y., 80
Weisser, D. C., 24, 117
Williams, C., 330, 352
Wills, J. S. C., 422
Wilson, M. T., 495
Wollnik, H., 466, 505
Wutte, D., 312, 384, 566

X

Xialing, G., 80
Xie, Z. Q., 312, 384, 566
Xu, X. J., 312

Y

Yongliang, J., 80
Yoshida, T., 152
Yueming, H., 80

Z

Zhang, T., 341
Zhiren, Y., 80
Zinkann, G. P., 279
Ziomi, L., 228
Zisman, M. S., 466
Zouloumian, P., 104
Zviagintsev, V., 185